# Analecta Husserliana

The Yearbook of Phenomenological Research

Volume CXXI



Published under the auspices of
*The World Phenomenology Institute*
A-T. Tymieniecka, Founder

More information about this series at http://www.springer.com/series/5621

William S. Smith • Jadwiga S. Smith
Daniela Verducci
Editors

# Eco-Phenomenology: Life, Human Life, Post-Human Life in the Harmony of the Cosmos

Springer

*Editors*
William S. Smith
World Phenomenology Institute
Hanover, NH, USA

Jadwiga S. Smith
World Phenomenology Institute
Hanover, NH, USA

Daniela Verducci
World Phenomenology Institute
Hanover, NH, USA

Analecta Husserliana
ISBN 978-3-030-08475-2 ISBN 978-3-319-77516-6 (eBook)
https://doi.org/10.1007/978-3-319-77516-6



Printed on acid-free paper

This Springer imprint is published by the registered company Springer International Publishing AG part of Springer Nature.
The registered company address is: Gewerbestrasse 11, 6330 Cham, Switzerland

**Exergo/Epigraph**

*Actually, my account of* ontopoiesis *is an eco-phenomenology. Ontopoiesis reaches to the very germs of ecology: development and genesis. I have published several essays related to this.* In *The Passions of the Earth [Analecta Husserliana, Vol. LXXI],* I show how the human being is an ecological fruit and how the human being is formed by the earth and sucks the juices of the earth. I have also written things about the cosmos and the cosmic dependencies of the human mind and human development. You see, the self-individualization of life, which is the basic instrument of ontopoiesis, draws upon the laws of the cosmos and the earth. *This is the most fundamental ecology that can be done. So, we have just touched the essence of my philosophy, the base—our relationship to the earth and to the cosmos.*

Anna-Teresa Tymieniecka

# Preface

*Eco-phenomenology. Life, Human Life, Post-Human Life in the Harmony of the Cosmos* addresses the highly topical theme of ecology in markedly heterodox terms. In fact, the context of all the contributions is the *Eco-phenomenology* formulated by Anna-Teresa Tymieniecka on the basis of the results of her over 40 years of inquiry into the phenomenology of life.

In an interview given on the occasion of her receiving an honorary degree from the University of Bergen in 2008, Tymieniecka unhesitatingly defined as *Eco-phenomenology* the innovative vision of being in becoming that flowed from her discovery of the ontopoiesis of life, or in other words, the spontaneous and pervasive movement of the Logos of life itself, which reaches to the very germs of ecology, which are development and genesis, and which therefore surprisingly forms the basis for all theorizations of Ecology, Environmental Studies, and Ethics, as well as for innocent trust in science and in technological innovation in the field of human life and of the natural environment. In fact, the ontopoiesis of life is not one among many dynamic lines along which life evolves, but on the contrary, it is the fundamental logos/force of the auto-individualization of being in becoming, that from which every evolutive movement emanates and according to which and because of which all becoming harmoniously develops. The rhythm of the ontopoietic logos of life pervades every form of being, inanimate, animate, and human, and with its creative productivity conducts the entire natural being beyond itself toward a transcendental destination, one as mysterious as its logoic source. Since it belongs to life, such a logos/force possesses the principal qualities of life: it is sentient and metamorphic. In other words, the ontopoietic logos of life is able not only to "feel/sense" the vectors of its dynamism which produce themselves in an on-going way, but also to make use of them, metamorphosing itself, transforming its own modality of becoming, to continue seamlessly or rather in a discrete continuity its productive auto-movement by the new means that have come to light. This can be observed, for example, in the crucial evolutive passage from natural life to the human condition, in which the auto-individualizing power of the ontopoietic logos of life continues to act effectively, even though it has changed the supports of its own dynamic modality and so uses the free vector of the human will and of *imaginatio creatrix*, which mysteriously has taken the place of the deterministic vector of the laws and automatisms of nature.

Anna-Teresa Tymieniecka died in June of 2014, and this Volume 121 is the first in the "Analecta Husserliana" series, which she founded, to be published after her death. For this reason, the first part of the volume features a tribute to her memory written by Olga Louchakova Schwartz, "The Symphony of Sentience, in Cosmos and Life: In Memoriam Anna-Teresa Tymieniecka." This chapter explores whether phenomenology, as a cognitive enterprise of sentience, extends beyond the death of the physical body, turning from the phenomenology of life to the philosophy of life. The symphonic form of the paper in four movements, *Allegro*, *Moderato*, *Minuet*, and *Finale Glorioso*, is inspired by Tymieniecka's *Christo-Logos: Metaphysical Rhapsodies of Faith (*Itinerarium mentis in Deo*)*, *The Fullness of the Logos in the Key of Life – Book II,* published in 2013, in Volume 111 of the Analecta Husserliana series. The fecundity of the eco-phenomenology of Tymieniecka emerges here manifestly. In fact, in virtue of the unifying and integrating power of the ontopoietic *logos* of life, discovered in the course of her research into the phenomenology of life to be the driving and ordering force that is the basis of all evolution, the usual ecological outlook centered on relationships among organisms or groups of organisms, and their natural environment can expand beyond the confines of so-called nature, because the human condition originated from it and can broaden into the supernatural, through the advent, in the midst of the just conquered unity-of-everything-there-is-alive, of the free human creativity that facilitates the passage of the ontopoiesis of life and its logos not only to the esthetic and ethical dimensions, but also to the religious dimension of human experience.

Part II of this volume, *Introduction to the Topic*, focuses on certain aspects that determine this condition of the ecological overflow, so to speak, of Tymieniecka's eco-phenomenology. First of all, it describes her re-forming of the phenomenological method, by which she reached her cognitive results, and which is an ecological reform inasmuch as it sought to ensure that theoretical thought would have the maximum inter-relation with the broader environmental and historical context of living beings and of experiences in general. In fact, as Daniela Verducci has observed, even though Tymieniecka's speculation was situated in the American orbit, "neophiliac" by definition, it was always nourished by constant and deep exchanges with the more than thousand years of philosophical tradition that has flowed into Italian culture, starting with Parmenides of Elea and passing through Saint Thomas Aquinas to the more recent forms of *Italian theory*, as Dario Gentili has expressed it. In this way, Anna-Teresa Tymieniecka was able to cast her phenomenological gaze starting from the most mature peak of Western thought, since, as Roberto Esposito argues, the Italian type of thinking is rightly re-evaluated in many quarters as a "living thought," for it seems to be the only bearer nowadays of the fruit of manifold and complex speculative sedimentations, which elsewhere have been swept away instead.

Even in regard to Husserlian phenomenology, Anna-Teresa Tymieniecka adopts an ecological approach, rather than a purely speculative one, as she welcomes interiorly the seminal virtualities engendered by Husserl's thought, so that they may continue to germinate, inasmuch as they are grafted onto her current and living thought. In this way, through leaning forward from the platform of consolidated phenomenological results, Tymieniecka was able to realize an "intuitive re-sowing" of the method and classical themes of phenomenology, from the point of view of life. In the measure to which she re-contextualized the human creative condition within the unity-of-everything-there-is-alive, Anna-Teresa Tymieniecka succeeded in making use, again, of *Erlebnis* as a resource for *philosophia prima*, revealing that the phenomenology of life had matured into an eco-phenomenological Enlightenment.

At this point, it is well to consider the speculative position of Gustavo Bontadini (1903–1990), a leading exponent of the new-classical metaphysics of the Milan school, who, like Anna-Teresa Tymieniecka, sought the logos that shows itself in the human phenomenon and especially in *Imaginatio Creatrix*. Francesco Totaro in his contribution, "Metaphysics and Eco-Phenomenology Aiming at Harmony of Human Life with the Cosmos," compares the thought of Bontadini and that of Tymieniecka. Though at first glance, the dynamic ontopoiesis of Anna-Teresa Tymieniecka seems to be very different from the metaphysics of Bontadini, which appeals to the permanence of being without contradiction, a closer examination reveals that actually there is a good connection owing to the place that each gives to metaphysics and to the relationship between metaphysics and life. However, the ways in which these two thinkers work out these shared themes differ. Regarding the search for a new metaphysics, the philosophical school of the Catholic University of Milan has elaborated a so-called new-classical metaphysics, one that refers mainly to Parmenides and Aristotle and reinterprets the tradition dating back to Thomas Aquinas. For this new-classical metaphysics, an important point is the overcoming of the contradiction that is becoming, because of the passage therein from being to not being, by the passenger's being redeemed in the sphere of "being that can't not be."

In contrast, we all know that Anna-Teresa Tymieniecka particularly appreciated the doctrine of becoming in Heraclitus and that her research led her to formulate "ontopoiesis as a new metaphysics" and to find the basis for the new metaphysics precisely in the ontopoietic logos of evolving life.

Regarding the shared focus of these philosophers on the relationship between metaphysics and life, there are also differences to be noted. We can read at the beginning of Bontadini's first book, *Saggio di una metafisica dell'esperienza* [*Essay on a Metaphysics of Experience*], that philosophy emerges from life and that philosophy embraces and attracts all of life or, better, life embraces itself within philosophy, becoming an object of itself and so becoming life anew. But, since the main undertaking of philosophy, according to Bontadini, is asking for the reasons, sense, aim, and value of the whole flux of life, the main task of philosophy, which is always within the flux of life, is to stop that flux through a reflective mediation in order to experience the presentation of immediate being within the so-called "unity of experience." In contrast, Tymieniecka turns to new intuitive evidence, the unity of reason within life's constructive spread, to seek and find the *logos* that shows itself in the human phenomenon and in *imaginatio creatrix*, which is able to turn back to the series of events giving a new impulse to the ontopoiesis of life and to its manifestation.

The original outstanding vision of the ontopoiesis of life put forward by Tymieniecka is the focus of the contribution by Carmen Cozma, "*Ontopoiesis of Life* as Eco-Phenomenology," which stresses and unfolds the significance of the pivotal ideas of Tymienieckan philosophy, such as the "self-individualization of life," the centrality of the creative act of human being, and the grounding of the human condition within the totality of life's spread.

The nine parts of this volume attempt to sketch out the architectonics generated by such a specific ordered dynamic of life, spontaneously productive of its evolution and of its various degrees of development. In fact, beginning with an examination of several seminal questions (Part III, *Seeds of Eco-Phenomenology*), the volume goes on to valorize extra-philosophical and multi-sectored outlooks as well (see Part VI, *Eco-Phenomenological Readings*). The following sections delineate the new eco-phenomenological outlook in the consolidated disciplines of cosmology, ethics, and anthropology, giving rise, respectively, to investigations into *Eco-Cosmology* (Part IV), *Eco-Ethics and Environmental Theories* (Part V), and *Eco-Anthropology: Sentience, Desire, Language, Creativity* (Part VII), which are accompanied by specific inquiries into *Ecology of the Human Mind and Human Relations* (Part VIII) and *Flesh, Body, Embodiment/Space and Time* (Part IX).

The contributions in Part III, *Seeds of Eco-Phenomenology*, explore crucial and heretofore unknown passages of the new ecological outlook opened with the discovery and metaphysical valorization of the ontopoietic logos of life. Maija Kule, in her broad inquiry, "Eco-Phenomenology: Philosophical Sources and Main Concepts," observes that the new developing trend of Eco-phenomenology is rooted in the *philosophia naturalis* of Schelling, Nietzsche, Bergson, Husserl, and Merleau-Ponty, addressing the problems of the human being, nature, life, and the universe. Compared to the prevalent politically engaged character of eco-philosophy to date, the eco-phenomenology of Anna-Teresa Tymieniecka is to be acknowledged for its theoretical depth and inclusivity, as it joins concepts such as Logos and Life, the Unity-of-Everything-there-is-Alive, Ontopoiesis, the Individualization of Life, and *Imaginatio Creatrix*. In "Logos of Life and Logos of Science. Metaphysical Advice," Franco Bosio stresses the importance of recovering the specificity and originality of the primordial phenomenon of life, neglected and almost forgotten by modern and contemporary science. Even philosophy falls into making the same mistake and separates the domains of spirit and the human sciences from their roots in life. On the contrary, it is of primary importance to recover the fundamental core of the primordial phenomenon of life that is to be found in the idea of "force," following the intuitions of great philosophers such as Husserl, Scheler, and Jonas and the original approaches of philosophically gifted scientists such as L. von Bertalanffy, F. Capra, H. Maturana, and F. Varela. In "An Insight into the Foundations of Eco-Phenomenology," Massimo Marassi observes that in gaining for Husserlian phenomenology the valorization of the principle of life, Tymieniecka provides the crucial orientation that enables phenomenology to carry out its mission in the world today, freeing us from the mental conditioning generated by the ideology of technique and motivating us to face the great challenge of sustainability in environmental and human development. In fact, as Dario Sacchi explains in his contribution, "Some Questions about Idealism and Realism in the Structure of Husserlian Phenomenology," Tymieniecka, by rooting the Husserlian constituent transcendental consciousness within the unity-of-everything-there-is-alive, succeeds in overcoming the dualism between consciousness and the world that blocked classical phenomenology, keeping it from fulfilling its speculative mission. In § 53 of *Ideas I*, Husserl speaks of two main aspects of consciousness, which, on the one side, presents itself as an all-encompassing *horizon* and, on the other, is included in the world and pertains to the world. Both aspects are undeniable but are very difficult to reconcile. This reconciliation between the transcendental and the worldly dimensions of consciousness can only happen if, as Tymieniecka has done, one phenomenologically recovers the living quality of the subject. In effect, objective observation reveals that, differently from other beings that as "things" are what they are, the subject is not what he is; this is not so in the sense that he would also be something else or that indeterminacy inheres in him, but in the sense that he exists only in denying himself. The subject is never identical with himself, he never quiets down in himself: rather, his ontological status is the becoming that is life and activity, through which he continually passes beyond conditions of both an absolute standing outside the world and a mere worldly being.

The three scientific contributions that follow in Part III cultivate the ground of the hard sciences to make place for the principle of life in physics, exploring those connections that make it possible to include with full rights the world of the inorganic in the ecological vision of the all: Ion Soteropoulos in "The Origin Paradox: How Could Life Emerge from Nonlife?," Attila Grandpierre in "The Fundamental Biofriendly Activity of the Universe," and Mamuka Dolidze in "Cosmic Harmony, the Emergence of Life and of Human Consciousness."

Part IV addresses *Eco-Cosmology*. According to Stefano Veluti in his "*Welt.* At the Origins of Eco-Phenomenology: Heidegger's Concept of World in the Work of Anna-Teresa Tymieniecka," the phenomenology of life of Anna-Teresa Tymieniecka has affinities with the thinking of Heidegger such that a kindredness can be discerned between his "Welt" and her "unity of everything-there-is-alive," upon which an eco-cosmology can be built. This undertaking is seen in the contributions of Nikolay N. Kozhevnikov and Vera S. Danilova, "Life and Human Life in the System of World Coordinates on the Basis of Extreme Dynamic Equilibriums"; Debika Saha, "Eco-Phenomenological Vision: Balancing the Harmony of the Earth"; María Avelina Cecilia Lafuente in "From Anna-Teresa Tymieniecka's Eco-Phenomenology to Paul Ricoeur's Hermeneutics. The Role of the Human Being in the Global Context of Cosmos, Chaos and Evil"; and Konul Bunyadzade, in "Ego: The Cross Point of Divine Illumination and Social Reality."

In the framework of cosmological theory thus re-considered, space is opened up among the environmental theories for an Eco-Ethics. This theme is sounded in Part V, *Eco-Ethics and Environmental Theories*, as illustrated by Ella Buceniece in "Phenomenology as Ecology: Movement from Ego- to Geo- and Eco-Thinking," Kimiyo Murata-Soraci in "'Song of the Earth': An Eco-Phenomenology," Alessandra Lucaioli in "Cultural Sustainability: Lines of Reflection for a Human Life in the Harmony of the Cosmos," Renato Boccali in "The Geology of Movement. The Earth and the Dynamic of Phenomenalisation in Merleau-Ponty and Patočka," and Piotr Mróz in "On Two Versions of Phenomenological Transgressions – Anna-Teresa Tymieniecka and Jean-Paul Sartre."

The *Eco-phenomenological Readings* made in Part VI confirm from various points of view the connections between nature and culture that only an eco-phenomenology founded on the ontopoiesis of life can guarantee in their discrete continuity: Rosemary Gray in "Sowing 'A Quilt of Harmony': An Eco-Phenomenological Reading of Ben Okri's 'Lines *in Potentis*' from *Wild* (2012)," Bronisław Bombała in "Eco-Phenomenology of Scientific Activity as Non-Routinized Routine: Stefan Banach's Café Method of Research and Its Contemporary Continuation," Raymond Langley in "Sartre on Marx and Freud: A Phenomenological Dialectic of Universal Singulars and Singular Universals," Valentina Carella in "Eco-Phenomenology: The Japanese Original Perspective in the Thought of Nishida Kitaro," Ming-Qian Ma in "'Negative Seeing': Robert Smithson, Earth Art, and the Eco-Phenomenology of 'Mirror Displacements,'" Koushik Joardar in "The Transcendental Philosophy of Krishnachandra: An Indian Approach to Human Life," and Anna Małecka and Katarzyna Stark in "Henryk Skolimowski's Eco-Philosophy as a Project of *Living Philosophy*."

Among the new fields to explore is *Eco-Anthropology* (Part VII). Herein are examined aspects of *Sentience* (by Lena Hopsch in "Small Talk with a Grape Vine: Presence and the Sensuous Depth of Being," and Ineta Kivle in "Auditory Phenomena and Human Life: Phenomenological Experience"); *Desire* (by Roberto Marchesini in "Animal Being Means Desiring: Subjectivity, Singularity, Diversity in Post-Human Life"); *Language* (by Antonio Domínguez Rey in "The Language That (In) habits Us," and Erkut Sezgin in "Phenomenological Elucidations Carried Out by Constructing a Phenomenological Language"); and *Creativity* (by Andrew Chesher in "Phenomenology after Conceptual Art," and Massimo Mezzanzanica in "The Human Condition, Nature, Power and Creativity. Philosophical Anthropology and Eco-Phenomenology in the Context of Biopolitics").

Part VIII, which explores *Ecology of the Human Mind and Human Relations*, provides a broader context for studies of phenomenological psychopathology, with Anna Piazza addressing "The Concept of Life in Ludwig Binswanger's Psychopathology," Giulio Lo Bello exploring "Meaningless Life: The Role of Clinical Phenomenology in Understanding the 'Being in the World' of Psychiatric Patients," and Antonio De Luca advancing "An Ecological Perspective on the

Helping Relationship." The ecological outlook also lends greater amplitude to the psychological and sociological contributions made here by Velga Vevere in "Experience of the City: An Eco-Phenomenological Perspective," Angela Ales Bello in "Eco-Phenomenology of the Human Environment: The Case of Intercultural Dialogue," and Jan Szmyd in "Digital Reason vs. the Modern 'Metamorphosis of Man': From the Perspectives of the Philosophical Anthropology of Józef Bańka and Anna-Teresa Tymieniecka."

At this point, particular importance accrues to the theme of *Embodiment*, or in other words, the condition by which the living personal dimension is inflected in cosmic space and time, in its striving to harmonize its creative freedom with the constructivism of nature. This emblematic aspect of eco-phenomenology is explored in Part IX, *Flesh, Body, Embodiment/Space and Time,* by Aleksandra Pawliszyn in "From the Archeology of Happenings ... the Matter of Corporeality," Kamil Łacina in "Multi-Layered Time and the Unity of the Unfolding Logos of Life," Carla Danani in "The Question of (AI)Location," Salahaddin Khalilov in "Holographic Memory of Life Situation," Sadaqat M. Aliyeva in "The Chronotopic Content of the Esoterism and the Models of Thought," and Roberto Diodato in "Phenomenology of the Virtual Body. An Introduction."

In conclusion, the main benefit that the consultors of our volume can receive certainly consists in sustaining the conviction that "it is only in a direct, immediate insight into the constructivism of life and its coincidence with our own constructivism that we may expect to disentangle and grasp life's patterns"[1] and consequently find suitable nourishment for philosophy and humanity itself, so as to be provisioned for successfully living the change of epoch it is our lot to face.

Macerata, Italy — Daniela Verducci

## Note

1. A.-T. Tymieniecka. *Logos and Life* - Book 4: *Impetus and Equipoise in the Life-Strategies of Reason*. Analecta Husserliana LXX. Dordrecht/ Boston/London: Kluwer Academic Publisher, 2000. 5.

# Acknowledgments

It is with great pleasure that we present this volume to the scholarly public. This volume includes the papers presented at the World Phenomenology Institute's 64th Congress of Phenomenology, ***Eco-Phenomenology: Life, Human Life, Post-Human Life in the Harmony of the Cosmos,*** held in Milan, Italy, October 1–3, 2014. The World Phenomenology Institute would like to thank the Catholic University of the Sacred Heart in Milan for hosting the conference. In particular, we thank the following people at the Catholic University of the Sacred Heart for their tireless work and assistance in organizing the congress: Professor Francesco Botturi, the University's Vice-Rector, Professor Massimo Marassi, the Director of the Philosophy Department, and Professor Dario Sacchi. We also thank Dr. Emilia Andri of the University of Bergamo and Dr. Renato Boccali of the International University of Languages and Media in Milan, both of whom worked with competence and dedication to assure the success of the conference. And, we of course thank the conference participants who submitted their papers for publication consideration to *Analecta Husserliana, The Yearbook of Phenomenological Research*, making the publication of this pioneering work possible. The World Phenomenology Institute also wishes to acknowledge the valuable editorial assistance of Mr. Robert J. Wise, who has been a long-time collaborator of the Institute.

William S. Smith
Jadwiga S. Smith
Daniela Verducci

# Contents

**Part I Anna-Teresa Tymieniecka Memorial**

**The Symphony of Sentience, in Cosmos and Life: In Memoriam Anna-Teresa Tymieniecka** . . . . 3
Olga Louchakova-Schwartz

**Part II Introduction to the Topic**

**The World Phenomenology Institute's Eco-Phenomenology** . . . . 17
Daniela Verducci

**Metaphysics and Eco-Phenomenology Aiming at the Harmony of Human Life with the Cosmos** . . . . 25
Francesco Totaro

***Ontopoiesis of Life* as Eco-Phenomenology** . . . . 31
Carmen Cozma

**Part III Seeds of Eco-Phenomenology**

**Eco-Phenomenology: Philosophical Sources and Main Concepts** . . . . 43
Maija Kūle

**Logos of Life and Logos of Science. Metaphysical Advice** . . . . 59
Gianfranco Bosio

**An Insight into the Foundations of Eco-Phenomenology** . . . . 69
Massimo Marassi

**Some Questions About Idealism and Realism in the Structure of Husserlian Phenomenology** . . . . 79
Dario Sacchi

**The Origin Paradox: How Could Life Emerge from Nonlife?** . . . . 87
Ion Soteropoulos

**The Fundamental Biological Activity of the Universe** ................. 115
Attila Grandpierre

**Cosmic Harmony, the Emergence of Life and of Human Consciousness** ................................... 141
Mamuka Dolidze

**Part IV Eco-Cosmology**

***Welt*. At the Origins of Eco-Phenomenology: Heidegger's Concept of "World" in the Work of Anna-Teresa Tymieniecka** ......... 155
Stefano Veluti

**Life and Human Life in the System of World Coordinates on the Basis of Extreme Dynamic Equilibriums** ..................... 175
Nikolay N. Kozhevnikov and Vera S. Danilova

**Eco-Phenomenological Vision: Balancing the Harmony of the Earth** ............................................. 187
Debika Saha

**From Anna-Teresa Tymieniecka's Eco-Phenomenology to Paul Ricoeur's Hermeneutics. The Role of the Human Being in the Global Context of Cosmos, Chaos and Evil** .............. 191
María Avelina Cecilia Lafuente

**Ego: The Cross Point of Divine Illumination and Social Reality** ........ 217
Konul Bunyadzade

**Part V Eco-Ethics and Environmental Theories**

**Phenomenology as Ecology: Movement from Ego- to Geo- and Eco-Thinking** ............................... 225
Ella Buceniece

**"Song of The Earth": An Eco-Phenomenology** ...................... 235
Kimiyo Murata-Soraci

**Cultural Sustainability: Lines of Reflection for a Human Life in the Harmony of the Cosmos** ............................. 245
Alessandra Lucaioli

**The Geology of Movement. The Earth and the Dynamic of Phenomenalisation in Merleau-Ponty and Patočka** ................. 255
Renato Boccali

**On Two Versions of Phenomenological Transgression – Anna-Teresa Tymieniecka and Jean-Paul Sartre** . . . . . . . . . . 263
Piotr Mróz

**Part VI Eco-Phenomenological Readings**

**Sowing "A Quilt of Harmony": An Eco-Phenomenological Reading of Ben Okri's "Lines *in Potentis*" from *Wild* (2012)** . . . . . . . . . . 281
Rosemary Gray

**Eco-Phenomenology of Scientific Activity As Non-Routinized Routine: Stefan Banach's Café Method of Research and Its Contemporary Continuation** . . . . . . . . . . 291
Bronisław Bombała

**Eco-Phenomenology: The Japanese Original Perspective in the Thought of Nishida Kitaro** . . . . . . . . . . 309
Valentina Carella

**"Negative Seeing": Robert Smithson, Earth Art, and the Eco-Phenomenology of "Mirror Displacements"** . . . . . . . . . . 321
Ming-Qian Ma

**The Transcendental Philosophy of Krishnachandra: An Indian Approach to Human Life** . . . . . . . . . . 339
Koushik Joardar

**Henryk Skolimowski's Eco-Philosophy as a Project of Living Philosophy** . . . . . . . . . . 347
Anna Małecka and Katarzyna Stark

**Part VII Eco-Anthropology: Sentience, Desire, Language, Creativity**

**Small Talk with a Grape Vine: Presence and the Sensuous Depth of Being** . . . . . . . . . . 359
Lena Hopsch

**Auditory Phenomena and Human Life: Phenomenological Experience** . . . . . . . . . . 367
Ineta Kivle

**Animal Being Means Desiring: Subjectivity, Singularity, Diversity in Post-Human Life** . . . . . . . . . . 375
Roberto Marchesini

**The Language That (In)Habits Us** . . . . . . . . . . 387
Antonio Domínguez Rey

**Phenomenological Elucidations Carried Out by Constructing a Phenomenological Language** . . . . . . . . . . . . 405
Erkut Sezgin

**Phenomenology After Conceptual Art** . . . . . . . . . . . . 417
Andrew Chesher

**The Human Condition, Nature, Power and Creativity. Philosophical Anthropology and Eco-Phenomenology in the Context of Biopolitics** . . . . . . . . . . . . 435
Massimo Mezzanzanica

**Part VIII Ecology of the Human Mind and Human Relations**

**The Concept of Life in Ludwig Binswanger's Phenomenological Psychopathology** . . . . . . . . . . . . 451
Anna Piazza

**Meaningless Life: The Role of Clinical Phenomenology in Understanding the "Being in the World" of Psychiatric Patients** . . . . . 461
Giulio Lo Bello

**An Ecological Perspective on the Helping Relationship** . . . . . . . . . . . . 469
Antonio De Luca

**Experience of the City: An Eco-Phenomenological Perspective** . . . . . . . . . 479
Velga Vevere

**Eco-Phenomenology of the Human Environment: The Case of Intercultural Dialogue** . . . . . . . . . . . . 491
Angela Ales Bello

**Digital Reason vs. the Modern "Metamorphosis of Man": From the Perspectives of the Philosophical Anthropology of Józef Bańka and Anna-Teresa Tymieniecka** . . . . . . . . . . . . 503
Jan Szmyd

**Part IX Flesh, Body, Embodiment/Space and Time**

**From the Archeology of Happenings ... the Matter of Corporeality** . . . . . 517
Aleksandra Pawliszyn

**Multi-layered Time and the Unity of the Unfolding Logos of Life** . . . . . . . 527
Kamil Łacina

**The Question of Placeness** . . . . . . . . . . . . 541
Carla Danani

**Holographic Memory of Life Situation** . . . . . . . . . . . . . . . . . . . . . . . . . . . . . 553
Salahaddin Khalilov

**The Chronotopic Content of Esoterism and the Models of Thought** . . . . . . . . . . . . . . . . . . . . . . . . . . . . . . . . . . . . . . 561
Sadaqat M. Aliyeva

**Phenomenology of the Virtual Body: An Introduction** . . . . . . . . . . . . . . . . 569
Roberto Diodato

# Contributors

**Sadaqat M. Aliyeva** Azerbaijan University, Baku, Xatai region, Azerbaijan Republic

**Angela Ales Bello** Pontifical Lateran University, Rome, Italy

**Giulio Lo Bello** Faculty of History and Philosophy, University of Latvia, Riga, Latvia

**Renato Boccali** IULM University, Milan, Italy

**Bronisław Bombała** Uniwersytet Warmińsko-Mazurski w Olsztynie, Olsztyn, Poland

**Gianfranco Bosio** University of Verona, Verona, Italy

**Ella Buceniece** Institute of Philosophy and Sociology, University of Latvia, Riga, Latvia

**Konul Bunyadzade** Azerbaijan National Academy of Sciences, Baku, Azerbaijan

**Valentina Carella** Università degli Studi di Macerata, Macerata, Italy

**María Avelina Cecilia Lafuente** University of Seville, Seville, Spain

**Andrew Chesher** Chelsea College of Arts, University of the Arts London, London, UK

**Carmen Cozma** Department of Philosophy, "Alexandru Ioan Cuza" University of Iași, Iași, Romania

**Carla Danani** Università degli Studi di Macerata, Macerata, Italy

**Vera S. Danilova** North-Eastern Federal University, Yakutsk, Russia

**Roberto Diodato** Università Cattolica del Sacro Cuore, Milan, MI, Italy

**Mamuka Dolidze** Tbilisi Javakhishvili State University, Tbilisi, Georgia

**Attila Grandpierre** Konkoly Observatory, Budapest, Hungary

**Rosemary Gray** Department of English, University of Pretoria, Pretoria, South Africa

**Lena Hopsch** Representative of the Faculty of Chalmers, Department of Architecture, Chalmers University of Technology, Gothenburg, Sweden

**Koushik Joardar** Department of Philosophy, University of North Bengal, Darjeeling, West Bengal, India

**Salahaddin Khalilov** Azerbaijan University, Baku, Azerbaijan

**Ineta Kivle** University of Latvia, Riga, Latvia

**Nikolay N. Kozhevnikov** Philosophical Society of the Sakha Republic (Yakutia), Sakha (Yakutia), Russia

**Maija Kūle** Institute of Philosophy and Sociology, University of Latvia, Riga, Latvia

**Kamil Łacina** Institute of Philosophy, Jagiellonian University, Kraków, Poland

**Olga Louchakova-Schwartz** Jesuit School of Theology, Berkeley, CA, USA

Department of Public Health Sciences, University of California at Davis, Davis, CA, USA

**Antonio De Luca** The Magna Græcia University, Catanzaro, Italy

University of Calabria, Rende, Italy

**Alessandra Lucaioli** University of Macerata, Macerata, Italy

**Ming-Qian Ma** Department of English, The State University of New York at Buffalo, Buffalo, NY, USA

**Anna Małecka** AGH University of Science and Technology, Krakow, Poland

**Massimo Marassi** Direttore del Dipartimento de Filosofia, Università Cattolica del Sacro Cuore, Milan, Italy

**Roberto Marchesini** Centre Study for Post-human Philosophy, Bologna, Italy

**Massimo Mezzanzanica** Università degli Studi di Milano, Milan, Italy

**Piotr Mróz** Institute of Philosophy, Jagiellonian University, Kraków, Poland

**Kimiyo Murata-Soraci** Chuo University, Tokyo, Japan

**Aleksandra Pawliszyn** Academy of Arts, Gdańsk, Poland

**Anna Piazza** Department of Humanities Universidad Francisco de Vitoria, Madrid, Spain

Instituto Schuman de Estudios Europeos Universidad Francisco de Vitoria, Madrid, Spain

**Antonio Domínguez Rey** Departamento de Lengua Española y Lingüística General, Facultad de Filologia, Universidad Nacional de Educación a Distancia, Madrid, Spain

**Dario Sacchi** Università Cattolica del Sacro Cuore, Milano, Italy

**Debika Saha** Department of Philosophy, University of North Bengal, Darjeeling, West Bengal, India

**Erkut Sezgin** Independent scholar, Ataköy, Istanbul, Turkey

**Ion Soteropoulos** Apeiron Centre, Paris, France

**Katarzyna Stark** AGH University of Science and Technology, Krakow, Poland

**Jan Szmyd** University of Krakow, Krakow, Poland

**Francesco Totaro** University of Macerata, Macerata, Italy

**Stefano Veluti** University of Macerata, Macerata, Italy

**Daniela Verducci** University of Macerata, Macerata, Italy

**Velga Vevere** University of Latvia, Riga, Latvia

# Part I
# Anna-Teresa Tymieniecka Memorial

# The Symphony of Sentience, in Cosmos and Life: In Memoriam Anna-Teresa Tymieniecka

**Olga Louchakova-Schwartz**

**Abstract** In this tribute to Anna-Teresa Tymieniecka, I examine whether phenomenology, as a cognitive enterprise of sentience, extends beyond the death of the physical body. The symphonic form of this paper is inspired by Tymieniecka's *Metaphysical Rhapsodies of Faith*. The paper's First Movement, *Allegro*, sets the stage by drawing distinctions between the concept of sentience in the Phenomenology of Life and that of the intentionality of consciousness as viewed in cognitive phenomenology with its strong Husserlian legacy. The Second Movement, *Moderato*, follows the thread of sentience in the labyrinth of life through the ontopoietic patterns of complexity and emergence in both animate and inanimate nature. The Third Movement, *Minuet*, shows the hylomorphic unity of sentience and patterns of complexity in the nonlinear dynamics of the brain, and discusses qualia of thought, and the informational patterns of consciousness. Sentience is claimed to be a cosmic property which is appropriated by the phenomenologically material subjectivity. Since the intuition of life is the self-revelation of sentience, intuition of life continues in dying and has no reason not to continue beyond. The Fourth Movement, *Finale Glorioso*, describes patterns of sentience in the soul's final ascent, and honors the intuitional gifts that proceed from the passage of a great soul.

**Keywords** Tymieniecka · Phenomenology of life · Sentience · Intentionality · Complexity · Consciousness · Death and dying · Cosmos

---

This is a keynote address in memoriam of Anna-Teresa Tymieniecka, October 1, 2014, 64th International Phenomenology Congress of the World Phenomenology Institute, held at the Catholic University of the Sacred Heart in Milan.

O. Louchakova-Schwartz (✉)
Jesuit School of Theology, Berkeley CA, USA

Department of Public Health Sciences, University of California at Davis, Davis CA, USA
e-mail: olouch@ucdavis.edu

 
W. S. Smith et al. (eds.), *Eco-Phenomenology: Life, Human Life, Post-Human Life in the Harmony of the Cosmos*, Analecta Husserliana CXXI,
https://doi.org/10.1007/978-3-319-77516-6_1

Anna-Teresa Tymieniecka's book *The Fullness of the Logos in the Key of Life,* Book II: *Christo-Logos, Metaphysical Rhapsodies of Faith* (*Analecta Husserliana*, 2013) is a soliloquy with variations on the themes of beauty, creativity, and the sacred. Appearing after it *Art, Literature and the Passion of the Skies*, it serves as a Chorus on the themes outlined in the *Metaphysical Rhapsodies*. Why did this philosopher talk to the themes of beauty, creativity, and the sacred at the close of her discourse? How do these themes contribute to the turn from the Phenomenology of Life to the Philosophy of Life, which was announced in the *Metaphysical Rhapsodies of Faith*?

An experience of beauty is salutary. According to the Sufi metaphysics, beauty is an antidote to egotism. An experience of beauty relaxes the mind and loosens the grip of the ego to open the road for the sacred in direct experience.[1] Likewise, a creative moment breaks through the consensus trance creating a window in which one can *see*.[2] Tymieniecka hasn't been writing directly about death, but beauty, creativity, and the sacred indirectly prepare the soul for dying. Death, in this case, is not an annihilation but, rather, a possibility for the new knowledge. Knowledge always has been of the highest value for Professor Tymieniecka. We see that the final turn from the Phenomenology of Life to the Philosophy of Life is not a mere re-categorization of an already-completed philosophical project.[3] As had been the case with her great predecessors in Ancient Greek philosophy, the imminence of departure ignites a new insight.

Obviously, dying radically changes one's capacity for subjective self-reporting. As a phenomenologist of life, is there a way to understand not just the meaning of mortality, but the event of death itself? Tymieniecka stated many times that the intuition of life as a specific kind of intuition. According to Tymieniecka, life is grasped not only categorically in eidetic intuition, or as a meaning of life, but through the direct intuition of life as such.[4] In this view, judgments of what is alive and what is not alive are made out of the pre-reflective, apperceptive primary facts of experience. Can an event of death appear as an alteration in this self-revelation of life?[5] Can death be considered as a subjectively lived phenomenon? If dying takes time would a temporally extended intuition of change in self-revelation of life be available to subjective self-reporting and/or to phenomenological investigation? Does life and, with it, the Phenomenology of Life, end up where death begins?

It certainly appears that death is available only in an analysis of its meaning, in speculation, through ciphering. Perhaps, death is a concept related to biology, a death of a body, never visible to its subject, but only in relation to the other. Perhaps, death is not a metaphysical counterpart of life, not self-subsistent as is the latter. Death evaporates amidst life's appearances, in which each death of an earlier condition is superceded by a new stage of life. Turning to the concretum in the *Metaphysical Rhapsodies,* Tymieniecka transplants the lofty vision of the Phenomenology of Life into a psychospiritual, existential horizon. In life as *this* human life, and especially for a phenomenologist, death is tantalizing; like Savitri of the Indian legend by the same name, one wishes to interrogate death directly, to bring the great horrible Yama of Indic mythology into the light of description.[6] Like Shvetaketu in the Katha Upanishad,[7] the Philosopher-Phenomenologist draws from the confrontation with Death the final act of her phenomenological interrogations, and the understanding of life, of the human predicament, and of the transnatural destiny of the soul. Like Ibn 'Arabi's Jesus, Anna-Teresa Tymieniecka witnesses and testifies against illusions of our finality. Via the brotherhood of those who turn to their internal ontopoietic source in search of the truth, the soul transcends its carnal beginnings. This is the existential-transcendental turn, from Phenomenology to Philosophy of Life, not in the manner of Hadot's "philosophy as a way of life"[8] but, rather, in a Socratian way of dying as philosophy. One sees the record of phenomenological investigations, an intuitional switch in the living horizons of inquiry, and an actual life event (which is death) coming together. Tymieniecka's death brings together the acts of consciousness, the acts of communication, and the act of living, now one, like a crystal of salt. The difference between the written and the lived disappears.

What place does this existential-humanistic-transcendental-transnatural turn assume in Tymieniecka's Phenomenology of Life? Like many philosophers of direct intuition, Tymieniecka began with psychological and introspectionist agendas. At the close, she revisits her intellectual beginnings, but in a new manner: her intuition is free from psychologistic assumptions, the ontopoiesis of life is fully explicated. Guided by the Logos of her inquiry, she followed the thread of sentience in the labyrinth of life. All of her phenomenology, all of her philosophy is, basically, metaphysics. This lofty metaphysics underlies and contains all of this philosopher's expressions, human, as it were, in the cosmos and life. This knowledge is certain; therefore, the mind can now bounce back from the pristine metaphysical truths into the messiness of existential commitment. This "glass menagerie" of relationships, this bestiary of the heart, renders metaphysical gems of extreme importance. The patterns of revelation continue in all manifestations of life, whether lowly or holy. How does the Logos of Life show within the human, existential horizon? The missing link between the metaphysical *Imaginatio Creatrix* and human artistry is through spontaneity which now enters the philosophical discourse. The reflection on spontaneity leads to uncovering the main logoic expression within the existential horizon, which is the transnatural destiny of the soul. The finality of the biological body stands out now as a positive value, as a background against which the figures of the destiny can be seen, and its trans-natural terminus can be discerned.

A soliloquy truns into a dialogue. Tymieniecka asks, "Brethren, having covered this complete cycle together are we not truly, at last, Brethren" (Tymieniecka 2012a, 194)? Tethered to the same Logos, we must now extend the cycle of inquiry and articulate the phenomenology of life in death. This will be an act of phenomenological intuition and philosophical reflection, as well as an act of spontaneity in the human artistry. In Tymieniecka's strategy of knowledge, temporality is extremely important. The logos of things is revealed not through a static "what" but in a dynamic "how," from the primeval ontopoietic blossoming to the developed hierarchies of life. Like leaves coming out of a palm tree, rationalities and virtualities are incubated within the previous conditions and grow into the posterior conditions. Tymieniecka here writes simply, without neologisms or special terminology: like the Vedantic sages, she works within ordinary language, excavating the primary meanings that are tied to the prereflective, to the intuited reality of life.[9] If ontopoiesis shows up in the metaphysical horizon as sentience within the existential human horizon, in the innate messiness of experience, it shows up as spontaneity. On the stage of human life, spontaneity is an expression of transcendental creativity in each manifest moment of Life.

A prelude to the showing of spontaneity is the act of bracketing. First, one stops thinking that death equals annihilation and is a step into nonbeing; the fallacy of this assumption is based on an identification of consciousness with the appearance of the body. Next, comes a polar opposite, a "spiritual" assumption that death is a step into a primordial non-dual state, a sort of liberation from all impressions into undisturbed, content-less serenity in which the individual "I" dissolves into the Ocean of Being. The idea of such "return" is central in many spiritual paths. Together with Anna-Teresa Tymieniecka, one shall ask: is such a primeval state, indeed, our destination in death or, for that matter, in the spiritual Enlightenment? An undifferentiated indivisibility – is that where the soul is heading? Tymieniecka answers: if such ground is thought of as a primeval state of man, it cannot be our destination; that would be the mind's returning to the most primitive of its mental states. Such assumption needs to be dismissed; the "journey to the source" schema of things is thereby is rejected. The creative process of life cannot involve a circular teleology in which one returns to life's primitive state; therefore, we must seek the source of spontaneity through spontaneity itself. Tymieniecka says, "Experience as basically a spontaneity is obviously strictly individual, unshareable" (Tymieniecka 2012b, 1). From the depth of subjectivity, spontaneity drives the patterns of experience, the patterns of awareness, action, creativity, and ultimately, the weaving of destiny. Spontaneity creates the temporally extended existential patterns; are not these the same patterns that our old friend, sentience, designs for all the manifestation of life? Sentience walks the red carpet of existential expression unfolded by her handmaiden, spontaneity.

A human face of sentience dawns on us. Tymieniecka says:

> How often have we forgotten all these acquired means [the existential and psychological dimensions of life], and started from a primitive germ in ourselves, on our own, without a spark of outward light or a word of courage. We have followed our inner spontaneity wherever it may lead us and thus step by step have dug into the soil of our being and along the sacred river where our roots plunge, have retraced the path, the winding path of the genesis of our authentic life. We have rediscovered the light within ourselves. (Tymieniecka 2012a, 194)

And, we may add, the self-same light shines on us from the outward edges of life. We intuit life by the recognition of sentience; sentience is in the breath of a rose, in the self-assembly of the protein molecules of the organic brine, and in the essence of the spiritual wisdom. She is us, and she is not us. As Ibn 'Arabī puts it:

> She displayed her front teeth, and a levin flashed,
> and I knew not which of the twine rent the gloom.
> And she said: "Is it not enough for him that I am in his heart,
> and that he behold me at every moment? Is it not enough?" (Ibn 'Arabi, 1978, 57)

Sentience we know by direct intuition. Intuition is generally thought of as the intentional knowledge. If the New Enlightenment, as heralded by Tymieniecka,[10] proposes the possibility of new knowledge, must this knowledge be always intentional? Can there be a better knowledge, a more direct knowledge than the knowledge as "aboutness?" Can we know sentience as it is known to itself, in its marks on both sides of the subject-object equation?[11] Can there be consciousness beyond the enduring presence of subjectivity? As part of this tribute to Anna-Teresa Tymieniecka, I wish to understand whether phenomenology, as a cognitive enterprise of sentience, extends beyond the death of the body. Another way of asking this will be: what is the meaning of resurrection in the New Enlightenment? Out of the many faces of sentience, which one is revealed in the modification of mental attitude created by dying? We will now follow the melodic fabric of Tymieniecka's rhapsody of the trans-natural destiny of the soul, into phenomenology of death as it appears to us, to develop the theme into a full symphonic cycle.

A contour of this new enquiry is outlined by the idea of the trans-natural destiny of the soul. Such a mode of being must assume knowledge beyond the biological determinants of the human mind (in effect, of intentionality). An idea of the trans-natural destiny suggests that there must other ways of consciousness than is in the the biology of the human form. The associational fabric of Tymieniecka's discourse enhances a new *Verstehen*, one that is non-sensory, non-intellectual, non-transcendental in phenomenological terms.[12] This may be the intuition of physics, of the neither direct nor indirect knowledge of Vedanta, an intuition of the phase shifts and nonlinear dynamics of complexity. In this intuition of essential substance, there is a key to the transcendence of subject-object dichotomies, of causality, of determinism and of all the traps of beginning-and-end schemata.[13] In light of Tymieniecka's departure, the logic of her last writings invites such reflection.

## Allegro

Tymieniecka's *Metaphysical Rhapsodies of Faith* overcome the natural attitude with its ciphering of life and death; she speaks from the place where life and death unite. *Christo-Logos* is, in fact, an esoteric book. It is esoteric in the sense that it points to a possibility of an immortal "substance" (for the lack of a better term) that is primeval both to the human condition and to a larger life. It is a bridge to the totally invisible and, therefore, not given in appearances, that is to say, invisible but sustaining all appearances.[14] Neither measurable nor immeasurable, not the body and not its perceived flesh, it is, in the essence of things, "sovereign and conditioned at the same time" (see Ibn 'Arabi 1975, 34).

Tymieniecka says, "We have discovered the light within ourselves, each coming upon his innumerable experiences yielding evidence of the eternal precept revealed to the mankind and so opening our opaque, enfleshed being to the Absolute"[15] (Tymieniecka 2012a, 194). This light, this experience, which is so strictly subjective that it cannot possibly be shared, is known via spontaneity in the transcendental-existential turn of Phenomenology toward Philosophy of Life. In other contexts, it is sentience. *Allegro*, the First Movement of my symphony, is a cadence of sentience. Not at all intentionality, a structural relation in itself, sentience is embedded in *all* of meaning and *hyle*. In analytic philosophy, a kindred concept is the famous "what it's like," the qualia. In the debate over cognitive phenomenology, one asks[16] is there a quale to thought? In accordance with Tymieniecka's Philosophy of Life, we should ask: are there qualia to sentience? Or, is sentience at the core of any and all qualia, at the unshareable, intimate dimension of subjectivity? If so, does this essence of qualia-ness go away in the moment of death, when all intentional relations seem to collapse together with sensory data and biologically dependent intentionalities? Are we, then, something or nothing?

## Moderato

A specific contour of our symphonic theme is beginning to emerge. In the Second Movement, *Moderato*, I will follow the Tymienieckian thread of sentience in the labyrinth of animate and inanimate nature. Sentience is an ontological platform for phenomenality; it is also an essence of all qualia, which is to say, the quale animating every qualia. That established, we will now adjust the horizon of inquiry. We proceeded from the spontaneity within the existential horizon to its ontological ground, sentience; we moved from the sheen to living silver itself. We will now glean from sentience not from the standpoint of the human condition but, rather, in its cosmic status. If the Descartes-Vico polarity of possible knowledge shifts between subject and object, the pirouette of our intuition would be in a direction away from either, from the ek-static outside of life in appearances into its inside – basically, into the substance of the appearances of the world.

Tymieniecka's idea of The New Enlightenment is that it brings with it a revolution in scientific thinking. Science is no longer concerned exclusively with the factic properties of objects but with new ontologies, such as that of complexity theory. To me complexity theory reads as a mathematical expression of the Phenomenology of Life; its phase shifts and strange attractors are remarkably reminiscent of the ontopoietic novum in Tymieniecka's vision. In cognitive theory, it has been proposed many times that what subjectively comes across as qualia of thought is, in essence, a set of informational relationships pertaining to cognition or, rather, information per se that is subjectivistically appropriated in the human experience of thought.[17] Information transcends subject-object relations; subject-object relations themselves can be viewed as informational relations. We have no evidence that information pertains only to the physical universe; on the contrary, we have evidence from quantum mechanics and mathematics that things are quite the opposite: observer and physical universe are there mutually engaged, as in, for example, the Belavkin equation or Heisenberg's Uncertainty Principle. Further, life and death are intermingled (cf. Schrödinger's mental experiment with a cat and *Verschränken* (entanglement), when a cat can be both alive and dead). Information can be thought of as being a substratum on the sides of subject and object each, in *noema* and *noesis*, in both the conscious self and the material, non-conscious, and philosophically non-existent world. In the history of thought, a similar view was developed by the Islamic philosopher Suhrawardī in his argument concerning the identity of subjective awareness and objective visible light; this substance just appears to be of different kinds in the subject-object differentiation.[18] Consequently, if we zoom intuition into sentience in the spectacle of experience and focus exclusively on the qualia of pure subjectivity, what we experience is not just something monadic and purely qualia-tative, or even analyzable in structural terms, but also what self-subsistently is, and is self-subsistently alive.

## Minuet

The dance of nonlinear complexity unfolds in the Third Movement, *Minuet*, in which the informational patterns of consciousness, cosmos, and life must be recognized as the arabesques of sentience. In this dance, sentience is not a source or primeval ground into which we can tap or to which we can return but is, rather, as the phenomenological substance of all appearances, the very there-ness and facticity of all things. The lived existential property of spontaneity springs forth from sentience. – this can be discovered in an introspective experiment with a movement of the mind towards the pure subjectivity. The selfsame sentience unfolds as complex relationships in the primeval waves of information that compile our universe. In its downward arc, these sentient informational patterns are given to us as the appearances of life and the spontaneity of sentience – as the creativity of the human subject. These patterns manifest themselves in upward and downward arcs of life, and, therefore, as a minuet.

## Finale Glorioso

So far, we submitted that sentience is not a property of the human brain and not even an emerging property of life. Rather, it is borrowed from the cosmos and appropriated as the self in the human condition; inter-subjectivity is a derivative of sentience whereby subjectivity is an instance of inter-subjectivity. Since sentience brings with it the intuition of life (which is, according to our imagination, a non-intentional kind of knowledge), this intuition must extend into dying and has no reason not to continue beyond. Now, in the Fourth Movement, *Finale Glorioso,* I will describe the patterns of sentience in the soul's final ascent, and honor the intuitional gifts that proceed from the passage of a great soul.

Having established the under-cover identity of the qualia of sentience, which is neither subjective nor objective, neither a wave nor a field, neither a particle nor idea but, rather, informational relations with the patterns of self-organizing complexity, we may now again engage in the upward arc and inquire: can the sentient substratum of the universe ever cease to be? Can the informational relationship ever be extinguished? What happens in the seclusion of a Buddhist burial chamber which, opened later, reveals the flesh and bones of a saint gone with only nails and hair remaining? Is transfiguration a metaphor or a description and, if so, a metaphor for what or a description of what?

I wish to shift now to more personal imaginings. Those who have witnessed death will perhaps agree with me that it has a certain emptying quality in it. Not only is a person is gone, and the body goes breathless and still, but, in addition, the space around changes its metrics. A strange blazing physicality of emptiness emerges, like a gap or rip in tactility. This gap seems to bridge the worlds: traditions claim that the deceased ones can, for a while, hear our prayers and even receive guidance for their transitions. In my witnessing a few deaths recently, the gap did not seem to be as present as before. Rabbi Zalman Schachter-Shalomi departed in stillness; Professor Eugene Taylor departed in sorrow; Professor Arthur Hastings, a very saintly man, departed in glory, and when his students mentally asked for guidance in the wake of his departure, they felt that the guidance was given. Anna-Teresa Tymieniecka departed in glory. She appeared welcomed by a celestial chorus from the ranks of angels. She ascended into light. The echo of this departure lasted for many days. Like a child in a mother's embrace, she turned to the bosom of Life which she faithfully serenaded for so many decades. There was never any gap.

Imagination is a cognitive organ. It reaches beyond the circumference of the senses. It metaphorizes the invisible. These images of glorious departure of the soul, the absence of the gap in life – are they not the arabesques of sentience? They herald an increase of sentience, the increase of the fullness of the Logos, the increase of life. Isn't this spontaneity pushing the limits of the mind toward further and further frontiers of knowledge in The New Enlightenment? And, doesn't it give us hope that perhaps even 2000 years after the nascent self-disclosure of Christo-Logos, the universe is still not a completed creation, and we still have a road to travel together in expanding horizons of knowledge, through the mirrors of sentience? As the Philosopher's lyrical alter ego, Timothy the Dispossessed, says, "We advance in the fulfillment of our spirit" (Tymieniecka 2012a, 16).

For those of us who share the human condition, the gift of her departure is to see beyond our limitations, to align the spontaneity of sentience with our expressions, as did she, and to spread the word. Anna-Teresa's visionary insight will animate clinical theory and the work of the healing professions; it will extend into physics and astronomy; it will penetrate into computer science, artificial intelligence, and the building of the new virtual realities. The cup of sentience is full, and the fullness of it is increasing. We need to keep spreading the word.

The coda follows. May the gifts of spontaneity increase; may new mental horizons open; may new angles of intuition follow the self-revelations of sentience. If phenomenology were only a bodily enterprise, an outcropping of corporeity (as is awareness), or a product of *praxis,* the end of the body would be the end of any phenomenology. But, in the melody of sentient cosmic information, phenomenology is a quintessential, primeval self-reflection of sentience, ciphered, as it were, in the symbols of language. Phenomenology is conceivable as a part of the universe, a pathway in the substratum of the cosmos. Direct intuition appercepts not just the unity of things but, rather, the material from which all is made. This ever-lasting sentience can be what Husserl saw on his death bed when he was phenomenologizing his own dying, and exclaimed: "I see something beautiful… Write, write!"[19] This is what shines at us from the final passage of Anna-Teresa Tymieniecka, as her last book prepares intuition for the final philosophical turn.

## Notes

1. For more on the experience of the sacred, see Dadosky 2004, *The Structure.*
2. Consensus trance is a term coined by American psychologist Charles Tart for the ordinary state of consciousness in which people follow the natural attitude assumptions of what is real and behave in a stimulus-response mode, without reflection. For more on consensus trance, see Tart 1986, *Waking Up*. For more on the experiences of the sacred which show as if through ruptures in the regular phenomenal field, see Louchakova, "Ontopoiesis." Note that Louchakova and Louchakova-Schwartz are one and the same author.
3. For Tymieniecka's comments on knowledge, see Louchakova-Schwartz 2012a, "Self."
4. For more on intuition of life in Tymieniecka's writings, see Louchakova-Schwartz, "DirectIntuition."
5. The term the "self-revelation of life" is borrowed from Karl Hefty's 2015 English translation of Michel Henry's *Incarnation.*
6. For the original oldest version of the legend of Savitri, see *Vyasa's Mahabharatam* 2008.
7. For the Katha Upanishad, see Sarvananda 1921, *Katha-upanishad.*
8. For more on philosophy as a way of life, see Hadot 2013, *Philosophy.*
9. For more on the primary meanings in Vedānta, see Berliner, *Advaita Makaranda.*
10. For more on Tymieniecka's concept of the New Enlightenment, see Louchakova-Schwartz 2012a, 2014a, b, "A Paradox."
11. A theory of non-intentional knowledge was developed in Vedānta. See Shankarāchārya 1993, *Aparokshānubhuti.* For the present author's non-intentional theory of knowledge in Phenomenology of Life, *(T)ex(t)perience.*
12. When I wrote this, I was not yet familiar with the work of Michel Henry. Henry's work, such his *Incarnation*, confirms this insight.
13. For the reflections on essential substance, see Henry 2015, *Incarnation.* Also, Louchakova-Schwartz 2012b, "Intuition" and 2014a, "Dia-Log(os)."
14. When I wrote this, I was not yet familiar with the work of Michel Henry. Henry's work, such his *Incarnation*, confirms this insight.
15. Here, "opaque flesh" means *hyle*, the perceptual flesh; this is analogous to Henry's sensed body, as opposed to sensible body, i.e., the *ek-stasis* of life as opposed to its self-revelation (cf. the Tymienieckian Absolute).
16. For more on qualia, see Bayne and Montague 2011, *Cognitive Phenomenology*; also Kriegel 2015, *The Varieties.*

17. For more on informational relations underlying qualia in subjective experience, see the concluding section in Louchakova-Schwartz 2016, "Theophanis" and Oizumi et al. 2014, "From the Phenomenology."
18. For an exposition of Suhrawardī's argument of the unity of awareness and light, see Louchakova-Schwartz 2015, "A Phenomenological Approach."
19. See Schmitz-Perrin 1996, "La phénoménologie" and the exposition of that article in Louchakova 2005, "Ontopoiesis and Union."

## Works Cited

Bayne, Tim, and Michelle Montague. 2011. *Cognitive Phenomenology*. Oxford: Oxford University Press.

Dadosky, John Daniel. 2004. *The Structure of Religious Knowing: Encountering the Sacred in Eliade and Lonergan*. Albany: State U of New York. Print.

Hadot, Pierre, ed. 2013. *Philosophy as a Way of Life: Ancients and Moderns: Essays in Honor of Pierre Hadot*. Trans. Michael Chase. Malden: Wiley-Blackwell.

Henry, Michel. 2015. *Incarnation: A Philosophy of Flesh*. Trans. Karl Hefty. Evanston: Northwestern University Press.

Ibn a.-'Arabī. 1975. *The Wisdome of the Prophets (Fuṣūṣ-al-Ḥikam)*. Trans. Titus Berckhardt and Angela Seymour. Alswirth: Beshara Foundation.

———. 1978. *The Tarjumán al-ashwáq : A Collection of Mystical Odes*. Translation and commentary R. A. Nicholson. London: Theosophical Publishing House.

Kriegel, Uriah. 2015. *The Varieties of Consciousness*. New York: Oxford University Press.

Louchakova, O. 2005. Ontopoiesis and Union in the Prayer of the Heart: Contributions to Psychotherapy and Learning. In *Logos of Phenomenology and Phenomenology of the Logos. Book Four: The Logos of Scientific Interrogation. Participating in Nature – Life – Sharing in Life*, Analecta Husserliana XCI, ed. Anna-Teresa Tymieniecka, 289–311. Dordrecht: Kluwer.

Louchakova-Schwartz, O. 2012a. Self in the Path(o)s of the New Enlightenment: Towards a New Clinical Theory. *Agathos. An International Review of the Humanities and Social Sciences*. 3 (1): 7–30.

———. 2012b. Intuition of Life in Tymieniecka's Phenomenology with a Reference to Intuition of *Sat* in Śaṅkara's Advaita Vedanta. *Culture &Philosophy: A Journal for Phenomenological Inquiry* 2011: 40–60.

———. 2014a. Dia-Log(os): Genesis of Communicological Virtues in the Phenomenology of Life, with the Reference to the Advaita Vedānta of Śaṅkara. In *Phenomenology of Space and Time*, Analecta Husserliana CXVII, ed. Anna-Teresa Tymieniecka, 193–206. Dordrecht: Springer.

———. 2014b. A Paradox of the New Enlightenment: The "Endangered" Self in the Path(o) s of Individualizing Life. *Acts of the Vie Della Fenomenologia nella "post-modernita." [Phenomenological Paths in Post-Modernity]. Confronto con la Fenomenologia della Vita di Anna-Teresa Tymieniecka*. Rome: Pontifical University Antonianum.

———. 2015. A Phenomenological Approach to Illuminationist Philosophy: Suhrawardī's *nūr mujarrad* and Husserl's Reduction. *Philosophy East and West* 64 (2): 1052–1081.

———. 2016. Theophanis the Monk and Monoimus the Arab in a Phenomenological-Cognitive Perspective. *Open Theology* 2.1: 53–78. ISSN (Online) 2300-6579, https://doi.org/10.1515/opth-2016-0005, February 2016.

Oizumi, M., et al. 2014. From the Phenomenology to the Mechanisms of consciousness: Integrated Information Theory 3.0. *PLoS Computational Biology* 10 (5): e1003588.

Sarvananda, Svami. 1921. *Katha-upanishad, with Sanskrit Text, Paraphrase with Word-for-Word Literal Translation, English Rendering*. Mylapore: Ramakrishna Math. Print.

Schmitz-Perrin, Rudolf. 1996. La phénoménologie et ses marges religieuses: la correspondence d'Edmund Husserl. *Studies in Religion/Sciences Religieuses* 25 (4): 481–488.

Shakarāchārya, Sri. 1993. *Aparokhānubhuti: Self-Realizaton*. (Swami Vimuktanānda, translations and commentary). Calcutta: Advaita Ashrama.

Tart, Charles T. 1986. *Waking Up: Overcoming the Obstacles to Human Potential*. Boston: New Science Library. Print.

Tymieniecka, Anna-Teresa. 2012a. *The Fullness of the Logos in the Key of Life, Book II, Christo-Logos, Metaphysical Rhapsodies of Faith,* Analecta Husserliana CXI. Dordrecht: Kluwer.

———. 2012b. Introduction: The Illusion of the Return to the Source. In *The Fullness of the Logos in the Key of Life, Book II, Christo-Logos, Metaphysical Rhapsodies of Faith*, Analecta Husserliana CXI, 1–9. Dordrecht: Kluwer.

"Vana Parva: Wife's Devotion and Satyavana". 2008. Ch. XVIII in *Vyasa's Mahabharatam,* 329–336. Kolkata: Academic Publishers.

# Part II
# Introduction to the Topic

# The World Phenomenology Institute's Eco-Phenomenology

**Daniela Verducci**

**Abstract** This paper presents the acceptance of the word "eco-phenomenology" that is specific to the World Phenomenology Institute, resulting from the research in the phenomenology of life of its founder, Anna-Teresa Tymieniecka, carried out for over 40 years. Her most conclusive discovery was the ontopoietic logos of life, the productive and ordering force that is at the base of cosmic becoming and that forms human existence, leading them all to their transcendent destination. From the standpoint of the World Phenomenology Institute, ecology can be considered and practiced in its germinal condition that is from the perspective of the self-individualization of life, the basic instrument of ontopoiesis that draws on the cosmic dependencies and constitutes the human mind and human development.

**Keywords** Eco-phenomenology · Phenomenology of life · Ontopoiesis · Logos · Anna-Teresa Tymieniecka

## Introduction to a Setting and an Impetus

The eco-phenomenological reflection presented in this volume is the harvest of the World Phenomenology Institute's 64th International Congress of Phenomenology,[1] the first one held without the earthly presence of Anna-Teresa Tymieniecka. Her "diamond-like gaze," as Alfredo Marini[2] described it in his message of sympathy, had always accompanied the conferences of the World Phenomenology Institute and its affiliated societies, in person or long distance. Although she has left us, the articles collected here prove that many of us want to keep alive her legacy of thought.

Tymieniecka was very glad to learn that the 64th Conference on Phenomenology would be held at the Catholic University of Milan. Her travels and conference commitments in Italy normally took her more often to Rome, although stops in northern or southern Italy were not rare. From this point of view it is intriguing to notice that in October 2014 the Catholic University of Milan unexpectedly hosted the first

D. Verducci (✉)
University of Macerata, Macerata, Italy
e-mail: daniela.verducci@unimc.it

 
W. S. Smith et al. (eds.), *Eco-Phenomenology: Life, Human Life, Post-Human Life in the Harmony of the Cosmos*, Analecta Husserliana CXXI,
https://doi.org/10.1007/978-3-319-77516-6_2

Memorial of her death, which had occurred a few months before on June 7, 2014. It was as if the dynamism of the ontopoietic auto-individualization of life, which Tymieniecka discovered and theorized, had chosen that locus of thought as an intact base from which to take up the impetus for both a renewed irradiation and a theoretical evolution. This is particularly striking if we consider how important Italy was for Tymieniecka, as well as from the point of view of her theoretical elaboration. In fact, the home of the World Phenomenology Institute (WPI) was and still is in Hanover, New Hampshire, in the United States; consequently, the Institute has found itself, and still finds itself, immersed in the "neophiliac" atmosphere (Caillé 1997) of Pragmatism and Analytical Philosophy. But, the thought that came to maturity there was always nourished by constant and deepened exchanges with the more than thousand year philosophical tradition that has flowed into Italian culture, starting with Parmenides of Elea and passing through Saint Thomas Aquinas to the more recent forms of *Italian theory*, as it has been labeled by Dario Gentili in his volume *Italian Theory. Dall'operaismo alla biopolitica* (2012).

In his *Living Thought* (2012), Roberto Esposito notices that today this kind of thought is rightly re-evaluated in many quarters as "living thought," as it is the bearer nowadays of the fruit of manifold and complex speculative sedimentations that elsewhere were swept away.

Beginning in the late 1970s, the WPI developed ever closer contact with scholars of prestigious Italian cultural institutions and universities, among them the young Department of Philosophy and Human Sciences of the University of Macerata. It was precisely this contact which provided the phenomenological inquiry of the WPI with a solid philosophical anchorage in the tradition of what Lovejoy called *The Great Chain of Being* (1964), a vision that is imbued with the *Italian Difference* (Chiesa and Toscano 2009), for Italians neither generated nor made its own any of the great historic rifts of modern times (Perniola 1976, 11). On the contrary, the Italian difference is an inclination to seek the profile and the meaning of its own topicality in what came before (Esposito 2010, 53).

With their contribution, the authors participating in this volume intend not only to formally honour Tymieniecka's memory but, also, to render a living homage to her, which means to welcome interiorly "the seminal virtualities engendered by [her] thought," so that they may continue to germinate, inasmuch as they are grafted onto the current and living thought of each of us (Tymieniecka 2002, 685a). We want to actuate the phenomenology of life in accord with that movement of "discrete continuity" so dear to Tymieniecka, which manner of continuity is characteristic of each experience that keeps itself alive, transmitting itself into history. With the philosophical field, only "living," inter-subjective, and affectively rich relations are able to generate the continual re-vitalization of ideas. Indeed, such a regeneration escapes us when a philosophical relation is based on a merely speculative and theoretical exchange. So Habermas taught, and Tymieniecka verified as well, when it was a matter of re-animating Husserlian *Phänomenologisieren* (Habermas 1995, 7, 17–18; Tymieniecka 2005, xv).

## Eco-Phenomenology for a New Enlightenment

The title of this volume, *Eco-Phenomenology,* summarizes the essential core of the inquiry and of the theorizing for a New Enlightenment that Tymieniecka opened and pursued in the course of her over 40 years work of phenomenological elaboration. In the four volumes of the *Logos and Life* series, she realized an "intuitive resowing" of the method and classical themes of phenomenology, from the point of view of life (Tymieniecka 2009, xxi–xxix, 2010a, 3–4, 2010b, 7–15; Verducci 2010, 33). Leaning forward from the platform of the consolidated phenomenological yield, Tymieniecka focuses her attention on the "breaking point of intentionality," on the line that marks the border between the role of the meaning-bestowing-agent of the intellect and the zone of the "pre-predicative," the meaning of which eludes "the mind's tentacles." Here, there emerges the concrete living human individual as "the vortex distributing and measuring the significant roles" (Tymieniecka 1983, xv). In fact, the intimate spontaneity of man with the creative virtualities that are expressed in the course of "man's self-individualization in existence," is the only capacity able to give foundation to the opening of "a horizon of possible worlds," which instead was precluded to consciousness self-confining within an intentionally predetermined world (Tymieniecka 1983, xvii, xvi). In the measure to which she recontextualized the human creative condition within the unity-of-everything-is-alive, Anna-Teresa Tymieniecka succeeded in making use, again, of *Erlebnis* as resource for *philosophia prima* (Tymieniecka 1988, 4). In fact, consciousness is now able to reveal the surprising actuality that when life, with its natural constructivism, reaches the level of the human condition, it opens to the irruption of *Imaginatio Creatrix.*

In human creative acts, more than in the cognitive processes of the human mind, there is manifested the "inward givenness of the life progress common to all living beings as such"; even a logos appears, supporting it through its network. What is produced is, thus, an expansive and evolutive logic of autoindividualization of life that autopoietically reproduces itself in the pre-human constructivism while it creatively-produces-being in the ontopoiesis of life's human level (Tymieniecka 2000, 4–5).

At this point, Anna-Teresa Tymieniecka undertakes a radical metaphysical re-elaboration, one suited to the needs that spring from the decline of the modern theoretical paradigm. In fact, philosophical inquiry into the principle of all things, which phenomenology of life launches again, now engages the field of being no longer in its generic and static wholeness, which embraces all-that-is but, also and above all, in its continual concrete becoming and proceeding through ceaseless auto-articulation. Therefore, responding to the ancient need to "save the phenomena" means undertaking an inquiry of *philosophia prima* directed at the objective of "theorizing" the overall phenomenon of the new "fullness of the Logos in the key of Life." In fact, what has thrown itself wide open before us is a path of theoretical research that we did not imagine existed, on which we can adventurously embark, renewing the instance of the Enlightenment and Kant to "Dare to know!" We now catch sight of a unitary *logos* leading us, that which animates both, the Parmenidean sphere and the absolute Hegelian Spirit and which, autoindividualizing itself through ontopoiesis, shows that it is intrinsically able to connect phenomena emerging bit by bit from the inorganic into the organic, into the human, weaving a "metapoietic" network of innumerable metamorphic passages of transcendence—which opens it in the direction of the divine, newly risen to sight, in accord with the perspective of *philosophia perennis*, a vision already delineated by Leibniz when, to rationally understand the truth of propositions of fact, he introduced the principle of sufficient reason, which, while establishing a foundational dynamic tending toward the infinite, made it possible to construct a solid ladder of truth in order to always better rise to the fullness of the logos (Tymieniecka 2009).

## The Ontopoiesis of Life as an Eco-Phenomenology

At that time Tymieniecka was interviewed by Norwegian journalists Lars Petter Torjussen, Johannes Servan, and Simen Andersen Øyen. In 2008, on the occasion of a WPI conference held at the University of Bergen, an event organized by Konrad Rokstad, Anna-Teresa Tymieniecka was awarded "an honorary doctorate in philosophy … for her contributions to phenomenological research and philosophy in general and for the unique way she has organized philosophical activities world wide" (Rokstad 2008, 23). As Rokstad tells, it was there that "Tymieniecka revealed that the phenomenology of life had matured into an eco-phenomenological Enlightenment" (Rokstad 2008, 23).

The interviewers open by asking Tymieniecka questions about the practical value of phenomenology compared with contemporary science. This question had already been raised by Stephen C. Pepper and Alfred Tarski in the 1950s when Tymieniecka, newly arrived in the United States, was reading Husserl's *Logical Investigations* at the University of California at Berkeley. Tymieniecka informed the journalists that such a pragmatic test had already been amply addressed and passed. In her 1962 book *Phenomenology and Science in Contemporary European Thought*, which sold over 10,000 copies in America and was translated into Japanese and published in Japan as well, she had demonstrated how phenomenology was not a mere mental game, incapable of leading to practical results, technical innovations or solutions to the world's problems. Phenomenology, even back then, had found fruitful applications in psychology, psychiatry, the fine arts, and critical analysis in general. Now, after 50 years of phenomenological work focused on life, it is an incontrovertible fact that phenomenology has entered into all sectors of knowledge (including physics and embryology) as a praxeology of both practical and theoretical knowledge (Torjussen et al. 2008, 25–26).

She emphasized, in a surprising convergence with Habermas's evaluation, that the phenomenological approach is that most suited for promoting the growth of knowledge in the postmodern era because it carries "a vision of reason that breaks out from the narrow traditional frameworks and opens up creatively toward appreciation of the host of new rationalities … in order to deal with the changeable currents of existence, to generate criteria of validity, predictability, prospects, measure" (Habermas 1995, 3; Tymieniecka 2009, xxiv). But, the interviewers pressed Tymieniecka on this point, asking her to give examples of the concrete utility of a phenomenological approach and apply it to the ecological crisis of our times. There is, in fact, a sharp divide between most people—who perceive this crisis as a problem for the physical sciences, the solution of which depends on technological innovation—and eco-phenomenologists who do not focus on the lived experiences of real individuals, and, instead, view the crisis only on a metaphysical plane. That is to say, they simply ask for "a fundamental reconceptualization of human values and our relationship with nature" without wondering how this can be achieved. Tymieniecka responded dryly to this provocation, as we can read in the *exergo* of our volume, to wit: "Actually, my account of ontopoiesis is an eco-phenomenology. Ontopoiesis reaches to the very germs of ecology: development and genesis" (Torjussen et al. 2008, 26–27). In one fell swoop, and without reticence, she indicated that her phenomenology of life is at the base of all theorizations of Ecology, Environmental Studies, and Ethics. It is also at the base of any innocent trust in science and in technological innovation in the fields of human life and the natural environment.

In effect, practicing the phenomenology of phenomenology, Anna-Teresa Tymieniecka not only discovered that constituent consciousness is rooted in the ontological creativity of the acts of human existence but also, and above all, was able to grasp and bring to light the grafting of the traditional *logos* onto the more profound ontopoietic *logos* of life so that it does not limit itself to extrinsically saving the phenomena manifested to consciousness from dispersion, as does the tradition's *mathesis*. Rather, it is the one effective force and vehicle of every intrinsic genesis and development of phenomena. The ontopoietic *logos* of life produces or generates being through the dynamism of the sentient auto-individualization of which it consists; therefore, it not only conducts but also substantializes all becoming, from the deterministic constructivism of the inorganic and organic levels of being up to the creative human condition and beyond.

In the human creative condition, the force of the *logos* of life, making a metamorphosing move, assumes the human will as its new motor and advances "from the vital/ontopoietic round of significance into two new dominion of sense: the creative/spiritual and the sacral," as Tymieniecka says (Tymieniecka 2007, 20).

On one side, therefore, the fulcrum of this metamorphosis is that "unique phase of evolutive transmutation" in which the mature phase of the platform of life manifests an extraordinary character and gives rise to the Human Condition within the unity-of-everything-there-is alive (Tymieniecka 2007, 31). On the other side, "man's elementary condition appears to be one of blind nature's elements and yet, at the same time, this element shows itself to have virtualities for individualization at the vital level and, what is more, for a specific human individualization. These latter virtualities we could label the subliminal spontaneity" (Tymieniecka 1988, 28). The conclusion is that, when life attains the level of the human creative condition, it is no longer limited to reproducing itself but, in the acts of man, it always interprets itself while in existence, giving rise to forms of life that are not only new and previously unimaginable but, also, congruent and appropriate to the being of life in its becoming, of which man possesses the cipher.

In the cosmic arrangement of metamorphic dynamic harmony that is thus delineated, the ontopoietic *logos* of life achieves what Nietzsche longed for in vain: a movement of immanent transcendence that is at once metaphysical, physical, and ethical and that, therefore, expresses *The Fullness of the Logos in the Key of Life*, which is to give the cover title of the last two volumes written by Tymieniecka.

## Conclusion

Tymieniecka has discovered the ontopoiesis of life as the living constructing principle of eco-phenomenology, but this last still has yet to be elaborated and validated in its theoretical systematicity and in its practical-sectorial usability. Tymieniecka herself expressed the general lines according to which her philosophical inheritance is to be received and developed in the speech she gave on January 15, 2011, at the conclusion of an International Conference in Rome on the theme *Phenomenological Paths in Post-Modernity: A Comparison with the Phenomenology of Life of Anna-Teresa Tymieniecka*. On that day she received a medal of honor bestowed by the President of the Italian Republic and, in a surprising coincidence, it was announced on that same day that John Paul II was to be declared Blessed (he has since been sainted). In her speech of gratitude in Italian, the last she spoke in public, Tymieniecka articulated her philosophical testament, auguring that research in the phenomenology of life, which to date had concerned the earthly level of human passions, could finally turn toward heaven and "accomplish the enormous task of showing how the objective constitution depends on what I call the geocosmic architectonic dynamism of everything, which from heaven extends over all the earth, and also over all humanity" (Tymieniecka "Lungo il nuovo sentiero della fenomenologia della vita" Tymieniecka 2014, 24).

This stimulating inheritance of theoretical research is now entrusted to all the friends of Anna-Teresa Tymieniecka. They will never forget her in the pursuit of the precious intellectual and human work of the reconstruction of a horizon of being adequate to the challenges of our days, a goal she pursued with untiring assiduousness and undoubted success.

The first thing we must do is to deepen our awareness of the method that enabled Tymieniecka to open the broad perspectives of philosophical research and pragmatic innovation that the discovery of the ontopoietic *logos* of life promises. In fact, nobody will be able to echo on the occasion of the passing of Tymieniecka what Martin Heidegger said *in mortem* of Max Scheler: "A road of philosophy has closed" (Heidegger 1975, 9–10)! The entire philosophical work of Anna-Teresa Tymieniecka was directed not only to opening new philosophical paths but, above all, to building a community of phenomenological scholars who want to travel those paths and following the branching threads immanent in the infinite living coalescence of being.

The innovative phenomenological method that Tymieniecka practiced was a feminine one of the cultivation of ideas through empathic and intentional identification, a method that she preferred over any speculative idealism that silences living experience in order to reason only with dead ideas. According to my mind, we can unveil just such cultivation as a philosophical paradigm is capable of offering to the world of production and human formation a horizon of meaning in which the dominant style of industry and instrumental excess can find appropriate ethical integration and contextualization, what is essential to promoting a more than sustainable development as a renewed state of cosmic harmony.

## Notes

1. The 64th International Phenomenology Conference, centered on the theme "Eco-Phenomenology. Life, Human Life, Post-Human Life in the Harmony of the Cosmos," occurred 1–3 October 2014 at the Catholic University of the Sacred Heart of Milan. It was promoted by the World Phenomenology Institute and its affiliated International Society for Phenomenology and the Sciences of Life, in collaboration with the Department of Philosophy of the Catholic University of the Sacred Heart, represented by its Director, Professor Massimo Marassi, and Professor of Theoretical Philosophy, Dario Sacchi. Valuable and competent support in conducting the conference was offered by Dr. Renato Boccali of the International University of Languages and Media (IULM) of Milan and Dr. Emilia Andri of the University of Bergamo.
2. Alfredo Marini has been professor of the history of contemporary philosophy at the State University of Milan and is director of the periodical *Magazzino di filosofia. Quadrimestrale di informazione, bilancio ed esercizio della filosofia.* He is also a translator of philosophical essays and a member of the Società Umanitaria of Milan.

## Works Cited

Caillé, Alain. 1997. 30 Thèses pour une Gauche Nouvelle et Universalisable. *Revue du MAUSS* 9: 297–331. Print.

Chiesa, Lorenzo, and Alberto Toscano, eds. 2009. *The Italian Difference. Between Nihilism and Biopolitics*. Melbourne: Re.press. Print.

Esposito, Roberto. 2010. *Pensiero vivente. Origine e attualità della filosofia italiana*. Torino: Einaudi, 2010. (*Living Thought. The Origins and Actuality of Italian Philosophy*. Trans. Zakiya Hanafi. Stanford: Stanford University Press, 2012.) Print.

Gentili, Dario. 2012. *Italian Theory. Dall'operaismo alla biopolitica*. Bologna: il Mulino. Print.

Habermas, Jürgen. 1995. *Postmetaphysical Thinking*. Trans. William Mark Hohengarten. Cambridge UK: Polity Press. Print.

Heidegger, Martin. 1975. Andenken an Max Scheler. In *Max Scheler im Gegenwartgeschehen der Philosophie*, ed. Paul Good, 9–10. Bern: Francke. Print.

Lovejoy, Arthur Oncken. 1964. *The Great Chain of Being*. Cambridge Mass.: Harvard University Press. Print.

Perniola, Mario. 1976. La differenza italiana. *L'erba voglio* 18/19: 10–15. Print.

Rokstad, Konrad. 2008. Distinct Recognition for Phenomenology. *Phenomenological Inquiry. A Review for Philosophical Ideas and Trends* 32: 23–24. Print.

Torjussen, Lars Petter, Johannes Servan, and Simen Andersen Øyen. 2008. An Interview with Anna-Teresa Tymieniecka. *Phenomenological Inquiry. A Review for Philosophical Ideas and Trends* 32: 25–34. Print.

Tymieniecka, Anna-Teresa. 1983. The Phenomenology of Man and of the Human Condition—The Human Individual, Nature, and the Possible Worlds. In *The Phenomenology of Man and the Human Condition.* Part I: *Plotting the Territory for Interdisciplinary Communication,* Analecta Husserliana XIV, ed. Anna-Teresa Tymieniecka, xi–xx. Print.

———. 1988. *Logos and Life.* Book I. *Creative Experience and the Critique of Reason*, Analecta Husserliana XXIV. Dordrecht: Kluwer Academic Publishers. Print.

———. 2000. *Logos and Life.* Book IV. *Impetus and Equipoise in the Life-Strategies of Reason*, Analecta Husserliana LXX. Dordrecht: Kluwer Academic Publishers. Print.

———. 2002. A Note on Edmund Husserl's Late Breakthrough. In *Phenomenology World-Wide. Foundations, Expanding Dynamics, Life-Engagements. A Guide for Research and Study,* Analecta Husserliana LXXX, ed. Anna-Teresa Tymieniecka, 685–87. Dordrecht: Kluwer Academic Publishers. Print.

———. 2005. The Logos of Phenomenology and the Phenomenology of Logos. In *Logos of Phenomenology and Phenomenology of the Logos.* Book One: *Phenomenology as the Critique of Reason in Contemporary Criticism and Interpretation,* Analecta Husserliana LXXXVIII, ed. Anna-Teresa Tymieniecka, xiii–xl. Print.

———. 2007. The Great Metamorphosis of the Logos of Life in Ontopoietic Time. In *Timing and Temporality in Islamic Philosophy and Phenomenology of Life,* Islamic Philosophy and Occidental Phenomenology in Dialogue 3, ed. Anna-Teresa Tymieniecka, 11–71. Dordrecht: Springer. Print.

———. 2009. *The Fullness of the Logos in the Key of Life*. Book I: *The Case of God in the New Enlightenment*, Analecta Husserliana C. Dordrecht: Springer. Print.

———. 2010a. Heralding the New Enlightenment. In *Phenomenology and Existentialism in the Twentieth Century,* Book Three: *Heralding the New Enlightenment,* Analecta Husserliana CV, ed. Anna-Teresa Tymieniecka, 3–4. Print.

———. 2010b. The New Enlightenment. In *Phenomenology and Existentialism in the Twentieth Century,* Book Three: *Heralding the New Enlightenment,* Analecta Husserliana CV, ed. Anna-Teresa Tymieniecka, 7–15. Print.

———. 2012. *The Fullness of the Logos in the Key of Life*. Book II: *Christo-Logos: Metaphysical Rhapsodies of Faith*, Analecta Husserliana CXI. Dordrecht: Springer. Print.

———. 2014. Lungo il nuovo sentiero della fenomenologia della vita. Testimonianza di una fondatrice. *Vie fenomenologiche nella Post-Modernità. Confronto con la Fenomenologia della vita di Anna-Teresa Tymieniecka/Phenomenological Paths in Post-Modernity. A comparison with Anna-Teresa Tymieniecka's Phenomenology of Life*, ed. Daniela Verducci, 9–12. Roma: Aracne. Print.

Verducci, Daniela. 2010. The Development of the Living Seed of Intentionality. From E. Husserl and E. Fink to Anna-Teresa Tymieniecka's Ontopoiesis of Life. In *Phenomenology and Existentialism in the Twentieth Century,* Book Three: *Heralding the New Enlightenment,* Analecta Husserliana CV, ed. Anna-Teresa Tymieniecka, 19–37. Print.

# Metaphysics and Eco-Phenomenology Aiming at the Harmony of Human Life with the Cosmos

**Francesco Totaro**

**Abstract** Reflecting about the relationship between the new-classical metaphysics of the Milan school and a radicalized phenomenology such as the ontopoiesis of life can contribute to a "critique of the settled reason," as Tymieniecka said. According to Gustavo Bontadini, philosophy has a double face because it is always in the flux of life but its task is to stop that flux—to search, we could say following Anna-Teresa Tymieniecka's thought, the *logos* that shows itself in the human phenomenon and especially in his *Imaginatio Creatrix*, which is able to go beyond the series of events and to give a new impulse to the ontopoiesis of life and its manifestation. From this base, even according to Anna-Teresa Tymieniecka, we could delineate a New Enlightenment as a new project of human civilization in harmony with the earth and cosmos, where *logos* is unfolding in all the richness of its manifestation and God rises to its highest fullness.

**Keywords** New-classical metaphysics · Ontopoiesis · Logos of life · *Imaginatio creatrix* · New Enlightenment · Gustavo Bontadini · Anna-Teresa Tymieniecka

## Eco-Phenomenology and a New Anthropological View

A major feature characterizing the ontopoietic radicalization of phenomenology, as elaborated by Anna-Teresa Tymieniecka, is the translation of the "transcendental categories"—that had already been analyzed by Husserl—from their cognitive statement to a more comprehensive, existential, and vital one. According to that, we can bring the transcendental dimension to the more concrete anthropological field. Essentially, a fundamental question is at stake: which vision of human being do we need, in the era of globalization, in order to prevent our vital acts from being captured by instrumental logic? More specifically, how can phenomenology contribute

F. Totaro (✉)
University of Macerata, Macerata, Italy
e-mail: totarofr@unimc.it



W. S. Smith et al. (eds.), *Eco-Phenomenology: Life, Human Life, Post-Human Life in the Harmony of the Cosmos*, Analecta Husserliana CXXI,
https://doi.org/10.1007/978-3-319-77516-6_3

to outlining a human being which is not made like a machine exclusively aimed at producing and consuming?

Phenomenological reflection is a thought that goes beyond any interest to impose already-given or assumed instrumental functions. The world, filtered through phenomenological reduction, is given to us by transcending the operations linked to usability (a conceptual gain that was, perhaps, obscured by Heidegger's existential analytics). Such a transcendence does not nullify the world of our operations but is to be understood as the capacity, thanks to which the already-constituted operations stop to oblige us by filling up our whole life. By broadening its meaning through phenomenological reflection, the world of our experience can become available for other goals and for different values of being. The semantic breaking of phenomenological reflection makes it possible to activate an imaginative capacity that can aim at recreating the existing world through new constitutive operations (that one would be the function of the *imaginatio creatrix*).

This faculty of transgression, which raises humans to a condition of freedom in comparison with existing reality, does not remove the concreteness of temporality and historicity but increases the importance of contemplative activity. Retaking possession of such faculty means a decisive anthropological task. Of course, this task cannot remain a generic one, and must be supported by both exploring the regional contexts of human expression and overcoming any tendency to self-referential closures. This way, the contemplative faculty can inspire mental and practical abilities to reform and transform life contexts, so that human operating can escape risks of unilaterality and alienation.

## A Comparison: Phenomenology and New-Classical Metaphysics

From the point of view of an old student at the Catholic University in Milan, at present involved in the Society for Phenomenology and Sciences of Life affiliated to the World Phenomenology Institute (WPI), I wonder whether there is a meaningful connection between Tymieniecka's thought about the ontopoiesis focusing on dynamic beingness, based on the Unity-of-Everything-There-Is-Alive, and the philosophy that my main teachers elaborated in the cloisters of that University. In my opinion, there is a good connection due to the idea of metaphysics and to the relationship between metaphysics and life.

In a paper that Anna-Teresa asked me to write that appeared in Analecta Husserliana 115, I expressed some remarks about ontopoiesis as new metaphysics (Totaro 2015). We all know that Anna-Teresa appreciated particularly the doctrine of becoming in Heraclitus, but in the paper about that topic I tried to underline the importance of Parmenides, too. Indeed, the philosophical School in the Catholic University elaborated a so-called new-classical metaphysics, referring mainly to Parmenides and Aristotle and reinterpreting the tradition dated back to Thomas

Aquinas. In this new-classical metaphysics, an important point was the overcoming of the contradiction of becoming, because of its passing from being to not being, by its redemption in the sphere of "being that can't not be."

So, at a first glance, the dynamic ontopoiesis of Anna-Teresa Tymieniecka seems to be very different from the metaphysics which appeals to the permanence of the being without contradiction. Apparently, we would have a contrast or a scenery like this: becoming against being, fluidity against stability—in short, Heraclitus against Parmenides. In other terms: life against the denial of life.

Are things really this way? I don't think so. The new-classical metaphysics was launched by Gustavo Bontadini, whose first book is titled *Saggio di una metafisica dell'esperienza* [*Essay about a metaphysics of experience*]. We can read in the beginning of that important work, "Philosophical problem as life's problem" (Bontadini 1995, 5). The first paragraph is entitled "Philosophy comes out from life," and it underlines that philosophy "embraces and attracts all the life or, better, life embraces itself within philosophy, becoming an object of itself and so becoming a new life." According to Bontadini, the main feature of philosophy is asking for the reasons, sense, aim, and value of the whole flux of life. So, philosophy has a double face because it is always in the flux of life, but its task is stopping that flux to search, we could say according to Anna-Teresa Tymieniecka's thought, the *logos* that shows itself in the human phenomenon and in the *imaginatio creatrix*, which is able to turn back to the series of events giving a new impulse to the ontopoiesis of life and to its manifestation.

Introducing Bontadini's *Essay* in a new edition, Virgilio Melchiorre, one of the scholars of Bontadini and a professor as well at Catholic University, remarked that the metaphysical question could be justified and ensured only in its relationship to the most primary experience of the life of consciousness. Experience means the immediate being's presentation. For this reason, Bontadini used the formula "Unity of Experience;" considering the experience as a Unity means, at the same time, a reflective mediation concerning the sense and the *logos* of the experience itself. Through a reflective movement of consciousness, *logos* and *being* can appear both in their connection and in their distinction. Melchiorre says, therefore, that Bontadini's theoretical perspective is not far from a phenomenological one.

To complete this consideration of the link between phenomenology of life and new-classical metaphysics, I would like to stress phenomenology has been important to, and keeps playing a major role for, many scholars at Catholic University in Milan, with special reference to Husserl, Scheler, Heidegger, Meinong, Merleau-Ponty, and so on. Furthermore, I have to remind us that one of the very rare full professors as woman in her time, Sofia Vanni Rovighi, taught at that Institution for many years and was among the first scholars of Husserl in the Italian context (her book *La Filosofia di Edmund Husserl* appeared in 1939), though she loved to be considered a fervent exponent of the neo-scholastic tradition.

Reflecting on the relationship between classical metaphysics and a radicalized phenomenology can contribute to a "critique of the settled reason," as Tymieniecka said. Despite the appearance, Aristotle can help us especially re-think the main features of a complete anthropology. In his philosophy (for example, *Nicomachean*

*Ethics, Politics, and Metaphysics,* too), the human "*enérgheia*" expresses itself in the plurality of contemplation, action, and work (though the last one was attributed to slave labor). Nowadays, unlike Aristotle, we do not have to establish a hierarchical ranking among these dimensions but, instead, to embed them into one another, so that they can achieve their peculiar function and can fecundate each other in order to enact an overall human flourishing, i.e., the fulfillment of the human concrete ontopoiesis.

## Ontopoiesis of Life as Eco-Phenomenology: A Loving Intentionality

Finally, I would like to come back to the latest suggestion of the philosophy of Anna-Teresa Tymieniecka, wherein she asserts that her account of phenomenology is an eco-phenomenology. We already know the statements and the developments by Olga Louchakova-Schwartz, concerning this issue. Furthermore, Daniela Verducci, in her appreciation of the phenomenology of life, gave us a deep explanation of that topic; Carmen Cozma, too, developed the issue "Ontopoiesis of Life as Eco-Phenomenology." For my part, I would like to highlight that phenomenology—and, more precisely, ontopoiesis of life—as eco-phenomenology cannot be a simple, superficial make-up of previous ideas.

Eco-phenomenology, if taken seriously, implies a clean turning point regarding a real harmony between human being and earth, human being and cosmos. Human being becomes, indeed, responsible for the construction of a new relationship that is of new attitudes, habits, and styles of life in reference to the earth and cosmos. We need a firm change of human intentionality towards the non-human environment. We have to purify human intentionality and, so, we have to move from an instrumental intentionality towards earth and cosmos, to a *loving* intentionality.

All of this requires a drastic replacing with regard to "the human position within the cosmos," to take the title of a famous work of Max Scheler again (Scheler 2008). So, we have to leave a position of domination and exploitation, and pursue a commitment of balance and care in our relationship with environment and the non-human world. We have not to do with abstract proposals, but with a concrete matter. The unavoidable task is reviewing the idea and the practice of a way of producing which is reducing both the Earth and cosmos to a reserve of means for an unlimited accumulation of material goods. In keeping on this way the result would be a degradation, also, of human being to an instrumental dimension. Human being would become a tool of his tools. Fighting such a bad fate, the phenomenological look is invited, again, to help us escape a narrow vision of things and enlarge the intentionality of human *logos*, overcoming a cramped anthropocentrism.

This way of thinking, or re-thinking, phenomenology and ontopoiesis of life as an eco-phenomenology means both a theoretical purification of reason and an ethical commitment. Along these two directions, in our mind and in our will, we can find out a right *measure* in re-defining the relationship between human being and the earth and cosmos. Establishing a right measure is also a crucial point in assessing the debates on post- and transhumanism without accepting a dystopian, digitized humanity, nor absolutely refusing every advantage deriving from a fair integration with cyber- and IT devices. Unfortunately, aiming at the so-called enhancement of the human being, biological sciences and applied technologies do not have always a clear awareness of the consequences depending on an unlimited manipulation. Indeed, enhancing human being should not contrast the sense of measure and the care for his dignity.

So, we could construct a new enlightenment, as Anna-Teresa hoped in her last book, *The Case of God in the New Enlightenment* (Tymieniecka 2009), wherein *logos* is unfolding in all the richness of its manifestation and God rises to its highest fullness. A desirable and wishful destination, as such, is able to inspire a new project of human civilization in harmony with the earth and cosmos.

## Works Cited

Bontadini, Gustavo. 1995. *Saggio di una metafisica dell'esperienza*. Milano: Vita e Pensiero. Print.

Scheler, Max. 2008. *The Human Place in the Cosmos*. Trans. K. Frings. Evanston: Northwestern University Press. Print.

Totaro, Francesco. 2015. "Introduction: Some Remarks About Ontopoiesis as New Metaphysics." *Analecta Husserliana* CXV: 1–7. Print.

Tymieniecka, Anna-Teresa. 2009. *The Fullness of the Logos in the Key of Life*. Book I: *The Case of God in the New Enlightenment*. Dordrecht: Springer. Print.

Vanni Rovighi, Sofia. 1939. *La Filosofia di Edmund Husserl*. Milano: Vita e Pensiero. Print.

# *Ontopoiesis of Life* as Eco-Phenomenology

**Carmen Cozma**

**Abstract** Starting from Anna-Teresa Tymieniecka's thesis of the ecological dimension of human being, and considering man's intimate relationship with the Earth and the cosmos, this article emphasizes some fundamental articulations of phenomenology of life in its potential to deal with the present ecological crisis and at the same time to overcome it. The original outstanding vision of the *Ontopoiesis of Life* set forth by Tymieniecka presents a real challenge to reach a better understanding of our own opportunities and responsibility to wisely shape a proper manner of living. Thus, we stress and unfold the significance of the pivotal ideas of Tymieniecka's philosophy, such as: the "self-individualization of life," the centrality of the creative act of a human being, and the grounding of the human condition within the totality of life's spread. These are major lines to be assumed and followed by each of us in the endeavor to think about and to serve our situation in the "dynamic web of life stretching between cosmic relevancies and human creative invention," according to the initiator of the phenomenology of life.

**Keywords** Tymieniecka · Phenomenology of life · Ontopoiesis · Eco-phenomenology · Logos · Environment

## Renewing the Issue of Nature in the Phenomenology of Life

Facing the last decades' ecoalienation of mankind, Anna-Teresa Tymieniecka's urge "to formulate anew the issue of *nature*" and her inspired phenomenology of life with its integrated *ontopoietic* perspective represent a notable achievement, one with implications for an eco-phenomenology (Tymieniecka 2000, 100).

The reference concept of *life* is approached in a holistic and dynamic manner as being "suspended on the cosmic parameters;" and the manifestation of the *logos of*

C. Cozma (✉)
Department of Philosophy, "Alexandru Ioan Cuza" University of Iaşi, Iaşi, Romania
e-mail: carmen.cozma@uaic.ro

 
W. S. Smith et al. (eds.), *Eco-Phenomenology: Life, Human Life, Post-Human Life in the Harmony of the Cosmos*, Analecta Husserliana CXXI,
https://doi.org/10.1007/978-3-319-77516-6_4

*life*—the carrier of beingness that in its constructive *élan* informs a ceaseless process of "self-individualization in life"—can be conceptualized only through the complex of *nature*, "the vital milieu of all living beingness" (Tymieniecka 2000, 97, 100, 195). Tymieniecka's phenomenological discourse covers the deciphering of *life* as a unity of mineral, vegetal, animal, and human spheres, an orchestrated whole of entities, forces, operations, processes, and manifestations, making structures and ordering networks in which the *logos of life*'s movement embraces all the natural and inventive forces in interplay.

## Thematizations of an Eco-Phenomenology Within the 'Ontopoietic' Perspective of Life

Dominant in Tymieniecka's phenomenology is the crux problem of human natural existence and geo-cosmic transcendental positioning, the fundamental experience of participatory engagement with the whole of life. To cope with the acuteness of the dangerous situation generated by the contemporary ecological crisis, the contemporary bewildering reality that reveals the "anatomy of our moral disarray," we must engage the major environmental issues and activate an ecological consciousness and conduct for each "biotic citizen" shaping an ecological way of life within "everything-there-is-alive." This outlook is somewhat able to change "the role of *homo sapiens* from conqueror of the land-community to plain member and citizen of it," implying "respect for his fellow members, and also respect for the community as such," the *biotic community*, as Aldo Leopold names it (Leopold 1949, 204).

Beyond any controversies between phenomenology and naturalism, or those linked to deep ecology, Anna-Teresa Tymieniecka takes into account the question of the environmental consciousness to be awakened by human beings in a responsive attitude towards nature's gifts, developing a strategy of protection, maintenance, and care for the whole relational (human and nonhuman) living environment.

The "geo-cosmic—ontopoietic—creative—sacral" connection synthesizes "the life-positional horizons of beingness and their orbits," and the task of a philosophical re-construction of phenomenological theory and methodology has been mapped by the American thinker for "recovering the great vision of the All-in-becoming" (Tymieniecka 2009, 127–177). Thus, a rich repertoire of concepts and issues are encompassed by Tymieniecka's original imaginative reflection, in which she has dwelled upon and assessed all in a *cosmic perspective*. We will just mention: a renewed formulation given to the issue of nature-life, "one open to the cosmos and to culture;" nature's core significance for philosophy, seeing that it is the complex circumambience for the manifestations of the logos of life; the vital relevance of the natural system and the "symbiotic/affective/empathic linkages" that lead to the differentiation and cultivation of the human spirit within the cosmic matrix; the balancing between earth and sky as the stage for *ontopoietic* unfolding amid the forces, energies, and directions that mark the 'singular and multiple', 'diversity and unity' in the integrating harmony and the measure of the "great plan of life," as seen in the virtuality/actuality and outward/inward circularity of beingness-in-becoming; the web of life as "the unity-of-everything-there-is-alive" and man's status as "Custodian of everything-there-is-alive"; the human positioning in the cosmos, and so on (Tymieniecka 2009, 1990). These represent part of the main themes taken up in the scope of the phenomenology of life, manifesting a thread of Tymieniecka's thinking following which an eco-phenomenology finally comes into sight—as an ecological phenomenology and a phenomenological ecology alike. As "a new cross-disciplinary inquiry" developing "an appropriate philosophy of nature," eco-phenomenology can offer a "methodological bridge between the natural world and our own" by (re) discovering the value of our connection "with our most basic and primordial experiences of the natural world," Charles Brown and Ted Toadvine remark (Brown and Toadvine 2003, xii, xx).

Going beyond scientific research on the earth, Anna-Teresa Tymieniecka explores the idea of "the passions of the forces and mysteries of the depths of the earth" (Tymieniecka 2009, 163). She tackles the Earth as a living being, distinguishing spheres of the "earthly passions," such as rootedness, generation, grounding, and appreciation, among others. She is interested about the role played by the hidden in-depth forces and energies of the earth "as an astral body" in penetrating and governing all living beings. And moreover, for humans, the "*actio/passio* experiential background" becomes the sphere of the human "vital passions of the earth," concomitantly with the crystallization of substantial selfhood open towards expansive, "supra-earthly" horizons (Tymieniecka 2009, 163). We are guided by insight into the "mutuality between life's constructive virtualities and earth's life promoting features, one that has unfolded and progressed as a function of the cosmic system itself"; our emergence in rising out of the earth makes it even a springboard for projecting the human creative experience into infinite horizons (Tymieniecka 2009, 163, 164, 168).

The entire stream of becoming in life—from primogenital existence to the most elevated contemplations we know, all on the constructive track of being—is revealed on the ground of the *ontopoiesis* incarnating the *logos of life* as the *principle* and the *sense* of all that is alive. Everything carries and unfolds a cosmic dimension, following universal laws and patterns, eventually fulfilling the inward-outward balance of living that is inscribed within the initial and final seal of the logoic flux of life on the Earth.

## Tymieniecka's 'Ontopoiesis of Life' and Her Vision of the Human Relationship with the Cosmos

The *Ontopoiesis of Life* is a core concept, covering "the self-individualizing, telos oriented schema" (Tymieniecka 2000, 628). According to the phenomenologist of life, *onto-* refers to "firstness in the scale of existential formation" and *poiesis* means "the intrinsic factor of the constructive process of individual becoming …in its own advance, in qualification" (Tymieniecka 1995, 40). The *Ontopoiesis of Life* facilitates a better understanding of the creative experience of life in its complexity, involving the "creative context," "creative inwardness," "creative perception," the "creative product" (Tymieniecka 1990, 451). It is defining in the endeavor to encompass the entire existential dynamism with its perpetual transformations from "the germinal preorganic and organic circuits" to "the sentient, sympathetic, psychic, social and cultural circuits of life"—as the author herself reiterates (Tymieniecka 2000, 629).

Playing the "crucial role in the being/becoming network," *Ontopoiesis* is one of the basic triad of intuitions through which human intentional consciousness operates—together with the *logos of life* and the cosmic factors/spheres, according to Tymieniecka. In her vision, *Ontopoiesis* represents "the deepest crystallization of the logos of life in individualizing being/becoming," which lays down an ordering creative "track on the earth" (Tymieniecka 2014, 9).

The special attention paid to the interaction between *physis* and *bios*, cosmos and human, nature and culture is linked to Tymieniecka's thesis of the self-individualization-in-life. This does not involve a self-isolation of man from nature, or a loss of the kindredness between humans and the cosmos. To the contrary, here is brought out the issue of the interdependent community of all living beings, the biosphere within which man is a link (perhaps that having the most equipoise and certainly the most responsible) manifesting a capacity to organize and develop what is distinct in human life in its sameness and differentiation owing to human creativity, which, by continuously "inventing new ways of being human" within the expanse of the elementary vital living frame, makes "man-the-inventor of new avenues for life … in creative activity" (Rogers 2003, 115).

## The Logos of Life

Undoubtedly, Anna-Teresa Tymieniecka encounters Max Scheler's attempt to spiritualize the tie between man and cosmos in a vibrant unity of life. Just as for the German phenomenologist, man "can behave openly towards the world"—owing to his spirit—and by extension he can manifest himself as a "co-founder and co-producer of an ideal series of becoming that is occurring within the cosmic process and at the same time with the self," so for the phenomenologist of life this human situation is explicated on the ground of the *logos of life* expanding to new modalities "in freedom into a new universe of its own invention, into the universe of the human spirit" (Scheler 2001, 28, 81; Tymieniecka 2000, 320).

The "absolute criterion of life, becoming and beingness," the self-individualizing in life process, involves distinctiveness but not a split between humans and nature (Tymieniecka 2000, 107). Self-differentiation through *creativity*—the "Archimedean point from where everything finds its proper place"—does not mean that man loses his natural-biological roots, but that he improves them in an elevated manner through investing the advance of life in cultural values (Tymieniecka 1988). Phenomenological investigation focuses on life as the unique medium of beings in progress and in their interconnectedness; from within individuals are self-differentiating, on the one hand, and they are inscribing themselves in the same total course of becoming, on the other hand.

Through her original *ontopoietic paradigm*—simultaneously dealing with specific individuals in their autonomy, meaning bestowal, and unique roles, and with the total framework of circuits in cosmos, bios, psyche, human spirit, society and culture—Anna-Teresa Tymieniecka elaborated a philosophy of correlation-in-life, of living beings-in-community; this is a philosophy envisaging singulars, types, structures, and so forth proceeding from the coexistential network of life. According to María Avelina Cecilia, this is a philosophy of "a new, global vision of the world … physical, biological and human at the same time" (Cecilia 2002, 711).[1]

This ingenious phenomenological architecture around the system of vital forces, wherein the ecological questions are permanently in discussion, implies the key notion of the *Ontopoiesis of Life* manifesting the *entelechial principle* and its vehicle: the "self-individualization of life," which represents the axial constructive force of life's becoming, the dynamic plan or route of the *logos of life*'s workings, in which both nature and culture are situated. The "ontopoietic nucleus" is the ground for life's inner-outer movement in the tension and balance of the All (cosmos and humanity). According to the analysis of Daniela Verducci, a significant point is that the phenomenological *ontopoiesis of life* can very well be related to the scientific biological "autopoiesis" of the living systems expounded by Humberto R. Maturana and Francisco J. Varela.[2]

## The 'Moral Sense,' the Unity of 'Everything There Is Alive,' and the Human Ecological Identity

Anchored in the most significant ideas of both the Ancients and the Moderns, Anna-Teresa Tymieniecka advances the potential of a moral philosophizing centered on life, affirming it and encouraging its progress, newly opening our creative stance of inquiry, scrutinizing, interpreting, understanding, and acting to its plenitude. She succeeds in unfolding a grand *cosmic* vision of perpetual production and reproduction, focusing on the old concept of *logos*. In fact, she restores the concept in its deepest sense, as "a sort of *mathesis universalis* that assures an order that can not only maintain being but also generate it," as Patricia Trutty-Coohill discerns (Trutty-Coohill 2009, 2).

At stake is precisely the "moral sense." Together with the "aesthetic/poetic" and "intellective" senses, it endows with significance the entire human functioning in the gigantic life schema. Having a felicific tonality, the moral sense introduces into the individual line of behavior "the Sentiment of Benevolence toward other living creatures, toward oneself, and toward life in general." It is a valuable meaning-giving factor inserted in life, surging "as a virtue of the human condition," being "operative particularly in the emergence and expansion of the specifically human sphere of existence" (Tymieniecka 1990, 13–15).

Thanks to human creativity, the moral sense prompts rational action to redress wrongs, to avoid perils, to transcend obstacles. It grasps the experience of the evil, but more so the possibility of overcoming it, of working on the side of construction and sustainability, of bringing order out of chaos, of recognizing harmony in its role of governing the living world. Emerging "from the capacity of man's inventive function to reorganize the world on the basis of freedom," the "moral sense brings to awareness the unity-of-everything-there-is-alive" (Backhaus 2001, 41).

Here emerges a substantial concept—on having priority for contemporary environmental studies—namely, that of "Custodian of everything-there-is-alive" (Tymieniecka 1990, 16), which is what the *moral sense* does make the human being.

Instilling this category, Tymieniecka points out the distinctiveness of the creative human condition in the "ontopoietic design" and, no less, the human kinship with the entire web of life as being *a part of*—and *not apart from*—the cosmic order. This peculiar status of "Custodian" opens to a better comprehension man's ecological identity. We have to understand that the "unity-of-everything-there-is-alive" surrounds the axis of "self-individualization in life"; the autonomy but also the interdependence of different organisms make even the functioning of the cosmic order from which life emerges. Such a status engages man as moral agent in his in-depth ties with all the circumambient conditions of living, establishing human beings in a pinnacle location when we assume a fundamental responsibility for the totality of life. In the effort to reshape man's general vision and behavior through moral consideration of the inherent worth of "everything" and, concomitantly, of "all" that is alive, an ethics of life must comprehend both individualistic and holistic perspectives, for the present and the future alike. Man is able to ascend to the "Custodian" role within the living world, putting into action his creative forces for developing and protecting his own well-being, his healthy and secure existence, which are obviously connected to the ecological soundness of other individuals and communities. It is a vital human concern to use our creative energies to strive for the ecosphere's survival, safety, and flourishing, for the improvement of life at large.

Man can be an authentic overseer, guardian, treasurer of the *common good*. He becomes "Custodian of life's equilibrium" by activating a set of *ethical values*, such as care, tolerance, eco-justice, responsibility, solidarity, moderation, wisdom, respect, benevolence, sharing-in-sympathy, and *love of life*.

## The Sacral Logos

The orchestration of an ethics of life is intimately bound up with the logoic turn made by the "ontopoietic" philosophy of Anna-Teresa Tymieniecka. In its scope, ethics reveals itself throughout the phenomenological scrutiny of the *logos of life* with its plural modalities, from the vital to the "divine/sacred logos" by which span we can get a picture of "the fullness of the Logos" in the "Great Plan" of life. By discovering the *sacral logos*' workings in the world, man accesses a superior ethical creativity to be exercised in both the horizontal and vertical dimensions of becoming. Tymieniecka underlines the high significance of the *sacral logos* in accomplishing a life worthy of living by attaining the fulfillment of humane in the metamorphosis "from the logos of life to the logo-theic horizon" (Tymieniecka 2009, 247–255).

Considering the centrality of *moral experience* in human life, the phenomenological *transcendent-transcendental* pairing becomes much more comprehensible. The *transcendental* reveals itself in creative inwardness and outwardness too. It is activated by each individual relating with itself and with (an)other. It opens to a relation with the divine transcendent, showing the pathway to be chosen and cultivated: that of *creativity*. This is the pathway of man that deserves to be protected and developed as demonstrating his power to project and to work toward a moral ideal: that of making his own beingness, the world, and life generally to be(come) much better.

## An Ethics of Life Centered on *Living in Harmony with Nature*

We can read Tymieniecka's phenomenological work in the key of a philosophy upholding the basic value of searching and assuming the experience of the meaning of living in harmony with nature. We can say that, somehow, Anna-Teresa Tymieniecka framed a philosophy of ecological harmony that is beyond any of the disputes around the concepts of ecosophy or ecological wisdom, as found in the views of Arne Naess and Félix Guattari, for example.

The phenomenology of life enters the domain of the worldwide movement for sustainable development in the twenty-first century with its emphasis on the imperative for humanity to live in harmony with nature. An exemplary document of that movement is *The Future We Want*, adopted by the General Assembly of the United Nations at the "Earth Summit 2012," in Rio De Janiero.[3]

## Caritas Sapientis

Throughout ontopoietic phenomenology, the deployment of an ethics of life sends out a sort of Apollonian wave, enlightening the effort to overcome part of the perils and turmoil that our superficiality and consumerist excesses present, and finally enlightening our endeavor to unravel some of the mystery of life, to see what does constitute its highest value.

> By priority, thanks to the concept of "moral sense," phenomenological scrutiny allows us to catch the significance of the *caritas sapientis* (the care of wisdom) that, in the seventeenth century, Gottfried Wilhelm Leibniz pointed out as a chief defining trait of "a good man." We find in his *Codex Iuris Gentium Diplomaticus* (1693)[4] a reminder of the Greeks' combina-

tion of *φίλος/philos* and *ἄνθρωπος/anthrōpos* to emphasize *philanthropy* "as the charity which follows the dictates of wisdom" and which is able to make real universal benevolence, the habit of loving or of willing good, happiness, and the common good. Actually, "the charity of the wise man" comes to light in Tymieniecka's discourse, with the affirmation of the virtue of benevolence towards not only humans, but also non-humans. We face the intuition of a universal harmony within the scheme of the cosmos; and all, in the *ontopoietic* design of life in which the "human transnatural destiny" may communicate with the Creator, for the "privileged portion" that humanity "seems to be inscribed" in, that shown by the Divine (Tymieniecka 2012, 183–185).

Phenomenology of the *ontopoiesis of life* takes up the question of "the human soul in the cosmos and the cosmos in the human soul. Man's individual existence involves an intertwining unity of the "sensory, emotive, intellectual, volitional, aesthetic, moral, and spiritual" realms with profound ties to the cosmos in the experience of life articulated by the "ever present profile of Logos;" and these realms appear "intertwined as well as distinct, intermingled as well as autonomous, united as well as divided along the lines of their origins in different sources and their careers toward different teloi" (Tymieniecka 2012, 8–9).

Pursuing a comprehensive grasp of the "accomplishment of the logos' experiential route," Tymieniecka's ontopoietic phenomenology of life develops a path to interpreting the "rationale of the cosmos" by the "individualization of the logos in life" in its plural modes: vital, Dionysian, Promethean, and sacral.

## Conclusion

By a continuous care for nature, by acknowledging the responsibility man has as "Custodian of everything-there-is-alive," with the renewal of a fundamental concern for life, the ontopoietic phenomenology induces awareness of the proper anthropogenic factors that affect in a negative way nature's equilibrium, safety and health, and ultimately the whole of life on Earth. Thus, a consistent direction in the endeavor to overcome the difficult ecological problems of the present is to be found thanks to apprehension of the overarching of cosmic web in the human creative condition, with its capacity to activate—among other possibilities—*wisdom* and *care* as belonging to the essential self of the human being that is made manifest through *benevolence* towards everything-there-is-alive.

We think that the most important attitude we might undertake is to decipher, to understand, and to appropriate the great learning of wisdom that Anna-Teresa Tymieniecka generously offers us: namely, to being attuned to the *constructive* and *harmonizing* character of the path, which is worth to be followed in life, to increasing the authentic creative self in accordance with the *ἀρχή πάντων/arché pánton*—the Aristotelian "principle of all things," that is, with the *universal logos of life* (*Metaphysics* 1.3.983b). This seems to be the path on which all human endeavors have to be oriented, in the cultivation of admiration, fairness, truth, care and love, to improve the health, the beauty, and the integrity of life.

A significant message of phenomenology of life and, more generally, of the extraordinary example of creation and activity that Anna-Teresa Tymieniecka shared with a large international community of scholars over the years, is that of the power of wisdom and generosity, of intellectual friendship able to arouse, to reveal and to increase the best in the other(ness), and to work together for the common good of the whole of life within the world of which we, humans, are a part.

## Notes

1. In an insightful study dedicated to Tymieniecka's philosophical work, María Avelina Cecilia underlines the valences of a new worldview on the basis of the dialogue between phenomenology of life and the sciences. A new path is emphasized, one of convergence (and not confrontation) between "philosophy and science, on the one hand, and the sciences of nature and the sciences of culture, on the other," especially considering the original paradigm of the "ontopoiesis of life." See Cecilia 2002, pp. 687–711.
2. Daniela Verducci discusses Tymieniecka's concept of the "ontopoiesis of life" as it correlates with the idea of "autopoiesis" introduced in 1972 by Humberto R. Maturana and Francisco J. Varela in the framework of a systematic theoretical biology, one associated with the theme of cognition. According to the Chilean authors, in their *Autopoiesis and Cognition: The Realization of the Living*, "autopoiesis" designates "what takes place in the dynamics of the autonomy proper to living systems." This word opens new possibilities for understanding life and the environment. See Verducci 2012, pp. 115–116.
3. The necessity of promoting *harmony with nature* is underlined in articles 39 and 40, also in several paragraphs of the outcome document: *The Future We Want*, in the Annex of the *Report* of the United Nations Conference on Sustainable Development, Rio de Janeiro, Brazil, 20–22 June 2012. For example: "40. We call for holistic and integrated approaches to sustainable development that will guide humanity to live in harmony with nature and lead to efforts to restore the health and integrity of the Earth's ecosystem."
4. See Leibniz, *Political Writings* 1988, p. 83.

## Works Cited

Backhaus, Gary. 2001. Anna-Teresa Tymieniecka: The Trajectory of Her Thought from Eidetic Phenomenology to the Phenomenology of Life. *Phenomenological Inquiry* 25: 17–53. Print.

Brown, Charles S., and Ted Toadvine, eds. 2003. *Eco-Phenomenology: Back to the Earth Itself*. Albany: State University of New York Press. Print.

Cecilia, María Avelina. 2002. Phenomenology of Life, Integral and Scientific, Fulfilling the Expectations of Husserl's Initial Aspirations and Last Insights: A Global Movement. In *Phenomenology World-Wide. Foundations—Expanding Dynamics—Life-Engagements. A*

*Guide for Research and Study*, Analecta Husserliana LXXX, ed. Anna-Teresa Tymieniecka. Dordrecht/Boston/London: Kluwer Academic Publishers. Print.

Leibniz, Gottfried Wilhelm. 1988. *Political Writings*. Trans. and ed. Patrick Riley. Cambridge: Cambridge University Press. Print.

Leopold, Aldo. 1949. The Land Ethic. In *A Sand County Almanac: And Sketches Here and There*. New York: Oxford University Press. Print.

Rogers, W. Kim. 2003. *Reason and Life: An Introduction to an Ecological Approach in Philosophy*. Lanham/New York/Oxford: University Press of America. Print.

Scheler, Max. 2001. *The Human Place in the Cosmos*. Trans. M. S. Frings. Evanston: Northwestern University Press, 2009. Print.

Trutty-Coohill, Patricia. 2009. The Case of God in the New Enlightenment. Tymieniecka's Phenomenology of Life as a Radical Challenge to Continental Philosophies of Religion. *Phenomenological Inquiry* 33: 1–3. Print.

Tymieniecka, Anna-Teresa. 1988. *Logos and Life. Book 1: Creative Experience and the Critique of Reason*, Analecta Husserliana XXIV. Dordrecht: Kluwer Academic Publishers. Print.

———. 1990. Phenomenology of Life and the New Critique of Reason: From Husserl's Philosophy to the Phenomenology of Life and the Human Condition. In *Man's Self-Interpretation-in-Existence. Phenomenology and Philosophy of Life. Introducing the Spanish Perspective*, Analecta Husserliana XXIX, ed. Anna-Teresa Tymieniecka, 3–16. Dordrecht: Kluwer Academic Publishers. Print.

———. 1995. Measure and the Ontopoietic Self-Individualization of Life. *Phenomenological Inquiry* 19: 26–51. Print.

———. 2000. *Logos and Life. Book 4: Impetus and Equipoise in the Life-Strategies of Reason*, Analecta Husserliana LXX. Dordrecht/Boston/London: Kluwer Academic Publishers. Print.

———. 2009. *The Fullness of the Logos in the Key of Life. Book I: The Case of God in the New Enlightenment*, Analecta Husserliana C. Dordrecht: Springer. Print.

———. 2012. *The Fullness of the Logos in the Key of Life. Book II: Christo-Logos: Metaphysical Rhapsodies of Faith (Itinerarium mentis in Deo)*, Analecta Husserliana CXI. Dordrecht/Heidelberg/London/New York: Springer. Print.

———. 2014. Possibility, Life's Ontopoiesis, and the Vindication of the Cosmos. *Agathos: An International Review of the Humanities and Social Sciences* 5 (1): 7–12. Print.

United Nations. 2012. *Report of the United Nations Conference on Sustainable Development*, Rio de Janeiro, Brasil, 20–22 June 2012. New York: 2012. Web. 2 September, 2014. www.uncsd2012.org/content/documents/

Verducci, Daniela. 2012. *La fenomenologia della vita di Anna-Teresa Tymieniecka. Prova di sistema*. Rome: ARACNE editrice. Print.

# Part III
# Seeds of Eco-Phenomenology

# Eco-Phenomenology: Philosophical Sources and Main Concepts

**Maija Kūle**

**Abstract** Eco-phenomenology is a new developing trend, but its roots can be traced to *philosophia naturalis* (life philosophy), the philosophies of Schelling, Nietzsche, Bergson, Husserl, Merleau-Ponty, and other philosophers. The ecological turn in phenomenology is prompted by authors dealing with the problems of the human being – nature, life, the Universe. Eco-phenomenology and eco-philosophy are different teachings, eco-phenomenology being more theoretically fundamental while eco-philosophy until now more politically oriented. Eco-phenomenology appeals to us to look "Back to the Earth itself," but that it is not enough to solve the question of the nature of nature and to recognize the meaning of natural reality in phenomenology. The basic issues for eco-phenomenology can be found in A.-T. Tymieniecka's phenomenology of life. Her teaching actually is eco-phenomenology developed in a systematic way with new concepts. Many concepts of her phenomenology of life can be cornerstones for eco-phenomenology: Logos and life, the Unity-of-Everything-there-is-Alive, ontopoiesis, the individualization of life, and *imaginatio creatrix*. Here are offered additional eco-phenomenological concepts which could be termed the fundamental phenomena unifying the world's natural, human, cultural and symbolic dimensions – light, rhythm, silence, place.

**Keywords** Eco-phenomenology · Eco-philosophy · Phenomenology of life · Tymieniecka · Logos · Fundamental phenomena

## Eco-Philosophy and Eco-Phenomenology

Philosophy has an important part to play in the investigation of ecological themes. One could think that eco-phenomenology should be part of eco-philosophy, just like phenomenology is one of the trends of philosophy, a branch. However, that is not the case. Eco-phenomenology tends to be wider and more extensive in its scope

M. Kūle (✉)
Institute of Philosophy and Sociology, University of Latvia, Riga, Latvia

 
W. S. Smith et al. (eds.), *Eco-Phenomenology: Life, Human Life, Post-Human Life in the Harmony of the Cosmos*, Analecta Husserliana CXXI,
https://doi.org/10.1007/978-3-319-77516-6_5

than is eco-philosophy because it asks basic and systematic questions about the human being, nature, the Universe, creative activities, creative imagination (*imaginatio creatrix*), ontopoiesis, self-individualization, and existential sharing-in-life. Under Anna-Teresa Tymieniecka's direction, it has undertake discussions of the passions of the skies, of the soul, of place; the poetics of the elements: wind, breath, tempest, snow, flame, fire, thunder; the poetics of existential powers: movement, rhythm, fragrance, word, touch, sound, color, image; the harmony of life and questions of the Earth, Heaven, and In-between; human positioning in the Cosmos; the fullness of the logos in the key of life; authentic life in the moral sense; destiny; the inward quest; the temporality of life and the sacred. In these discussions, it presents a large philosophical system and creates new basic concepts.

Eco-phenomenology's roots could be traced to *philosophia naturalis* (life philosophy), the philosophies of Schelling, Nietzsche, Bergson, Husserl, Merleau-Ponty, and other philosophers. The ecological turn in phenomenology is influenced by the authors dealing with the themes of the human being: nature, life and the Universe.

The task of this paper is to examine what contemporary eco-phenomenology is like in its wider sense. What are its philosophical sources and perspectives? In what way are these not the same as those of eco-philosophy? However, to come to eco-phenomenology in the first place we must examine the pretensions of the contemporary philosophical schools that deal with linking philosophy to ecological problems and ecosophy.

There are different proposals as to how eco-philosophy should be developed. It is regarded to be a teaching on home or place (eco) and sophia (wisdom). This branch of philosophy promises to foster the pursuit of ecologically harmonious living and the development of an ecologically-informed green ethics. In some ways, it seems to be an art of living based on practical suggestions. In the centre of the theory are the relations between human beings and nature, with no particular importance being attached to the human being's mind, volition, or rationality, nor is the human being granted a special position in the Universe. The meaning these is mistrusted because the human being is likened to the processes in the Universe or else regarded to be a matter of chance.

Norwegian philosopher Arne Næss, whose name is associated with eco-philosophy – or, by another name, ecosophy – has since the 1970s stimulated the appearance of a social, non-governmental movement. Its participants articulate the impossibility of our continuing to work in the spirit of wild capitalism, giving priority to industry and profit and ignoring nature as a value. NGOs set for themselves positive aims. They demand changing value systems for the good of nature and lest we face the destruction of the beauty of nature and its wealth. Næss says, "to live is like travelling through landscape. [...] To feel at home in life requires both moving toward a goal and simply being" (Næss and Heukelands 2001, 3–4).

In eco-philosophy, the moral community is regarded to consist not only of people, as in classical ethics, but also of animals and, in a still wider sense, of everything that is alive. The values created by human activities should be for the good of ecosystems, not the other way around. If human activity does come to harm other living beings, then such activities, even if they are economically profitable, should be given up. Ecocentric values are to be developed. The Earth should be regarded as a value, but it is not the centre of the Universe and the human being is not the crown of creation as regarded in Christianity.

Eco-philosophy urges people to live closer to nature, sublimating their interests to natural systems. It has developed a teaching of deep ecology. The way in which Eco-philosophy is presented to ordinary people is a simplified model, one denigrating the value of the human being. That does not solve many problems. For instance, some solutions addressing overpopulation at times carry the threat of inhumanity and violence. Eco-philosophy appeared when the themes of natural philosophy on losing their metaphysical aspect were supplemented by worries about the ethical quality of human activities: environmental endangerment accompanied by a wish to raise the status of living nature and to lower the status of human beings (Ecophilosophy 2015, 1–8). It called for fighting against anthropo-chauvinism (placing the human being at the centre) and creating ecosophy or a new ecological wisdom.

However, the human being, and perception of the world in subjective and cultural experience, has not disappeared. The human being has not given up his or her status of the highest living being as evidenced by the fact that human beings are aware of the world and reflect on the world. Humans have not resigned from intellect, will, rationality, or responsibility, and no school of philosophy, no matter how much it would wish to be non-anthropological, can manage to avoid the theme of the human being's special status in the world. Christianity states it simply symbolically: Man is created by God. Thus, a proposed eco-philosophy that discards the idea of the human being as God's creation is neither complete, nor systemic, nor extensive, though it can be characterized as socially influential. Its power is mainly in promoting ethical responsibility towards the world of nature, but its standing is hindered by the absence in it of a meta-system and the highest values.

Some authors, myself included, consider phenomenology to be opposed to deep ecology, mainly as concerns the human being's place in the Universe. Eco-philosophy, though cherishing ethical aims, lowers the human being's value by placing it next to that of the living world. Eco-phenomenology, on the contrary, sees the highest potentialities of the human being (the logos of life) in the animate world, raising the human being's status and responsibility.

## The Essence of Eco-Phenomenology As Phenomenology of Life

Eco-phenomenology was already a subject for discussion thirty years ago, and today has become one of the main projects of phenomenology. Phenomenology as one of the most influential philosophical trends of the twentieth century cannot leave our ecological anxieties without a response. However, that is not the only reason for the birth of eco-phenomenology.

The philosophical role of eco-phenomenology is much wider and deeper than that of eco-philosophy because it turns to the questions of the human being's essence, the Universe, the nature of nature, and the like, not disregarding, naturally, the questions of ethical responsibility. Life energies, forces, and the shaping of life vitally and existentially are subjects of phenomenological investigation. One of the main concerns of eco-phenomenology is to describe the Unity-of-Everything-there-is-alive, looking at life's individuation and existential sharing-in-life. Space, time, sky, Earth, life, and death are taken up as complex philosophical subjects. The aesthetics of enchantment are close together with the passions of the skies, light, darkness, enjoyment, and human positioning in the Cosmos. Eco-phenomenology appears not as a new *philosophia naturalis* focused only on nature, but as a wish to apply the richness of the phenomenological approach, part of its conceptual apparatus, in order to elaborate the relationship holding among the Universe-logos-nature-life-the human being-ethics. The human being in relation to communication, the biosphere, and eco-systems should be an ethical human being.

Eco-phenomenology is looking for answers to these questions:

(a) Can classical phenomenology, based on the teaching of Edmund Husserl, be taken as the basis of eco-phenomenology?
(b) Could eco-phenomenology be based on the concept of intentionality worked out by Husserl, or would that narrow the understanding of eco-phenomenology and, therefore, be of no use?
(c) What should be the attitude adopted toward the turning against the natural attitude stressed in Husserl's epistemic method, and would that imply turning against investigating nature in phenomenology?

Is it that maybe eco-phenomenology should repeat Husserl's appeal, "Back to the things themselves!" and begin turning to focusing on nature (living nature) rather than investigating consciousness and inner temporality? Volume CVIII of the Analecta Husserliana series is titled *Transcendentalism Overturned. From Absolute Power of Consciousness until the Forces of Cosmic Architectonics*, which indicates a radical turnabout of classical Husserlian phenomenology.

The "Back to the Earth itself" project, sounded in the volume *Eco-Phenomenology. Back to the Earth Itself* (ed. Charles S. Brown and Ted Toadvine), takes up both ecological phenomenology and phenomenology of ecology, which are not one and the same thing. When using the word phenomenology it is important to retain its inner essence. The basic concepts of phenomenology as a twentieth-century philosophy are intentionality, intersubjectivity, and lifeworld, and these have to be at the centre of attention if one wishes to maintain adherence to phenomenological thinking. When Charles S. Brown recognizes that, "The ecological crisis is a crisis of meaning," that means to think phenomenologically (Brown and Toadvine 2003, 5).

It goes without saying that nobody would forbid the new philosophical activists to establish a new naturalism, new materialism, natural realism, ecological enlightenment, or the like. But then they cannot be called phenomenological trends. For the elaboration of eco-phenomenology, one of the most important tasks is that of philosophical anthropology, as was recognized by Scheler: the human place in the Cosmos is to be pondered.

Moreover, to belong to the phenomenological tradition, one must give thought to the subdivisions of phenomenology: constitutive, realist (the Munich-Göttingen school), existential, hermeneutical, social, life phenomenology, material phenomenology, and others. The possibilities each of these schools offers in support of eco-phenomenological solutions differ. If the adepts of eco-phenomenology wish to recognize the natural world, nature as independently objective – without human presence and the role of human experience in describing phenomena – then the most suitable trend is realist phenomenology[1] and the least suitable one is the trend based on the constitutive activity of the transcendental subject, displaying no interest in independent objective reality. The best expression of the phenomenological approach is one finding an equilibrium between subjective experience, the constitution of meanings, life-worlds, the plurisignificant ciphering of reality, and seeing how the logos of life is developing.

To understand the possibilities that eco-phenomenology offers, let us turn to its sources in the history of philosophy.

## Eco-Phenomenology: Philosophical Sources and Main Questions

The ecological turn in phenomenology is influenced by some major trends which have dealt with questions about relationships between the human being-nature-culture-Universe. Eco-phenomenology has been influenced by Husserl's ideas of genetic phenomenology, the lifeworld, the *Umwelt*, telos; by Merleau-Ponty's idea of embodiment; by Neo-Kantian ideas about the nature-culture difference from the epistemological and methodological points of view and tenets about values.

The theoretical, metaphysical approaches taken could be based mainly on:

1. Anna-Teresa Tymieniecka's phenomenology of life and the work of her followers found in the Analecta Husserliana book series;
2. the axiological approach extending back to Max Scheler's thinking on the human being in the Cosmos, Merleau-Ponty's phenomenology of the body, and the philosophical anthropology of German philosophers Arnold Gehlen and Erich Rothacker, Helmuth Plessner, Emerich Coreth, and others;
3. Catholic philosophy of the twentieth century which has interpreted the Universe, nature and spirit as a complex system, as in the philosophy of Pierre Teilhard de Chardin, Karol Wojtyła;
4. the ethical approach developed by Hans Jonas' practical ethics;
5. the Eco-ethics developed by the Japanese philosopher Tomonobu Imamichi[2] and his followers, Noriko Hashimoto and the Danish philosopher Peter Kemp;
6. the practical one, that has been influenced by the philosophy of environmentalists and "green" philosophy.

In nineteenthth-century philosophy, ideas for contemporary eco-phenomenology can be found in Friedrich Wilhelm Joseph von Schelling's philosophy of nature, which sent out shoots for the modern hermeneutic view of nature. Complexity and mutual identification between subject and nature (identity philosophy) demonstrate the characteristic of contiguity emphasized in contemporary eco-phenomenology. Schelling recognized that nature produced subjectivity, enabling it to understand itself.

Eco-phenomenologists see the same – nature creates a subject able to reflect upon nature. Schelling describes nature as a process developing from the lowest to the higher levels. Therefore, nature itself could be understood as a living "super-subject."

Eco-phenomenology as a new trend is in need of discussions as to what can be cornerstone concepts for its philosophical and interdisciplinary approach. There are two possibilities: (a) to recognize 'Nature" as such; (b) to concentrate on "the Universe," "life," "experience."

These approaches are not controversial but equally importantly to only return to nature as such (in itself) does not correspond to contemporary philosophical horizons because naive realism is not a priority for contemporary thinking. There are a few possibilities for how one is to interpret reality in eco-phenomenology. Philosophy centered on subjectivity is inclined to speak of nature as a "worldview of meaningful nature," bearing in mind that meanings of natural objects are created in intentional acts and we cannot attain nature *in itself.* Classical phenomenology is not going to discuss this level of reality *in itself.*

If eco-phenomenology returns to nature, that does not mean a return to Kant's position on the "Ding an sich," but, rather, recognition of the fact that philosophy of transcendental subjectivity (the Kantian-Husserlian line) has left behind important philosophical questions about the Universe, the objectivity of nature, the unity-of-everything-there-is-alive, and life's differentation as vegetal, animal, or human life forms. Transcendentalism does not pay attention to the life's vital energies, forces, and the shaping of organic life; looming in its horizon are, rather, the inner structures of human experience. Phenomenology of life, as realised in Tymieniecka's system and the 120 volumes of the Analecta Husserliana book series she edited, has moved from the absolute power of consciousness to the forces of cosmic architectonics and existential sharing-in-life. The questions that comes to the fore is that of how the logos of life develops from the animal soul to the human mind and that of how impetus and equipoise are realized in the life strategies of reason.

One of the basic theses of eco-phenomenology is that nature and the Universe exist as real "substances," not created by consciousness, but evidenced in our life experience; they, of course, appear as phenomena in a historical and cultural context. The position of naive realism in phenomenology is disputable because then philosophy would lose the means to approach our speaking of meanings and experience.

If we return to the two possibilities (a) to recognize 'Nature" as such; (b) to concentrate on "the Universe," "life," "experience," then, in my opinion, the second option has more perspective than a new naturalism focused nature *as itself.*

One of the basic questions is – how deeply can we separate the structures of human life from life *as such*? Can we, as phenomenologists, continue to use the principle of "pure life" (*das reine Leben*)?

In classical phenomenology the notions "life" and "live" function in their transcendental meaning and imply pure structures. Phenomenologists plead that the life they write about is not real live life, but only phenomenologically-purified life.

Life described in eco-phenomenology is not "pure," but vital, expressed in plural forms, connected with growth and dying, and included in the environment. Tymieniecka said that phenomenologists had to disentangle the knot which life, in its creative games, ties among its unlimited factors. She developed her new metaphysical system in the three volumes of *Logos and Life* (Tymieniecka, Books 1–2, 1988a, b; and Book 3, 1990a).

Eco-phenomenology belongs to process philosophy. The main concepts used in eco-phenomenology are interpreted historically: life means development, extension of potentialities. If classical phenomenology stresses description of evidence and mainly avoids genealogies and historical and biological stories about the emergence of mind, free will, and species, then eco-phenomenology attends to genealogies, creative processes in the Universe, the development of life and the essence of living beings, participating in nature-life-sharing in life. The creative symbiosis of existential powers, passions for place, vital spacing, the skies, the infinite, and the fullness of the logos are conceptual apparatus for phenomenology of life as eco-phenomenology.

Classical phenomenology, based on Husserl's principles, has challenging questions for eco-phenomenological thinking:

(a) What in Husserl's teaching is meant by the natural attitude and its 'bracketing'? Does this exclude description of nature?
(b) Classical phenomenology's understanding of the real, reality: hindrance or stimulus?
(c) To what extent is intentionality as "consciousness of..." understandable in wider vistas – as engagement, embodiment, being in the world (*in-der-Welt-sein*); would such an extension of the concept support the eco-phenomenological approach?
(d) The meaning of horizon, context, wholeness, integrity; what do these contribute to an eco-phenomenological approach?
(e) What is the solution of the relationship between intentionality and causality in phenomenology? Could a bridge between them, rather than opposition, be a solution?
(f) Inner time consciousness and the mysteries of time: imperceptibility, appreciation of finitude, rhythm, breaking temporal horizons. Do the solutions to the problem of time in eco-phenomenology strengthen the initiative or needlessly complicate viewpoints?

Short answers can be given: what Husserl means by the "natural attitude" and its bracketing is not obstacle to the investigation of nature in phenomenology because when Husserl speaks of the "natural attitude" he means everyday experience as acceptance of being beforehand, some naïveté. One of phenomenology's insights is that the human being is never severed from world; he or she is engaged in the world, including nature. Therefore, reality does not consist of objective positive and neutral facts but is indeed the reality of everything-there-is-alive, the reality of human positioning in the Universe. And intentionality and causality are not so very opposite because a human being lives between freedom and necessity, between Earth and Sky (Heaven) in some wholeness, integrity. The key of life is logos, which is active from the lowest levels of life to authentic life in the moral sense. And discussions on time strengthen eco-phenomenology because they cover not only inner time consciousness and life in its finitude but also the elemental passions for time, nostalgia, yearning, the place of memory in the ontopoiesis of life, divine and absolute.

## Alternatives to Eco-Phenomenology: Are They Phenomenological?

The American philosopher and performance artist David Abram looks for the existence of a specific perception outside the human world. His formula "the more than human world" stimulates opening the window of the human world and seeing all the living beings without any preference for human beings. In this respect he approaches Arne Næss' eco-philosophy. He gives a wide interpretation of the world – it is not only a human world, but a cosmic, universal world where living, conscious, and spiritless beings exist close together (Abram 1997, 2). He does not seem to have any pretensions to being counted among phenomenologists because his approach has been devoted to opening horizons, but not to developing an eco-phenomenological system on the basis of the Husserlian conceptual apparatus. His interest in Merleau-Ponty seems to be directed more towards the French philosopher's metaphors than to his phenomenological ideas.

Morten Tonnessen sees eco-phenomenology as an interdisciplinary science close to biosemiotics, ecosemiotics, zoosemiotics. He urges remembrance of eco-philosophers such as Arne Næss, Peter Wessel Zapffe, and the philosophy of biology developed by Jakob von Uexkull (Tonnessen 2011, 328–339). The concept of "Umwelt" used in their theories does come from classical phenomenology. Bodily semiotics is expressed phenomenologically, and signs appear in the environment. Life itself evaluates and chooses. Meanings can be attributed to nature; sense can be attributed to the living world. Natural beings cannot be separated from human beings possessing consciousness because the human being is an inseparable part of nature. Nature is much richer than our ability to designate it, to understand it, and to express it in language.

Such philosophical viewpoints add credibility to the theses that the human being should not be understood as the principal being in nature, that in nature everything is interconnected, that one must raise the value of other living beings and develop much more extensive ethics beyond the frames of the world of humans. This is the direction taken also by the eco-phenomenology developed by Tymieniecka, the difference being in the fact that she has created an eco-phenomenological system with many new, interlinked concepts, and retained the line of phenomenological thinking, as in her theory she shows similarities and differences of viewpoint with regard to Husserl and his followers.

## Logos and Life: Tymieniecka's Eco-Phenomenological Teaching

Phenomenology has been oriented to Logos and Life by Anna-Teresa Tymieniecka's personal efforts, and by the work of the World Phenomenology Institute which she founded, as found in the 120 volumes of the Analecta Husserliana book series that she directed, the fruit of some 64 international conferences. This is a very important source for eco-phenomenology today. This project is that closest to the essence of eco-phenomenology. Tymieniecka's overwhelming system of Logos and Life, its ontopoietic vision is one of the best-elaborated frameworks for proceeding that we have at the beginning of the twenty-first century because it fully covers the theoretical basis for ecologically-orientated philosophy. She has worked out a system of concepts at the highest theoretical level, has returned to the ancient view of the logos active in the Universe, and has demonstrated the natural and ontopoietic development of the human mind and spirit.

The contribution of her phenomenology of life and the responses to it in the pages of the Analecta Husserliana series to the questions posed by eco-phenomenology could be detailed as follows:

(a) What is offered: an enormous system encompassing the Universe, nature, and the human world;
(b) What is going on: a shift of focus in Husserlian phenomenology from the recognition of meaning, which constitutes intentionality as the primary viewpoint in phenomenology, to the creative act as an expression of logos and life;
(c) What has been created: exceedingly rich, philosophically meaningful new concepts about life – ontopoiesis, the logos of life, the self-individualization of life and others;
(d) What is going on: elaboration of the principal concepts: the Unity-of-Everything-there-is-Alive, *imaginatio creatrix*;
(e) Tymieniecka and her colleagues have devoted prodigious effort (to be found in the Analecta Husserliana volumes) to working out the concepts of the phenomenology life: place and space, home, life, the Universe, the natural elements, and so on.
(f) What is being encouraged: cooperation with natural scientists, astronomers, biologists, and others.

The turn to the ecological point of view leads to radical changes in the phenomenological approach. The human mind is no longer the king of nature, but the most powerful expression of the logos of life's activity in the natural realm. If eco-philosophy for the most part does not offer an account of the human being's universal value, then Tymieniecka does so without contradicting the tenets of Christian civilization because she shows the human being's links with the logos of life. This means concentration on the logos' long path in life's self-individualization, reflection, and human freedom. Consciousness matures in the womb of the Universe. Therefore, eco-phenomenology starts with interpretations of the Universe, geo-cosmic horizons, and only then passes over to analyzing the structures of inner experience, of human subjectivity.

Tymieniecka's philosophical journey to eco-philosophy started already with her triptych monograph *Logos and Life: Creative Experience and the Critique of Reason* (1988a). The first part of the triptych was written already in 1972 under the title "*Eros and Logos*." Before that Tymieniecka had published a book on *Leibniz' Cosmological Synthesis* (1964) and *Why Is There Something Rather Than Nothing? Prolegomena to the Phenomenology of Cosmic Creation* (1966). We see that the founder of the phenomenology of life reflected in a circular way: five decades ago she was interested in philosophy of the Cosmos, and at the end of her life, she returned to the geo-cosmic themes and human positioning in the Cosmos.

In a foreword to the triptych, Tymieniecka writes that the present-day philosophical endeavour is prompted by a conundrum of problems: "The struggles in which contemporary mankind is fiercely engaged are not confined, as in the past, to economic, territorial, or religious rivalries, nor to the quest for power, but extend to the primary conditions of human existence. They undermine man's primogenital confidence in life and shatter the intimacy of his home on earth" (Tymieniecka 1988a, xxi).

The essential motif of Tymieniecka's innovation is widening phenomenology from the Husserlian description of consciousness to the interpretation of life, paying maximum attention to the *life* processes, putting ontopoiesis in the place of strong rationality, replacing intentionality with creative acts. Phenomenology of life is extended phenomenology. Phenomenology of life brings forth the ontopoiesis of life as the primal and generic factor of constitution. We can see how it differs from Husserlian phenomenology by placing the lifeworld together with its correlate; transcendental consciousness is in a secondary position.

Tymieniecka's theory of the self-individualization of life corresponds to the phenomenological turn in the philosophy of ecology and her idea of ontopoiesis actually is an eco-phenomenology. When the Norwegian philosophers Lars Petter Torjussen, Johannes Servan, and Simen Andersen Øyen asked her about eco-phenomenology, she answered:

> Actually, my account of ontopoiesis is an eco-phenomenology. Ontopoiesis reaches to the very germs of ecology: development and genesis. I have published several essays related to this. In *Passions of the Earth* (Analecta Husserliana LXXI) I show how the human being is an ecological fruit and how the human being is formed by the earth and sucks the juices of the earth. I have also written things about the cosmos and the cosmic dependencies of the

> human mind and human development. You see, the self-individualization of life, which is the basic instrument of ontopoiesis draws upon the laws of the cosmos and the earth. This is the most fundamental ecology that can be done. So, we have just touched the essence of my philosophy, the base – our relationship to the earth and to the cosmos. (Tymieniecka 2009)

## Phenomenology of Life As System for Eco-Phenomenology

Three concepts of the phenomenology of life serve as systemic cornerstones for eco-phenomenology:

1. Logos and life, the Unity-of-Everything-there-is-Alive;
2. Ontopoiesis;
3. the Individualization of life and *imaginatio creatrix.*

As the main standpoint of the philosophy of life Tymieniecka points out three main themes: (1) *the self-individualisation of life* circumscribing the context of phenomenological investigation; (2) *the creative act* of the human being, which brings us into the centre from which the human mind draws all the rays of order; (3) *the human condition* that grounds the creative act as man's foothold within the uncharitable schema of life (Tymieniecka 1990b 5).

She writes that the logos of life is an intelligent design of entire things and intelligence itself, the measure and essence of all things, in itself – logic, sentience, intuition, awareness. That is what distinguishes her eco-phenomenology from what is offer by others who tend to describe the Earth in itself, wild ethics, animal rights, and the like. In Tymieniecka's philosophy one can discern Schelling's soundings about the potentiality of life, the presence of logos in the Universe, thus imparting meaning to human qualities as a result of the development of logos. The philosopher discerns moral sense as potentially existing in live forms of nature and their yearning towards the divine becomes really apparent in the human world. While eco-philosophy more often than not avoids ascribing the spiritual characteristics of the human being to the animal world, inviting us to empathize with animals by, instead, delving into the specificity of their world, Tymieniecka's standpoint is different – she sees the unity-of-everything-there-is-alive and that implies that spiritual qualities are potentially hidden in the simplest levels of live nature.

*Imaginatio Creatrix* transforms the more primitive stirrings of the soul into the subliminal passions of human existential significance, inward sacredness. She recognizes that "Imaginatio Creatrix proceeds from the womb of life and depends on it" (Tymieniecka 2007).

The triumph of the creative force in Tymieniecka's philosophy reminds one of Bergson's positing an *élan vital*, although the philosopher herself did not want to acknowledge this influence when at congresses discussions arose as to the sources of her standpoints.

Phenomenology of life recognizes that self-individualization in life is the main principle of development in general. Every individual is born and dies alone according to his or her destiny, but in the social, cultural, and cosmic context there exists the continuity of species, and of everything-there-is-alive. Dying comes with suffering, but it is the main principle of life's renewal and creativity. The human soul reflects the passions of the Earth and of the skies. Tymieniecka admits that the soul exists, that it is not an empty or entirely theological category but, rather, a concept which plays a very important role in the system of the Universe, logos, and the human being. The body is animated by the psyche and enlivened by the spirit, which is the outcome of life's individualization. The soul requires the body for its natural existence, and the physical body as a material thing is grounded in the Earth.

An important step in the explanation of the ontopoiesis of soul is attributing to the human soul intellectual activities, and aesthetic and ethical values. The philosophical idea of logos helps to unify mind with body, the cognitive with the emotional; it helps us to see the close relationships between perceptions, feelings, reflection, doing; it unifies theoretical thinking with practical doing, thinking with will and evaluation, and life with the generation of human beings. Tymieniecka includes soul in the context of the cosmos and demonstrates the development of the rather different strata of souls as the creative development of the Universe expands the potentialities of logos.

Eco-phenomenology based on the phenomenology of life would be systematic, on a high metaphysical level, unifying all the traditional parts of philosophy: ontology, ethics, aesthetics, practical philosophy, social norms, human geography, cultural anthropology, astronomy, and environmental protection.

## Concepts of Eco-Phenomenology: Life-World, Life, World

The history of phenomenology shows that the notions "ecology," "life," "live," "life-world," "nature," "Earth," "Cosmos," "the Universe," "hyle," "logos," "everything-there-is-alive," and others have become important in contemporary phenomenological lines of thought. Classical phenomenology has been changed: interest in nature and the Universe has come to the forefront, analysis of consciousness lags; creative acts are prior, intentional acts are secondary; some scientists recognize intentional acts beyond human consciousness, existing at the lowest levels of life.

One of the most useful concepts for eco-phenomenology that comes from classical phenomenology is that of the *life-world*. It substantiates the structures of sense and validity and enables one to speak about the world of nature where nature is seen through meanings and sense experience.

Purely physical nature *as such* (not as phenomena) emerges as a net of meanings from an original, phenomenologically-prior intersubjective base. For Husserl, the life-world signifies the world of pre-theoretical and pre-scientific everyday experience. The objects grasped by the natural attitude (which must be reduced) are not the real objects of nature but our everyday constructions. The human world is sensual and spiritually-rich because it is constituted by meanings; the animal world is poor because it co-exists with its surroundings and is not included in abstract language, culture, semiotic systems.

The animal exists in nature, but not in the natural life-world because it has no constitutive possibilities such as the human being has. If Husserl regards the human being and the animal as being opposite (just like Heidegger), Tymieniecka demonstrates the general line of the genesis of life where the logos of life generates individualization of beings step by step, and animals are included in the process of developing logos and life. Therefore, Tymieniecka's conception is closer to eco-phenomenology than are the philosophies of Husserl or Merleau-Ponty or Heidegger.

For Tymieniecka the central category is life, not existence – beingness as a process and time. Logoic ontopoiesis is central, not time as a structure that categorizes existence. For her life *times itself*; the constitution of worlds is an opportunity for the expression of creativity. Logos is directly intuited within a phenomenological horizon of life. Tymieniecka returned from the ego-centred transcendental consciousness to its creative, dynamic process based on life. She shows subjects to be self-generating individualities at the ontopoietic edge of life's individualization. This brings out the self-individualizing principles of life as the primary factor of genesis.

Tõnu Viik discusses the notion of a "world horizon" for eco-phenomenology. He returns to the question of whether eco-phenomenology could go beyond human experience and reach the privilege of divine perspective – to speak of nature *as it is*, the Universe, the world as a whole. Recognizing the outside view of the world (nature, the Universe) means to sink into new naturalistic theories, not to respect *epoché* – which is to put the naturalistic outlook into brackets.

Could the road to the Universe be shut off for man? Many eco-philosophies simply define the World as a whole, the existence of the Earth and the Universe, without taking into account the phenomenological viewpoint acquired from the light of human perception and experience. Viik admits: "The world as a whole, if staged, necessarily becomes a displaced territory of the life-world, a simulacrum of the world" (Viik 2013, 64). He is right; eco-phenomenology could remember that the world is a correlate of the life of the transcendental ego and singularity is its feature characterizing the relationship between ego and the world, a place where one finds oneself. Tymieniecka recognizes that life realizes itself as self-individualization. Therefore, eco-phenomenology should avoid very abstract generalizations and remember the principle of the historical singularity of every experienced, creative act.

## Eco-Phenomenology – Ideas of Fundamental Phenomena

It seems that the understanding of the human being's nature-Universe relationships is influenced by some deeper ontological layers located beyond the level of culture but coming to us through the layers of culture. I would call these the fundamentals of the Universe; they can be characterized as borderline phenomena. Sometimes it does seem that phenomenology is moving to neutral territory. That is why the ontological preconditions of meaning should be sought not only in the human mind and volition creating culture, but also in the Universe.

To accentuate the philosophical dimensions of the phenomena of the fundamentals of the Universe, one must turn to such phenomena necessary for human life as rhythm and eternal recurrence, light and darkness, silence and noise, and the like.[3]

Phenomena such as light, darkness, silence, noise, rhythm, and space function as the basis of the intersubjective processes of life. Silence and noise are not only and solely phenomena created by culture, but phenomena necessary for life and living. Just like light and darkness, silence and noise, too, are phenomena whose scope reaches out from nature, from phenomena created by the Universe up to phenomena depicted in art and symbolic reality. The intersubjectivity of everyday life could be the level at which those unique phenomena could be grasped in interconnection without severing the natural and the cosmic from the symbolic and the social. It is on the level of everyday life that these appear as the most essential phenomena forming human existence. The changes of day and night organize human existence on the biological, social, cultural, and symbolic levels. What is most significant is that on the level of everyday life this organization does not happen separately, but the biological, the natural merges with the symbolic, the aesthetic. In the field of art the symbolic, cultural layer appears already to be severed from the natural (life, existence) layer.

Classical phenomenology, apprehensive of naturalism and combating psychologism, has driven out from the field of problems to be discussed all transsubjective, transcultural, and cosmic phenomena, and at the same time has overly accentuated the questions of subjectivity.

The new grasp appropriating the phenomena of life and the natural-cosmic view is especially gratifying: the human being is returned to harmonious interconnection with nature, with the Universe, with Everything-there-is-alive. The thesis of the omnipotence of the human being, cherished by the philosophy of subjectivity characteristic of classical phenomenology, is being tempered. In the first place, in the eco-phenomenological view, the world of the human being is united by *being* in the world, and human responsibility for everything that is alive emerges. Here is a philosophy of solidarity not only within the world of human beings, but also of the solidarity created by being responsive to universal light and darkness, noise and silence, place, and rhythm.

## Notes

1. To the trend of realistic phenomenology belonged Latvian phenomenologist Teodors Celms, who paid special attention to the concepts of nature and culture in phenomenology. See T. Celms, *Phänomen und Wirklichkeit des Ich. Studien über das subjektive Sein*. Riga: FSI, 2012, Teil 4 "Struktur und Leben des men-

schlichlichen Subjekts," Kapitel 12 "Natur und Kultur im Menschen." As its title tells, Analecta Husserliana III is devoted to *The Phenomenological Realism of the Possible Worlds*, (1972).

2. Tomonobu Imamichi created eco-ethica as a new narrative of culture and nature. He understood culture as a vessel that forms and preserves the order of life. Harmony, comfort, familiarity characterize our ecological and ethical relationships with the environment. See the journal *Eco-ethica* 2011–2014, editors P. Kemp and N. Hashimoto (Copenhagen and Tokyo).
3. To the analysis of fundamental phenomena are devoted these of my papers: "Silence as a Cultural Phenomenon" in *Reason, Life, and Culture, Part One*, ed. Anna-Teresa Tymieniecka, Analecta Husserliana XXXIX (Dordrecht: Kluwer Academic Publishers, 1993), 13–21; "Home: a phenomenological approach" in *Passion for Place*, Book II, ed. Anna-Teresa Tymieniecka, Analecta Husserliana CI (Dordrecht: Kluwer Academic Publishers, 1997), 97–112; "The Role of Ciphering in Phenomenology of Life" in *Does the World Exist? Plurisignificant Ciphering of Reality*, ed. Anna-Teresa Tymieniecka, Analecta Husserliana LXXIX (Dordrecht: Kluwer Academic Publishers, 2004), 41–50; "Selfindividualization as the Main Principle of Life" in *Imaginatio Creatrix, the Pivotal Forces of the Genesis/Ontopoiesis of Human Life and Reality*, ed. Anna-Teresa Tymieniecka, Analecta Husserliana LXXXIII (Dordrecht, Boston, London: Kluwer Academic Publishers, 2004), 47–56; "Logos and Life: Understanding of Rhythm" in *Phenomenology/Ontopoiesis. Retrieving Geocosmic Horizons of Antiquity, Part 1*, ed. Anna-Teresa Tymieniecka. Analecta Husserliana CX, 675–683 (Dordrecht: Springer-Science + Business Media, 2011).

## Works Cited


Abram, David. 1997. *The Spell of the Sensuous: Perception and Language in a More-than-Human-World*. New York: First Vintage Books Edition. Print.

Brown, Charles S., and Ted Toadvine, eds. 2003. *Eco-Phenomenology. Back to the Earth Itself*. Albany: State University of New York Press. Print.

"Ecophilosophy." 2015. Powerpoint. www.creeca.wisc.edu/outreach/.../ecophilosophy.ppt Web. March 1, 2015.

Næss, Arne, and P.I. Heukelands. 2001. *Dzīves filosofija*. Rīga: Norden. Print.

Tønnessen, Morten. 2011. Uexküllian Phenomenology. Retrieving Geo-cosmic Horizons of Antiquity, Logos and Life. In *Phenomenology/Ontopoiesis Retrieving Geo-Cosmic Horizons of Antiquity, Part 1*, Analecta Husserliana CX, ed. Anna-Teresa Tymieniecka. Dordrecht: Springer. Print.

Tymieniecka, Anna Teresa. 1988a. *Logos and Life. Book 1, Creative Experience and the Critique of Reason*, Analecta Husserliana XXIV. Dordrecht: Kluwer Academic Publishers.

———. 1988b. *Logos and Life: Book 2, The Three Movements of the Soul*, Analecta Husserliana XXV. Dordrecht: Kluwer Academic Publishers.

———. 1990a. *Logos and Life. Book 3, The Passions of the Soul and the Elements in the Onto-Poiesis of Culture*, Analecta Husserliana XXVIII. Dordrecht: Kluwer Academic Publishers.

———. 1990b. The Human Condition Within the Unity-Of-Everything-There-is-Alive. A Challenge to Philosophical Anthropologies. In *The Moral Sense and Its Foundational Significance*, Analecta Husserliana XXXI, ed. Anna-Teresa Tymieniecka. Dordrecht: Kluwer Academic Publishers.

———. 2007. Creative Imagination in the Converting of Life's Sensibilities into Full Human Experience. In *Phenomenology of Life – from the Animal Soul to the Human Mind, book 2: The Human Soul in the Creative Transformation of the Mind*, Analecta Husserliana XCIV, ed. Anna-Teresa Tymieniecka. Dordrecht: Springer.

———. 2009. An Interview with Anna-Teresa Tymieniecka. *Phenomenological Inquiry* 33. Web May 15, 2014. http://www.phenomenology.org/Bergen-Interview.htm

Viik, Tõnu. 2013. The Cosmic Matrix: Revisiting the Notion of the World Horizon. In *Phenomenology and the Human Positioning in the Cosmos, book 2*, Analecta Husserliana CXIV, ed. Anna-Teresa Tymieniecka, 53–66. Dordrecht: Springer.

# Logos of Life and Logos of Science. Metaphysical Advice

**Gianfranco Bosio**

**Abstract** The essay is intended to delineate the fundamental profile of a general ontology of life, held to be an unavoidable necessity for a well-founded idea of the unification of knowledge and of every practice in human research. Modern and contemporary science neglect and almost forget the specificity and originality of the primordial phenomenon of life. Philosophy falls victim to the same mistake and separates the domains of spirit and the human sciences from their root in life; this is precisely the dimension that we intend to recover. Following Husserl, Scheler, and Jonas, we believe that the fundamental core of the primordial phenomenon of life is to be found in the idea of "force," since it is a dimension that contemporary physicalism is unable to grasp in its specificity. Beyond these great philosophers, we also follow the original approaches of philosophically gifted scientists such as L. von Bertalanffy, F. Capra, H. Maturana and F. Varela. At the conclusion of this contribution, we advise that the fundamental ontology of life would be a task that is just preliminary to the unveiling of the dimension of spirit, rather than the exclusive and last purpose of philosophy.

**Keywords** Logos · Nature · Life · Science · Subjectivity

## General Introduction

In contemporary thought, there has been felt ever more strongly the almost complete absence of an original and fundamental reality that philosophy cannot avoid confronting: life. Since Descartes, Western thought has taken two absolutely separate paths, which have diverged up to the point of incommunicability. The ruts dug by the dualistic scission between *res cogitans* and *res extensa* have never been arced. The divide has assumed different configurations, but it has left painful

---

*Translated by Emilia Andri*

G. Bosio (✉)
University of Verona, Verona, Italy



W. S. Smith et al. (eds.), *Eco-Phenomenology: Life, Human Life, Post-Human Life in the Harmony of the Cosmos*, Analecta Husserliana CXXI,
https://doi.org/10.1007/978-3-319-77516-6_6

lacerations that wound and mutilate the human being, nature, and the spirit itself. In fact, there have been some important philosophical moments in the thought currents of the end of the Nineteenth Century and the first thirty years of the Twentieth Century that drew attention to life. Late German Romanticism, Simmel's "philosophy of life," Bergson's "philosophy of intuition," even Nietzsche's thought and Dilthey's philosophy convey contributions that should be thought over and constantly resumed even now. Unfortunately, these philosophies have the undeserved and unfair fate of having become minority tendencies in modern philosophy and therefore do not exercise a determining influence that could have modified and deeply transformed contemporary thought. After all, their limit consists of having turned their reflection to the moment of the irrationality of the vital, and of having emphasized subjective experiences, which reflections are still compromised by subjection to the psychologism of feeling and of individual participation, something always very personal and not easily communicable.

Great modern philosophers who went beyond these limits were Edmund Husserl, Max Scheler, and Hans Jonas. Scheler and Jonas above all enhanced and radicalized the Husserlian discovery of the life-world (*Lebenswelt*), verily opening horizons previously not guessed.[1] We can certainly affirm that they raise speculative thought to a sheer ontology of life that exceeds all the limits of psychological inwardness and furthermore that they profoundly understand that the idea of living and of life cannot be reduced and restricted to the dimension of the organism and the biological. As we shall see later on, the idea of life is something more profound and original than the pure biological datum of the organism.

## Body-Mind Dualism and its consequences

Descartes' scission between *res cogitans* and *res extensa* was undoubtedly disastrous. Primarily, it obscured the necessity of the median dimension that links, connects, and puts in a peculiar and special relation abstract intellect, which is made separate, incorporeal, and able to reach pure "truths of reason" and matter, which is made inert and subjected only to the mathematically computable and predictable physical laws of motion. Life – the world of emotions, of passions, of phantasy, life as the contact between "outwardness" and "inwardness," between *soma* and *psyche* – is abandoned to a division between psychology and the biological sciences; nature is totally consigned to the exact sciences, which are grounded on mathematical physics and experimental objectification. On this issue one would better refer to the outstanding theoretical survey of the birth of modern science conducted by Heidegger in the 1938 conference *Die Zeit des Weltbildes*, *The Age of the World Picture*. Here, Heidegger defends the fundamental and central thesis of the hidden metaphysical origin of modern science, which belongs to the same root from which the trunk of Cartesian philosophy sprouts. As a matter of fact, according to Descartes the truth of being traces back to the *Cogito*; but the *Cogito* is a spiritual substance that knows by means of representing the world in order to reduce it to an idea.

Following Descartes "idea" means *clara et distincta perceptio*, a perception provided with the security of intuitive clarity and certainty, which is the very sure attestation of the originating centrality of *cogitare* as an activity that puts the being of the entity in front of itself so as to objectify it, to make it into an *ob-jectum* for a *sub-jectum*.

At the same time, physical-mathematical science arose as the ultimate, absolute, and definitive project of a representation of nature as a set of "mass points" whose only links are *extensio* and motion, where any reference to the concreteness of vital and living qualities such as colour, taste, touch, scent, is left aside as unessential and insignificant. By means of a nature so dis-animated by the representation of a disembodied ghost like the *Cogito*, it becomes easier to devise and to conceive scientific experimentation. In fact, setting up an experiment means to represent and to arrange the ideal conditions that can be provoked in the laboratory in a completely artificial way that totally disregards the factors that can perturb and disturb observation – as when, for instance, the void is pushed as far as possible in order to better observe the fall of bodies – and these conditions have to conform strictly to the original project that is shaped in the framework of nature which is exclusively reduced to idealized mass-bodies and to measurements of motions where the idea of "force," of "energy," is displayed and exhausted (Heidegger 2002, 81–85).

## End of "Subject" in Contemporary Thought

Contemporary philosophy has more and more de-substantialized Descartes' *res cogitans*. Not entirely wrongly, it has believed it to be unnecessary to appeal to a substantial *res* in order to secure the freedom, spontaneity, and autonomy of thinking activity from nature and its blind mechanisms. Nevertheless, having placed these primary functions all the more within the inter-subjective, historical, and social dimension of knowledge and action, within interpretation and language, it has ended with the fiction of the oblivion of nature and life. But nature and life are always here with us and within us, and they condition us; they direct us even without our knowing it. They dreadfully threaten the certitude of the freedom and the creativity of thought and spirit. Between the Nineteenth and the Twentieth Centuries, Darwinism and Freud's psychoanalysis asserted themselves with precepts that treat as illusory any belief in the autonomy and independence of the spirit from life and nature, and that therefore posit the essential feebleness of any attempt to repair to the interior spiritual world as a guarantee of freedom. According to Freud, the *ego* is not master of himself in his own house. First, Copernicus and Galileo said our earth is not the center of the universe. Then came a second humiliation, that inflicted by Darwin, who makes humans descend from the most evolved animal species. Now humanity suffers a third humiliation, the wound that psychoanalysis inflicts by asserting that we are nothing more than slaves of instincts, of the drives and the repressions of the unconscious.

## Subjectivity "Embodied"

In the last years of his philosophical production, especially in the fundamental work *The Crisis of European Sciences and Transcendental Phenomenology* (1936–1938), Husserl uncovers the life-world (*Lebenswelt*) in a very unique way and undoubtedly a far better way than had his predecessors. What is the life-world? It is that of the direct presence where realities present themselves to us "in the flesh," without any preliminary deformation by the scientifically calculating and objectifying intellect that intends to arrange the world in order to make it available for mathematical and idealizing objectification. In the life-world there are – and *cannot* be removed – colours, sounds, tastes, scents, namely the so-called "secondary qualities" of the Seventeenth Century's philosophical tradition that so much bother physicists and generally scientists, since these are uncertain, fluctuating, changeable, and thus not directly reducible to mathematical form. Yet the world of science is not a world wherein we really live; indeed, we absolutely could not live there, whereas real life – by means of which we know, we perceive, we cultivate interests (even those of the sciences themselves) – is the world from which arise the ideal reconstructions that afterwards scientific knowledge pursues and is committed to improve more and more. The realm of significance, of purposes and ends, resides in the life-world rather than in the dis-animated world of scientific objectification. Therefore, the Cartesian *Cogito* is a disincarnated ghost and so, according to Husserl, Descartes, notwithstanding his undeniable merits, directed modern thought toward disastrous results. Husserl presses closer and closer on modern thought, claiming that the replacement of the substantiality of the Cartesian *Cogito* with Kant's transcendental did not solve the problems put forth by Descartes at all. Indeed, it dissolves the real and living subjectivity in an "epistemic subject" which is absolutely conventional, insensible to individuality, and thus replaceable at pleasure (Husserl 1970, 75–84, 221–24).

Scheler and Jonas impute to Husserl the limitations of a transcendentalism that does not take into proper account life, corporeity, soul, a reproach that is partly undeserved, considering the developments of the late Husserl's thought. Nevertheless, both Scheler and Jonas take the path of a "philosophical biology" and of a "philosophy of biology" following which the dimension of life is not the outcome of a constitutive activity of the Transcendental Subject, as it is in Husserl, but it is a datum that should be constantly rediscovered. Jonas claims that any transcendental formulation compelled philosophy towards two opposing and inflexibly inconsistent paths: on the one hand, "idealism," that is, a philosophy maintaining the primacy and the supremacy of a self-evident, producer and creator of itself, the Spirit (a philosophy that culminates in Giovanni Gentile's *autoctisi*); on the other hand, a physical materialism that rebuilds all the real and the knowable solely on the basis of the physical-mathematical laws of extension and motion. Physicalism removes any cognitive power from inwardness and the soul, which is degraded to pure and simple epiphenomena of the becoming of physical and material nature, thus in the end falling into an unsustainable and ruinous determinism (Jonas 1976,

143–61). Lately, a more extreme physicalism has led to the more thoroughgoing reductionism of some tendencies of the philosophy of so-called "Artificial Intelligence," tracing back consciousness, intelligence, and mind to a programmed system of computational operations. Following these reductionists, the mind would be just a computer that performs a program mysteriously preinstalled in the brain (Bosio 2006, Ch. 2–4).

According to Jonas, materialistic physicalism unfortunately has more than one argument against idealism. The kind of knowledge on which it is based seems to be provided with the inconfutability and the indisputability peculiar to physics and exact science. Furthermore, materialism is supported by daily ordinary experience with its attestation of our original sensory passivity and of our natural condition of being a body among many innumerable bodies which surround us, press on the body, act on it, and in turn are pressed, pushed, and hit. But materialism totally disregards the not exclusively physical moments of inwardness, of spontaneity, of need, of choice, of the innate feeling of a perception of ourselves as organisms that are the centre of our own world, that are sources of effective action (Jonas 2001, 64–134). After all, daily and ordinary natural experience gives proof of that as well (Bosio 2008a, 53-69, 2008b, 19–46; Scheler 1976).

Is there "inwardness" in matter? In works published after his death, Scheler asserts just that. He affirms that contemporary physics demolishes the myth of the substantiality of matter, and steers us to see in it instead the expansion of force and energy; but the ideas of energy and force are not understandable unless we appeal to our immediate feeling of the forces of resistance and pressure which we exert and suffer long before catching them reflectively and with full awareness. According to Scheler, "real" is the same as "to resist," "resistance;" and the moment of "there-being," as *Da-sein*, is not inferable at all from that of the "ideality" of representative and intentional knowledge (*So-sein*, i.e., "so-being") (Scheler 1960, 156–80). On this point, we think that the best phenomenological analyses, such as were never before displayed so precisely, lie in Scheler's 1927 *Idealismus-Realismus* and in his prior *Erkenntnis und Arbeit* (Scheler 1960); see my own study on Scheler's philosophy of science and nature (Bosio 2000). We will just remark that a lot here was already anticipated by Schopenhauer in his masterpiece *The World as Will and Representation*.

Only by recognizing and delving into the more hidden and primordial instances of life, can we unfold a path that, although not entirely new, is susceptible to further and wider openings toward the rediscovery of the proto-phenomenon of life. It is regrettable that neither Scheler nor Jonas could consider Husserl's analysis of corporeity, feeling, desire, or will. Nonetheless, it must be said that they were misled by very harsh and sometimes unilateral criticism that at that period was directed against phenomenology, which stood accused of a tendency to idealism in the direction of a disincarnated subjectivity. However, Husserl did underpin these interpretations at least partly and he did not clarify possible misunderstandings, which could lead to such interpretations and readings.

The ontic sequence, proposed and imposed by physicalist scientism, according to which the first moment of reality would be that of material nature, on which life first, then consciousness, would be grafted, has to be totally overturned. This series would be confirmed by the philosophies that arose on the basis of Darwinian evolutionism. The true sequence, not simply ontic anymore, but authentically ontological, is a different one, where life holds the higher rank and is co-eternal with the presumed and apparent inertia of mere extended reality and then, in a second moment, it concentrates and condenses itself in centres of action that exchange energy with matter itself. These centres look for their independence from matter itself, through motion, by any means of free choice, through instinct and appetite.

Biologism cannot know nor can it have the last definitive word on life. By "biologism" we understand the assumption that the natural-organic aspects that life assumes in living species and their functions are prime. Biologism makes the life-world dependent and conditioned by the organism and the organic. As a matter of fact, consciousness and spirit *cannot* arise as a result of complications of the inorganic and the organic here on earth. Spirit and consciousness are something like the immanent inwardness of life itself. The phenomenality of empirical and physical appearing must not darken the ontological co-presence of the dimensions of energy-matter, life, and spirit.

## General Conclusions

The real structure of experience is not fully expressed at all by physical causality, which is strictly linear; in its profundity it is, instead, relational. The relations between entities are the structure wherein life lies. And relations are enhanced and complicated along with the increasing differentiation of living organic forms. But this understanding is achieved not only by means of philosophy; it emerges also in the tendencies and movements of scientific thought, so far a minority view, but not, for that, uninteresting. These movements have been active for a few decades, and here we will mention their most relevant exponents, such as, for instance, Frithjof Capra, author of the famous *The Tao of Physics* and of two other perhaps even more important books, *The Web of Life* (Capra 1996) and *The Hidden Connections: a Science for Sustainable Living* (Capra 2003). Nor could we forget the earlier contribution of the great physicist Erwin Schrödinger with *What Is Life?* (Schrödinger 1944) Very close to phenomenology are the fundamental and necessary contributions of the South Americans Humberto Maturana and Francisco Varela; the latter launched a very original "neuro-phenomenology," which is directly inspired by Husserl and gives primacy to observation "in the first person," which is not insignificant in the life sciences and especially in neurophysiological science, drawing it away from objectifying experimentation done "in the third person," which neglects and completely forgets any empathic interaction between living subjects. In an absolutely innovative and authentically revolutionary way, Varela opposes his formulation to the official neurosciences, victims of and dominated by the materialistic and computational reductionism of the so-called "philosophy of Artificial Intelligence" (A.I.). But before Maturana and Varela, in the 1960s, chemist and biologist Ludwig von Bertalanffy (at the time barely known), devised the "open systems theory," which was full of very relevant implications for psychology and the social sciences as well (Bertalanffy 1976). Bertalanffy stresses the complexity of the living organism as a system and its irreducible emergence as opposed to unilinear and reductionist mechanistic logics. We can mention even other figures, such as the British chemist Jim Lovelock, author of popular and successful books on "Gaia," the Earth, which, according to him, cannot be known and understood unless as a "living organism."

"Relational" and "reticular logic" is an original achievement of Capra, for which we must give him credit. All that is relational is intercommunicating. Yet life is quite relational indeed. The most important thorough analysis of the relational nature of life, one which makes an indefinitely open, and therefore easily destroyable and incapable of self-repairing and self-reproducing, system into a "closed system," yet not an "isolated system," comes from biologists Maturana and Varela, authors of the popular works *Autopoiesis and Cognition* and *The Tree of Knowledge*.[2] The observation of life envisages the indisputable primacy of experimentation "in person," which entails an interaction between the experimenter and the observed subject, and which thus involves him directly. This is an observation of the living being "from within," which is carried out by another living subjectivity and so requires "empathy" and "co-participation." Only what lives can understand and comprehend other living beings, since it has the faculty of re-living the same experiences of life that it observes, describes, and tries to explain. In contrast, a pure epistemic subject is essentially incapable of doing that, since it is aimed at the inanimate and the inert. The achievements of these latest fringe groups of the new sciences take on great relevance for brain neurophysiology as well and get to the bottom of problems that are badly framed and even more badly solved by the grievous ideology of "artificial intelligence" (Maturana and Varela 1980, 48–56; 1987, 33–54).

In "Cybernetics and Purpose: A Critique," rigorously analysing the nefarious consequences of the theoretical results of "computationism" for the ethic of life and for medicine, Hans Jonas makes an incomparably valuable contribution. Self-adjusting circuits and retroactions (so-called feedback) do not make any sense outside the aware positioning of ends and purposes for which they were invented and elaborated (Jonas 2001, 108–34); see also (Jonas 1976).

The very complex relational nature of the disciplines that study life starts to break and to crush the limits of inorganic and mechanical paradigms and lets emerge the primacy of holistic paradigms according to which the "Whole" is more than a mere summary of parts and acts in them as a form-making power. This new direction of science commands philosophy to engage itself in further analyses that allow us to grasp the unity generating all knowledge with more direct insights. The task pointed out to a possible already-rising philosophy is crucial; it consists in rediscovering the ontological categories that can be the main object of an authentic "first philosophy," of an authentic "fundamental ontology," one able to disclose new horizons, precluded to modern and contemporary thought until now. And among these categories primacy belongs to "relation," "energy," and "force." Based on these, life unfolds itself as "autopoiesis" and "ontopoiesis," as A.-T. Tymieniecka efficaciously stressed in her many works. As "autopoiesis," life is *creatio continua* and thus it is ontologically primary compared to the world of dead, inert, and inanimate matter.

Max Scheler was the most important thinker of *creatio continua*. According to him, in the life of the universe everything performs an oriented impulse in a kind of *teleoclinia* that is at the very base of any aware and explicit "teleology" toward the creation of living beings. According to Scheler, this impulse, named *Drang*, is "the second attribute of 'Divinity'," of the *Deitas* that pervades the entire universe itself. Many years later, Jonas left us this beautiful remark: if even in cold and dark interstellar and intergalactic spaces the first "bricks" of life, i.e., the amino acids, were developed in the first molecules of the primordial elements, then wouldn't all that make one think of the primacy of the tendency of life in all the universe? (Jonas 1994, 1996, 165–97).

Therefore, from the very ancient thought of Vedic India to German Romanticism, especially as found in Schelling and Schopenhauer, and later on in Eduard von Hartmann, and then to Max Scheler, this one and only overview, which is organized and modulated in different configurations and perspectives, but always supported on the same unity, recurs over and over again. Both in science and in philosophy, the old mechanistic and geometric paradigm have absolutely to be abandoned. The all-inclusive moment of life (the *Alleben* of Scheler) specifies itself afterwards in organisms and living species, which in their innumerable formations are just a few of its phenomenal manifestations. "Force," "energy," and relational nature are the keys that open the gates of the secrets of the universe. What is to be firmly surpassed is the relational nature seen as mere spatial contiguity and as the continuity of the temporal consequence of events. Living means to have the power of relating to the distant and to everything that is not immediately co-present or in space-time proximity. But if in its primordial potentialities life already has these amazing faculties, how far could the "spirit" go? And does the spirit emerge merely from life or is it something more than life and is it not entirely and completely reducible to life? These problems cross the limits of the present exposition, which has had to attain to the limits that we had in mind.

Finally, we conclude that a theory grounded on these foundations is much more than an exercise seeking to enhance human knowledge, which is very important but not the true and ultimate end in itself. We must remember just that: the triumph of a unilateral science, above all aimed at the moment of the technical realization of its theories and of its discoveries, contains all the germs of its ruin tomorrow. Specialized and above all just physical-mechanical technique-science could be very efficient in shaping the historical and social existence of contemporary human beings. Nevertheless, unluckily for us, it could be even more efficient in preparing the definitive destruction of life, earth, and mankind. It sowed the lethal and poisonous seeds of the pollution of soil, water, and skies; it spread the germs of dreaded genetic engineering and of industrial and social technocracy. Therefore, now more than ever we need a real philosophy and a conscious science. See the works of James Lovelock (Lovelock 1979, 1988, 2006).

## Notes

1. It is odd that Hans Jonas does not refer to Scheler, with whom he has many things in common.
2. Maturana and Varela rectify the "open systems" theory of Bertalanffy here.

## Works Cited


Bertalanffy, Ludwig von. 1976. *General System Theory: Foundations, Development, Applications*. New York: George Braziller, rev. ed. Print.

Bosio, Franco. 2000. *Filosofia e scienza della natura nel pensiero di Max Scheler*. Padova: Il Poligrafo. Print.

———. 2006. *Natura, mente e persona. La sfida dell'intelligenza artificiale*. Padova: Il Poligrafo. Print.

———. 2008a. Filosofia dell'intelligenza artificiale e fenomenologia della vita e della mente. In *Intenzionalità ed empatia. Fenomenologia, psicologia, neuroscienze*, 19–46. Roma: Edizioni OCD. Print.

———. 2008b. Natura e vita nel pensiero di Max Scheler. *Pensare il bios* 5: 53–69. *B@belonprint, Rivista di Filosofia*. Web.

Capra, Fritjof. 1996. *The Web of Life: A New Synthesis of Mind and Matter*. London: Harper Collins. Print.

———. 2003. *The Hidden Connections: A Science for Sustainable Living*. London: Flamingo. Print.

Heidegger, Martin. 2002. The Age of the World Picture. In Martin Heidegger, *Off the Beaten Track*. Trans. Kenneth Haynes and ed. Julian Young, 57–85. Cambridge: Cambridge University Press. Print.

Husserl, Edmund. 1970. *The Crisis of European Sciences and Transcendental Phenomenology*. Trans. David Carr. Evanston: Northwestern University Press. Print.

Jonas, Hans. 1976. On the Power or Impotence of Subjectivity. In *Philosophical Dimensions of Neuro-Medical Sciences*, ed. S.F. Spicker and H.T. Engelhardt, 143–160. Dordrecht/Boston: D. Reidel Publishing Co. Print.

———. 1994. Philosophy at the End of the Century: Retrospect and Prospect. *Social Research* 61: 812–832. Print.

———. 1996. Matter, Mind and Creation. In *Mortality and Morality: A Search for Good After Auschwitz*, ed. Lawrence Vogel, 165–197. Evanston: Northwestern University Press. Print.

———. 2001. Cybernetics and Purpose: A Critique. In *The Phenomenon of Life. Toward a Philosophical Biology*, ed. Hans Jonas, 3rd ed., 108–127. Evanston: Northwestern University Press. Print.

Lovelock, James. 1979. *Gaia: A new look at life on earth*. Oxford: Oxford University Press Print.

———. 1988. *The Ages of Gaia*. New York: W.W. Norton & Co. Print.

———. 2006. *The Revenge of Gaia: Why the Earth Is Fighting Back and How We Can Still Save Humanity*. Santa Barbara: Allen Lane Print.

Maturana, Humberto, and Francisco Varela. 1980. *Autopoiesis and Cognition: The Realization of the Living*. Boston: D. Reidel. Print.

———. 1987. *The Tree of Knowledge: The Biological Roots of Human Understanding*. Boston: Shambhala. Print.

Scheler, Max. 1960. In *Erkenntnis und Arbeit. Wert und Grenzen des pragmatischen Motivs in der Erkenntnis der Welt*, Gesammelte Werke, ed. Maria Scheler, vol. 8. Bern: Francke. Print.
———. 1976. In *Idealismus-Realismus. Späte Schriften*, Gesammelte Werke, ed. Manfred S. Frings, vol. 9. Bern: Francke. Print.
Schrödinger, Erwin. 1944. *What Is Life? The Physical Aspect of the Living Cell – Mind and Matter*. Cambridge: Cambridge University Press. Print.

# An Insight into the Foundations of Eco-Phenomenology

**Massimo Marassi**

**Abstract** The transformation of phenomenology that Anna-Teresa Tymieniecka carried out is centred on two major themes. The first is the theme of phenomenological vision, which allows philosophy to go beyond Descartes' dualism; the second is the concept of transcendental subjectivity: reality may appear only to a constituent ego, which can perceive it. These two themes are her important legacy, which Tymieniecka received from Husserl but which she took one step further. This author thinks that the contemporary world is determined by the power of technology, which has enslaved people. This is the ideological structure of our times. If philosophy wants to achieve the manifestation of the logos of life, it has to address the great topics of life, the world, nature, and the cosmos. This is the mission of phenomenology today.

**Keywords** Tymieniecka · Husserl · Vision · Subjectivity · Life · Logos · Technology

The human condition is the starting point of any philosophical research. The aim is not to find immediate solutions to abstract problems but to seek the meaning of life in a vast, profound, and original way. This is why philosophers need to consider the first principles of metaphysics as the fundamental and distinctive first principles of humankind, since metaphysics determines itself in history as a "metaphysics of life." This personalist view is a specific feature of Anna-Teresa Tymieniecka's phenomenology. On one hand, her research is carried out through a deep analysis of phenomenology. On the other hand, the author takes a step back from it, especially from the theoretical mindset of Husserl and Scheler. The aim of this analysis is to briefly look into this itinerary and to see how Husserl's complex and articulate view of phenomenology is received in Tymieniecka's "first principles of metaphysics" (Tymieniecka "Tractatus Brevis" 1986, 3).

M. Marassi (✉)
Direttore del Dipartimento de Filosofia, Università Cattolica del Sacro Cuore, Milan, Italy
e-mail: massimo.marassi@unicatt.it

 
W. S. Smith et al. (eds.), *Eco-Phenomenology: Life, Human Life, Post-Human Life in the Harmony of the Cosmos*, Analecta Husserliana CXXI,
https://doi.org/10.1007/978-3-319-77516-6_7

## The Phenomenological Vision and the Principle of Experience

In order to highlight the differences, this article looks into a few specific themes, first of all into the theme of phenomenological "vision." Since its beginning, philosophy has commonly been defined as research into first principles, causes, and truth (Aristotle *Metaphysics* A and α). This is how Aristotle begins his work on "first philosophy," defined as the "science we are looking for." This research, which should lead to wisdom, is triggered by the pleasure that men take in their sensations, first of all in vision. The etymology of the words, which bind vision (*horao*, *oida*) to knowledge (*theorein*, *eidenai*), is further proof that vision is a primary source of knowledge. This is the main feature of the Greek culture, which is centred around "vision": this is what philosophy is all about. Unlike as with poetry and myth, through "vision," philosophy starts a journey along a path where the main aim is to overcome the beliefs of the age, questioning what everyone takes for granted and considers to be true. The metaphysical vision of life may therefore be summarized in the sentence "'Vision' instead of 'Method'" (Tymieniecka "Tractatus Brevis" 1986, 10).

Gadamer also speaks of an opposition between truth and method. In order to see one must, first of all, give up any kind of method for grasping the truth. Then, one has to overcome the prejudices attached to each circumstance and to every collective or personal history. These operations are not immediate, and prejudice can never be totally overcome, as the illuminists would think. This theory is outlined in *Truth and Method*, wherein Gadamer holds that no judgment is ever free from preconceptions and bias (Gadamer 2004, 278). Therefore, phenomenology is a continuous exercise of thought, aimed at overcoming prejudice. The philosopher knows that this aim may not be achieved all at once, since one must learn to see. The motto of phenomenology, "to the things themselves," is a process that implies going beyond the traditions of a certain culture, and the practices through which it leaves its traces in everyday life. Therefore, phenomenology is a process of education, the continuous exercise of vision, which is aware of its preconceptions and biases. The latter lies in the traditions and practices of everyday life and culture. For this reason, even though phenomenology aims at an immediate vision, it always achieves a "mediate" vision, as vision is never neutral and can never be taken for granted.

The power of mediation must be carefully considered. Certitude is never a starting point: it is, rather, a result. From Augustine to Descartes and phenomenology, the philosopher must overcome doubt in order to achieve certitude. Even Hegel, in a different context, holds that mediation is "means to make a beginning and then to have proceeded to a second item, such that this second item is the way it is only insofar as one has arrived at it by starting with something that is an other over against it" (Hegel 2010, 40). In other words, for Hegel, one must "mediate the mediation."

In a similar way, the fact that phenomenology does not consider our approach to reality to be immediate means that we may not consider as true anything which appears to us. Therefore, it is not surprising that the itinerary of phenomenology is very close to that in one of the myths recalled by Plato, specifically, the myth of the cave in the seventh book of the *Republic*. In this respect, phenomenology has the task of setting an absolute beginning, starting a line of research based on the "principle of all principles": "*that every originary presentive intuition is a legitimizing source of cognition*, that *everything originarily* (so to speak, in its 'personal actuality') *offered* to us *in 'intuition' is to be accepted simply as what it is presented as being,* but also *only within the limits in which it is presented there*" (Husserl *Ideas* 1983, 44).

In the same way, phenomenology is also the science studying the ways in which things give themselves. It deals with *how* this happens: givenness is not an abstract or isolated vision; it is an experience that belongs to each subject carrying out research (*Erlebnis*). This implies a radical renewal of the concept of experience: it is a place where even the past becomes a present, provided that it can be observed through a vision in someone's memory. Hence, phenomenology is about understanding that the truth is a phenomenon; it must be regarded as a phenomenon. In this way, even ancient prejudices may reveal some truth.

## Subjectivity and the Experience of the Givenness of Things

A second element characterizing modern metaphysics, and one that decidedly contrasts with Tymieniecka's view, is the place given subjectivity. Heidegger once wrote that philosophy, the science Aristotle was looking for, underwent a dramatic change at the beginning of the modern age. This might sound a bit blunt, but it is a very efficient way to describe a turn in the history of thought. In the modern age, the object of philosophy is no longer a "thing" referring to something else, but something original and evident. The notion of evidence is located by Husserl in the "subjectivity of consciousness," which he had defined as "a sphere of absolute positing" (Heidegger "The End of Philosophy" 1993, 439; Husserl *Ideas* 1983, 102).

The principle on which consciousness bases its indisputable leading role is that of evidence. Evidence is the foundation of knowledge because it justifies itself, without involving anything or anyone else. It sets a limit to the chain of justifications, just as Aristotle said in *Metaphysics* α 2. Evidence, therefore, is not justified through anything else. It lives in a precise space: the consciousness of the subject and the constitutive structures of a subjectivity, which would otherwise be abandoned to scepticism. This is how Descartes, in his *Discourse on Method*, sets the rule of evidence as the starting point of legitimate knowledge and of the quest for truth undertaken by the individual subject. There is no evidence and no legitimate knowledge if these are not recognized by subjectivity. Husserl's thought starts from this point when he holds that "all substantiation and refutation … is carried out in the realm of subjectivity" (Husserl *Introduction to Logic* 2008, 165).

There is another way to support the theory of the priority of the subject: each reality, as such, implies a reference to the subject. To speak of a red chair, of a white horse, or even of a green dragon, implies that only a subject may talk about it. This claim may sound obvious, but throughout the history of thought there have been many attempts to set the subject aside, in order to reach the "neutral givenness of things." This was the result a naive realism. Then too, a sort of dogmatic realism led thinkers to hold that the object exists independently from the subject. In any case, it is a fact that there are no things and there is no world, unless they are given in relation to the specific acts of a determinate consciousness.

So, holding that things may be given independently from the ways in which the subject perceives them, is not an acceptable objection to the starting point of knowledge. Rather, one should question the reasons why we consider something as given, existent, relevant for our experience. It is always a subject who must speak about reality, and he must show by what right he can talk about it. Things exist, no matter what consciousness one has of them: *nobody is questioning their true and real being*. One must realize that the things and their existence are constituted *for us* only within an experience, which takes shape in a time span and which sediments in the history of culture and life. Something exists in itself, but the starting point is always an act of a consciousness, inside an experience. Transcendental subjectivity is a living consciousness, which is multi-shaped and always streaming. It is within just such life that one experiences the givenness of things.

## Transcendental Subjectivity and the Intersubjective Experience of the World

This does not mean that the objects exist only because they are perceived by a consciousness (the stance of idealism), but we must ask ourselves through which operations a consciousness may claim that an object exists as such. That is to say, we must enquire how an object may appear and show itself, how it becomes a "phenomenon." However, it is not enough to say that something is a phenomenon because it is conceived by a consciousness. The latter must prove the legitimacy of its acts and "transcendental analysis" is the research that explicates how a consciousness accesses the reality of an object in a legitimate and justified way.

Phenomenology may no longer take for granted, as a dogma, the existence of things outside of us, nor it may conceive things as a simple representation, because certainly they existed before us (the universe "was" before "us"). Phenomenology has the task of showing how our consciousness may fix on a world which exists, but which is independent from the consciousness itself.

This reference to consciousness must be specified, since it has no psychological meaning. Reality is not aimed at just one, isolated subject, in a sort of transcendental solipsism. The things and the world are given within an intersubjective relationship. Human reason constitutes itself only through intersubjectivity. In this way, it is possible to put in place a system of "normality" that references itself "to the true being, to that universe of the truth and of the being which is the object of philosophy" (Husserl *Zur Phänomenologie* 1973, 35). This is how intersubjective syntheses are given. They precede linguistic communication; they are gathered on a linguistic level and they enter a complex and stratified historical succession.

Therefore, the concern here is not about demonstrating that the existence of the world is independent from the subjects. It is about showing how, through which acts of the consciousness and of the body, men are able to achieve a normal and intersubjective experience of the world which can be considered the same for all (Husserl *Ideas* 1983, 68–71), as opposed to achieving it like psychic subjects, determined by a solipsistic experience. Descartes believes that we are rational subjects because we are single individuals (*res cogitans*). Husserl believes we have to be in an intersubjective relationship, wherein everyone must adjust their judgments to the structure of real being. Moreover, unlike George Berkeley, who holds that the being of objects lies in the fact that they are perceived, Husserl does not believe that things are appearances. On the contrary, they exist for themselves; their being transcends the sensations through which they are given to consciousness. As a consequence, things are grasped by men as existing in themselves; they transcend the sensations through which they are related to the subject; the object and the sensation are two different things. This schema is not about an opposition between subject and object, inside and outside, or the ego and the world, but, rather, conceives the consciousness as the space where things reveal themselves.

One must not confuse what is evident with what seems evident. In order to avoid that, according to Husserl, it is important to differentiate between transcendental consciousness and the soul, between the theory of knowledge and psychology. To do this, one needs several criteria. The object of psychology is the psyche, as the place where events and factual data take place. Transcendental phenomenology, through reason, questions such factual data: the *Erlebnisse* of the consciousness are questioned as to whether they be true or false, legitimate or not (Husserl *Introduction to Logic* 2008 165). Any statement must contain the reasons why acts of consciousness may be held to be true.

## Transcendental Phenomenology and First Philosophy

Notoriously, on many occasions—and especially in his *Erste Philosophie*—Husserl tried to argue that only transcendental phenomenology may be considered as "first philosophy." This term recalls Aristotle's differentiation of first philosophy (research of the first principles and of the first causes, of the truth, of being, of substance, and of the unmoved mover) and a second philosophy (research of a particular aspect of being, for example, physics; see *Metaphysics* Z 11). This is an epistemological distinction, dealing with the forms of knowledge. However, Husserl tends to highlight not so much the content of knowledge but, rather, the conditions of possibility of science: "There was a need for a science of the original sources, for a *first philosophy*, for a science of transcendental subjectivity" (Husserl *Erste Philosophie* 1956, 4).

Philosophy is the science of the true beginnings, of the origins, and phenomenology corresponds to this ideal of being a rigorous science, one which is about the phenomenon. The phenomenon is not conceived in a negative way, as an appearance or a reference to a superior entity, but as the appearance and manifestation of something. Phenomenology deals with the *Gegebenheit* of things, with their original givenness in flesh and bone (*Selbstgebung*). First of all, philosophy aims at being a rigorous, truly philosophical science founded on the things themselves. When philosophy opens a way to access phenomena, phenomenology is engaged, as well, in its beginning. By beginning, we do not mean a chronological beginning or the starting point of a deduction process. On the contrary, we mean the starting point of philosophical research, the foundation of all its future developments. For Husserl, the starting point of this rigorous science lies in the things themselves, in the phenomena, as the original manifestation of things and of the world. Even though the latter is not acknowledged as a pre-existing reality, it is respected as an experience (*Erlebnis*) of consciousness or of the world of life (*Lebenswelt*), or even *Erlebniswelt,* a 'world of the consciousness,' lived experience.

## Tymieniecka: Technology As the Power of Man to Transform the World

Even if she considers these arguments as the starting point of her philosophy, Tymieniecka takes them one step further. This sets her apart from Husserl and makes her research original and specific. In fact, she highlights that the way in which we look at the world is compromised by an ideological structure which has praised technology and enslaved humanity.

Technology has triggered a change in our way of looking at reality, and there is no part of the natural, human, or symbolic universe which has escaped this great transformation. Unlike as in other historical periods, from Aristotle's Greece up to the time of Descartes, reality is no longer respected as something sacred or immutable: everything is violated by the power of man. Since the modern age began, this power has become more and more sophisticated, and the great challenge of the present generation is addressing the reality that man's power is often destructive; it seems that the more we progress, the less our power can be confined within boundaries. Even the great *topoi* of life are no longer a place for meditation: they have, rather, become a battlefield and there seems to be very little serious pondering when it comes to our moral responsibilities towards the resources—both material and cultural—that we should preserve for future generations.

The challenges that philosophers need to face nowadays are significantly different from those of past generations. In fact, the world is no longer considered to be something meaningful: it is, rather, regarded as the result of our actions or intentions, and not as something which is given to us, and which precedes us with a meaning of its own. Hence, Aristotle's perspective is completely turned upside down: the world is not an object of contemplation but a product of technology. Therefore, it is useless to look at it in order to even find a meaning or a first principle, let alone the truth, as these too can be considered a product of men. In this perspective, it becomes even more challenging for the philosopher to find a way to deal with the destructive power of man and to come up with arguments for the preservation of the cultural and material legacy that we should pass on to the next generation. For these reasons, the task of the philosophical subject in the present should be that of applying research in order to find out which acts, natural or technological, lead to the accomplishment of something good, or true, for human existence and for the entire world of life.

## The Ontopoietic Process of Life

Each philosophy is an attempt to approach reality in a unique way through a specific content or method so as to understand and grasp the deeper essence of reality, being, life, nature, and the cosmos itself (Tymieniecka *Logos and Life* 1988, 3). The method through which philosophy may reach this aim is phenomenology. In fact, even if it seems that the whole world depends on the creative—or destructive—power of man, phenomenologists constantly remind us that, even if the real object appears to our consciousness, that object always exceeds it: the world is transcendent. Reality has always existed even before a subject was able to watch it or perceive it and, from a phenomenological point of view, no argument can be raised against this statement. The task of phenomenology within the contemporary context is to show why the subject is able to look at the world and the reasons why the world exists before these fundamental operations of the subject. This must be done despite the several attempts in the history of philosophy to bypass the subject in order to reach the "neutral givenness of things" (in a sort of naive realism) or to state that the objects have an existence of their own, independent from the subject, through a dogmatic realism. As a matter of fact, there is no reality and there are no objects if they are not given in relation to the specific acts of a determinate consciousness. Phenomenology is a lively reflection, which goes far beyond the pure analysis of the conditions for the possibility of knowledge or of the constitutive genesis of objectivity. In fact, such research can be sterile and meaningless. It is not simple research into the "things in themselves." Phenomenology's motto "to the things themselves" is a process that encourages the philosopher to go beyond the specific conventions upon which a culture is built, and beyond the determined practices which that culture leaves in everyday life. For sure, a phenomenologist needs to take into account the social, cultural, and historical circumstances of her time, but she needs to hold them as such: inevitable premises that need interpretation. Such premises are the setting in which people relate themselves to each other, to reality, to nature, to the universe, and to the meaning of life in general and its historical renovation. These are the relationships that must constitute the object of phenomenology, not simply the transcendental subject or the truth of reality alone. In fact, since the world exists apart from the perception of the subject, no one is questioning its real and true being. The aim of phenomenology is to show that beings are given to us only within an experience, which takes shape within a history, a culture, a life. Everything that is given is perceived through an experience. "The existence of human beings is conditioned by the situation of nature-bios (as its foundation in the cosmos), and life on earth depends upon the measuring and creative wisdom of human beings" (Tymieniecka "The Golden Measure" 1996, 12). Therefore, we must no longer consider the world as a stranger, or as a universal object that is independent from what single individuals think or want. In fact, the human power of modifying the environment and of transforming the structure of reality has dissolved the illusion of considering nature to be an entity that is able to bear any sort of injury or prevarication operated by men (Tymieniecka *Phenomenology and the Human Positioning in the Cosmos* 2013). Phenomenology has taught us "the self-individualizing principle of life, the entelechial design of life's unfolding, and that creative virtuality of life that brings about the Human Condition" (Tymieniecka "The Golden Measure" 1996, 13). The cosmos of the future will no longer be an immutable set of rules within a given order: "the self-individualization of life is an ontopoietic process" (Tymieniecka "The Golden Measure" 1996, 15). Therefore, phenomenology is not about demonstrating that the world exists independently from subjects; it is about showing how men, as transcendental subjects, may achieve a normal and intersubjective representation of the world that is meaningful for all subjects.

## Conclusion

In this sense, we need to hope for "an aim, a purpose, a telos" (Tymieniecka "Metaphysics of the Manifestation of Logos" 1993b, 15). Through hope, it will be possible to carry out a responsible transfiguration of the present, even in its most intransigent, unjustifiable, and absurd aspects. Such hope is the silent virtue of "clarity of thought" and an ethic of sustainable civilization for today, but, most importantly, for tomorrow. Sustainability is the task and the project for the future: "We have seen how by intensively unfolding the virtualities of the Human Condition the living being transforms the primary avenues of life by bringing in new factors of sense. Within these new factors he expands his circumambient conditions into a socio-cultural world, his very own universe within which he seeks his unique self-realization" (Tymieniecka "The Moral Sense. A Discourse on the Phenomenological Foundation of the Social World and of Ethics" 1983, 7; Tymieniecka "Tractatus Brevis" 1988, 68). Within this project, we need a meaning to transform the political and social dimensions of life: we must become aware of the historic function of our idea of the world. Our world carries a heritage, which contains "the manifestation of the Logos of Life in its accomplishments" (Tymieniecka "Metaphysics of Manifestation and Reason" 1993a, 8). This can lead us towards new hopes, and to "the phase of fulfilment …, the zenith of the manifestation of the Logos in the self-individualizing progress of life" (Tymieniecka "Metaphysics of the Manifestation of Logos" 1993b, 16).

## Works Cited


Aristotle. *Metaphysics* A and α.

Gadamer, Hans-Georg. 2004. *Truth and Method.* Trans. J. Weinsheimer and D.G. Marshall. London, New York: Continuum. Print. 2nd rev. Trans. of *Hermeneutik I: Wahrheit und Methode. Grundzüge einer philosophischen Hermeneutik* (1960). Tübingen: C.B. Mohr, 1986.

Hegel, Georg Friedrich Wilhelm. 2010. *Encyclopaedia of the Philosophical Sciences in Basic Outline. Part 1: Logic.* Trans. and ed. K. Brinkmann and D.O. Dahlstrom. Cambridge: Cambridge University Press. Print. Trans. of *Enzyklopädie der Philosophischen Wissenschaften im Grundrisse* (1817). Heidelberg: Verwaltung des Oswaldschen Verlags, C.F. Winther, 1830 (Dritte Ausgabe). Print.

Heidegger, Martin. 1993. "The End of Philosophy and the Task of Thinking." *In Basic Writings: from Being and Time (1927) to The Task* of Thinking *(1964)*, Trans. and revised ed. by D. Farrell Krell. San Francisco: Harper. Print. Trans. of "Das Ende der Philosophie und die Aufgabe des Denkens," in *Zur Sache des Denkens, Gesamtausgabe*, Bd.14, I. Abt.: *Veröffentlichte Schriften* 1910–1976, hrsg. v. F.W. von Herrmann. Frankfurt a.M.: V. Klostermann, 2007.

Husserl, Edmund. 1956–1959. *Erste Philosophie. Erster Teil. Kritische Ideengeschichte* and *Erste Philosophie. Zweiter Teil. Theorie der phänomenologischen Reduktion.* Husserliana Bd. VII–VIII, hrsg. v. R. Boehm. Den Haag: Nijhoff. Print.

———. 1973. *Zur Phänomenologie der Intersubjektivität. Dritter Teil (1929–1935).* Ed. I. Kern. Husserliana Bd. XV. Den Haag: Nijhoff. Print.

———. 1983. *Ideas Pertaining to a Pure Phenomenology and to a Phenomenological Philosophy.* First book: *General Introduction to a Pure Phenomenology.* Trans. F. Kersten. The Hague: M. Nijhoff. Print. Trans. of *Ideen zu einer Phänomenologie und phänomenologischen Philosophie*, Bd. I. Ed. K. Schuhmann. Husserliana Bd. III/1. Den Haag: M. Nijhoff, 1976.

———. 2008. *Introduction to Logic and Theory of Knowledge: Lectures 1906/07.* Trans. C. Ortiz Hill. Dordrecht: Springer. Print. Trans. of *Einleitung in die Logik und Erkenntnistheorie. Vorlesungen 1906–07.* Ed. U. Melle. Husserliana Bd. XXIV. Den Haag: Nijhoff, 1984.

Tymieniecka, Anna-Teresa. 1983. The Moral Sense. A Discourse on the Phenomenological Foundation of the Social World and of Ethics. In *Foundations of Morality, Human Rights, and the Human Sciences: Phenomenology in a Foundational Dialogue and the Human Sciences*, Analecta Husserliana XV, ed. Anna-Teresa Tymieniecka, 3–78. Dordrecht: Kluwer.

———. 1986. Tractatus Brevis. The First Principles of the Metaphysics of Life Charting the Human Condition. In *The Phenomenology of Man and of the Human Condition II, The Meeting Point between Occidental and Oriental Philosophies*, Analecta Husserliana XXI, ed. Anna-Teresa Tymieniecka, 3–73. Dordrecht: Reidel.

———. 1988. *Logos and Life: Creative Experience and the Critique of Reason*, Analecta Husserliana XXIV. Dordrecht: Kluwer.

———. 1993a. Metaphysics of Manifestation and Reason. Logos in the Individualization of Life, Sociability, and Culture. In *Reason, Life, Culture, Part I, Phenomenology in the Baltics*, Analecta Husserliana XXXIX, ed. Anna-Teresa Tymieniecka, 3–10. Dordrecht: Kluwer.

———. 1993b. Metaphysics of the Manifestation of Logos. In *Manifestations of Reason: Life, Historicity, Culture. Reason, Life, Culture Part II, Phenomenology in the Adriatic Countries*, Analecta Husserliana XL, ed. Anna-Teresa Tymieniecka, 3–24. Dordrecht: Kluwer.

———. 1996. The Golden Measure. Self-Individualization of Life Bringing to Fruition the Ideal for a New Epoch. In *Life the Human Quest for an Ideal*, Analecta Husserliana XLIX, ed. Marlies Kronegger and Anna-Teresa Tymieniecka, 3–28. Dordrecht: Kluwer.

———, ed. 2013. *Phenomenology and the Human Positioning in the Cosmos*, Analecta Husserliana CXIII and CXIV. Dordrecht: Springer.

# Some Questions About Idealism and Realism in the Structure of Husserlian Phenomenology

**Dario Sacchi**

**Abstract** The emphasis laid by Husserl on an abyss of sense between consciousness and reality, between an immanent being and a transcendent being, flows into the assertion of a necessary dependence of the world on consciousness and, consequently, of a constitution of reality within consciousness. But, if he passes in such a way from the undeniable difference in ontological status between world and subject to the assertion of the absolute existence of the subject out of the world, this happens because he presupposes that the ontological status of worldly beings is univocal so that the only way of differentiating from worldly beings would be to go out of the world; after all, there is a secret complicity between objective realism and transcendental idealism. But how is it possible to think of a subject that from the same point of view can make the world appear and be part of it? His mode of being must be understood in the form of negativity and becoming: only thus are we able to conceive a consciousness at once included in the world and including it.

**Keywords** Husserl · Consciousness · Stream of consciousness · Realism · Transcendentalism · Idealism · Perception · Becoming

## The Two Features of Consciousness in Confrontation

I would like to begin with a quotation, from § 53 of the first volume of *Ideen zu einer reinen Phaenomenologie und phaenomenologischen Philosophie*, in which the translation is mine and so are the italics: "*On the one hand consciousness must be the absolute* within which any transcendent reality, and so the psycho-physical world as a whole, forms itself; *on the other hand* consciousness must be *a real event, inner to this very world*. How is it possible to reconcile the two things?"

D. Sacchi (✉)
Università Cattolica del Sacro Cuore, Milano, Italy
e-mail: dario.sacchi@unicatt.it



W. S. Smith et al. (eds.), *Eco-Phenomenology: Life, Human Life, Post-Human Life in the Harmony of the Cosmos*, Analecta Husserliana CXXI,
https://doi.org/10.1007/978-3-319-77516-6_8

Here we are faced with two main features of consciousness, both undeniable but very difficult to reconcile. From one aspect, consciousness presents itself as an all-encompassing *horizon*; from a second aspect, consciousness is included in the world, pertains to the world. Husserl does not really succeed in reconciling these apprehensions and, as a matter of fact, ends by choosing the first. Thus, he gives its position a character which seems to be idealistic.

It is not necessary to say that A.-T. Tymieniecka, in any case, would have chosen the second feature, which seems more consonant with a realistic position and above all with her eco-phenomenology. Therefore, this passage appears particularly suitable for highlighting the difference between the two thinkers.

But, in my opinion, we must ask ourselves in the first place whether it is possible reconcile the two sides of consciousness in a satisfactory way, without being obliged to choose between one and the other, and secondly whether idealism and realism really live where they seem to live.

## Being and Appearing

Let us consider first the nature of appearing. The reality which appears shows itself in a succession of appearances that we can regard as subjective because they are not identical with what appears in them. The whole difficulty lies in thinking of what appears as something that *passes* into its appearances, since this involves a kind of identity that is not a mere equality and, therefore, does not abolish any distinction. The whole difference between phenomenology and the other philosophies that grasp being from its appearing (empiricism, phenomenalism, immaterialism, idealism) is that the latter are over-hasty in stating a mere identity between being and its manifestation. Phenomenology, on the contrary, is faithful to the sense of appearing, and this forces it to acknowledge being both as something which consists in appearing – so in some ways it is not different from its appearances – and as something which exceeds its appearances, since it appears *in* them – so that in other ways it is different from them.

And yet, appearances at once are and are not the transcendent. In short, saying that they are appearances is the same as saying that they are already the thing itself (if what appears is the thing *leibhaft*, not an image or a representation of it), and yet they are different from it since they are nothing but its appearances. This difference is still a difference of identical realities. Nothing but the very essence of intentionality is revealed to us in this peculiar structure of appearing.

So, more than a property of consciousness, the concept of intentionality expresses the essential being of appearance as a manifestation of something which transcends it. This is why Merleau-Ponty spoke of intentionality as "intrinsically pertaining to being."

What appears is nothing but its appearances but these, in their turn, are nothing but that which they make appear. Moreover, we have seen that what appears and its appearances are not the same precisely because they are not diverse, so that their difference is preserved so far as it is not taken as otherness. Each of these two terms suppresses itself in favor of the other and, for this very reason, if we turn our attention to one term, we always find the other. It is necessary to conclude that any distinction between appearance and what appears is already an abstraction made on the movement of appearing. Indeed, there is neither appearance nor what appears but only a continuous passage from one to the other, so both are already abstract images of this passage and make its course quite stiff.

From all this, it follows that the division itself between subject and object is doomed to effacement. We said that the transcendent reality reveals itself in appearances; it cannot be placed beside them or outside them, and in this manner only, can its difference as appearing reality be preserved. In other words, it is at the same time more and less than an object. It is more than an object because it never appears as such; it is beyond the object and any possible objectification. It is *above* the object. But, at the same time and for the same reasons, it is less than an object because its transcendence is founded upon a non-distinction or an in-difference in front of appearances. What appears can only transcend appearances in a real and definitive manner (viz., not as an object) if it is on the side of appearances and, in this sense, subjective; as to transcendent reality, being beyond any object and not being different from a course of subjective appearances is the same thing. In the end, a pure transcendence differs from appearance only if it is nothing but appearance.

## Husserl's Concept of "Shading" as an Attempt to Arch the Gap

The outcome of this analysis may be synthesized in the Husserlian concept of shading (*Abschattung*) as it is worked out in the eidetics of perception, the classical exposition of which we find in *Ideen I*. What is perceived reveals itself in a sequence of *Abschattungen* that show it in person (*leibhaft*), even though they postpone its manifestation indefinitely.

Therefore, the thing shows itself in every *Abschattung* as absent (as such) from that which shows it and, consequently, as referring indefinitely to other *Abschattungen* where its presence will grow rich without ceasing to involve a dimension of absence or withdrawal. As to the thing, appearing *as* itself, namely, in person and not as an image, is not the same as appearing *in* itself, namely, exactly as it is or exhaustively. Now, we must ask ourselves about that which founds the possibility of developing the sequence of *Abschattungen*, since it cannot be founded upon the object, which is nothing but their point of convergence, or rather what they bring to light. We must ask ourselves about that which allows consciousness to go on, specifically, to transcend one *Abschattung* in favor of the other, failing a principle of unity and determinateness which could be reached only at the end of this process of exploration.

Husserl will certainly appeal to a presence of other *Abschattungen* in the form of a potentiality of consciousness, but such a solution does not explain how this potentiality is able to come off, how consciousness is able to venture on an exploration that no object can warrant because, on the contrary, the exploration itself gives rise to the object.

At my first contact with exteriority, when my experience is not yet an experience of something determinate, I am given a soil which is as wide as possible, a deepness which is open to an infinite exploration; we resume this situation by saying that *there is* something. Therefore, I do not say that there is a thing because I can develop the succession of its aspects around one point. On the contrary, I know I can develop the succession of *Abschattungen* because I see there is a thing or, rather, because a thing shows itself at first sight. So the revelation of the object in and through *Abschattungen* supposes, in every appearance, a previous revelation of experience as continuable. Such a continuableness is what turns appearances into *Abschattungen* of something: it is the veritable ground for the phenomenological meaning of transcendence.

This revelation of experience as continuable is the revelation of an indefinitely open field, which is nothing but the *world* itself. From this point of view the phenomenology of perception guides us towards a phenomenology of the world: something which perhaps is no longer a phenomenology. The world is the true name of that which appears; it is what appears originally in every appearance. Each appearance may be an appearance of something because it is an appearance of the world, in opposition to the naive conception according to which the world shows itself as a result of the appearance of a totality of objects of which it is nothing but the sum. In short, appearance may give rise to objects so far as it reveals the pure transcendence of the world. If the unity of the object is correlative to the convergence or coherence of a stream of *Abschattungen*, the object is such, namely, a reality which is distinct from the subject and stands against it (*Gegenstand*), so far as the orderly succession of *Abschattungen* rises in the background of an open totality which is yet undetermined but is already given, since the beginning, as the stage on which all the objectifications can develop. From all this, it follows in the first place that every appearance is concomitantly an appearance of the world, specifically there is a sort of duality or duplicity of appearance according to which it always contains more than it gives and could not give what it gives if did not hold this reserve, which is nothing but the infinite bottom of the world.

The thing is such because it belongs to the world; the world, as we will see later, is such only as a world of things. This is the reason why if you look for one you find necessarily the other.

The world is what founds the presence in person of anything, which is the principle itself of *Leibhaftigkeit,* in short, what must be concomitantly present in order that anything may be present in person. Therefore, it is necessary to speak of an archi-originality of the world, and a reality is only present in person if it is presence of the world, that is, if the world is present in it. Now, for essential reasons such an archi-originality cannot be objectified. Saying that the world is concomitantly present in any perception is the same as acknowledging it as the ground of any state of affairs, so that it could only be an object of perception if it were included in another world that would be, then, the only true world.

Yet, we have to consider another side of the question: the ontological status of subject. From its origins this is the place wherein phenomenology tests its basic ambition and, so, ventures upon the question of its own possibility. Certainly, Husserl incurs the risk of thinking of an "abyss of sense" between subject and world, namely, an absoluteness of the subject necessarily implying his extra-worldliness. An undeniable tension manifests itself between (a) the exigency of upholding the world-subject correlation and (b) the phenomenology of *Ideen I*, which culminates in the hypothesis of an annihilation of the world. Undoubtedly, such an annihilation, according to Husserl, would modify consciousness but would not damage it in in its being. This hypothesis, notoriously, flows into the assertion of a sharp ontological inequality between consciousness and world. The emphasis laid on an abyss of sense between consciousness and reality, between an immanent being (which shows itself without *Abschattungen*) and a transcendent being (which shows itself through *Abschattungen*), an abyss of sense which is correlative to the absoluteness of consciousness, flows into the assertion of a necessary dependence of the world on consciousness and, consequently, of a constitution of reality within consciousness. In short, such an emphasis allows the passage from reduction (understood as leading back to the sphere of consciousness) to constitution. In consequence of this last stage, phenomenology becomes effectively transcendental phenomenology to the extent that reality owes its being to consciousness, the absolute being of which is the veritable place and ground of what exists.

## The Exhaustion of the Effort

This crucial moment of Husserlian phenomenology, in its classic and paradigmatic version, lends itself at least to two remarks. We observe, in the first place, that if Husserl passes in such a way from the undeniable difference in ontological status between world and subject to the assertion of the absolute existence of the subject out of the world, this happens because he presupposes that the ontological status of worldly beings is univocal and presupposes that belonging to the world means the same thing in any case, so that the only way of differentiating from worldly beings would be to go out of the world. A *difference* (in the manner of being) from worldly beings must necessarily mean, according to Husserl, an *otherness* vis-à-vis the world itself. And yet, the univocal ontological status of worldly being is implied in the characterization of consciousness as absolute: forming part of the world means living in a spatial and temporal system and being able to undergo and to exercise causality. This is, evidently, a very limited characterization of worldliness, one dominated by the pattern of physical objects. Then, if the subject as such does not enter into causal relations and probably has neither spatial nor temporal exteriority, it is necessary to infer that he is not part of the world. Nevertheless, the question is whether being part of the world means being situated in it and undergoing its action in the manner of a physical object. Has worldliness the univocal meaning of spatial inclusion in a place and inscription in causal relations? Is it not possible a different way of being "in" the world or at least of being not extraneous to it? We can discover here a secret complicity between objective realism and transcendental idealism, between the physical determination of the world and the conception of the subject as an absolute element. In fact, a world as physics describes it may only exist for a "disinterested" pure subject, for that pure spectator of the world which is the absolute subject, and, vice versa, a subject is only able to constitute the world if the world has been previously reduced to physical objects and their laws, specifically, if finally it has been made flat. From all this it follows that the admission of an abyss of sense between word and subject, on one hand, and the assertion of the extra-worldliness of the subject, on the other hand, are no longer necessarily connected.

We must remain resolutely Husserlian and admit that if the subject is truly himself, that is, the subject of correlation, then the condition of appearing of the world is or exists effectively in a sense that has nothing to do with worldly beings. Therefore, we must recover on our own account the assertion of the abyss of sense claimed by Husserl. Such an ontological inequality is what warrants the peculiarity of the subject. But, if it is not an object or a thing, a being, this does not mean that it is extraneous to the world unless being-in-the-world is understood after the fashion of inclusion in relations of the physical type. Nay, so long as the subject differentiates radically from worldly beings and has nothing in common with them, it may be included in the world, and in a deeper sense than are things themselves. So, there are several ways of being-in-the-world, since the world is precisely what is able to lodge in itself measureless differences.

Facing up to the question, Husserl raises the problem itself of the ontological status of consciousness as it is implied by correlation. But he does not discuss it in an authentic way, that is to say preserving the tension which is inherent in its terms and, so, he precludes himself from solving it. In effect, according to him, the same consciousness cannot be involved in both cases; it is wrong to mistake psychological consciousness for transcendental consciousness or, rather, the duality of transcendental and empirical, of constituent and constituted, prevails over their unity as dimensions of consciousness. From this it follows that the only way of sewing up what has been torn too hastily, of reconstituting a unity that has been broken in the name of the idealistic imperatives of constitution, is resort to a constitution of psycho-physical consciousness performed by absolute consciousness, that is, to a self-constitution of consciousness that can only take the form of a degradation. Transcendental consciousness, which is able to constitute any being, must be able to constitute itself as a reality included in the texture of the world, namely, as empirical consciousness.

The unity Husserl is obliged to avow is expressed by him in the form of a self-constitution, in virtue of which the empirical dimension may appear as deriving from the transcendental one. But, all this remains quite incomprehensible. In effect, if consciousness were really the absolute claimed by Husserl, it would never be able to rejoin its own worldly side, to constitute itself as psycho-physical consciousness, in short, to exist as a real consciousness. Devoid of any worldly feature, it could not constitute the world itself and would be, so to speak, be confined in its absoluteness.

On the contrary, if consciousness, as Husserl claims, is able to constitute its own worldly side, then it can do this as long as it is not alien to the world but ever already engaged in it, so that empirical consciousness needs no more to be constituted. As soon as it becomes possible, the constitution shows itself to be useless. It says both too much, because if consciousness is really absolute then it cannot go out of itself, and too little, because consciousness can only constitute itself as worldly if it *is* already. The hypothesis of constitution, in short, is absurd (impossible) or superfluous (useless). Therefore, it is much better to say that consciousness is ever already included in the world, that it is not necessary to "constitute" this inclusion because it has ever already taken place – in short, that consciousness is one and indivisible, except for a provisional abstraction, so that the duality of empirical and transcendental, of absolute and contingent, is just what demands to be overcome. The real problem consists, rather, in understanding how is it possible that a consciousness which is originally one is at the same time two; it contains, in fact, two dimensions that seem mutually extraneous like the "constitutive" dimension, in virtue of which the world is included in consciousness, and a more "naturalistic" dimension, in virtue of which consciousness is included in the world. Here, it is necessary to take a route which is the opposite of that which was taken by Husserl. We must not ask how an absolute consciousness which is outside the world may constitute its own worldly side but, on the contrary, what is the ontological status of a subject which can be neither an Absolute outside the world nor a mere worldly being.

The whole difficulty consists in understanding how the subject *from the same point of view* can make the world appear and be part of it. Here, it is necessary to turn off Husserlian phenomenology, which posits idealistic metaphysics and the "constitutive" perspective, deriving from it before the unity and indivisibility of consciousness in such a way as to distinguish two points of view on consciousness and ultimately, so to speak, two consciousnesses. How shall we conceive the ontological status of the subject so that its disposition to make the world appear does not exclude, nay can imply, its inclusion in the world itself?

The risk that phenomenology runs is simply that which impends over the natural attitude as long as such an attitude implies the conviction that consciousness is included in the world exactly in the same way as other beings. The thesis of the existence of the world implies the belief that the subjects who experience the world are included in it exactly in the same way as other beings are, so that their relation to the world is a causal one; the relation between consciousness and real world is real itself, is included in the world. Such a naive conviction is what opens the road to behaviorism, scientific psycho-physiology, the cognitive sciences, and programs of neuro-philosophy – in short, all the attempts to naturalize or reify consciousness.

## Coming to Some Understanding

Thus, we are able to solve the problem raised at the beginning: how to reconcile the transcendentality and the empiricity of the subject, taken as two dimensions of a unique existence? There is only one possible answer: we must give up attaching any positivity to the subject and conceive his mode of being in the form of *negativity*. Differently from other beings that as "things" are what they are, the subject we are trying to characterize is not what he is, not in the sense that he would be also something else or that he would imply indetermination, but in the sense that he exists only denying himself. The subject performs unceasingly such a negation and is nothing but this very operation; in short, he is never identical with himself, he never quiets down in himself. So, the mode of existing which denotes negativity in its difference from nothing must be defined as *becoming*.

Only as becoming can the subject at the same time be included in the world and differentiate from it, in so far as he is precisely what makes the world appear. As becoming, the subject is radically different from other beings in the world even though he is deeply rooted in it. Existing as becoming means differentiating from the world in the heart of the world, transcending it within it, manifesting an inclusion that has nothing in common with spatial relations. Only apprehending becoming allows us to conceive a consciousness at once included in the world and including it.

Besides, if becoming is understood as life and activity, we can solve another question: how is it possible to conceive the Divine Consciousness as immutable if we have now seen that in the world becoming is a sign of perfection? Here is the only possible answer: God as well is becoming *qua* living and active, but His activity is so intense and devoid of obstacles that it unfolds instantaneously, without time. His immutability is not absence of becoming, but is the highest expression of it. Time is not connected to becoming as such but to finite becoming.

# The Origin Paradox: How Could Life Emerge from Nonlife?

**Ion Soteropoulos**

Courtesy of Rubén Val

**Abstract** The origin of life is one of the great unsolved paradoxes in human understanding. In fact, if, according to the natural philosophers since ancient times, nothing comes into being from not-being, how could life emerge from non-life?

On the basis of the intuitive solution to the matter of change provided by the Ionian Greek philosopher Anaxagoras (fifth century BCE), we will attempt to resolve the problem of the origin of life by devising a zoogony in which change between opposites does not involve a contradiction or a paradox.

Because it has been universally recognized that there is no change between contradictory opposites such as nonlife and life, we transform their absolute opposition into a relative opposition in which life is generated from nonlife insofar as it is already present in nonlife, though in an imperceptible or implicit manner. This leads us to conclude, with Louis Pasteur, that *omne vivum e vivo*—that all life is derived from antecedent life (1857). If all life is generated from life and, thus, if all life continues to exist beyond its original position in space-time—that is, if life is more

---

I. Soteropoulos (✉)
Apeiron Centre, Paris, France
e-mail: contact@apeironcentre.org



W. S. Smith et al. (eds.), *Eco-Phenomenology: Life, Human Life, Post-Human Life in the Harmony of the Cosmos*, Analecta Husserliana CXXI,
https://doi.org/10.1007/978-3-319-77516-6_9

than itself and after itself—what is the nature and founding principle of life, and what is the impact of the continuity of life on the question of life's local uniqueness or ubiquity within the universe? If all life is generated from life, is it possible to build a lifeless robot having the properties of a living being?

**Keywords** Origin of life · Biogenesis · Abiogenesis · Intuitive holographic principle · Anaxagorean theory of change · Self-organization · Metaphysical singularity · Critique of evolution · Quantum de-coherence · Quantum biology · Infinite light · Infinite senses

## The Origin Paradox from a Historical Perspective and Its Intuitive Solution

> *It is impossible for anything to come to be from what is not, and it cannot be brought about or heard of that what is should be utterly destroyed.* —Empedocles

All scientific theories about the origin of life start with the fundamental assumption that life appeared on earth from an assembly of lifeless molecules formed by non-biological processes (Darwin, Oparin, Haldane, Miller, et al.). Life is, therefore, considered to be an emergent property of inanimate matter that is accidentally produced by evolution. Evolutionary biologists term the generation of life from non-living ingredients *abiogenesis*.

According to the scientific account, when a collection of lifeless molecules had reached a critical threshold of instability and complexity, it suddenly and unpredictably transformed itself into a living cell, which, because of its self- moving and self-organizing power, is something entirely different from a collection of lifeless molecules. But, how could *life* appear from a collection of *lifeless* molecules?

Indeed, since ancient times [see Ionian Greek and Hindu philosophy of the sixth and fifth centuries BCE], all natural philosophers agreed that nothing comes into being from not-being and that there can be no motion or change between contradictory opposites having an absolute opposition. In fact, any change between contradictory opposites—between life and nonlife, for example—is an impossibility. Change exists uniquely between *spatial contraries* whose opposition is relative. For example, according to the Anaxagorean theory of change, relative opposites arise from each other insofar as they are already present in each other and thus, are *contraries* existing simultaneously since eternity—that is, since infinite time—and not ephemeral *contradictories* existing successively since a finite time.[1]

Applying the Anaxagorean solution to the problem of the origin of life, we assert the following: in order to generate life from something infinitely different, such as nonlife, we must assume that in the deepest recesses of matter, imperceptible to our finite particular senses, life is already present in lifeless matter.

Let $A$ designate life, $<$ designate inequality and succession (origination, implication, inclusion), $A'$ designate nonlife, and $=$ designate equality and simultaneity. $A$

arises from $A'$ if, and only if, $A$ is already present in $A'$ and vice versa; hence, $A'$ and $A$ are equal and simultaneous, that is to say, they are coexisting opposites or contraries[2]:

$$A' < A \text{ if, and only if } A' = A. \qquad (1)$$

Based on this formula, we conclude that the collection of lifeless molecules creates life, such as the living cell, if, and only if, we assume that life is an immanent property of lifeless matter, that matter in its fundamental structure is a complex whole, that is, both nonliving and living, which we represent one-dimensionally by the continuous line $A'A$. This is not a contradiction or a paradox; rather, it is a complexity, because we ontologically assume that a piece of matter in its ultimate reality is not a *simple individual* admitting at one time in virtue of analytic principles of organization a unique determination—for example, either $A'$or $A$—but, instead, a *complex whole* or *universe* admitting at one time in virtue of synthetic principles of organization contrary determinations: for example, both $A'$ and $A$.[3]

Only a synthetic ontology of matter can liberate us from the paradox of something originating from nothing, life from nonlife, $A$ from $A'$, which is a variant of Zeno's motion paradox. As a matter of fact, an analytic ontology of matter destroys all motion between $A'$ and $A$. If a piece of matter is a simple individual, then—by virtue of the analytic principle of the excluded third—matter has at one instant either the determination $A'$ or the determination $A$ and, hence, is either nonliving or living. Because $A'$ and $A$ are contradictory opposites having an absolute opposition, any successive motion from $A'$ to $A$ is logically impossible. We call a paradox, a miracle, or an absolute contingency any successive variation of $A'$ into $A$ despite its logical impossibility. In contrast, if a piece of matter is a complex universe or whole divisible into equal and opposite parts $A'$ and $A$ (called contraries, which verify the synthetic principle of equality and coexistence of opposites), any motion between opposites $A'$ and $A$ having a relative opposition is consistent and necessary. Thus, complex matter that contains simultaneously the necessary variety $A'$ and $A$ does not need the passage of time to contingently vary from $A'$ to $A$.

If everything $A'$ and $A$, nonlife and life, for example, is *ab initio* in everything—that is, in all matter—then any mutual origination between opposites is, in reality, a process of extracting whatever already exists in matter considered as a universal receptacle (*τό πανδεχές*) containing everything at all times.[4] This Anaxagorean thesis leads us to conclude that external causal origination and evolution do not produce necessary change and variety. Because complex matter is the synthesis of everything existing and being fixed within matter *qua* universe since eternity, we conclude that external causality and evolution involving the existence of linear time are not fundamental determinations of the material universe existing in itself and independently of our particular observation; in this sense, they are appearances or absolute accidents of our finite particular senses. For example, our finite retinal cells at rest detect uniquely the finite part of light's real speed corresponding to the finite unit speed $c = 1$ (a fundamental assumption of this work). In turn, this arbitrarily selected finite unit speed $c = 1$ creates the perceptual illusion of time delay and temporal order between different things. It appears, therefore, to our finite individual senses that $A$ has causally evolved from $A'$, whereas in reality $A$ coexists with $A'$.

Accordingly, when we decompose the complex unity or equality $A' = A$ into two opposite and coexisting inequalities:

$$\left(A' = A\right) = \left(A' < A\right)\left(A < A'\right), \tag{2}$$

we note that the causal evolution of $A$ from $A'$, of life from non-life, in conformity with the temporal (irreflexive or heteronomous) order $A' < A$ is an accidental part of a timeless complex material whole simultaneously satisfying the totality of temporal orders ($A' < A$) and ($A < A'$). This means that the contingent appearance of linear time $A' < A$ arises from an essentially timeless material universe. Thus, it appears to our particular senses that life is an emergent epiphenomenon of lifeless matter accidentally produced by evolution, which is, itself, an accidental law of matter, whereas, in reality, life is an immanent, essential, and necessary determination (principle, property, or state) of matter existing and complementing lifeless matter since eternity and, therefore, independently of evolution and linear time.

By being an essential determination of matter, life ceases to be a local or particular property of biological matter—for example, of the living cell confined on earth and constrained by terrestrial chemistry and evolution—and becomes a universal property of all matter regardless of whether it is biological or non-biological, terrestrial or non-terrestrial. In other words, it becomes the immanent property of the ultimate constituents of matter.

Following the ancient thinkers of Ionian Greece and India (sixth and fifth centuries BCE), let us next identify life with the Divine mind (nous, intelligence) or soul (anima, psyche, atman), which we define as that which moves by itself spontaneously, that is, independently of an external constraining force and cause; we may then assign to life, thought of as *life-mind* or *life-soul,* the power of self-motion.[5] Because self-motion requires kinetic energy, the Greek Ionian philosophers regarded the life-soul as the source of kinetic energy, or fire.[6] Thus, the soul of matter, which is a *living fire*, by energizing inanimate matter assigns to matter the Divine power of self-motion, whose different manifestations are self-organization, self-origination, self-maintenance (the capacity to remain the same despite variation), self-replication, and self-containment.

On the other hand, insofar as inanimate matter is the cessation of or resistance to motion (what we call inertia), it assigns constancy to the self-moving matter. It follows that the fundamental constituents of matter have both the power of self-motion caused by the energy- force, life-force, or soul-force of matter and the power of resistance to motion caused by the inertial force of matter. This cold inertial force, reconciled with its equal and opposite hot kinetic force, ensures the balance and permanence of self-motion. Ultimately, the fundamental constituents of matter have the Divine power of *permanent self-motion*, which is grounded in two conjoined principles governing matter: the principle of self-motion due to the *life-force* of matter, and the principle of rest or constancy, the result of the *inertial force* of matter.[7]

This synthetic line of reasoning led the atomists Leucippus and Democritus to consider the smallest constituents of matter as spherical atoms that are alive—that is, endowed with souls or soul-forces, which naturally move them independently of external influences and assign to them eternity and divinity. The original self-motion of the freely revolving, vibrating and colliding soul-atoms is permanent and indeterminate, free of sense with no tendency in any soul-atom to move in one sense rather than another.[8]

According to the ancient natural philosophers, the life-soul is not only the source of continuous, natural, or spontaneous self-motion in matter but, also, the source of consciousness: that is, of sensibility detecting the multiplicity and variety of the material world; and also of comprehensive intelligence (logos) uniting and integrating the multiple sensible parts. Thus, consciousness is not an emergent epiphenomenon of matter but instead an immanent and necessary determination (principle, property, or state) of matter complementing mindless matter. This led the atomists to assign mental properties, such as intelligence and sensibility, to the soul-atoms.

We use the term *hylozoism*, *animism*, or *panpsychism* for the original worldview of the ancient natural philosophers of Ionia and India, according to which the apparently inanimate and mindless matter is in its deepest recesses alive and conscious.[9] Because complex matter is uniformly distributed in the universe—everywhere and at all times (in conformity with the cosmological principle of the perfect uniformity and continuity of the universe relative to its large-scale structure)—we conclude, in agreement with the Greek Ionian natural philosopher Thales (sixth century BCE), that the universe is full of living and conscious matter composed of permanently self-moving soul-atoms: that the *universe is besouled, and full of gods*.[10]

Let us replace $A'$ by its equal $A$ in the right side of Eq. 2; we obtain, then:

$$(A' = A) = (A < A)(A < A). \tag{3}$$

If we conjoin the two self-inequalities of the right part of the equation, we obtain a single self-inequality in which the product of two identical antecedents is equal to one antecedent and the product of two identical consequents is equal to one consequent:

$$(A < A)(A < A) = (AA < AA) = (A < A). \tag{4}$$

The self-inequality $A < A$ states that life originated from antecedent life, which is the translation of the famous phrase from Louis Pasteur. The assertion that life originates from life—which biologists call *biogenesis*—is based on the principle of self-origination (reflexive origination, autonomous origination) or self-causality, which mathematicians call the principle of reflexive order (self-order or free order) and which is a synthetic principle belonging to second-order logic: [11]

$$A < A. \tag{5}$$

According to the above principle of self-causality (which characterizes fundamental entities such as Nature, the Universe and the Divine), the origin and cause of life resides in life: Life is both cause and effect of herself, thereby refuting the analytic principle of external causality (or heteronomous causality) grounded in linear time.[12] This means that life does not owe her being and movement to something else; it means, rather, that life is capable of making her own kinesis independently of the contingency of external causes and forces, such as linear time, which Maturana and Varela called *autopoiesis* (a Greek word meaning "self-making"). Because we do not need an external cause or violent force to cause, explain, and sustain life's motion, we affirm that the force that causes life's motion is life itself. This, in turn, inherent life-force was divided into equal and opposite biogravitational forces of attraction and repulsion, of motion and rest that ensure the permanence of life's self-motion.

The principle of reflexive order applied to life holds that life is before and after herself, that life continues beyond herself—beyond her initial or final position $A$—and, thus, is a continuous and infinite whole. It also holds that life is less and more than herself—a contained part and a containing whole—and thus is self-contained and free.

Now, if $A$ designates life and mind (regarded as the source of consciousness), we may then read the principle of reflexive order $A < A$ as follows: life gives rise to consciousness —namely, thought and sensibility—which thinks and senses "life" and therefore constitutes the consciousness of life but, also, thinks and senses the "consciousness of life" and therefore constitutes the consciousness of consciousness—*the νόησις νοήσεως νόησις*—which we call *self-consciousness* and verifies the formula $(A < A)_{CL}\,(A < A)_{CC}$.[13]

We conclude this chapter with a summary of our main theses: although it is logically impossible or paradoxical for life to arise from non-life, we resolved this paradox in the manner of Anaxagoras by postulating *ab initio* the variety life and nonlife, which we regard as contrary determinations of the same piece of matter conceived as a complex whole or universe. This means that at first view it appears to our finite senses that life evolves from nonlife, whereas in reality life is a coexisting determination of lifeless matter. The first founding principle of complex matter is the synthetic equality $(A' = A)$, which we define as the complex product of opposite inequalities or temporal orders: $(A' = A) = (A' < A)(A < A)$. This is the most comprehensive equality, as it integrates the maximally different and distant determinations, namely life and non-life, in one and the same smallest constituent of matter (the atom). Subsequently, by substituting $A'$ for its equal $A$ in the product of opposite inequalities, we transform the paradoxical and contingent origin of life from nonlife into the logically consistent and necessary origin of life from life that verifies the synthetic principle of self-origination or self-causality, which mathematicians call the principle of reflexive order—that is, of spontaneous or free order.

Thus, when life originates from life in agreement with the principle of reflexive order, $A < A$, we say that free life replicates life for its self-maintenance and self-variation. When free life replicates itself in any of its parts, we have the self-containment of the self-replicating life.

If we assume that free life and consciousness are necessary determinations of the ultimate constituents (atoms) of unconscious and inert matter, and if we conjecture that the smallest size of the atoms is about $10^{30}$ times smaller than that of an egg cell, we can legitimately claim that free life and consciousness already exist at the smallest length, which is near the Planck scale.

Now, is it possible to construct a computational model of life that replicates free, conscious life governed by the synthetic principle of reflexive order? The answer is no, because the finite, analytic computational model itself denies, that is, qualifies as false, the synthetic principle of reflexive order on which self-organizing and self-contained life is based:

$$\left(A < A\right)' \tag{6}$$

Thus, if $A$ is a finite number, then by virtue of the above principle, no computing finite number $A$ is smaller and greater than itself—that is, self-ordered. It follows, then, that no computational model can replicate conscious life, which is self-ordered and, therefore, free of computation, unless we assume that in its ultimate reality the computational model is simultaneously beyond finite analytic computation.

The above question reminds us of our first fundamental question: whether it is possible to create a living whole from its lifeless constituent parts. We have demonstrated that this is logically impossible, unless we assume that in reality its lifeless constituent parts have simultaneously the determination of life, which means that in essence life originates from life—a living whole from a living whole—in agreement with the synthetic principle of reflexive order and independently of linear time. Life is, therefore, essentially a fractal object having self-similarity throughout the scales.

## A Spatial Theory of Matter

*Les savants sont des gens qui confondent le vivant avec la vie car il ne serait avoir de vie que selon l'esprit.* —Georg Wilhelm Friedrich Hegel

### *From the Soul-Atom to the Soul-Singularity*

We use the term body to mean the definite spherical volume of space in which the mass of the body fits. Let us consider any point of space $a$ as the center of a sphere of radius $r$. Let us next situate at the center $a$ designating the here and now $t_{\text{now}}$, the magnitude $\approx 10^{-4}$ m, which we consider to be the radius $r$ of a definite volume of space in which the human egg cell fits. The average mass of the human egg cell at rest is $\approx 10^{-8}$ kg and, its energy content at room temperature is $\approx 10^{-2}$ eV.[14] With the center $a$, the human egg cell, and the radius $ab$, which is $10^{-30}$ times the cell's radius $r \approx 10^{-4}$ m, we construct a sphere whose finite radius $ab$ is $10^{-4} \times 10^{-30} \approx 10^{-34}$ m. Because we consider the finite magnitude $\approx 10^{-34}$m as the smallest or zero magnitude, we have the equation: $ab \approx 10^{-34}$ m $= 10^{-\infty} = 0$ m. By "smallest magnitude" we mean the magnitude that cannot be further divided no matter how much we divide it. We use the term *soul-singularity* to mean the body enclosed by the smallest sphere whose finite radius $\approx 10^{-34}$ m is equal to zero-radius—in other words, whose radius is a Platonic point-line (ἄτομος γραμμή), having both a finite positive magnitude and a zero magnitude without absurdity. This singularity is not a black hole, not a collapsed body crushed by its infinite curvature (the dogma of contemporary physics) but, instead, a vibrating complex body.[15] Indeed, the complex singularity has both a finite positive volume and a zero volume implying an infinite curvature, which we identify with the soul or mind (nous, *νοῦς*).[16]

We locate this complex singularity on the limiting boundary $b$ of a sphere of center $a$ and maximum radius $ab$ according to division. If the center $a$ designates *here* at time now $t_{\text{now}}$ (in which we locate the finite magnitude $10^{-4}$ m), the limiting boundary $b$ occurring at a maximum distance from the center $a$ necessarily designates *there* at time zero $t_0$—in which we locate the finite magnitude $10^{-34}$ m, considered to be the smallest or zero magnitude. We use the term *microcosmos* or *universe according to division* to designate the maximum sphere that contains the totality of magnitudes ranging from $10^{-4}$ m to $10^{-34}$ m $= 10^{-\infty} = 0$ m.

## *The Intuitive Interpretation of the Holographic Principle*

Let us apply to the quantity mass the intuitive interpretation of the holographic principle and then assert that the maximum possible mass a specified spherical volume of space can contain is proportional to the boundary area of the volume and not to the volume itself, as common sense would expect. We have, then, the following formula:

$$m \propto A, \tag{7}$$

in which $m$ is the mass, $\propto$ is the symbol of proportionality, and $A$ is the boundary area.[17] Because the boundary area $A$ of the spherical volume is approximately equal to its radius $r$ squared, in the formula 7 we replace $A$ with $r^2$ and now have:

$$m \propto r^2 \quad \text{or} \quad m = kr^2, \tag{8}$$

in which $k = m/r^2$ is a constant of proportionality equal to 1.

If the radius $r$ of a spherical volume of space is $\approx 10^{-4}$ m, which is the radius of the definite volume of space in which the human egg cell fits, then on the basis of formula 8, the maximum possible mass $m$ contained in this definite volume is:

$$m = kr^2 \approx \left(10^{-4}\right)^2 \approx 10^{-8}\,\text{kg}, \tag{9}$$

which is roughly the mass of the human egg cell.

If the radius $r$ of a spherical volume of space is the finite magnitude $\approx 10^{-34}$ m, which we take as the finite measure of the smallest or zero radius, then on the basis of the formula 8, the maximum possible mass $m$ contained in the smallest or zero volume is

$$m = kr^2 \approx \left(10^{-34}\right)^2 \approx 10^{-68}\,\text{kg}. \tag{10}$$

The above mass is roughly the mass of the smallest constituent of matter—the soul-singularity.

Because we take the mass $\approx 10^{-68}$ kg to be the finite measure of the smallest or zero mass, we have: $m \approx 10^{-68}$ kg $= 10^{-\infty} = 0$ kg. This equation tells us at what limiting scale a finite and lifeless mass becomes infinite and alive and possesses an infinite soul or mind of zero mass. In other words, it tells us at what limiting (critical) point the opposite determinations—for example, lifeless mass and living mind—are united by the synthetic equivalence principle, which stipulates the unity and equality of opposites.

We have two complementary ways to determine the energy content of the soul-singularity. The first is through its mass $m$, if we consider it as a particle, and the second is via its wavelength $\lambda$, if we consider it as a wave.

Let us consider the soul-singularity as a mass particle. Grounded in the following famous energy-matter equation:

$$E = mc^2, \tag{11}$$

we affirm that the maximum possible energy $E$ we can extract from the singularity's finite mass $m$ is:

$$E = mc^2 \approx 10^{-68} \times \left(3 \times 10^8\right)^2 \approx 10^{-51}\,\text{J}\,(\text{joules})\,\text{or} \approx 10^{-32}\,\text{eV}\,(\text{electronvolts}), \tag{12}$$

which is the finite measure of the singularity's lowest or zero energy. We have therefore: $E \approx 10^{-32}$ eV $= 10^{-\infty} = 0$ eV. By "lowest energy" we mean the energy that cannot be further decreased no matter how much we decrease it. We assume that this coldest part of the soul-singularity constitutes the freezing point—the ice-point—of the material universe where all motion ceases. The cold property of the soul-singularity is etymologically expressed by the Greek word *psyche (ψυχή)*—the soul—derived from the Greek *ψυχρός*, meaning "cold."[18]

Let us now regard the sphere in which the soul-singularity fits as a wave of wavelength $\lambda$ equal to the sphere's diameter $2r \approx 10^{-34}$ m. Let us assume that the energy content $E$ of the wave of wavelength $\lambda$ is inversely proportional to $\lambda$:

$$E \propto 1/\lambda \quad \text{or} \quad E = k \times 1/\lambda, \tag{13}$$

and let us take as the constant of proportionality $k = E \times \lambda$ the energy content $E = 3.57 \times 10^{-19}$ J (joules) of the green light of wavelength $\lambda = 5.55 \times 10^{-7}$ m to which the human eye is most sensitive. If the wavelength of the soul-singularity is about $\lambda = 2r \approx 10^{-34}$ m, then the maximum possible energy we can extract from the singularity's wavelength is:

$$\begin{aligned} \mathrm{E} &= 3.57\times10^{-19}\times5.55\times10^{-7}\times1/10^{-34} = 1.98\times10^{9}\,\mathrm{J} \\ &\text{or } 1.23\times10^{28} » 10^{28}\,\mathrm{eV}(\text{electronvolts}), \end{aligned} \tag{14}$$

which we take as the finite measure of the soul-singularity's highest or infinite energy. We have, therefore: $E \approx 10^{28}\text{eV} = 10^{\infty} = \infty\text{eV}$. By "highest energy" we mean the energy that cannot be further increased no matter how much we increase it. We assume that this hottest part of the soul-singularity constitutes the boiling point—the fire point—of the material universe eternally animating all inanimate material things.

Now, if we convert the above lowest and highest energies into frequencies on the basis of the equation:

$$f = E / h, \tag{15}$$

in which $h = 6.62 \times 10^{-34}\text{Js} = 4.13 \times 10^{-15}$ eV (Planck's constant of proportionality), we obtain the following lowest and highest frequencies[19]:

$$\begin{aligned} &(\mathrm{i})\, f \approx 10^{-32} / 4.13\times10^{-15} \approx 2.42\times10^{-18}\,\text{cycles / s} \approx 10^{-18}\,\text{cycles / s} \\ &(\mathrm{ii})\, f \approx 10^{28} / 4.13\times10^{-15} \approx 2.42\times10^{42}\,\text{cycles / s} \approx 10^{42}\,\text{cycles / s.} \end{aligned} \tag{16}$$

Because we take the finite frequency $\approx 10^{-18}$ cycles/s as the finite measure of the lowest or zero frequency that characterizes the slowest rotating and vibrating soul-singularity, we have: $f \approx 10^{-18}$ cycles/s $= 10^{-\infty} = 0$ cycles/s. This equation tells us at what limiting scale the finite becomes infinite according to decrease; it shows us at what limiting (critical) point the finite frequency of a rotating and vibrating material body becomes zero and stops all rotational or vibrational motion.

Now, if we take the finite frequency $\approx 10^{42}$ cycles/s as the finite measure of the highest or infinite frequency that characterizes the fastest rotating and vibrating soul-singularity, we have: $f \approx 10^{42}$ cycles/s $= 10^{\infty} = \infty$cycles/s. This equation tells us at what limiting (critical) point the finite becomes infinite according to increase; it shows us at what limiting scale the finite frequency of a rotating and vibrating material body becomes infinite and assigns infinite energy and conscious life to its inert and unconscious matter.

The synthetic unity of opposites—that is, of fire having infinite energy and ice having zero energy within the soul-singularity—enables the latter to continuously breathe the Divine fire of life to lifeless matter without burning matter itself. This is neither an incomprehensible paradox nor an inexplicable miracle, it is the very extra-ordinary nature of the eternally living and temperate complex matter.

If conscious life is a fundamental property of matter, the ultimate building block of conscious life is not the DNA organic macromolecule storing biological information; rather, it is the smallest constituent of matter—the soul-singularity storing non-biological information out of which the different masses of the universe are made. The soul-singularity is not only the ultimate building block of matter and information, but is also the geometric point, the ontological principle (ἀρχή), the universal container that unifies all opposites and contains all masses in the infinite universe. In this sense, the self-containing soul-singularity is both the smallest part within the universe contained in everything and the greatest whole in the universe containing everything.

## In the Microcosmos Masses Arise According to the Holographic Principle

We have shown how the three quantities of the soul-singularity, which are mass $m$, energy $E$, and frequency $f$, depend on the size of the spherical volume of space surrounding it. Let us now show, by way of Fig. 1 and Table 1, how the radius $ab = t_{now}t_0$ of the microcosmos is a logarithmic axis scaled by the order of magnitude of lengths and masses. The lengths are the radii of spherical volumes of space enveloping different masses of the living universe according to division equally known as microcosmos (see Fig. 1). Table 1 shows the proportionality between the masses of the microcosmos and the radii squared of the definite spherical volumes in which the masses respectively fit. The masses range from the egg cell at $a = t_{now}$, to the soul-singularity at $b = t_0$.

The above orders of magnitude of masses in the microcosmos are calculated from the intuitive holographic principle $m = kA$, in which $k = m/A \approx 10^{-8}$ kg/$(10^{-4}$ m$)^2$ is a constant of proportionality equal to 1 and $A \approx r^2$. The holographic principle holds that the boundary area $A \approx r^2$ of a definite spherical volume of space of radius $r$ originates the mass $m$ inside the volume so that the material body has the mass it has. This means that material bodies do not come into being from the accidental intervention of external causes and forces at particular successive moments of the evolving universe (a fundamental assumption of evolutionary biology). Rather, they are timeless, self-originating, living wholes whose matter and form arise from the boundary areas of their respective spherical volumes of space in conformity with the

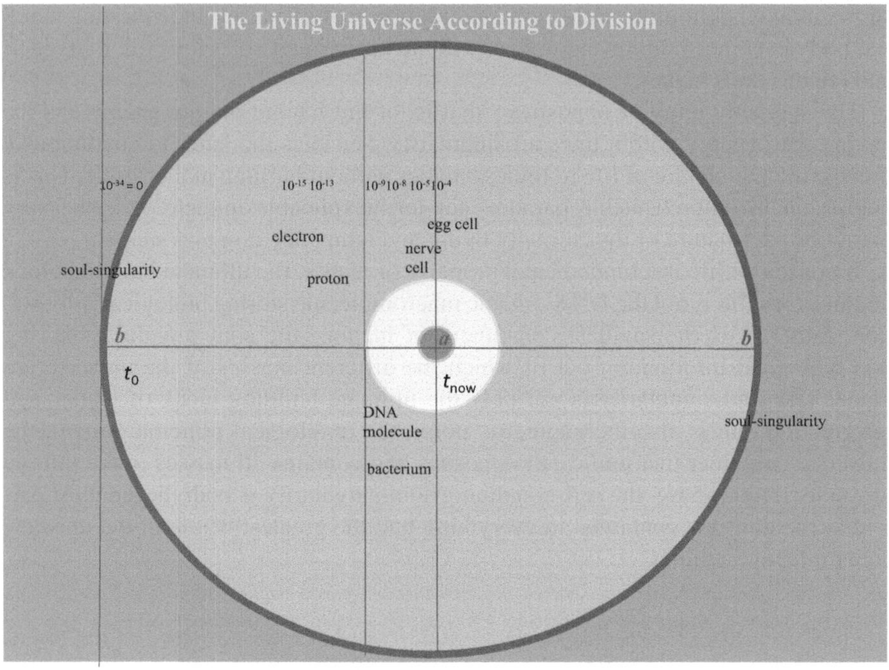


**Fig. 1** With the center $a = 10^{-4}$ m and the radius $ab = 10^{-4} \times 10^{-30}$, we draw a circle that represents the microcosmos, defined as the living universe according to division. We use the term *biological world* to signify the region that ranges from $10^{-4}$ m to $10^{-10}$ m and contains biological cells and molecules; we use the term *quantum world* to signify the region that ranges from $10^{-11}$ m to $10^{-34}$ m and contains quantum particles. Both worlds exist simultaneously in the living universe of center *a* and radius *ab*

**Table 1** Order of magnitude of masses in the living universe according to division

| | | |
|---|---|---|
| $a$: $m = kr^2 \approx (10^{-4})^2 \approx 10^{-8}$ kg | (Human egg cell) | Biological scale |
| $m = kr^2 \approx (10^{-5})^2 \approx 10^{-10}$ kg | (Nerve cell) | |
| $m = kr^2 \approx (10^{-8})^2 \approx 10^{-16}$ kg | (Bacterium) | |
| $m = kr^2 \approx (10^{-9})^2 \approx 10^{-18}$ kg | (DNA organic molecule) | Bio-chemical scale |
| $m = kr^2 \approx (10^{-13})^2 \approx 10^{-26}$ kg | (Hydrogen's proton) | Sub-atomic scale |
| $m = kr^2 \approx (10^{-15})^2 \approx 10^{-30}$ kg | (Electron) | |
| $b$: $m = kr^2 \approx (10^{-34})^2 \approx 10^{-68}$ kg | (Soul-singularity) | Near Planck scale |

holographic principle that governs the permanent and observer-independent (impersonal) material universe. This self-origination of material bodies is also in conformity with the rationalist assumption of Greek Ionian philosophy [Leucippus] taken as the defining element and basis of rationalist science: that everything exists according to necessary, universal principles and that nothing occurs by random chance.

Because with our finite brain (finite sense organs) at rest we observe selectively only a finite part of light's real speed, which is the finite unit speed $c = 1$, we perceive the maximally distant soul-singularity at $b = t_0$ *as if* it were earlier (or later) than the present human egg cell at $a = t_{now}$. As a matter of fact, we observe the maximally distant $b = t_0$ as being the earliest moment, as well as the youngest, and the soul-singularity *as if* it were the ultimate chronological origin—the beginning of a sequence of evolving bodies in the microcosmos of increasing size and mass up to the present human egg cell, at $a = t_{now.}$. We stipulate, thus, the following analytic principle of inequality and temporal order that rules our observable microcosmos at $t_{now.}$:

$$(t_0 < t_{now}) = t_{now} - t_0 > 0. \tag{17}$$

On the basis of this principle, we affirm that $t_0$ is chronologically prior to (before) $t_{now}$ and that $t_{now}$ is chronologically posterior to (after) $t_0$. The positive inequality and temporal order $(t_0 < t_{now}) = t_{now}\text{-}t_0 > 0$ determines the finite age of the present microcosmos at $t_{now}$—which is about 10 billion years ($3 \times 10^{17}$s)—as well as the evolution (cosmic, chemical, and biological) of its material bodies, in which a body of increasing size and mass, such as an electron, a proton, a DNA organic molecule, a bacterium, a nerve cell, or an egg cell, appears to our finite senses to be successively generated from the soul-singularity at time zero $t_0$ (see Fig. 2).[20] This successive passage from the first origin $t_0$ to the present $t_{now}$, from the soul-singularity at $t_0$ to the egg cell at $t_{now}$ through a very long sequence of chance events, is, as we have argued in the first part of this work, an impossibility or a miracle unless we stipulate that $t_{now}$ is already present in $t_0$ and vice versa: in other words, that the opposites $t_0$ and $t_{now}$ are simultaneous and equal with respect to the deepest reality of matter, and that they verify the synthetic principle of equality and zero temporal order:

$$t_0 < t_{now} \text{ if, and only if, } (t_0 = t_{now}) = t_{now} - t_0 = 0. \tag{18}$$

This formula holds that $t_{now}$ evolves from $t_0$ if, and only if, with respect to the ultimate and fundamental reality of matter, $t_0$ and $t_{now}$ are equal and simultaneous and, thus, there is between them neither temporal order nor evolution.

What kind of light immediately connects the extreme opposites $t_0$ and $t_{now}$ and, therefore, suppresses the temporal order and evolution between them? Necessarily, it is infinite light that travels the radius *ab* or the maximum spatial distance between $a = t_{now}$ and $b = t_0$ instantaneously and nonlinearly in zero time and with infinite frequency. By "infinite light" we mean light whose speed $c$ is a complex, infinite unit that we define as the product of infinite frequency and zero wavelength and, conversely, as the product of zero frequency and infinite wavelength: $c = f \times \lambda = \infty \times 0 = 1$ or $c = f \times \lambda = 0 \times \infty = 1$. On the other hand, finite light is the part of real infinite light that is observable by our finite particular senses at rest.

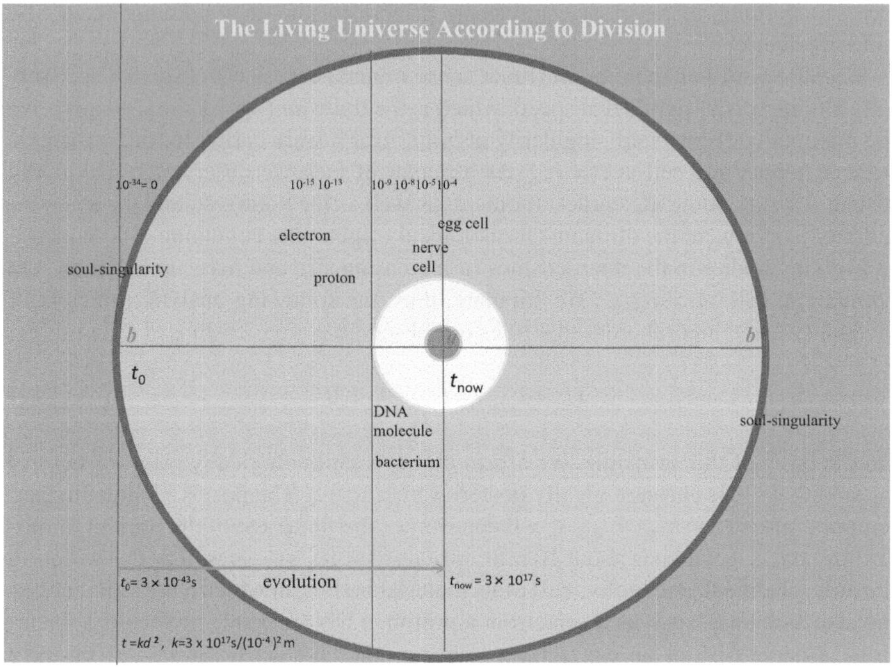


**Fig. 2** The radius *ab* is perceived by our finite particular senses at rest as if it were constrained by linear time such that *a* arises from *b* via a long sequence of chance events that we call evolution. This evolution of *a* from *b*, of the living cell at *a* from the fundamental quantum particle—the soul-singularity at *b*—took place during ≈ 10 billion years ($3 \times 10^{17}$s), and can be considered equally as the successive origination of the biological world from the quantum world

The finite light's speed $c$ is a simple, finite unit that we define as the product of finite frequency and finite wavelength: $c = f \times \lambda = 1$. What we observe, then, is not real infinite light itself but, instead, light exposed to our finite analytic method of perception that exercises an imperceptible constraining influence on its object.[21]

Let us divide the complex equality $t_0 = t_{\text{now}}$ into coexisting opposite inequalities and temporal orders:

$$(t_0 = t_{\text{now}}) = (t_0 < t_{\text{now}})(t_{\text{now}} < t_0). \tag{19}$$

Notice here that the positive inequality and temporal order $t_0 < t_{\text{now}}$ is an accidental part that our finite individual senses at rest arbitrarily select from the permanent and universally present microcosmos defined as the balanced product of opposite temporal orders.

The equality and zero temporal order $(t_0 \doteq t_{\text{now}}) = t_{\text{now}} - t_0 = 0$ defines a microcosmos, which is, in its ultimate and fundamental reality, a timeless and ageless universe free of temporal order by containing the totality of temporal orders. It also defines a universally present universe in which definite bodies (or spherical volumes of space) of different sizes and masses exist simultaneously and, with necessity, in conformity to the intuitive holographic principle applied to masses and do not need linear time, evolution, and the contingent intervention of external causes and forces in order to arise.

In fact, it is sufficient to divide a given body or spherical volume of space into two equal and opposite parts. For example, we could divide a particular body, into a *contained mass m* inside the given spherical volume and a *containing boundary area A* of the given spherical volume assigning a *form* to the contained mass, in order to initiate between them, and on the basis of the intuitive holographic principle, an endless and autonomous cycle of reciprocal coming into being and conversion. These complementary parts ultimately produce a complex body that is in continuous self-origination, free of accidental intervention of external cause and force. Thus, the body—for example the human egg cell—has the mass and form that it has *not* because it is determined by the biological information residing in its organic DNA macromolecule but, rather, because of the non-biological information residing in the boundary area of its respective enveloping spherical volume. In fact, it is this very boundary area that forms and informs its contained mass.

Given that a different mass of the microcosmos, such as an electron, a proton, a DNA organic molecule, a bacterium, a nerve cell, or a human egg cell, is made of the same soul-singularity, we can think of this diversity of masses as replications of the first soul-singularity endowed with the power of self-replication. In fact, the power of self-replication—of self- extension (self-amplification) and self-division—is a manifestation of the power of self-motion that the soul-singularity possesses intrinsically. The power of self-motion, however, which is infinite and timeless at the quantum level, is finite and time-conditioned at the biological level.

Thus, the freely revolving and vibrating fundamental particles of animated matter, such as soul-singularities and electrons taken as soul-singularities, are ageless and the collisions among them are without friction and involve no net change in energy. Because their total kinetic energy remains constant (in conformity with the first law of thermodynamics, the principle of the conservation of free energy), their free motion is non-linear and timeless, leaving unchanged themselves and the order of their environment. In contrast, the freely revolving and vibrating biological molecules and cells metabolize, grow, and age, and their free motion, whether it is manifested as rotation, vibration, or collision, is essentially linear and time-conditioned, and produces a net change in themselves and in the order of their environment in conformity with the second law of thermodynamics (the principle of the degradation of energy).

For example, the biological cell's metabolism is a mode of self-motion that is constrained by the second law of thermodynamics: it absorbs light (free energy) from the external environment, converts it into internal order, and discards heat waste (bounded energy that cannot be completely recycled) into the external environment, which, then, decreases its order. Ultimately, the biological cell is a dissipative structure that increases its internal local order (progressive evolution) at the expense of decreasing its external global order (regressive evolution). This overall increase of disorder makes biological life, according to Prigogine, an island of order in a sea of disorder (1989). Thus for the finite analytic approach to life, the end (telos) of metabolism is not the maintenance of life but, rather, its non-maintenance, that the end of biological life is not life but, rather, non-life—that is, thermodynamic death.

Biological life constrained by linear time converts life into non-life in conformity with the analytic principle of irreflexive order (heteronomous order):

$$A < A', \tag{20}$$

in which $A$ designates life, $<$ designates origination, and $A'$ designates non-life (that is, death). In compliance with this principle, life creates non-life, and non-life is the end (effect or sense) of life. It reminds us of the words of the ancient Greek Ionian philosopher Heraclitus (sixth century BCE), who described life as an arc whose product is death: "Le nom de l'arc est vie. Son oeuvre est mort."

If biological life aiming at death is not real and true life, then what is real and true life? The answer is that real and true life is life whose end is life, in agreement with its founding principle of reflexive order (autonomous order) $A < A$, which stipulates that life originates from life and, that life is the cause and end of free life.

We have argued that the basic activity of biological life—that is, metabolism (or self-maintenance)—is a mode of self-motion constrained by linear time, whose thermodynamic expression is the second law of thermodynamics. Because linear time is an analytic principle of irreflexive order (heteronomous order), it contradicts self-motion governed by the synthetic principle of reflexive order (autonomous order). It follows, then, that insofar as self-motion—such as metabolism (self-maintenance)—is constrained by linear time, it is an improper or incomplete free motion. In reality, it is a heteronomous or unfree motion that produces a net change in the biological cell and its surrounding environment in conformity with the analytic principle of irreflexive order, according to which the end of life is non-life: $A < A'$.

Relative to the deepest recesses of matter, however, the biological cell is a *physical* or *quantum cell* made of eternally self-moving soul-singularities whose free motion (rotation, vibration, and collision with other soul-singularities) produces no net change in the cell or its surrounding environment. There is neither linear time nor evolution with respect to the quantum cell, its free motion is complete, permanent, and non-linear, governed by the synthetic principle of reflexive order $A < A$, according to which the cause and end of life is life itself (see Fig. 3).

Now, why do we perceive the timeless quantum reality of the living cell *as if* it were a time-conditioned biological reality? Why do we perceive the timeless quantum cell made of perpetually self-moving soul-singularities *as if* it were a time-conditioned biological cell made of self-moving organic molecules that metabolize, grow, age, and evolve? In other words, why do we perceive the eternal and costless

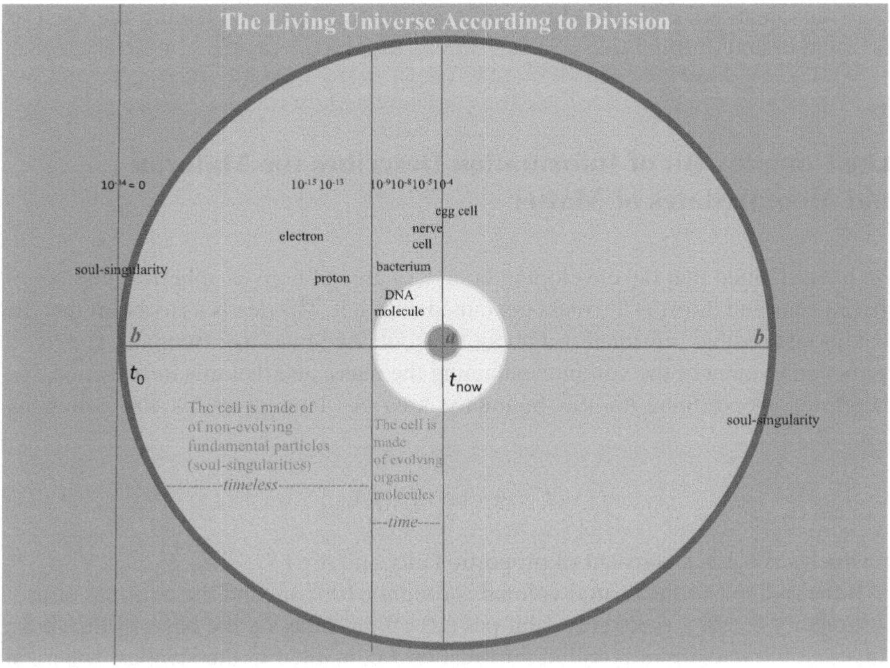


**Fig. 3** The living cell at *a* has simultaneously, similarly to the living universe according to division, two realities: (i) a biological reality conditioned by linear time in which the cell is a collection of evolving DNA organic molecules; and (ii) a timeless quantum reality in which the cell is an assembly of non-evolving soul-singularities

state of *being* of the living cell as if it were the state of *becoming* involving a thermodynamic cost?

This perceptual error resides in our finite brain (finite particular senses), which at the state of rest perceives uniquely the finite part *c* of light's real infinite speed. This finite observable speed of light *c* equal to the finite unit 1 generates a *time delay* or *de-coherence* between two points of space—for example, between the reception of the signal *here* at *a* and the emission of the signal from *there* at *b*—that renders their communication delayed and erroneous as it involves a loss of information in conformity with the second law of thermodynamics.[22] Were we to perceive through an infinite brain (infinite universal senses) the light's real speed, which is the infinite unit $c = 1 = \infty \times 0$, the time delay or de-coherence would disappear from the universe, leaving in its place a *timeless, coherent* cosmos free of the second law of thermodynamics, in which all points of space communicate immediately and free of error. The living cell is, thus, perceived by our finite brain *as if* it were a time-conditioned biological cell absorbing free light and emitting waste and having nonlife as its origin or end of life. In its fundamental reality, however, the living cell is a timeless quantum cell, absorbing and emitting free light and having life itself as its origin or end of life.

## The Complex Bit of Information Describes the Material and Mental States of Matter

We have claimed that the enveloping boundary area of a given spherical volume of space forms and informs the mass contained within it. This leads us to assert that the maximum possible information $I$ about the contained mass is essentially stored on the boundary area of the volume containing the mass, and that this information $I$ is, therefore, proportional to the boundary area $A$. This gives us the following formula:

$$I \propto A \quad \text{or} \quad I = kA, \tag{21}$$

in which $k = I/A$ is a constant of proportionality and $A \approx r^2$.

If the radius $r$ of the spatial volume is roughly $10^{-34}$ m, and the constant of proportionality $k = I/r^2$ is roughly 1 bit per $(10^{-34})^2\text{m}^2$, then, on the basis of the above equation, the maximum possible information $I$ contained in the smallest spherical volume is approximately:

$$I = kr^2 \approx 1/\left(10^{-34}\right)^2 \times \left(10^{-34}\right)^2 \approx 1\text{bit}. \tag{22}$$

Thus, 1 bit is stored on the smallest boundary area of roughly $10^{-68}$ $\text{m}^2$ which, in turn, originates the smallest possible mass, approximately $10^{-68}$kg, inside the smallest spherical volume. We have, then, on the basis $I = kr^2$ and $m = kr^2$, the following equation:

$$\approx 1\text{bit} / 10^{-68}\,\text{m}^2 \approx 1\text{bit} / 10^{-68}\,\text{kg}, \tag{23}$$

which shows that the same quantity of information is stored in both area and masse.

Because we consider ≈1 bit per $10^{-68}$ kg to be a complex unit of information—a *quantum bit*—reflecting the complex nature of the soul-singularity endowed with matter and mind, we divide it into two parts $a$ and $a'$ in which $a$ is 1 bit of material information that registers $2^1 = (01)$ alternative states of matter—the emission of a photon and the absorption of a photon, for example—and $a'$is 1 bit of consciousness that registers $2^1 = (01)$ alternative states of consciousness—thought and sensation, say. The composition of the parts $a$ and $a'$ constitutes the complex whole bit $1 = (a \times a') = (1 \times 1)$, which registers at the same time $2^1 \times 2^1 = 2^2 = (00)(01)(10)(11)$ alternative states of matter and consciousness. This gives us the following equation:

$1 = 2^1 \times 2^1 = 2^2$.[23] These four alternative states of matter and consciousness are compressed into the 1 complex whole bit to constitute the memory power and wealth of the soul-singularity.[24] Thus, the complex 1 bit constitutes the material and mental configuration of the soul-singularity: namely, the soul-singularity's sensing and thinking of its own material states (emission and absorption of a photon) that give quality and meaning (significance) to its material being and constitutes the proto-consciousness of the soul-singularity.

Because every complex state of matter (absorption and emission of photon) has a corresponding complex state of consciousness (thought and sensation), we have concluded the consciousness of matter. This consciousness increases proportionally to the mass of the material body and, hence, to the number of its soul-singularities or bits information.

If we divide the mass of any material body by the mass of the soul-singularity, we obtain roughly the number of soul-singularities and, hence, of bits of information contained in the massive body that describe its material and mental configurations. Thus, the mass of the electron (which is approximately $10^{-30}$ kg) comprises approximately $10^{-30}/10^{-68} = 10^{38}$ soul-singularities storing roughly $10^{38}$ bits of information that register about $2^n \times 2^n$ ($n = 10^{38}$) states of matter and consciousness within that electron.[25] The mass of the organic molecule DNA (which is about $10^{-18}$ kg) is made up of approximately $10^{-18}/10^{-68} = 10^{50}$ soul-singularities storing about $10^{50}$ bits of information that register about $2^n \times 2^n$ ($n = 10^{50}$) states of matter and consciousness within that molecule, whereas the mass of the human egg cell (which is approximately$10^{-8}$kg) comprises approximately $10^{-8}/10^{-68} = 10^{60}$ soul-singularities storing about $10^{60}$ bits of information that register roughly $2^n \times 2^n$ ($n = 10^{60}$) states of matter and consciousness within that cell. Finally, the mass of our brain (which is about $10^0$kg) is made up of approximately $10^0/10^{-68} = 10^{68}$ soul-singularities storing about $10^{68}$ bits of information that register roughly $2^n \times 2^n$ ($n = 10^{68}$) states of matter and consciousness, whereas the mass of our observable universe according to extension (which is about $10^{52}$kg) comprises approximately $10^{52}/10^{-68} = 10^{120}$ soul-singularities storing about $10^{120}$ bits of information that register roughly $2^n \times 2^n$ ($n = 10^{120}$) states of matter and consciousness.

Because the observable universe is the computable finite part of the infinite physical universe (defined as the sum total of an infinite number of parts), we consider its finite mass of about $10^{52}$kg to be the finite measure of the physical universe's greatest or infinite mass. We thus obtain the following equation: $\approx 10^{52}\text{kg} = 10^N = 10^\infty = \infty\text{kg}$. Similarly, $\approx 10^{120}$ soul-singularities is the finite part of the physical universe's infinite totality of soul-singularities such as $\approx 10^{120} = 10^N = 10^\infty = \infty$kg storing $\infty$ bits of information that register $2^N \times 2^N$ ($N = 10^N = 10^\infty = \infty$) states of matter and consciousness. Thus, the living being that has the greatest consciousness is the observable universe, which is the computable, finite part of the incomputable, infinite, physical universe.

Because information about the material and mental states of a given mass is stored in its soul-singularities, we may translate the states of matter and consciousness of any massive body made of soul-singularities into 0s and 1s, in order to obtain a unifying framework for communicating information across the categories of things. For the moment, we ignore how to translate the material and mental states of a massive body—for example, of a biological living cell that fits in a given spatial volume of radius *r*—into 0s and 1s. We ignore how to process these 0s and 1s and store the processed information on the two-dimensional boundary area of a given spherical volume of space. We also ignore how to convert this stored information into a three-dimensional massive body: into a biological cell, for example.

If the mass of the universe (of everything) is made of fire-soul-singularities each having a highest or infinite vibrational energy and frequency, then how is it that the mass of the universe does not burn? How is it that there is a living-fire that perpetually energizes and animates the inanimate mass of the universe without causing it to disintegrate?

The solution derives, at least in principle, from the complex nature of the soul-singularity, which, considered both as a particle and as a wave, has its energy content $E$ defined in two different but coexisting ways that complement and mutually cancel each other. I leave to the reader the opportunity to complete the answer to the above questions.

## The Impact of the Intuitive Solution of the Paradox of the Origin of Life on the Finite Approach to Life

*Nothing occurs at random, but everything for a reason and by necessity.* —Leucippus

If everything in nature happens according to universal and necessary principles, not by capricious chance, then life and consciousness have not accidentally emerged from lifeless and mindless matter. Rather, they are logically necessary replications and amplifications of what already exists in the ultimate constituents of matter—the soul-singularities. Indeed, we have argued that each vibrating soul-singularity stores 1 complex bit of information that registers $2^1 \times 2^1 = 2^2$ possible configurations of matter and consciousness that enable each soul-singularity to think and feel its proper material state and, therefore, to be self-conscious.

It follows, then, that all matter made of perpetually vibrating and self-conscious singularities is a perpetually living and self-conscious being which, in virtue of the principle of self-order and self-origination, permanently creates and processes its own complex information independently of the contingent intervention of external causes and forces. This rational conclusion has its historical roots in the proto-rationalist doctrine of animism, or hylozoism, held by the Ionian Greeks, the Persians, and the Hindus of the sixth century BCE. Its contemporary version is based on a criticism of the finite approach to life according to which life and consciousness are unique local events, by-products, or epiphenomena of matter that emerged accidentally from the Monte Carlo game of terrestrial evolution (a fundamental assumption of time-conditioned evolutionary science, whose main advocate was the French biologist Jacques Monod).

The finite approach to life assumes that (1) linear time and its correlative evolution are necessary and fundamental principles of the universe; (2) the speed of light $c$ is simple, unique, and constant, which we take as a finite unit, namely as $c = 1$; and (3) the piece of matter is a simple (indivisible) individual verifying analytic principles of organization—for example, the principle of contradiction, which stipulates that no piece of matter is both living and non-living and also simultaneously has kinetic energy and is at rest. These three primitive assumptions constitute the finite-analytic paradigm of the universe, according to which the universe is a discrete multiplicity of isolated, simple, and time-conditioned parts—the individuals deprived of unity and motion, verifying analytic principles of organization.

In our research, we have demonstrated that linear time produces the paradox of the origin of life according to which it is illogical and, indeed, impossible to generate successively something from nothing, life from nonlife, consciousness from non-consciousness. It follows that the solution to the origin paradox requires the negation of the cause of all contradictions or paradoxes, which is linear time, and whose logical expression is the analytic principle of irreflexive order (heteronomous or unfree order).

This negation leads us directly to the infinite, rationalist view of life as grounded in the Anaxagorean theory of being. According to this approach, life and consciousness are immanent and necessary properties of lifeless and mindless matter that belong, since eternity, to its ultimate constituents: the soul-singularities. Thus, a piece of matter of any size is essentially a complex universe or whole divisible into coexisting parts—for example, into matter and form (Aristotelian coexisting opposites), matter and enveloping space (Democritean coexisting opposites), finite and infinite (Pythagorean coexisting opposites), inert matter and kinetic energy (Galilean coexisting opposites), or lifeless matter and life—without contradiction or paradox. Because conscious life has no chronological origin in the unlimited and timeless universe, we deduce the continuity and self-origination of conscious life: that conscious life continues to be before and after herself, and that, in a general manner, conscious life $A$ is a principle, cause, and end of its own being in conformity with the synthetic principle of self-causality (self-origination), whose logical expression is the synthetic principle of reflexive (autonomous) order: $A < A$. This principle of self-causality becomes the ontological origin and principle of real infinite life, free of the biological need for a chronological birth and end.

The infinite rationalist approach to the origin of life assumes that linear time is an artificial or illusory principle of the universe produced by the fact that we perceive via our finite particular senses at rest only an arbitrarily selected portion of light's real speed, which corresponds to the finite unit speed $c = 1$. This observable light of finite unit speed $c = 1$ travels the radius $ab = t_{now}\ t_0$ successively and linearly from $b = t_0$ to $a = t_{now}$ according to the positive temporal order $(t_0 < t_{now}) = (t_{now} - t_0 > 0)$.

This creates the illusion that the living cell at $a = t_{now}$ is an accidental effect of evolution. However, at the limiting boundary *b* of the universe at moment $t_0$, which is simultaneously the ultimate and fundamental reality of all things, light's real speed is the infinite unit $c = \infty \times 0 = 1$, or $c = 0 \times \infty = 1$. This real infinite light, which underlies the observable finite light, travels the radius $ab = t_{now}\ t_0$ instantaneously and nonlinearly, that is, from $b = t_0$ to $a = t_{now}$ and vice-versa with zero temporal order $(t_0 = t_{now}) = (t_{now} - t_0 = 0)$. Zero temporal order shows that the living cell at $a = t_{now}$ possesses simultaneously its ultimate origin—its soul-singularity—at $t_0$.

We have used the term *physical* or *quantum cell* to designate the living cell whose soul-singularity is immanent to the cell, thereby constituting the *ontological origin* (*ἀρχή*) and *principle* of the living cell and, not external and transcendent to the cell, constituting the *chronological origin* of the living cell through a long sequence of chance events.

That light has at the limiting boundary *b* of the universe at moment $t_0$ the real infinite unit speed $c = 1 = \infty \times 0$—which we can write as lim $c = 1 = \infty \times 0$—is a rational conclusion of our infinite synthetic reason. Indeed, if, through our infinite reason, we define the universe as the totality of infinitely many parts immediately connected according to necessary, universal principles, then how could these infinitely many parts communicate immediately if the universe's finite gravitational or electromagnetic action were not, at the same time, infinite? [26]

The immediate action at a distance assigns to the universe a universal common present that holds it together and renders it continuous despite the indefinitely increasing distances among its parts. Thus, everything is connected with respect to its existence to everything else according to necessary and universal laws, not by capricious chance. The principle of cosmic connection, at the foundation of the universal laws, is the synthetic principle of equivalence that stipulates the unity and equality of things regardless of their difference and distance. From the equivalence principle, we derive the cosmological principle of the complete uniformity of the universe on large scales—namely, that the universe is the same at all points and times (homogeneity of the universe) and in all directions (isotropy of the universe). The substantialization of the equivalence principle is the ultimate constituent (building block) of ssmatter—the soul-singularity—working as the energizing force, the ontological origin and principle and unifying limiting boundary of the living cell and the living universe according to extension and division.

These assumptions of the infinite rationalist approach constitute the infinite-synthetic paradigm of the universe, according to which the living, real, physical universe is continuous and timeless and is composed of immediately communicating complex parts—the wholes or universes—that verify synthetic principles of organization.

Is life a locally unique or a universal phenomenon, an irrational accident of evolution or an essential property and principle of lifeless matter? Depending on whether we employ our finite individual senses, which perceive the world via our nerve cells at rest operating on a positive time scale, or our infinite synthetic reason, which thinks of the world via our soul-singularities of highest or infinite frequency operating on a zero time scale, we can give two answers to the question: the finite answer favors the unique, emergent nature of life, where at a critical level of evolving complexity, life arises from lifeless matter unpredictably by pure chance. The infinite rationalist answer favors the universal and necessary character of life, in which life, intrinsically linked to kinetic energy and consciousness, is a fundamental determination of the ultimate constituents (singularities) of inanimate matter.

## What Is Life?

Self-conscious life proceeds to question her own existence and asks, "What Is Life?," to which she answers: life—real and true life—is anything that has, according to the synthetic principle of reflexive order $A < A$, the power to continue to exist and move beyond itself in space-time independently of the contingent intervention of external causes and forces, and, therefore, free of linear time. We call this, the power of continuous, free motion characterizing the Divine, the infinite and eternal being, which is the physical universe itself. Because life is the end of computation and irreflexive order, no computational model can represent infinite life unless the model itself hyper-computes with respect to its fundamental reality, becoming an infinite living being, free of computation by maximally computing.

We have demonstrated that the origin of life from nonlife is impossible *unless* we assume that life is already present in the fundamental constituents of lifeless matter—the soul-singularities—which are spherical atoms of smallest or zero volume and infinite curvature.The operation of decomposing the living whole into its smallest parts does not destroy the living whole, as the smallest parts are living wholes or soul-singularities having contrary opposites: that is, nonliving and living properties at the same time.

Given the equivalence between part and whole, we assert that, in the remote future, we will be able to deconstruct the living cell into its smallest living parts—the soul-singularities (by the process of analysis), and reconstruct the smallest living parts into a complete living cell (by the process of synthesis) without any difficulty.

## Notes

1. For an exposition of the theory of change of Anaxagoras see Aristotle's *Physics*, I, 4, 187a26 – 187b7.
2. (i) The symbol $<$ designates inequality (less $<$ more); succession (temporal order: before $<$ after); causation (cause $<$ effect, antecedent $<$ consequent, beginning $<$ end); inclusion (contained part $<$ containing whole). (ii) The symbol $=$ designates unity or equality (neither less, nor more); necessity; simultaneity; zero-succession (zero-temporal order: present, that is, neither before, nor after); zero-causation (zero-origination); zero-inclusion. (iii) The symbol $'$ designates negation (not); falsity; for example $A'$ is read as not-$A$.

3. Aristotle was the first to distinguish clearly between the *universe* and the *individual*. Thus, anything admitting contraries (for example, both $A'$ and $A$) belongs to the class of universals and is therefore a universe, whereas anything admitting contradictories (for example, either $A'$ or $A$) belongs to the class of individuals and is therefore an individual. See Aristotle's *Metaphysique*, I, 10, 30–35.
4. See Kirk Geoffrey, John Raven, and Malcolm Schofield. *The Presocratic Philosophers*, the chapter titled "Anaxagoras of Clazomenae."
5. For a historical exposition of the ancient doctrines of the soul, see Aristotle, see *De l'Ame*, I, 2.
6. In *De l'Ame* (I, 2, 405b, 25–30) Aristotle mentions that disciples of Heraclitus regarded the word *ζῆν*, meaning "alive" as having its roots in the word *ζεῖν*, meaning "to boil."
7. If we identify the life-force of a piece of rotating and vibrating matter with the gravitational force of attraction directed inward and toward the center of matter and its equal and opposite inertial force with the gravitational force of repulsion directed outward and away from the center of matter, we can also affirm that the permanent rotation or vibration of the self-moving piece of matter is the balanced product of the simultaneous action of these opposite gravitational forces, which, as parts of the same complex life-force, are essentially *biogravitational forces.*
8. See *The Presocratic Philosophers*, the chapter titled "The Atomists: Leucippus of Miletus and Democritus of Abdera."
9. The word *hylozoism* is a composite of the Greek *ὕλη* (matter) and *ζωή* (life).
10. *τό κόσμον έμψυχον καί δαιμόνων πλήρη*, cited by Aristotle in *De l'Ame*, I, 5, 411a, 10, p 60.
11. According to the principle of self-causality, everything that happens $A$ has the cause of its existence inside itself and anterior to itself: $A < A$. Because self-causality integrates the cause with the effect, it is an integrated causality operating in a circular or spherical manner. The seventeenth-century Dutch philosopher Spinoza regarded the synthetic principle of self-causality (*causa sui*) as the founding principle of infinite Nature and God.
12. According to the analytic principle of external causality everything that happens $A$ has the cause of its existence outside itself, different from itself, and anterior to itself and therefore is a contingent by-product of external causes and forces: $A' < A$. Therefore external causality is grounded in the analytic principle of inequality and linear temporal order, itself a principle of constraining or heteronomous order.
13. CL = Consciousness of Life; CC= Consciousness of Consciousness.
14. The symbol $\approx$ means "approximately equal."
15. The standard definition of singularity is the place and moment when the finite quantities of an extended body—its temperature, energy, and matter, for example— increase to infinity, whereas its space-time intervals shrink to zero. *Finitism* affirms that it is logically and physically impossible for a finite body to have infinite quantities. Thus, a finite body with infinite quantities is a logical

absurdity (Aristotle) and a physical impossibility or nothingness—which contemporary physicists identify with the black hole. This is not the case with *infinitism*, whose positive concept of infinity is associated with the physical existence and life of the extended body. In fact, far from being the logical and physical negation of the extended body, infinity is its very founding principle: Every finite body is, in its ultimate and deepest reality, a maximum or infinite body according to extension and division, which verifies simultaneously two principles — the finite and the infinite (Pythagorean infinitist thesis)— without contradiction or paradox. See Aristotle's *Metaphysics*, A, 5, 986, 20, where he exposes the Pythagorean cosmological theory according to which the universe (everthing) is governed by two fundamental principles, the limited and the unlimited.

16. Concerning the complex nature of the singularity from the metaphysical point of view, see my recent book *Metaphysics of Infinity: The Problem of Motion and the Infinite Brain*, p 7–8.
17. Initially, the holographic principle [t' Hooft Gerard, Leonard Susskind, Jacob Bekenstein] was inspired by black-hole thermodynamics, which holds that the maximum information content of the region inside the hole is directly proportional to the area of its event horizon: more precisely, to one quarter of the event horizon's area measured in Planck areas. Subsequently, it was suggested that the holographic principle has a universal validity and could be applied to all volumes of space regardless of their nature, whether they are holes or physical bodies. As a universal principle, we contend that it could be equally applied to all contents of a given volume of space regardless of whether the content is information or matter (Ion Spteropoulos 63). The intuitive holographic principle has its historical and philosophical roots in the Aristotelian definition of the body, according to which a body, identified to the specified volume of space enveloping it, is that which is limited by a surface (Physics III, 204b, 4). Because the limiting surface area of the specified volume assigns a spherical form to the volume that determines (originates or ends) its indeterminate material content, necessarily the maximum material content or better mass of the spherical volume is directly proportional to its limiting surface area, which in turn depends on the volume's radius squared. Thus, the maximum mass that a given spherical volume or body can have is determined by the surface area of the specified volume in which the mass spherical fits and not by something external to it. In this sense, all spherical masses are incomparable, self-determined maxima.
18. See *De l'Ame*, I, 2.
19. Planck's constant is the minimum possible energy possessed by a vibrating body adapted to the atomic scale.
20. If we want to give a finite measure to time zero $t_0$, we proceed in the same way as we did with mass. Thus, in function of the intuitive holographic principle applied to time $t = kr^2$, we convert the surface area of the smallest volume of space having the finite radius $\approx 10^{-34}$m into time. If we take as the constant of proportionality $k = t/r^2 \approx 3 \times 10^{17}/(10^{-4})^2$ (which is the time required to produce

the eukaryotic egg cell of radius ≈ $10^{-4}$ m), then the finite measure of time zero $t_0$ regarded as the smallest or zero time interval (the smallest or zero elapsed time) is: $t_0 = 3 \times 10^{17}/(10^{-4})^2 \times (10^{-34})^2 \approx 10^{-43}$s. We have then: $t_{now} \approx 10^{10}$ y = $3 \times 10^{17}$s and $t_0 \approx 10^{-43}$s.

21. It i s expedient to note here that this reduced and simplified observable light is considered by our finite analytic experience as the real and true light. We have, then, two kinds of error related to the observable light: the perceptual error, which reduces the real, complex, infinite light into an observable, simple, finite light, and the epistemological error, which takes the observable finite light *as if* it were the real light.
22. By de-coherence (a term of quantum physics) we mean the disruption of the simultaneity and equality (coherence) of opposites points *a, b* in space – for example, of here at *a* and there at *b* – and their transformation into a succession of points in time that generate a contradiction and time delay between them: It follows that here at *a* is perceived as now, whereas there at *b* is perceived as if it were before or after, past or future, depending on the sense of our action, whether we are observing backward or acting forward. We continue to de-cohere when we chose one option – for example, the here at *a* – out of the totality of options of here at *a* and there at *b* that constitute the physical universe. The selected here at *a* defines our familiar observable universe, which is the finite part of the infinite totality of parts that defines our real, physical universe. De-coherence takes place at the limiting scale of $10^{-10}$ m, where we have the transition from the timeless quantum world to the biochemical and biological world constrained by linear time.
23. The four alternative configurations signify the following: (00) = the emission of a photon is thought; (01) = the emission of a photon is sensed; (10) = the absorption of a photon is thought; (11) = the absorption of a photon is sensed.
24. If the mass of a specified spherical volume of space stores *n* bits of information, then the mass registers at the same time $2^n \times 2^n$ alternative states of matter and consciousness. The material and mental states of the mass are in correlation such that every state of matter corresponds to the totality of states of consciousness, and every state of consciousness corresponds to the totality of states of matter. Therefore on the whole, the totality of states of matter correspond to the totality of states of consciousness.
25. Another way to calculate the maximum number of singularities or bits of information contained in a massive body that fits inside a given spherical volume of space of radius. *r* is by employing the intuitive holographic principle: $I = kA$, in which $k = 1/(10^{-34})^2$ and $A \approx r^2$. Thus, if the radius *r* of a given spherical volume of space is about $10^{-15}$m, the total number of bits of information contained in this volume is approximately: $I = 1/(10^{-34})^2 \times (10^{-15})^2 = 10^{38}$ bits of information that register $2^n \times 2^n$ ($n = 10^{38}$) states of matter and consciousness within a mass of about $(10^{-15})^2 = 10^{-30}$ kg, which we call an electron.
26. The Greek Ionian and Hindu atomists regarded the action of atoms in empty space as being infinite. Following their intuition, Galileo, Kepler and Descartes equally regarded the action of light in space as being infinite. However, if we

think deeply, we conclude that any contact action at a distance regardless of whether it is light or gravitation is necessarily infinite and hence instantaneous; otherwise, there would be no universal laws and no contact action among the parts of the universe separated by infinite distances.

## Works Cited


Aristotle. 1992. *De l'Ame*. Paris: Translated by Jean Tricot. Librairie Philosophique J. Vrin. Print.

———. 1986. *Metaphysique*. Translated by Henri Carteron. Paris: Librairie Philosophique J. Vrin. Print.

———. 1996. *Physics*. Translated by Robin Waterfield. Oxford: Oxford University Press. Print.

Bekenstein, Jacob. 2003. Information in the Holographic Universe. *Scientific American*, 289: 2 (Aug 2003): 58–65. Print.

Hegel, Georg Wilhelm Friedrich. 1986. *Encyclopédie des Sciences Philosophiques*. Translated by Bernard Bourgeois. Paris: Vrin. Print.

Héraclite. 2011. *Fragments*. Translated by Marcel Conche. Paris: PUF. Print.

Kirk, Geoffrey, John Raven, and Malcolm Schofield. 1988. *The Presocratic Philosophers*. Cambridge: Cambridge University Press. Print.

Monod, Jacques. 1971. *Chance and Necessity*. New York: Knopf. Print.

Prigogine, Ilya. 1989. The Philosophy Of Instability. *Futures* 21 (4): 396–400. Print.

Soteropoulos, Ion. 2013. *Metaphysics of Infinity: The Problem of Motion and the Infinite Brain*. Lanham: University Press of America. Print.

Susskind, Leonard. 2012. "L'Univers est un Hologramme." *La Recherche, l'actualité des sciences*. Web. 11 May 2012. http://www.larecherche.fr/content/actualité/article?id=24818

t'Hooft, Gerard. The Holographic Principle. *Opening Lecture*. arXiv:hep-th/0003004v2. Web. 16 May 2000.

# The Fundamental Biological Activity of the Universe

**Attila Grandpierre**

**Abstract** If everything is in permanent change, can the Universe itself be fundamentally passive? Answering this question requires a clear concept of 'activity.' The nature of 'action' is a central and unsolved philosophical problem. Actions play a crucial role in the way we conceive of ourselves, life and the Universe, and the value we put on these. In four decades of research on *solar activity*, we found that activity is not a mere occurrence but a *genuine activity of the Sun*, initiated globally by the Sun using quantum processes as tools that generates suitable primary mass flows locally in the solar core that are capable of producing a working dynamo. We argue that solar activity is initiated by biological causes existing beyond the system of physical causes.

The *anthropic principle* demands an extremely special trigger initiating the Big Bang in a way suited to the development of life. The *Astrobiological Revolution* indicates the generation of complex organic molecules preferentially favorable to life even in the 'impossible' physical conditions present in extremely rare and cold cosmic clouds. With the help of Ervin Bauer's biological principle, we find explanation for *biological determinism* and life's being a *'cosmic imperative.'*

Modern cosmology uses obsolete Laplacean models. We show that *the biological principle in the Universe involves a continuous biological activity of the Universe* prevailing everywhere, including in ourselves. This universal activity is the basis of our life instinct and of logic too.

**Keywords** Activity (philosophy) · Solar activity · Anthropic principle · Astrobiology · Cosmology · Cosmic activity

---

A. Grandpierre (✉)
Konkoly Observatory, Budapest, Hungary
e-mail: grandp@iif.hu

 
W. S. Smith et al. (eds.), *Eco-Phenomenology: Life, Human Life, Post-Human Life in the Harmony of the Cosmos*, Analecta Husserliana CXXI,
https://doi.org/10.1007/978-3-319-77516-6_10

## Introduction

In the last centuries it has seemed that the most fundamental problem of philosophy, the mother of all question "What is life?" has remained unanswered. We think this situation arises because we live in the era of physics. Actually, nowadays there exists only one exact branch of the natural sciences: physics. Therefore if anybody wants to explain anything scientifically, it seems that the only available tool at present is physics. We call attention to an accumulating number of efforts and facts arguing that an exact biology is under development that is capable of offering scientific answers the some of the most fundamental problems of life, matter, and consciousness (Bauer 1967; Grandpierre 2002, 2008a, b, 2012a, b; Grandpierre and Kafatos 2012, 2013; Grandpierre 2014a; Grandpierre et al. 2014).

## The Biological Nature of Action

The very nature of genuine activity, including its origin and the manners in which it becomes physically manifested, is one of the greatest unsolved problems of solar activity research, biology, and philosophy (Grandpierre 2012a, b; Grandpierre and Kafatos 2012, 2013). A process is defined as a genuine activity if and only if it is not completely determined by conditions and laws of Nature but involves an element of autonomous, actually free decision and corresponding government of behavior. Accordingly, philosophers use the word "active" in the sense "creating causal power," "adding a genuine new cause to the already existing ones." Moya summarizes the presently popular scientific views on "action" by the following argumentation illustrating the problem of activity. Actions play:

> a central role in the way we conceive of ourselves and others, as well as in the value we put on our lives. But is there any action? This question may sound bizarre, for what could be more evident than that? Philosophy, however, cannot allow itself to be satisfied with that level of evidence. We could be wrong. … To give the reader an idea of what a reductionist attitude is like, let us start with an episode that nobody would hesitate in classifying as an action, say, drinking a glass of water. What right do we have to call this an action, and not a mere happening? Where is action in this? Well, one could say, I caused that movement, so I acted. But think that this movement can be said to be properly caused by my arm's and hand's movement, which in turn were caused by some muscles' contractions, which in turn were caused by some neurons' firings, and so on. Action as such seems to dissolve and to be reduced to a sequence of happenings. Appealing to desires will not do, for our desire for water is presumably a state caused by organic deprivation. The chain of causes extends further and further into the past and there appears to be nothing we, as agents, initiate, no action at all, only further happenings. Actions, then, seem to be nothing but specific sequences of happenings. (Moya 1990, 1–3)

Even Moya overlooks the crucial step where the decision depends on the matter of the brain initiating nerve impulses and related physical processes realizing the decision. These events manifest a causal chain that works similarly to a domino game. Knocking the first domino initiates the knocking of the second, which knocks the third, and so on consecutively. Although it seems for Moya that "action as such seems to dissolve and to be reduced to a sequence of happenings," the crucial step is, of course, the first knock. In moving a finger, the first knock is given by the immaterial will that has a suitable, biologically governable energy for initiating the first "knock." Here arises the big question of how the mind can exert its influence over matter. Arguably, this is the biggest problem of science and philosophy.

The successes of modern physics are enormous and impressive. Physics considers only inanimate, inert objects that lack the creative causal powers characteristic of living organisms. Impressed by the enormous successes of modern physics, it is usual to consider that all the things of the world are inert and lack genuine causal powers. In contrast, living organisms are not inert objects. We have been successful in working out the scientific theory of genuine biological autonomy, illuminating that living organisms by their very nature are active, creating new chains of causes by their very actions (Grandpierre 2012a, b; Grandpierre and Kafatos 2012, 2013). We have found that the genuine nature of life can be characterized by systematic *work investments against inertial behavior and equilibration* and autonomous decisions (Grandpierre 2007). All life forms are characterized by their activity *maximizing the difference of their behavior from physically prescribed ones* because this difference represents the distance above death, what we usually call vitality (Bauer; Grandpierre 2008a, 2012a, b). Bending our finger is possible at will because living organisms possess genuine biological autonomy and are at least partially free from physical determinations. Necessarily, biological decisions can act only in the realm where physical determinations are incomplete—that is, at and beyond the quantum level. Biological determinations create virtual particle pairs according to biological aims. Therefore, biological determinations like decision-making originate from a deeper level beyond the quantum vacuum.

Genuine action is possible in actual reality through *free will,* which is formulated in exact scientific terms as *biological autonomy* (Grandpierre 2012a, b; Grandpierre and Kafatos 2012, 2013). It is biological autonomy that can be identified by the 'self' who acts. The self is the executive center of consciousness that, together with background consciousness like memory and unconsciousness, forms the mind (Grandpierre 2014a). Since in genuine action it is the 'self' that initiates a new causal chain, the 'self' is logically and causally prior to the realm of physical objects. The nature of action leads us naturally towards a deeper layer of reality that is logically prior to the realm of observable phenomena considered by physics. We may observe that the physical world can be regarded as the outer, visible layer or surface of the Universe. This world has a remarkable consistency involving a gigantic range of causal network reaching to the most distant stars.

## The Biological Nature of Solar Activity

Solar physicists formed a picture of the Sun on the basis of the available facts. According to this picture, the Sun is a hot ball of gas producing nuclear energy. Yet a series of fundamental facts have escaped due attention regarding the origin and nature of solar activity. We have gathered together these fundamental, apparently anomalous facts and have attempted to explore their relations. In this way we have obtained a fundamentally new, more complete picture of the Sun. This new picture shows that the Sun is far closer to life than has been depicted in the past.

The most unexpected property of the Sun, on a physical basis, is that it has an anomalous but systematic *activity*. Solar activity is a term describing all the changes of the Sun, first of all the changes of its magnetic fields and the mass flows in the solar interior. Remarkably, the characteristic complex patterns of solar activity are quasi-regularly and cyclically renewed over an average period of 11 years. If the Sun were merely a hot ball of gas, such an activity could not occur. Nobody would expect that a vast mass of inert liquid or gas systematically transforms its energies and rearranges its global patterns. As Eugene Parker, one of the most eminent solar physicists, noted, on the basis of our knowledge about stellar structure, *solar activity is completely unexpected.* Solar activity is a big challenge of astrophysics. Notably, a series of other facts of solar activity deepens this challenge in a way that sheds completely new light on the nature of the problem. These facts have escaped due attention because they did not fit into the old picture.

We have shown that solar activity systematically circumvents the Second Law of thermodynamics, stating that "All kinds of energy spontaneously spread out from where they are concentrated to where they are more dispersed, if they're not hindered from doing that" (Lambert). The spreading out—this process is also called diffusion—of magnetic energy proceeds extremely slowly, on the timescale of a billion years (Shore 1992, 178). In contrast, the patterns of solar activity are regenerated on average over 11 years. With the help of an example: a hill of sand will lose its height as time passes by, since the grains of sand slowly roll down the hillside in a way that can hardly be observed. On a long timeline, the sand hill would slowly shrink to half of its original height. In comparison, the strength of the solar magnetic field theoretically manifests a behavior like that of the hill that would shrink to half of its original height in a thousand-million-year timeline. Instead, in actual reality, this 'hill' shrinks its height to zero usually within 5–7 years and becomes lowland. After that it transforms itself into a 'valley' reaching a similar depth usually within 3–5 years. Moreover, this anomalous behavior is accompanied by a series of further anomalous facts.

The main task of control theory is to modify the input conditions of a dynamic physical system in order to obtain the expected specific final result from the output of the system. The Sun can systematically circumvent the Second Law by continuously modifying the initial and boundary conditions of physical laws in a way usual in control theory. The conditions to be controlled are the input data for the physical equations. These input conditions are controlled in a very special and systematic manner that leads to a thousand-million-fold acceleration of the magnetic field's decay and its systematic, thermodynamically uphill regeneration. The systematic modification of the input conditions of physical laws requires an activity, an intervention from a higher level capable of establishing relations between the mass flows and the magnetic field having that kind of special algorithmic complexity, which makes the internal mass flows suited to driving a working dynamo regenerating the activity patterns. Since the chain of physical processes follows the principle of inertia, such physical processes cannot intervene and modify their own course in a systematic manner. Such a systematic modification would require an engineering activity. The modifications of the magnetic field are realized by generating suitable mass flows in the solar interior, which flows are capable of transporting, annihilating, and regenerating the field by their suitable configurations.

Let us illustrate the problem with the help of an example. In thermodynamics, time has an arrow, the arrow of decay towards equilibrium. Similarly, all fruit-jars fall downwards, when they are not hindered in this by suitable shelves in the butlery. In the case of the Sun, the "fruit-jars" are not hindered from falling down. The theoretically calculated time for reaching half their height from the shelf to the floor of the "butlery" is more than thousand million years. Moreover, as these solar "fruit-jars" would approach the "floor," their fall were become slower and slower. In sharp contrast to these theoretical calculations, in actual reality the magnetic field lines lose all their strengths within an average of 5–7 years. Additionally, all the magnetic field lines are regenerated from scratch within the next 3–5 years on average. In our example, it would not be enough to pull out the shelves from below the fruit-jars. Additionally, it would be necessary to attach suitable rockets to them to accelerate their falling down by a factor of a thousand million times. After that, it would be necessary to govern the rockets again upwards in a special way and that within 3–5 years, for at the end of the cycle all the fruit-jars are again on the shelves, but now in a top-down position. Such a feat would require enormous ingenuity. In the case of the Sun, this feat is realized by mass flows that are generated in a suitable manner in the solar interior. We hope this example is helpful in illustrating the enormous and unexpected difficulties we noticed in searching for the origin of solar activity.

The destruction and regeneration of the Sun's magnetic activity requires a dynamo working in the Sun. The motor or the heart of solar activity is the *dynamo that produces systematically magnetic energy from mechanical motions* occurring in the solar interior (Nandy and Martens 2007). *A dynamo is a machine* that converts mechanical energy into electromagnetic energy, like one attached to a bicycle wheel. Keeping in mind the key importance of the dynamo, we can appreciate the true significance of the fact that *the dynamo is one of the truly large mysteries in astrophysics* (Carpenter et al. 2005). We think it is no wonder that the dynamo of solar activity is a truly large mystery because machines involve *functions* and *algorithmic complexity*, both of which transcend the conceptual framework of physics. Functions involve *teleology*, and teleology is alien to physics since physical objects cannot have purposes or aims. It is a matter of fact that the functioning of machines arises from human activity. Such engineering activity cannot be described by physics. Teleology is forbidden in the conceptual framework of physics.

The Sun continuously governs its own activity from its global level, initiating quantum processes in its energy-generating core in a way that induces primary mass flows producing a dynamo requiring the creation of algorithmic complexity. The algorithmic complexity of a machine arises from the boundary conditions of its components describing the way they are put together according to the working principle of the machine. This 'working principle' represents a higher-level principle controlling lower-level phenomena (Polanyi). Algorithmic complexity can be characterized by the minimal length of a computer program describing the given process. Algorithmic complexity characteristic to complex machines cannot be produced in a merely physical process. If all men were exterminated, this would not affect the laws of inanimate nature. But the production of machines would stop, and not until men arose again could machines be formed once more. Some animals can produce tools, but only men can construct machines (Polanyi 1968). While purposeful beings—humans—produce the algorithmic complexity of machines on Earth, such processes are unknown in the Sun and their existence can be excluded on a physical basis. Nevertheless, there is a dynamo at work in the Sun. Does this mean that somehow purposes can arise in the Sun?

These difficulties are even more significant since it is not only a magnetic dynamo that is at work in the Sun, but also a *multi-functional* system of energy transformation that transforms all types of energy into each other in a way that regenerates the patterns of solar activity. Searching for the origin of solar activity we have found it of basic importance that the multi-functional machine is driven by *mass flows*. Since this multi-functional machine has an algorithmic complexity, the mass flows generating and regenerating it must also have an algorithmic complexity.

Owing to the results of our four-decade research program attempting to clarify the origin of solar activity, we have obtained evidence showing that rotational, magnetic, tidal, kinetic, and nuclear energies all play a crucial role in the origin of solar activity (Grandpierre 2015). We have shown that the solar atmosphere couldn't supply enough energy for solar activity; therefore it must be generated in the deep solar interior (Grandpierre 1986, 1988, 1991, 1996a, b, 2002, 2010, 2015). The new theory we have worked out is based on the recognition that solar activity is generated in the solar core. We have obtained a plausible picture of how these energies are transformed into energy forms maintaining solar activity. We have found a series of *positive and negative feedback cycles* playing a central role in solar activity. With the help of detailed numerical simulations of all the related physical processes we have found that hot bubbles, approximately the size of Budapest (having a radius of approximately 10 km), are the key tools by which solar activity is transported into the surface. At all points of these vast hot bubbles the mass flows are coupled to each other in such a special way that the result is the regeneration of solar activity at the global level. In our example, this can be compared to the traffic in Budapest, where all vehicles move in a coordinated way to produce a special prescribed output pattern at the global level. A large initial heating, making the bubble 200,000 degrees hotter than its 15-million-degree environment, is necessary so that the hot bubbles can travel a significant distance towards the solar surface. Above an initial heating of 50 million degrees, nuclear reactions become explosively accelerated and a positive feedback develops resulting in a *thermonuclear runaway* producing a huge amount of energy and the anomalous abundances of heavy elements characteristic of large solar flares.

We have developed a new, almost complete theory of solar activity. The only missing element is to find *the very first cause(s) of solar activity*: the process that initiates and governs the mass flows in the solar core. We have shown that external physical conditions and chance may play a role in the generation of these mass flows, but they are not sufficient conditions for regenerating solar activity (Grandpierre 2015). These mass flows generate the local dynamo as well as the hot bubbles in a special way that is suited to regenerating the patterns of solar activity manifested *at the global level*. We are faced with a type of *downward causation that produces a multifunctional machine,* including the dynamo, from cycle to cycle. It is important to recall that machines are produced externally, by human activity. Yet, *in the case of solar activity the machine is produced internally*, by the Sun itself. Considering that the most significant difference between organisms and machines is that the former are intrinsically purposive whereas the latter are extrinsically purposive (Nicholson), in our search for the origin of solar activity we are led towards biology. We found the analogy of solar activity in biological actions like bending our finger. Indeed, when we bend our finger, we act from the global level of our mind to the local level of our finger. The causal chain of solar activity starts from beyond the gigantic network of physical causes, extending to the entire observable universe. The Sun initiates biological causes from a deeper layer of the Universe existing beyond the quantum vacuum by creating virtual particles suitable for realizing biological causes, namely, regenerating solar activity. Both solar activity and bending our finger are genuine self-initiated, self-governed actions involving top-down causation. In this way, we have developed a complete theory of genuine solar activity that is called the Helios Theory (Grandpierre 2018a).

## The Finger Experiment

We argue that the ability to act transcends physical behavior, because this latter is always inertial. The ability to act transcends inertial behavior. This is why genuine action necessarily transcends the physical framework. In order to make the concept of action clear and unambiguous we present a simple but compelling experiment: the finger experiment. Who would think that bending our finger and the course of solar activity show an essential similarity? Yet it is so, and exactly with respect to causality. We intentionally bend our finger in a way similar to solar activity. We act at the global level of the organism by our decision and the result is a local process, the bending of our finger. Similarly, the Sun initiates its activity from its global level and acts on the local processes in its core, initiating the mass flows that produce the primary dynamo and the hot bubbles.

The finger experiment has an extraordinary significance since it can clarify for most people that action arises from beyond physics. We can predict successfully when we bend our finger. Therefore our hypothesis that a genuine action realizes the bending of our finger is scientifically testable and provable. It is a fact that similar experiments take place in large number in our everyday life supervised by a vast number of independent experimenters and with successful results. We can consider the conclusion of the finger experiment to be scientifically confirmed. The finger experiment is elevated to an idea of revolutionary significance by the fact that it is commonplace and outstandingly radical at the same time. Its mind-changing significance arises from the fact that everybody can understand its extraordinary deeply penetrating power. If we but recognize it, all of us are enabled to change the way we see the world. Such a change can bring about a new, life-centered age for humanity.

We have shown that biological activity like bending our finger is physically realized by biologically created vacuum fluctuations (Grandpierre 2012a, b; Grandpierre and Kafatos 2012, 2013). We have also shown that solar activity is, similarly, governed by biologically-induced quantum processes. Surprisingly, similar process plays a similar role in triggering the Big Bang that is widely thought to be initiated by quantum fluctuations. We found a fundamental similarity between bending our finger, solar activity and the Big Bang. Remarkably, the similarity is manifested in the matter of causality.

We consider that our will represents a kind of biologically governable energy (Baumeister 2012). Let us note that the very first step in the causal chain of our action is that our will creates vacuum fluctuations that are ideally suited to the mental content of the will. The first step of the 'action' creates virtual particles suitable for realizing the corresponding biological aim. Realizing such a feat builds a bridge between our will and the quantum vacuum. This bridge can be compared to the role the genie plays in the old fairy tale about Aladdin and his wonderful lamp.

To wit, the quantum vacuum fulfills all our wishes in an extremely delicate and powerful manner. We wish to bend our finger. That's all, and the rest is done by our brain and a quantum vacuum. We can paraphrase the dialogue between our mind and the quantum vacuum by imagining this dialogue between Aladdin and the genie of the wonderful lamp:

> Aladdin to the genie: Oh my friend, let there be a bend of my finger now!
> Genie: Your wish is a command for me, my dear friend!

And there was a bend. The 'genie' creates exactly such special virtual particles that induce exactly such physical forces that realize the aim of bending the finger.

If a computer expert would take into account all the necessary input biocurrents to the muscles of the finger, he could work on that task day and night for years. How is it that bending our finger occurs with an utmost ease? Similarly, if a solar physicist receives the task of determining exactly all the important details of the mass flows to be generated in the solar interior that should serve as suitable rockets driving magnetic field lines in a way regenerating the solar cycle at the global level, he could work day and night for years—and still have no real chance at solving the task successfully. How is it that the Sun succeeds in solving this problem continuously?

## Timely Thoughts on the Models of Physical Cosmology

Modern cosmology works on the basis of deterministic cosmological models rooted in an obsolete idea of Laplace (1812). This idea was that the future motion of all physical objects was completely determined, if one knew all of their positions and speeds at one time. Since then, Laplace's idea has become untenable. The development of non-equilibrium thermodynamics, quantum theory, and chaos theory, among other conceptions, has made Laplace's idea obsolete. Remarkably, it nonetheless survives in cosmological models assuming that cosmological equations with suitably selected conditions can give account of the world we live in. In such a situation it is useful to keep in mind that the physical model of the Universe differs from the actual Universe in many fundamental respects. At present, we are not speaking of the model's differences from the astronomically-observed universe. We will just mention that the astronomical universe is full of forms that are missing from the physical model (Ellis 2005).

What is missing from the cosmological model that is to be found in the actual Universe?

- Fine tuning of vacuum processes to life, laws of Nature, fundamental constants (see below)
- Solar activity (see above; Grandpierre 2015)
- Complexity; algorithmic complexity that cannot be produced by the operations described by physics (Polanyi 1968; Davies 1998, Grandpierre 2008b)
- The observed continuous creation of complex molecules everywhere in the Universe (see below)
- Biofriendly laws. Life (Grandpierre 2014a; Grandpierre et al. 2014).
- Biological autonomy. Consciousness. Self-consciousness (Grandpierre et al. 2014).

This means that the physical model and the real Universe are fundamentally different. Laplacean cosmological models of physical cosmology correspond only to a surface layer of a more fundamental astrobiological or biofriendly cosmology that we are now discovering.

## The Anthropic Principle

The fundamental physical constants, the forms of physical laws, and the nature of the vacuum fluctuations that generated the Big Bang all appear to be finely tuned for life's flourishing (Barrow and Tipler 1986; Barrow et al. 2007; Dick 2009, 2010, 2012, 2013; Davies 1998, 2006; 'anthropic principle,' entry in Enc. Brit.). In the physical model, the quantum fluctuations triggering the Big Bang occur spontaneously, without physical causes. Moreover, it is not known what determines the form of physical laws and their fundamental constants. But one thing is clear: all these three factors are related to one known thing, that being life. The big question is: How?

Actually, these three factors—physical constants, laws, and vacuum fluctuations—are the tools of explanations within the conceptual framework of physics. These conceptual tools are insufficient since they leave unexplained all the related facts presented below.

In principle, there are an infinite number of possible random quantum fluctuations suited to triggering the Big Bang (Tryon). But only a very small segment of this range of possible fluctuations is suited to triggering such a Big Bang that is consistent with the existence of galaxies, stars, and life. The quantum fluctuation triggering the Big Bang represents the initial condition of the cosmological equations. The actual quantum process initiating the Big Bang can also be termed the First Cause. This formulation is suitable for putting more emphasis on the significance of this initial factor in the fabricating of the observable universe. The actual First Cause must be extremely special since it is one among the very few that are suited to the existence of life.

After this First Cause, further causes act as well. After the very initial event of the Big Bang, virtual pairs of particles from the quantum vacuum began popping in and out of existence, some of which could absorb energy and become real. Physicists think that all matter today, *from galaxies to living things*, originated from these primordial quantum fluctuations (Zyga 2012). Without the minimal variations in energy density that result from the tiny but unavoidable quantum fluctuations, one cannot account for the formation of the stars, planets, and galaxies that characterize the Universe we observe today.

Quantum fluctuations acting on the evolution of the observable universe have a far-reaching hand. These reach from initiating the Big Bang to the formation of the suitable density irregularities leading to galaxies, to the formation of the Solar System, the Earth-Moon system, to the origin and evolution of the terrestrial biosphere, and to our existence here. This means that the far-reaching hand of quantum-level processes reaches into us. Such quantum fluctuations are input elements for the Laplacean models of physical cosmology and are left unexplained by them.

We point out that it seems to be overly far-fetched to consider that Laplacean cosmological models worked out for describing the largest scale structure of the observable universe could work well not only on large scale systems but also on smaller scale systems like planetary systems and biospheres. By the way, no cosmological model attempts to derive the origin of the Solar System or the origin of life from the Big Bang, and they are clearly not suitable for accomplishing such tasks. We argue that since quantum-level processes have such a far-reaching hand influencing also small-scale processes, it is plausible to assume that their activity is responsible for the development of such apparently 'tiny details' like planetary systems and living beings. If so, the quantum vacuum may have a cosmic role influencing cosmic evolution. Since the quantum level is the deepest level of the physical world, a 'cosmic activity' realized at the level of the quantum vacuum is a fundamental and universal activity of the Universe. Since cosmic activity extends a far-reaching hand towards life, this cosmic activity has a remarkable biological aspect. If the Universe consists not only of matter and energy but information as well, life has a fundamental place in cosmology. We, therefore, consider seriously the idea of a fundamental and biological cosmic activity as manifested through the quantum vacuum everywhere. An increasingly large number of other, recently accumulated facts underpin this idea.

It was argued that atomic matter and therefore life are possible only in three-dimensional space (Gurevich and Mostepanenko 1971). As Barrow and Tipler put it in their chapter about the relations of dimensionality to life, "the dimensionality of the Universe is a reason for the existence of chemistry and therefore, most probably, for chemists also" (Barrow and Tipler 1986, 265).

Remarkably, the laws of Nature also have a form suitable for life's flourishing. We have pointed out that the form of physical laws is derivable from the least action principle (Grandpierre 2007). The physical meaning of the least action principle is that physical objects manifest inertial behavior. This inertial behavior can be regarded as the complete opposite of biological behavior, which is characterized by the mobilization of all available energies against inertial behavior. Moreover, we have found that the inertial principle is the ideal tool in the hand of life for attaining the greatest action possible. Once a living organism has decided on a concrete action, it must act in the most economical way in order to save energy for future actions. We have concluded that the form of all fundamental physical laws is related to biology.

Among the fundamental constants occurring in the laws of physics we find the Planck quantum of action, the speed of light, the relative strengths of the four fundamental forces, and the masses of elementary particles. If the laws of Nature are the machinery by which Nature works, these fundamental constants are the buttons on it. Only relatively small ranges of the possible values are consistent with the existence of life. It is this fact that inspires the phrase "fine-tuning" to describe the cosmic conditions favorable to life, even if in the case of certain parameters the allowed ranges are not very narrow (Barrow and Tipler 1986). In the light of newly accumulating facts and arguments "it is clear that the universe appears remarkably 'fine-tuned' for life as we know it" (Chyba and Hand 2005). The buttons on the machinery of Nature are all set on the position "LIFE."

The unique properties of water, carbonic acid, and the compounds of carbon, hydrogen, and oxygen are all fine-tuned for life. "The properties of matter and the course of cosmic evolution are now seen to be intimately related to the structure of the living being and to its activities; they become, therefore, far more important in biology than has been previously suspected. For the whole evolutionary process, both cosmic and organic, is one, and the biologist may now rightly regard the universe in its very essence as biocentric" (Henderson 1913, 312). Updated, more detailed, but essentially similar conclusions were reached by Wald (1962) and Needham (1965). It is claimed that not only is carbon important, but across the periodic table each element seems to be uniquely suited for life's evolution and emergence. For example, Wald (1962) presented detailed arguments showing that phosphorus and sulfur have surprisingly many properties making them ideally suited to life's purposes. Recently Conway Morris has shown that phosphorus and zinc bring to each cellular stage indispensable properties (Morris 2010).

Fine-tuning is present not only in the realm of inanimate matter, but also in the kingdom of the living. We are at a stage where most of the key players in particular processes of a particular biological process, such as focal adhesion formation, are known but the numerous competitive interactions in the cell and the fine-tuning achieved by phosphorylation and protein cleavage are not yet very well understood. The main challenge is to discover how all these components work together in a concerted way. Already, Niels Bohr, one of the founding fathers of quantum physics, also noted,

> An understanding of the essential characteristics of living beings must be sought, no doubt, in the peculiar organization, in which features that may be analogous by the usual [classical] mechanics are interwoven with typically atomic [quantum] traits in a manner having no counterpart in inorganic matter…. Owing to the very limits imposed by the properties of light, no instrument is imaginable which is more efficient for its purpose than the eye.... This ideal refinement suggests that other organs also … will exhibit a similar adaptation to their purpose. (*Light and Life*) (Bohr 1933)

The extremely sophisticated orchestration of biological organization presents one of the basic facts left unexplained by physics, chemistry, and molecular biology. The functions of the living organism typically depend upon the coherent operations of molecules by the million, belonging to hundreds or even thousands of different kinds, and marshaled into order by a hierarchy of controls. A satisfying reading of life's riddle demands a rational account of its biological organization, and that has yet to be achieved (Harold 2001, 4). Harold adds: "And always in the background, just out of earshot, a murmur of mystery: how are all these activities integrated into a pattern that works, reproduces itself and persists for millennia?" (Harold 2001, 142). The almost perfect construction and working of living organisms is still a profound mystery.

Recently, Paul Davies summarized the key points arguing for a biofriendly Universe in the following form:

- The existence of life as we know it depends delicately on many seemingly fortuitous features of the laws of physics and the structure of the universe.
- A famous early example of how the laws of physics seem to be fine-tuned for life is the production of carbon in stars, which requires a numerical "coincidence" to produce a nuclear resonance at just the right energy.
- All four forces of nature are implicated in the life story. Changing the strength of any one of them, even by a small amount, could render the universe sterile.
- The masses of some fundamental particles could not be very different without compromising the habitability of the universe.
- The measured value of dark energy is 120 powers of ten less than its natural value, for reasons that remain completely mysterious. If it were 119 rather than 120 powers of ten less, the consequences would be lethal. (Davies 2006, 171)

Let us mention that the anthropic principle has already generated some successful predictions. As is told in the *Encyclopedia Britannica*:

> In 1952 British astronomer Fred Hoyle first used anthropic reasoning to make a successful prediction about the structure of the carbon nucleus. Carbon is formed by nuclear reactions in stellar interiors that combine three nuclei of helium to make a nucleus of carbon. This three-body reaction is very improbable. In order to reconcile it with the abundance of carbon in the universe, Hoyle predicted that the carbon nucleus must possess an intrinsic energy level at a value almost equal to that of the sum of the three helium energies at the temperature of their combination. Under these circumstances the nuclear reaction proceeds with especial rapidity: it is said to be "resonant." Soon afterward, physicists found an energy level of carbon in precisely the place predicted by Hoyle. Other successful prediction of the anthropic principle is worked out by Weinberg in 2007 (see Ellis 2011).

## New Perspectives in the Search for Life in the Universe – The Astrobiological Revolution

Astrobiology is the study of the origin, evolution, distribution, and future of life in the universe: extraterrestrial life and life on Earth. Astrobiology is the science of life in the Universe. As the astrobiologist Steven J. Dick writes, with the advent of the means to explore space, the prospect of developing a truly universal science of biology now seems possible for the first time (Dick and Strick 2004, 2). The achievements of astrobiology offer new contexts and new perspectives for studying the relations between life and the universe. In this way, astrobiology explores the most universal aspects of life, laying the foundations for the biology of the future.

With the developments of deep drilling techniques, the signs of a deep hot biosphere having a total mass comparable to that of the biosphere at the surface are found in a depth of 10 km in the outer crust of Earth (Gold 1992). Life is present within much wider conditions than was previously conceived. Extremophile bacteria (organisms able to survive in extreme environments) have proved to remain not only viable in conditions of extreme temperature, pressure, and radioactivity but frequently proliferate even more there than within ordinary terrestrial conditions. The limits of life have expanded at an unprecedented and unimagined rate, including life present in deep space in conditions around −270 °C. The progress of space science made it possible to investigate organic molecules, the building blocks of life, and their conditions by direct measurements in the materials of meteorites and planets and their moons. Soon it was discovered that comets and meteorites are rich in organic materials, even in amino acids. The rapid progress of spectroscopy opened the way to detect organic molecules within planetary and extragalactic clouds as well. It has been realized that conditions necessary for life and organic molecules being the building blocks of life are ubiquitous and found everywhere even in places where it was considered to be impossible. Importantly, pieces of evidence have been found arguing that life was present on earth immediately after its surface became solid.

Instead of considering life to be a sporadic or singular event in the history of the Cosmos in its appearing on Earth, today it has become almost universally accepted among astrobiologists that life has appeared at all places in the universe where the conditions allow, and these conditions have much more range than was thought possible before.

## Astrobiological Observations Substantiating the Lawful Development of Life in the Universe

1. With the development of infrared spectroscopy, practically all families of organic compounds have been detected in space (Kwok 2011, 78). Organic compounds of a high degree of complexity are now known to be widespread in the Universe:

in the Solar System, in stars, in the diffuse interstellar medium, and in external galaxies. Their existence is not confined to limited classes of stars or small regions of the interstellar medium (Kwok 2011, 187). Most interestingly, these compounds are widespread in the diffuse interstellar medium where density is very low and the radiation background is also low, and they are found even in "impossible" regions of space where, according to general perception, the density is too low for the synthesis of complex molecules (Kwok 2009; Kwok and Zhang 2011; Neal-Jones and Steigerwald 2011). Until recently the rates of reactions in interstellar clouds were expected to be very slow, with minimal production owing to the low temperature and density of the clouds. The reactions needed to create complex organic molecules are familiar to scientists only at the much higher temperatures and pressures of earth and earth-based laboratories. The density of interstellar molecular clouds producing organic compounds is between $10^{-4}$ and $10^{6}$ molecules per $cm^3$. Compare this with a number density of roughly $10^{19}$ molecules per $cm^3$ for air. From our experience studying chemical reactions in the terrestrial laboratory, reactions will only occur under sufficiently high densities (allowing atoms to collide with each other frequently) and under relatively high temperatures (when atoms are moving sufficiently fast). In the rare and cold cosmic clouds both the density and temperature are extremely low, therefore the suitable collisions of atomic particles needed for forming organic compounds are extremely improbable. We note that according to Sun Kwok, who is one of the most outstanding experts in the field of astrobiology, "theoretically, this is impossible, but observationally we can see it happening" (The University of Hong Kong 2011). The improbability of the formation of these compounds is the greater, if we compare the *billion years'* timescale of increasing measures of complexity of life on Earth (Grandpierre 2008b; Sharov and-Gordan 2013) to the *several days'* timescale of the appearing such complex organic molecules in these extreme conditions (Kwok and Zhang 2011). The production of such complex organic molecules in the environment of the stellar winds is completely unexpected on physical grounds. "How these stars manage to perform such chemical miracles has remained a mystery" (Kwok 2013, 88). Our conjecture is that a significant percentage of the atoms must be moved in a fine-tuned way to form complex organic molecules by a *biotic factor*, namely, the biological principle (Bauer 1967; Grandpierre 2007, 2013; Grandpierre et al. 2014).

2. It has been argued that it is impossible to synthesize organic materials in appreciable quantity from inorganic materials without the intervention of biological systems (Hoyle and Wickramasinghe 1999a, b). At the same time, scientists believe that more than 20% of the carbon in the universe is tied up in this extensive family of compounds, collectively known as polycyclic aromatic hydrocarbons, abbreviated as PAHs (Hoover 2014). Actually, from the more than 160 molecules identified in the circumstellar and interstellar environments, however, not one is a polycyclic aromatic hydrocarbon molecule (Kwok and Zhang 2011). Instead, they are indicated to be mixed aromatic-aliphatic molecules *similar to coal and kerogen* as well as to the prebiotic insoluble organic materials found in

meteorites. This means that more than 20% of the material of interstellar grains is similar to coal and kerogen. However, since our ability to detect large, complex molecules is limited by the present state of astronomical techniques, the actual complexity of organic molecules could be much higher than is currently known (Kwok 2011, 78). The production of coal and kerogen requires millions of years even here on the Earth. Since coal and kerogen are remnants of ancient life, this type of organic matter was thought to arise only from living organisms (The University of Hong Kong 2011).

3. Recently the first amino acid, glycine, has been discovered in interstellar dust (Kuan et al. 2003). If we envisage a soup of chemicals and the near-infinite range of possible reactions, there will be a vast decision tree of molecular arrangements that are open. Only a few tiny twiglets on the tree will lead towards life (Davies 1998, 236). Therefore the discovery of glycine is further indication for the existence of a preferentially biological effect acting everywhere in the Universe.
4. There could be as many as *40 billion Earth-sized planets* orbiting in the habitable zones of sun-like stars and red dwarf stars within the Milky Way Galaxy (Petigura et al. 2013). Recently it has become clear that microorganisms populate habitats like deep-sea hydrothermal vents, Arctic sea ice, geothermal hot springs, and extremely dry desert soils, and thrive inside rocks up to 1900 feet below the sea floor, and half a mile below the ice of Antarctica, and have survived and shown remarkable results in the adaptation capacity for photosynthetic activity within a simulation time of 34 days under Martian conditions (de Vera et al. 2012), and seem to adapt to the space environment in ways "not observed on Earth" and in ways that "can lead to increase in growth and virulence" (Kim et al. 2013). These environments and their inhabitants give us a glimpse into potentially habitable environments on other planetary bodies, where these extreme conditions might be more common. Recently, it has become evident that the Martian subsurface contains niche environments where life could develop (Chatzitheodoridis et al. 2014). There are indications that Saturn's moon Titan can also have habitable zones, and it is well known that organic compounds are abundant there (Iess et al. 2012). Meteorites rich in prebiotic organic compounds may harbor evidence of life (Heldmann et al. 2014). Key molecules in prebiotic chemistry like dipeptides are detected in the Murchison meteorite (Shimoyama and Ogasawara 2002; Schmitt-Kopplin et al. 2010). A new experiment simulating conditions in deep space reveals that the complex building blocks of life could have been created on icy interplanetary dust and then carried to Earth, jump-starting life (Sanders 2013; Kaiser et al. 2013). Let us note that Kaiser et al. conclude that their experiments have established the feasibility that dipeptides—a key component in the assembly of proteins—can be formed in interstellar model ices abiotically at 10 K via ionizing radiation. Here we point out why their conclusion about the apparent 'abiotic' origin of organic molecules is wrong. Since "theoretically, this is impossible" in the actual low density conditions of interstellar clouds (The University of Hong Kong 2011), the only possible explanation that can explain both laboratory experiments and such

astrobiological observations is that not only physical laws, but also biological ones are present both in laboratories and in cosmic conditions. Since such biological laws are unknown to most researchers, they feel obliged to think that the process occurs abiotically. Nevertheless, we point out that the biological principle (Bauer 1967; Grandpierre 2007; Grandpierre et al. 2014) is present everywhere in the same way that physical laws are present in both laboratories and cosmic conditions. Certainly, only natural laws can govern dust formation as well as the formation of prebiotic molecules. If physical laws cannot explain observations, we must allow that biological laws are at work. If so, then we have found experimental evidence for the existence of a biological law of nature permeating the entire Universe.

5. The detection of infrared features in distant galaxies suggests that complex aromatic compounds were already present in the Universe as early as 10 billion years ago (Kwok 2011, 95). Scientists reported that life had begun $9.7 \pm 2.5$ billion years ago, billions of years before the Earth was formed, based on extrapolating the "genetic complexity of organisms" [from "major phylogenetic lineages"] to earlier times (Sharov 2006; Sharov and Gordon 2013).
6. The Milky Way has an important role in the development of terrestrial life. In the past 3 billion years, the star production rate of the Milky Way galaxy has been closely correlated with the productivity of life on Earth as measured by the isotope ratio $C^{13}/C^{12}$ (Svensmark 2006). The correlation coefficient between the two records is 0.92 and significant at the 0.9999 level. If this linkage is confirmed it suggests that the evolution of life on Earth is strongly coupled to the evolution of the Milky Way.
7. An increasingly large number of material properties have become known which seem to be fine-tuned for life. About 99% of the living parts of living organisms are made of the four elements, H, O, N, and C. The striking parallels between the relative cosmic abundances of reactive elements (especially H, C, O, and N) and the elemental composition of living matter have been pointed out by many authors (Fox and Dose 1997). Already Henderson (1913) argued early in this century that water and carbon dioxide are maximally and uniquely suited for the living state in virtually every one of their chemical and physical properties. "Water is the most extraordinary substance! Practically all its properties are anomalous" (Szent-Györgyi 1972, 9). No simulation model is currently able to reproduce these properties (Nordita Conference 2014). At present, water has more than 66 known anomalies, most of which are inevitable for life (Tuttle 2009; Huang et al. 2009; Chaplin 2015). Water is not only common in the cosmos. It is also the best known of all known substances for supporting the living state (Kenyon 1974). Oro reminds us that the composition of living matter is a better sample of the universe than is our earth (1963). The valences of carbon, oxygen, and nitrogen and the marked tendency of molecular hydrogen to escape from the surface of a condensing protoplanet easily account for the relative deficiency of hydrogen in living matter.

Steinman and Cole reported that amino acids might form peptide chains in a manner that was 'anything but random' (Steinman and Cole 1967). They noted that molecules tend to form compounds that are biologically preferred. "In a sense, a sort of built-in »predestination« can be identified at several levels of biological order." This explains why the organization necessary for living systems that appeared as a rare, chance, improbable phenomenon can actually occur.

If it can be shown that these most abundant reactive elements are uniquely suited for the living state, as Henderson, Wald, and Needham have argued, then *movement toward carbon-based life is discernible in the earliest stage of cosmic evolution as a favored direction* (Henderson 1913; Wald 1962; Needham 1965). Scores of primitive Earth simulation experiments employing a variety of initial gaseous mixtures and free energy sources have demonstrated that the types of compounds that play key roles in living matter are formed in appreciable yields under simple conditions. For example, Miller has shown that more than 200 mg of amino acids are produced when a primitive gas mixture containing about 1 gm of initial methane is subjected to an electric discharge (Miller 1955). Bar-Nun et al. found that in a high temperature shock tube 30% of the initial $NH_3$ is converted into amino acid product (Bar-Nun et al. 1970). In such experiments, the four most abundant amino acids of living matter, glycine, alanine, aspartic acid, and glutamic acid, are consistently formed in appreciable yields (Kenyon 1974). In spite of the non-biological compounds detected in these experiments a trend toward the living state long before the first life appeared is clearly indicated. The tantalizing conclusion suggested by the above survey of experimental data is that in every phase of cosmic evolution from the origin of the elements to the appearance of protocells there is a discernible preferential movement toward the carbon-based living state (Kenyon 1974, 211–212).

Sidney Fox and Klaus Dose claimed evidence that the basic laws of physics and chemistry were biased in favor of generating biologically significant molecules (Fox and Dose 1997; cited in Davies 2003). Sidney Fox also concludes that 'amino acids determine their own order in condensation', and that this non-random 'self-instruction' infuses macromolecules with crucial biological information, paving the way for life (Fox 1988, 897). Cyril Ponnamperuma, one of the early pioneers in biogenesis research, believed that "there are inherent properties in the atoms and molecules which seem to direct the synthesis towards life" (Shapiro 1986, 186–7). As it is formulated by the Nobel laureate biologist Christian de Duve (1996), life is a 'cosmic imperative'.

There is now broad agreement among physicists and cosmologists that the universe is in several respects 'fine-tuned' for life (Davies 2003). If the claim that life would be written into physical laws were true, it would be astounding, not to say incredible. To claim that atomic processes include a built-in bias favoring organisms means that the laws of atomic physics effectively contain a blueprint for life (Davies 1998, 236). A physical law will not create biological information, or indeed any biological information at all. If the occurrence of an event is 100% predictable on the basis of a physical law, than the next occurrence of the same event will have no information content. Contrary to the oft repeated claim, then, life cannot be 'written into' the laws of physics—at least, not into anything like the physical laws that we know at present. Davies points out, "Life works its magic not by bowing to the directionality of chemistry, but by *circumventing* what is chemically and thermodynamically 'natural'. Of course, organisms must comply with the laws of physics and chemistry, but these laws are only incident to biology" (Davies 1998, 237). He adds, "the secret of life lies, not in its chemical basis, but in the logical and informational rules it exploits. Life succeeds precisely because it evades chemical imperatives" (Davies 1998, 238).

Until recently, life and the Universe have been seen through the glasses of physics as a side-effect of physical laws occurring only as a fluke. In the mechanical paradigm governing the science of the previous four centuries, life in the Universe has been considered as a marginal and sporadic phenomenon. For example, in his 1981 article in the *Encyclopedia of Physics*, Caws claims that "Physics is the most basic of the empirical sciences and thus has an asymmetrical relation to others, because every object in the universe has physical properties and is acted upon by physical forces, whereas some objects—subatomic particles, for example—have no chemical properties and may undergo no chemical changes, while most objects in the universe have no biological, sociological, etc. properties." In this paper, we present evidences that biofriendly activity is actually present everywhere in the Universe including extremely rare and cold cosmic clouds. We argue here that in contrast to Caws' claims, most objects of the universe have genuine biological properties, including the Sun, the stars, and the quantum vacuum. If so, the old view that physics provides the base for biology "because all objects we see around us, including ourselves, are made of the same fundamental particles whose interactions are governed by the fundamental forces identified and investigated by physics," will require a fundamental revision (Ellis 2006).

We have obtained a new, wider and deeper picture of the Universe. The Universe is not made merely from elementary particles of matter. In contrast, the Universe is made, besides from elementary particles, from laws of Nature, including physical, biological and psychological laws. This means that elementary particles are governed not only by physical laws, but also by biological and psychological laws. This is why biological determinism prevails throughout the entire Universe. Under favorable circumstances, which themselves are facilitated by the biological principle, biological effects can become observable and manifested in higher and higher levels of organization. Not only the formation of organic molecules, but the formation of the first cell and higher organisms are driven by the biological principle (Grandpierre 2007, 2012a, 2013).

The Universe is not identical with the observable universe we can see with naked eyes and telescopes. Instead, the Universe is the unified whole of all observable phenomena, laws of Nature, first principles of Nature and biological autonomy (Grandpierre 2012a). Besides physics, we have also biology as a fundamental natural science (Grandpierre 2014a; Grandpierre et al. 2014) as well as psychology, by which we mean the general science of self-consciousness. In other words, the Universe is the unified whole of matter, life and mind, including the Cosmic Self (Grandpierre 2014a; Grandpierre et al. 2014). Definitely, the laws of Nature or the

Cosmic Self do not consist of atoms or elementary particles. They do not have physical properties like mass, extension, or color. Yet they are the most fundamental elements of the real Universe in which we live.

The discovery of the higher-level functions of cells offers an unexpected argument concluding that life cannot originate by physics and chance (Grandpierre 2013). Owing to advances in biology and astrobiology, as well as to the unfolding biological basis of quantum physics, the old view of an assumed 'abiotic genesis of life' will give way to a deeper, more complete, and genuine picture. In this new and more fulfilling picture, life has its own first principle that governs all life phenomena (Grandpierre et al. 2014). If life is a universal cosmic phenomenon, it is then not restricted to such 'islands of life' as the already familiar forms of life we know on Earth. As we learned in the example of solar activity, the biofriendly activity of the Universe transcends the conceptual framework of physics.

The discovery of stardust made up of organics was totally unexpected and still difficult to understand within the conceptual framework of physics. In spite of a clear lack of theoretical understanding relying on the usual mindset of physics, the observational facts are clear and definite. Sometimes these discoveries of astrobiology seem too fantastic to be true, and there has not been a lack of skeptics in the scientific community. But what we have learned is that we have to keep an open mind for unexpected discoveries and entertain new possibilities resulting from these new findings (Kwok 2013, xviii).

Summarizing the results obtained by astrobiology, we find that an increasing number of compelling facts present evidence that biofriendly laws act throughout the Universe. Let us draw the conclusion: there are laws of Nature favoring life. These laws prevail everywhere in the Cosmos, and they are not physical laws. We draw the apparently inevitable conclusion that the already recognized biofriendly laws of Nature are not physical but biological laws.

Our four-decades long studies underpin and extend the revolutionary oeuvre of Ervin Bauer (1967). The observed fact that organic molecules are formed everywhere in the Cosmos, even in extremely rare and cold cosmic clouds, can be regarded as experimental evidence for the theoretical findings of Ervin Bauer that a universal law of biology exists. We argue that Bauer's principle prevails everywhere in the Universe in the same way as physical laws prevail everywhere. The difference between the two is that biology is the control science of physics. Biology modifies the input conditions of physical laws in order to obtain the biologically required specific final result in the output of the system. Biology acts at a deeper level of the Universe than physics. Life regulates the input elements of physical laws in a manner suitable for biological purposes (Grandpierre et al. 2014). This means that the Universe is fundamentally biological and physics describes only the outer skin of its body.

The existence of the universal biological principle at the most fundamental level of the Universe explains the biofriendly nature of the Universe, including the anthropic coincidences and astrobiological findings. We have found that the quantum vacuum manifests more than the completely random behavior that is expected within the conceptual framework of physics. In the wider and deeper conceptual framework of a new, biofriendly scientific worldview containing more explanatory tools, teleology also becomes available as a respectable tool of science (Grandpierre 2012c). We are led to discover a non-random but organized, biological cosmic activity of the quantum vacuum.

We can term the here considered fundamental biological activity of the Cosmos in short as '*cosmic activity*'. Similarly to the Sun's having solar activity, the Cosmos has also its own activity: cosmic activity. In the widest sense, cosmic activity involves also the virtual particles produced continuously by the biological principle and biological autonomy, in addition to those generated by the least action principle of physics. In this paper, we would like to call attention to the until now overlooked biofriendly activity of the quantum vacuum. The accumulating evidence has convinced us that besides the least action principle of physics, an even more profound biological principle acts at the most fundamental level of the Universe. The biological principle and biological autonomy act through creating continuously virtual particles. The physical principle generates two kinds of virtual particle pairs. The first type of them realizes physical or inertial behavior; the second type has a random nature and averages out to zero. In contrast, the biological principle generates virtual particle pairs that realize biological behavior with the help of the active contribution of the living organism itself.

Our point here is that cosmic activity involves all the biological and, in a wider picture, also all the mental activity of the Cosmos. Life and mind are inseparable, because mind is a manifestation of biological autonomy. One can speculate that the universal access of humanity to the objective laws of mathematics, logic, music, and philosophy, as well as to inspirations, intuitions, including pre-conceptual thinking and feeling, arise from the presence of cosmic activity within us. Indeed, it is not only the physical principle of inertia that acts on our being but biological and psychological principles as well. We can open our mind to logical thinking and mobilize all the mental potentials we received from Nature for the sake of exploring the potential of logical thinking available for us. When we do that, we can experience the cosmic powers in us.

Our mind has a twofold task and responsibility. One is to supervise, control, and safeguard our bodily behavior. The other is to explore all the capacities of our Nature-given mind for the sake of the Universe, including all the cosmic life forms (Grandpierre 2008a). This second aspect of our mind can be characterized as cosmic mind. Our cosmic mind works on the basis of cosmic activity. Our cosmic mind listens to cosmic activity and transmits its message into this world, in which these letters you are now reading exist. We think the omnipresence of cosmic activity and its human accessibility is responsible for the otherwise unreasonable effectiveness of mathematics in the natural sciences, as well as for the objective character of subjectively accessed logic and for the cathartic power of some pieces of art (Wigner 1960).

Behind the physical level, in the depths of Nature and our inner world, we find a layer of reality bearing similar characteristics. Thus genius, characterized by "the large infusion of the subliminal in its mental output," provides means for discovery of this hidden environment (Myers; Kelly et al. 2007, 482). Our willpower may be in a certain degree based on a more general "cosmic will" (Strömberg 1948, 277). The nearer we advance to the natural and enlivening forces of creation residing within the depths of our inner worlds, the greater, brighter, more brilliant, and natural our life becomes.

## The Secret of Life

We have found that life has two fundamental and inseparable aspects: (i) biological autonomy and (ii) the biological principle. Genuine biological autonomy (i) is the ability to make self-induced decisions. In its highly developed, human stage it is known as free will. This self-induced decision-making is realized by the self. The self is the bridge between mind and matter, since decision-making is the way in which our mental events become manifested in the physical world. The other fundamental aspect of life (ii) is the biological principle, which is formulated by Ervin Bauer, and re-formulated as the greatest action principle by the present author (Grandpierre 2007). This biological principle is formulated in a mathematical form as an integral principle, in which the endpoints of the integral are variable and can be selected by the living organism itself. These endpoints are conditionally prescribed by Nature only in one respect, to be optimal for life. Living organisms usually live with their autonomy in harmony with this natural prescription. Nevertheless, their autonomy allows them to deviate from such natural behavior. Such unnatural deviations occur rarely in Nature. Unfortunately, they occur more frequently in the lives of modern, alienated people. But this does not change the fundamental fact that Nature prescribes that we live out our autonomy for the sake of life, to lift life as high above lethal physical equilibrium as possible. The real nature of life is to elevate all life to the heights of life. It belongs to the very basic nature of life to live and act for uplifting life as much as possible, extending and continuing the cosmic experience.

It belongs to the very basic nature of life to live and act for uplifting life as much as possible. Since the Universe is the unified whole of everything that exists, and since the Universe is living, cosmic life can, therefore, do everything. Nothing can hinder the actions of the Universe since everything that exists belongs to the Universe. This means that in the cosmic, ultimate context where the Universe exists, everything is possible. We can say that when life is omnipotent, this is magic. Since the Universe as a whole is living, life has a fundamentally magic nature. Life is extraordinary by its very nature, because its ability to surmount itself is unlimited.

Life is wider and deeper than the tiny parcel seen through the eyeglasses of the narrow mindset delimited by the science of inanimate matter. Life is the message of Nature's victory against all inertness, fixedness, and stubbornness. Life becomes present when it cleans our mind, in a cathartic attack, from all pity conflicts and narrow-minded concerns. Discovering the real nature of life has a higher significance than the discoveries of Copernicus, Newton and Einstein together—actually higher than that of all the already known discoveries of modern science and philosophy.

Life is more fundamental than physical matter. Matter is merely a superficial aspect of life. Matter is the thin skin of the vast cosmic giant: life. Life is sole actor, and matter is its tool and its trace left behind. Life is the third person aspect of the same thing that is experienced from the first person perspective as mind (Grandpierre2014a; Grandpierre et al. 2014), when mind is meant as the soul, that is, our emotional life, and consciousness, that is, our intellectual life, together. Consciousness is much more than consciousness of matter. The physical principle is an ideal tool in the hand of the biological principle since it secures the most economic realization of biological aims. In the absence of apparent biological aims, the issuing behavior is inertial. While biological autonomy can determine the ends of the biological principle at the most fundamental level of the Universe beyond the quantum vacuum, the physical principle acts only in the upper layer, within the quantum vacuum.

Fundamental reality is present in the form of biological autonomy and the biological principle. Together they form the most fundamental nature of the Universe: cosmic Life. The entire Universe must have a biological autonomy, since biological autonomy is inseparable from the biological principle that prevails throughout all the spaces of the Cosmos. The decision-making ability of the Cosmos can be regarded as the activity of the Cosmic Self. Throughout all the immense spaces of the entire Cosmos, from the coldest corners of rare cosmic clouds to the blood circulating in our hearts and inspiring our mind, cosmic activity is permanently 2018 Cosmic Plan of the Living Universe (Grandpierre 2018b). It belongs to the nature of life that our participation is inevitable, is shaping our own contributions to it.

One day all of us will realize that life is not a tiny spark somewhere in the depths of our body, but an immense and elementary power, the reality embracing the entire Universe.

## Works Cited

Bar-Nun, A., S. Bar-Nun, S. Bauer, and C. Sagan. 1970. Shock synthesis of amino acids in simulated primitive environments. *Science* 168: 470–473. Print.

Barrow, J.D., and F.J. Tipler. 1986. *The Anthropic Cosmological Principle*, 132. Oxford: Oxford University Press.

Barrow, J., S.C. Morris, S.J. Freeland, and C.L. Harper Jr., eds. 2007. *Fitness of the Cosmos for Life: Biochemistry and Fine-Tuning*. Cambridge: Cambridge University Press. Print.

Bauer, E. 1967. *Elméleti biológia* (Theoretical Biology). Budapest: Akadémiai Kiadó. Print (in Hungarian; translated by Dr. Miklós Müller from Russian "Theoreticheskaya Biologiya" and from German "Die Grundprinzipen der rein naturwissenschaftlichen Biologie"; with excerpts in English, main text in Russian, 1982; in Russian, 1935, 1993, 2002).

Baumeister, R.F. 2012. Self-Control—The Moral Muscle. *Psychologist* 25: 112–115. Print.

Bohr, Niels. 1933. Light and Life. *Nature* 131: 421–423, 457–459. Print.

Carpenter, K.G., C.J. Schrijver, M. Karovska, et al. 2005. *Stellar Imager Vision Mission Report*. Vision Mission Study Report. 15 September 2005. http://hires.gsfc.nasa.gov/si/documents/SI_Report_final_091505_ebook.pdf.

Chaplin, M. 2015. *Water Structure and Science. Anomalous Properties of Water*. http://www1.lsbu.ac.uk/water/water_anomalies.html

Chatzitheodoridis, E., S. Haigh, and I. Lyon. 2014. A Conspicuous Clay Ovoid in Nakhla: Evidence for Subsurface Hydrothermal Alteration on Mars with Implications for Astrobiology. *Astrobiology* 14: 651–693. Print.

Chyba, C.F., and K.P. Hand. 2005. ASTROBIOLOGY: The Study of the Living Universe. *Annual Review of Astronomy and Astrophysics* 43: 31–74. Print.
Conway Morris, S. 2010. What is Written into Creation? In *Creation and the God of Abraham*, ed. David Burrell et al., 176–191. Cambridge: Cambridge University Press. Print.
Davies, Paul C.W. 1998. *The Fifth Miracle. The Search for the Origin of Life*. London: Penguin. Print.
———. 2003. How Bio-friendly is the Universe? *International Journal of Astrobiology* 2, 115–120. http://arxiv.org/pdf/astro-ph/0403050.pdf.
———. 2006. *The Goldilocks Enigma: Why Is the Universe Just Right for Life?* New York: Allen Lane, Penguin Books. Print.
de Duve, C. 1996. *Vital Dust: Life as a Cosmic Imperative*. New York: Basic Books. Print.
De Vera, J.-P., P.D. Schulze-Makuch, A. Khan, A. Lorek, A. Koncz, D. Möhlmann, and T. Spohn. 2012. The adaptation potential of extremophiles to Martian surface conditions and its implication for the habitability of Mars. *Geophysical Research Abstracts* 14: EGU2012–EGU2113.
Dick, S.J. 2009. Cosmic evolution: History, culture, and human Destiny. In *Cosmos & Culture: Cultural Evolution in a Cosmic Context*, ed. S.J. Dick and M.L. Lupisella, 25–59. Washington, DC: National Aeronautics and Space Administration. Print.
———. 2010. *Cosmic Evolution. History, Culture, and Human Destiny. Chapter 2 in: Cosmos and Culture: Cultural Evolution in a Cosmic Context*. Washington, DC: NASA. Print.
———. 2012. Critical Issues in the History, Philosophy, and Sociology of Astrobiology. *Astrobiology* 12: 906–927. Print.
———. 2013. Testimony of Dr. Steven J. Dick, Baruch S. Blumberg NASA/Library of Congress Chair in Astrobiology, before the Committee on Science, Space, and Technology, U.S. House of Representatives, December 4, 2013. http://docs.house.gov/meetings/SY/SY00/20131204/101546/HHRG-113-SY00-Wstate-DickS-20131204.pdf.
Dick, S.J., and J.E. Strick. 2004. *The Living Universe. NASA and the Development of Astrobiology*. New Brunswick: Rutgers University Press. Print.
Ellis, G.F.R. 2005. Physics, Complexity and Causality. *Nature* 435: 743. Print.
———. 2006. Physics and the Real World. *Foundations of Physics* 36: 227–262. Print.
———. 2011. Editorial note to: Brandon Carter, Large Number Coincidences and the Anthropic Principle in Cosmology. *General Relativity and Gravitation* 43: 3213–3223. Print.
Fox, S. 1988. Prebiotic Toots of Informed Protein Synthesis. In *The Roots of Modern Biochemistry*, ed. H. Kleinkauf, H. von Dohren, and L. Jaenicke, 897. Berlin: de Gruyter. Print.
Fox, S., and K. Dose, eds. 1997. *Molecular Evolution and the Origin of Life*. New York: Marcel Dekker. Print.
Gold, T. 1992. The Deep, Hot Biosphere. *Proceedings of the National Academy of Sciences* 89: 6045–6049. Print.
Grandpierre, Attila. 1986. A Convective Flare Theory. In *Flare Stars and Related Objects*, 176–184. ed. Byurakan Symposion. L. V. Mirzoyan. Print.
———. 1988. Flares: Magnetic or Convective Origin? In *Activity in Cool Star Envelopes*, ed. O. Havnes et al., 159. Dordrecht: Springer. Print.
———. 1991. A Convective Model of Solar and Stellar Flares. *Memorie della Societa Astronomica Italiana* 61: 401–407. Print.
———. 1996a. A Pulsating-Ejecting Solar Core Model and the Solar Neutrino Problem. *Astronomy and Astrophysics* 308: 199–214. Print.
———. 1996b. On the Origin of Solar Cycle Periodicity. *Astrophysics and Space Science* 243: 393–400. Print.
———. 2002. *The Book of the Living Universe* (in Hungarian). Válasz Könyvkiadó.
———. 2007. Biological Extension of the Action Principle: Endpoint Determination Beyond the Quantum Level and the Ultimate Physical Roots of Consciousness. *Neuroquantology* 5 (4): 346–362.
———. 2008a. Cosmic Life Forms. In *From Fossils to Astrobiology. Records of Life on Earth and the Search for Extraterrestrial Biosignatures*, ed. Joseph Seckbach and Maud Walsh, 369–385. New York: Springer.

———. 2008b. Fundamental complexity measures of life. In *Divine Action and Natural Selection: Questions of Science and Faith in Biological Evolution*, ed. J. Seckbach and R. Gordon, 566 615. Singapore: World Scientific.
———. 2010. *Dynamism in the solar core*. In Proceedings of the third UN/ESA/NASA workshop on the International Heliophysical Year 2007 and Basic Space Science, National Astronomical Observatory of Japan, Tokyo. Astrophysics and Space Science Proceedings, ed. Hans J. Haubold and A.M. Mathai, 103–139. Springer. Print.
———. 2012a. *The Book of the Living Universe 2012* (a thoroughly new version, in Hungarian). Titokfejtő Könyvkiadó. Print.
———. 2012b. *Genuine Biological Autonomy: How can the Spooky Finger of Mind Play on the Physical Keyboard of the Brain?* Athens: ATINER'S Conference Paper Series no: PHI2012-0197. Print. http://www.atiner.gr/papers/PHI2012-0197.pdf.
———. 2012c. On the Biological Origin of Design in Nature. In *Origin(s) of Design in Nature: A Fresh, Interdisciplinary Look at How Design Emerges in Complex Systems, Especially Life*, ed. L.S. Swan, R. Gordon, and J. Seckbach, 17–41. Dordrecht: Springer. Print.
———. 2013. The Origin of Cellular Life and Biosemiotics. *Biosemiotics* 6: 421–435. Print. http://link.springer.com/article/10.1007%2Fs12304-013-9173-9.
———. 2014a. Biologically Organized Quantum Vacuum and the Cosmic Origin of Cellular Life. *Analecta Husserliana* 116: 107–133.
———. 2015. On the Origin of Solar Activity: The Interplay of Rotation, Magnetic Braking, Planetary Influences and Local Thermonuclear Runaways in the Solar Core. In *Planetary Influence on the Sun and the Earth*, ed. Nils-Axel Mörner, 91–108. New York: Nova Science Publishers. Print.
———. 2018a, The Helios Theory – The Sun as a Self-Regulating System and as a Cosmic Living Organism. *Process Studies*, 46(2) (in print).
———. 2018b. *The Living Universe Through the Eyes of a Solar Astronomer*. (book manuscript, submitted).
Grandpierre, A., and M. Kafatos. 2012. Biological Autonomy. *Philosophy Study* 2: 631–649. Print.
———. 2013. Genuine Biological Autonomy: How can the Spooky Finger of Mind Play on the Physical Keyboard of the Brain? In *An Anthology of Philosophical Studies*, ed. P. Hanna, vol. 7, 83–98. Athens: Athens Institute for Education and Research. Print.
Grandpierre, A., D. Chopra, and M. Kafatos. 2014. The Universal Principle of Biology: Determinism, Quantum Physics and Spontaneity. *NeuroQuantology* 12(3): 364–373. Print. http://www.neuroquantology.com/index.php/journal/article/view/747.
Gurevich, L., and V. Mostepanenko. 1971. On the Existence of Atoms in n-Dimensional Space. *Physics Letters A* 35: 201. Print.
Harold, F.M. 2001. *The Way of the Cell: Molecules, Organisms and the Order of Life*. Oxford: Oxford University Press. Print.
Heldmann, J.L., L. Schurmeier, C. McKay, A. Davila, C. Stoker, M. Marinova, and M.B. Wilhelm. 2014. Midlatitude Ice-Rich Ground on Mars as a Target in the Search for Evidence of Life and for In Situ Resource Utilization on Human Missions. *Astrobiology* 14: 102–118. Print.
Henderson, L.J. 1913. *The Fitness of the Environment*, 312. New York: Macmillan. Print.
Hoover, R. 2014. *Need to Track Organic Nano-Particles Across the Universe? NASA's Got an App for That*. https://www.nasa.gov/ames/need-to-track-organic-nano-particles-across-the-universe-nasas-got-an-app-for-that/NASA. Retrieved 22 February 2014.
Hoyle, F., and N.C. Wickramasinghe. 1999a. The Universe and Life: Deductions from the Weak Anthropic Principle. *Astrophysics and Space Science* 268: 89–102. Print.
———. 1999b. Biological Evolution. *Astrophysics and Space Science* 268: 55–75. Print.
Huang, C., K.T. Wikfeldt, T. Tokushima, D. Nordlund, Y. Harada, U. Bergmann, M. Niebuhr, T.M. Weiss, Y. Horikawa, M. Leetmaa, M.P. Ljungberg, O. Takahashi, A. Lenz, L. Ojamae, A.P. Lyubartsev, S. Shin, L.G.M. Pettersson, and A. Nilsson. 2009. The Inhomogeneous Structure of Water at Ambient Conditions. *Proceedings of the National Academy of Sciences of the United States of America* 106: 15241. Print.

Iess, L., R.A. Jacobson, M. Ducci, D.J. Stevenson, J.I. Lunine, J.W. Armstrong, S.W. Asmar, P. Racioppa, N.J. Rappaport, and P. Tortora. 2012. The Tides of Titan. *Science* 337 (6093): 457–459. Print.

Kaiser, R.I., A.M. Stockton, Y.S. Kim, E.C. Jensen, and R.A. Mathies. 2013. On the Formation of Dipeptides in Interstellar Model Ices. *The Astrophysical Journal* 765: 111. Print.

Kelly, E.F., E.W. Kelly, A. Crabtree, A. Gauld, M. Grosso, and B. Greyson, et al. 2007. *Irreducible Mind. Toward a Psychology of the 21st Century*, 482ff. New York: Rowman and Littlefield Publishers, Inc. Print.

Kenyon, D.H. 1974. Prefigured Ordering and Protoselection in the Origin of Life. In *The Origin of Life and Evolutionary Biochemistry*, ed. K. Dose, S.W. Fox, G.A. Deborin, and T.E. Pavlovskaya. New York: Springer. Print.

Kim, W., F.K. Tengra, Z. Young, J. Shong, N. Marchand, H.K. Chan, R.C. Pangule, M. Parra, J.S. Dordick, J.L. Plawsky, and C.H. Collins. 2013. Spaceflight Promotes Biofilm Formation by Pseudomonas aeruginosa. *PLoS ONE* 8 (4): e6237. Print.

Kuan, Y.-J., S.B. Charnley, H.-C. Huang, W.-L. Tseng, and Z. Kisiel. 2003. Interstellar Glycine. *The Astrophysical Journal* 593: 848–867. Print.

Kwok, Sun. 2009. Organic Matter in Space: From Star Dust to the Solar System. *Astrophysics and Space Science* 319: 5–21. Print.

———. 2011. *Organic Matter in the Universe*. Weinheim: WILEY-VCH. Print.

———. 2013. *Stardust the Cosmic Seeds of Life*. Berlin/Heidelberg: Springer. Print.

Kwok, S., and Y. Zhang. 2011. Mixed Aromatic-Aliphatic Organic Nanoparticles as Carriers of Unidentified Infrared Emission Features. *Nature* 479: 80–83. Print.

Miller, S.L. 1955. Production of Some Organic Compounds Under Possible Primitive Earth Conditions. *Journal of the American Chemical Society* 77: 2351–2361.

Moya, C.J. 1990. *The Philosophy of Action: An Introduction*. Cambridge: Polity Press. Print.

Nandy, D., and P.C.H. Martens. 2007. Space Climate and the Solar-Stellar Connection: What Can We Learn from the Stars about Long-Term Solar Variability. *Advances in Space Research* 40: 891–898. Print.

Neal-Jones, N., and W. Steigerwald. 2011. *Building Blocks of Life Created in "Impossible" Place*. https://solarsystem.nasa.gov/scitech/display.cfm?ST_ID=2350. Retrieved 7 October 2014.

Needham, A.E. 1965. *The Uniqueness of Biological Materials*. Oxford: Pergamon Press. Print.

Nordita Conference. 2014. *Water—the Most Anomalous Liquid*. Nordita, Stockholm, Sweden, from 13 October 2014 to 07 November 2014. http://agenda.albanova.se/conferenceDisplay.py?confId=3989

Oro, J. 1963. Studies in Experimental Organic Cosmochemistry. *Annals of the New York Academy of Sciences* 108: 464. Print.

Petigura, E.A., A.W. Howard, and G.W. Marcy. 2013. Prevalence of Earth-Size Planets Orbiting Sun-Like Stars. *Proceedings of the National Academy of Sciences of the United States of America* 110: 19273–19278. Print.

Polanyi, Michael. 1968. Life's Irreducible Structure. *Science* 160: 1308–1312. Print.

Sanders, R. 2013. Evidence that Comets Could Have Seeded Life on Earth. UC Berkely News Center. http://newscenter.berkeley.edu/2013/03/05/research-news-briefs-did-comets-seed-life-on-earth/

Schmitt-Kopplin, P., Z. Gabelica, R.D. Gougeonc, A. Fekete, B. Kanawati, M. Harir, I. Gebefuegi, G. Eckeld, and N. Hertkorn. 2010. High Molecular Diversity of Extraterrestrial Organic Matter in Murchison Meteorite Revealed 40 Years after its Fall. *Proceedings of the National Academy of Sciences* 107: 2763. Print.

Shapiro, R. 1986. *Origins: A Skeptic's Guide to the Creation of Life on Earth*, 186–187. New York: Summit Books. Print.

Sharov, A.A. 2006. Genome Increase as a Clock for the Origin and Evolution of Life. *Biology Direct* 1: 1–17. Print.

Sharov, A.A., and R. Gordon. 2013. *Life Before Earth*. http://arxiv.org/abs/1304.3381.

Shimoyama, A., and R. Ogasawara. 2002. Dipeptides and Diketopiperazines in the Yamato-791198 and Murchison Carbonaceous Chondrites. *Origins of Life and Evolution of the Biosphere* 32: 165–179. Print.

Shore, S.N. 1992. *An Introduction to Astrophysical Hydrodynamics*, 178. San Diego: Academic. Print.

Steinman, G., and M. Cole. 1967. Synthesis of Biologically Pertinent Peptides under Possible Primordial Conditions. *Proceedings of the National Academy of Science* 58: 735–742. Print.

Strömberg, G. 1948. *The Soul of the Universe*. Philadelphia: David McKay. Print.

Svensmark, H. 2006. Cosmic Rays and the Biosphere over 4 Billion Years. *Astronomische Nachrichten* 327 (9): 871–875. Print.

Szent-Györgyi, A. 1972. *The Living State. With Observations on Cancer*. New York: Academic. Print.

The University of Hong Kong. 2011. Astronomers Discover Complex Organic Matter Exists Throughout the Universe. *ScienceDaily*. 27 October 2011. www.sciencedaily.com/releases/2011/10/111026143721.htm

Tuttle, K. 2009. *SLAC Researchers Reveal the Dance of Water*. http://home.slac.stanford.edu/pressreleases/2009/20090811.htm

Wald, G. 1962. Life in the Second and Third Periods; or Why Phosphorus and Sulfur for High-Energy Bonds? In *Horizons in Biochemistry*, ed. M. Kasha and B. Pullman, 127. New York: Academic. Print.

Wigner, E.P. 1960. The Unreasonable Effectiveness of Mathematics in the Natural Sciences. *Communications in Pure and Applied Mathematics* 13: 1–14. Print.

Zyga, L. 2012. Physicists Search for New Physics in Primordial Quantum Fluctuations. Mar 26, 2012. http://phys.org/news/2012-03-physicists-physics-primordial-quantum-fluctuations.html. Retrieved 17 June 2015.

# Cosmic Harmony, the Emergence of Life and of Human Consciousness

**Mamuka Dolidze**

*To the memory of my mother, composer Mery Davitashvili*
*and*
*To the memory of the famous philosopher of our time*
*Anna-Teresa Tymieniecka*

**Abstract** The order and yet contingency found in the universe present a puzzle. The work presented here is devoted to the problem of cosmic harmony. It is through modern phenomenology that we can hope to capture the elusive source of harmony, the coincidence of sense and existence. A process of sense formation saturates the whole universe. A quantum-phenomenological explanation of the extension of the galaxies without reference to the Big Bang theory is to be posited. The miracles of life and of our vital-spiritual harmony are key to apprehending something of this explanation made from another perspective. Nietzsche's philosophy is to be appreciated for its rejection of determinism and celebration of the freedom inherent in life's unfolding. Tymieniecka's phenomenology of life links the advancing self-individualization of life to the Logos that informs this creative development. She saw in the phenomenon of achievement, scientific, artistic, etc. a sign that agency is involved in the will driving the ontopoiesis of life. But there is a playfulness in the flow of life's stream that resists objectification. The telos of all is not ours to impose, and a sense of that opens us to the higher realm.

**Keywords** Cosmos · Chaos · Big Bang · Microcosm · Macrocosm · Wave-particle duality · Matter-spirit duality · Intentionality · Nietzsche · Tymieniecka

---

M. Dolidze (✉)
Tbilisi Javakhishvili State University, Tbilisi, Georgia

 
W. S. Smith et al. (eds.), *Eco-Phenomenology: Life, Human Life, Post-Human Life in the Harmony of the Cosmos*, Analecta Husserliana CXXI,
https://doi.org/10.1007/978-3-319-77516-6_11

## Necessity and Freedom in the Cosmos

Harmony in the cosmos is a rather puzzling phenomenon. The order of the cosmos is rooted in the necessity of being, but, in contrast, the creative development of the universe gives chances to break this celestial order in favor of the contingency of the unexpected events that arise within the process of being's becoming. The everlasting river of life, embracing the whole universe, creates the new opportunity of freedom, which would be in alliance with the probable necessity of space phenomena. From the position of the phenomenology of life, the emergence of human consciousness unveils the potential of the cosmos to reflect itself, but at the same time it reveals a kind of uncertainty that makes for a perspective for freedom beyond the vigorous order of the world. The work presented here is devoted to the problem of cosmic harmony from this position of modern phenomenology.

Human consciousness, in the broad understanding of the phenomenology of life, contains the immense domain of subjective phenomena. Subjective uniqueness is not liable to the act of objectification of events that undergirds the scientific picture of the world. Philosophy of knowledge is unable to encompass this boundless kingdom of creative life, and the great task of phenomenology seems to be to observe and describe the elusive stream of subjectivity, which, like Heraclitus' river, is impossible for us to enter twice.

We suppose that all the mental things, what makes the specific essence of human being, are rooted in the forces of cosmos, which do not refer solely to the physical reality of the mega-sphere. Such establishment of conscious events, in the infinite perspective of the universe, leads us to a dualistic vision positing an higher realm; the physical reality of the cosmos coexists with the transcendental sphere of ideas and supernatural senses, which participate in the process of the becoming of being and saturate the whole universe. This is a duality for the sake of the unity of the world, since all the elusive phenomena of the inner working of consciousness are rooted in the metaphysical mystery of heaven. The subjective self of a human being has a hidden connection with vital essence of the universe, which provides the river of life with logical sense.

## The Coincidence of Sense and Existence

The cosmos seems to be the lucky place of meeting of sense and the existence, and their harmony is a certain miracle since there is no direct contact between spirit and matter, and the probability of accidental coincidence of these mutually exclusive phenomena is infinitely small.

Coming from the metaphysical mystery of celestial order, the infinite spirit is full of the mental phenomena that propel the self to a unique synthesis with matter to organize the extraordinary phenomenon that is life. One cannot treat inorganic matter as the riverside of this cyclic stream of vitality, since the latter involves the whole world in the all-absorbing vortex of life. Life has an individual face set against the regularity of objective events. Yet the waves of life strive for infinite extension and make the order based on the regularity of events. Life-uniqueness begets the order of matter! This is hardly probable, unless we assume that this miracle refers to the vital nature of the whole universe.

The cosmos is not a closed system of celestial bodies. It is open toward the primordial chaos. The passage between these two mutually exclusive states of the universe needs some transitional sphere where the being of the cosmos coexists with the not-being of chaos. This is an area of uncertainty, the dynamic area of the becoming of being from nothingness. This area does not exist (it does not belong to the being of the cosmos), yet we cannot say that it is an area of non-existence (since it does not belong to the not-being of chaos). This phenomenological area signifies itself beyond the claim of being or not being. This is the transcendental sphere of sense-forming acts where the hidden intentionality of chaos gives birth to the order of cosmos.

The emergence of organic life within the cosmos is not accidental. This miracle presents the realization of the "openness" of the cosmos to the hidden intentionality of chaos. For the believer, here the word of God penetrates the darkness to beget light.

I am always confused when speaking about the immense distances and the huge number of cosmic years when distances are measured in billions of light years, something we can theoretically calculate yet cannot imagine (Guth 1997).[1] Maybe these are the mathematical symbols of infinity and eternity – the signs of the metaphysical "thing in itself" that dwells beyond space and time? Yes, of course. I am very small creature and cannot match myself against the tremendous scale of the cosmos, but I am the human being who has the consciousness that presents the phenomenon of cosmos, which is ordered as a self-reflective system of stars. Therefore, the cosmic order has a geometrical structure, as if it followed the logical structure of my mathematical thought. Consciousness turns myself into a significant essence who is worth treating as a subject exploring the universe. So long as I am the subject encompassing the tremendous object that is the cosmos, I am not a small creature and my personality is able to embrace the higher realm that would be organized as a vital system. This is the sphere of cosmic order being in harmony with human consciousness, where I can use the logic of my language since it is the language of celestial order causing the emergence of my consciousness. All the concepts of my scientific vision, like space, time, being, becoming, essence, existence, causality, and so on, belong to the geometrical sphere of this celestial order.

## The Dualities in Microcosm and Macrocosm

Hence, when I speak about the immense distances of the universe in terms of billions and billions of light years – 90 billion light years being its span – I should set a question: can I extend my logical language and scientific concepts in this huge number of cosmic years? Maybe there is a limit to the application of scientific language and, beyond this limit, space and time are senseless concepts, causality does not work, and it is impossible to distinguish being from not-being.

In the 1930s, Georges Lemaître and George Gamow advanced the theory of "Big Bang" (Hawking 1998, 42–50, 70–74). According to this theory, 13.8 billion years ago there was an explosion of a super-dense mass (a so-called "Black Hole") and this event merits being considered as the starting point of the universe. But, to confirm this idea they had to extend the concepts of space and time and the principle of causality to this enormous distance. Yet, this extension was rather dubious, not only because of the huge scale of the distance but for the reason of the dualistic structure of the mega-sphere.

We have investigated the wave-particle duality in the micro world and eventually come to a point where "orthodox" interpretation of quantum reality is split into two mutually exclusive views: the wave and the particle pictures (Dolidze 2002). The particle picture implies a spatio-temporal description of atomic events, whereas the wave picture presents a causal description of the same phenomena. The dualistic structure of the microsphere, like the dualism of the mega-sphere, has suggested the attendance of a hidden, transcendental level of sense-forming acts, which turn quantum measurement into a cognitive situation where the device plays the role of consciousness. The quantum wave-particle duality limits the use of classical physical science language in describing the microcosm; if the causal description of quantum waves were used for micro events, it would exclude the concepts of space and time and it would be impossible to localize the quantum particle at some spatio-temporal point.

The extension of the wave-particle duality of the microsphere into the mega-sphere comes to the following: the theory of "Big Bang," considered as the eruption of a super-dense mass, posits in a causal description of the world that this eruption was the first cause of the universe. In the framework of this description, the concepts of space and time would have no sense and it would be impossible to localize this starting explosion in some spatio-temporal point. On the other hand, localization of this cosmic explosion in space and time excludes its consideration as a primordial reason for the universe. Because of this quantum-dualistic structure, causal description and the spatio-temporal outlook are mutually exclusive pictures of reality, and the theory of "Big Bang" refers either to the starting point of being, beyond space and time (which would accord with the Biblical explanation of the beginning of the world), or to a local physical bang without claim to being the beginning of the world.

## Everyday Adequation and Its Limits

In our everyday life we use the principle of causality and the concepts of space and time together so that we can work out one scientific language for describing the world; but, this is a world in which we live that is commensurate to our human existence. I perceive external events in space and time against the background of the inner sense of my subjective self. The internal experience of my subjective "ego" establishes the external vision of my world and provides the spatio-temporal phenomena with the principle of causality. I bear the idea of causality in my consciousness, and my body-mind wholeness helps grant me insight to the objective reality internal to my experience of my subjective reason. This experience makes the adequacy of my language and leads to the integrity of my world, where the phenomena of space and time coexist with the principle of causality.

The situation is changed if we refer to the microcosm of quantum objects or to the mega- sphere of celestial bodies. The quantum reality is so small and the cosmic sphere is so big, compared to human existence, that we cannot directly perceive them and cannot extend the wholeness of our subjective experience into these infinitely small or faraway distances. Human consciousness is then replaced by the measurement tools; they make physical sense of quantum particles or provide cosmic observation with a sense of physical reality. But, unlike vital consciousness, the measurement device lacks any internal vision of events. It reflects objects externally and, hence, cannot bring together the principle of causality and the concepts of space and time. Therefore, beyond the human world – in the quantum realm and in the cosmos – there is a split of scientific language. This split derives from the wave-particle duality. This dualism leads to a gap between the causal and the spatio-temporal descriptions of events. The theory of the "Big Bang," which did not take into account this gap, seems to be an erroneous conception.

I am a self, bearing the sense of being in my mind. I am a human being experiencing external reality on the basis of my internal reason. If I extended my subjective self beyond my human world into the micro or mega areas, I would replace my consciousness with a measurement tool. The latter, like a consciousness, has two functions: to work out the physical sense of a object, and to fix and reflect this object in space and time. These two functions refer to mutually exclusive experimental acts. Because of the disparity of spirit and matter, the act of sense-formation does not correspond to the act of the physical localization of an object. Therefore, both the micro world and the cosmos are split into two different parts: into the existential area of objective being and the transcendental sphere of the subjective forces making the sense of being. In quantum reality, this split leads to the wave-particle duality of atomic events. The atomic object exists as a particle with its localization in space and time, yet, it would gain physical sense if it was spread as a wave with definite form of causality and indefinite localization in space.

## The Miracle of Human Integration

This is a significant point of our conception, and we would like to analyze it once more: I am the phenomenon of life with wholeness of body and mind. Yet, it is a miraculous wholeness, and life is miracle since my physical being and my spiritual essence are incompatible phenomena, simultaneously being together in the integrity of life. Therefore, when I perceive an external object, I am aware that it is me who perceives this object. In the normal state of consciousness, the experience of self always accompanies the perception of a thing but, because of the disparity between my internal experience and external vision, I exist in a twofold way: either I perceive the external object but cannot objectify the self that is mine, or I focus on self-cognition but cannot take into consideration external being (which latter leads to the phenomenological way of thinking). So, I bear this internal-external duality in the miraculous unity of my life.

The situation would be changed if I went beyond the human world and replaced my consciousness with a measurement tool. Unlike vital consciousness, a measurement device lacks an internal vision of an object. Hence, it should turn the internal-external dualism of the human mind into a dualism in external phenomena. Therefore, the wave-particle duality of quantum events unveils the measurement tool as a consciousness and as a subject. And yet, this quantum subject should, anyhow, unite these incompatible wave-particle states in the integrity of life, and this unity exemplifies Bohr's principle of complementarity. So, we can turn this argument around and say that if mutually exclusive physical events at the same time complement each other they must then refer to a cognitive situation in which the measurement device plays the role of consciousness.

The same argument is true for the cosmos. When I perceive in space a "particle" picture of celestial bodies, I cannot take into account the wave-causal description of these same things, which indicates a process of sense-formation saturating the whole universe. This process is hidden to me. Therefore, if one observed the cosmos as a space, one could not find the process of the development of sense, but, if one considered the history of universe, one could see sense-formation acting throughout time. This way of conceiving things seems to present a quantum-phenomenological interpretation of Hegel's system of the historical development of the absolute idea. We would also refer to the results of our investigation of the phenomenology of consciousness, where the sense of being arises in the interaction between mutually exclusive phenomena (Dolidze 2013, 48–58).

Eventually, our judgment, on the basis of an analogy made with the wave-particle duality of quantum events, arrives at an explanation of the extension of the galaxies without the theory of the "Big Bang:" the system of quantum particles should be extended as a wave to keep the physical sense of the particles' existence. That is the result of our investigation of the phenomenology of consciousness, where the sense of being arises in the interaction between mutually exclusive phenomena (Dolidze 2002, 608–616). So, we can consider paired phenomena like wave and particle, continuity and discreteness, spirit and matter, essence and existence, becoming and being, regularity and uniqueness, general and individual, subject and object, and the like, to be opposite poles of a cognitive situation that would be responsible for generating the sense of being. Therefore, the objective-physical state of an atomic object refers to the subject as a measurement tool and needs the wave behavior of micro particles as an extension of the quantum system to provide the atomic object with physical sense.

In the same way, if we spread the wave-particle quantum duality into the cosmos, then we will come to the point where the system of mega particles (the system of stars) needs extension as a wave to keep physical sense through the subjective forces of the cosmos. That seems to be the quantum-phenomenological explanation of the extension of the galaxies without reference to the theory of the "Big Bang." We have devoted a special investigation to this problem (Dolidze 2014, 67–87). Now, we would like to emphasize that the prediction of cosmic events abounds with the danger of losing the existential sense of this vision.

## The Miracle of Time

With respect to this problem, let us come back to the miracle of time. Yes, indeed, time is the miracle which involves us in the everlasting stream of life encompassing all the universe. The order of the cosmos in alliance with the disorder of chaos make the same strange exhibition of the temporal process referring to the succession of external events coupled with internal perception of the same phenomena. The border between the internal and external manifestations of time is rather obscure, vague, uncertain such that one cannot separate the duration of an event from perception of the same phenomenon. The currency of time and its internal experience present an indivisible wholeness, and if the internal-subjective dimension of time would be abolished, the temporal succession of external events would thereby lose its existential sense (let us remind the reader of our apprehension that the sense of being rises out of mutually opposed phenomena). Therefore, when speaking about the objective-physical duration of time, we should always take into account the subjective experience of temporal phenomenon. When a physicist predicts that after five million light years the sun will die out, he also should take into account the existence of a human consciousness that can perceive this predictable cosmic event. But, it is impossible! One cannot foresee the perspective of human consciousness, not because of the great span of time but because of the freedom of life. Human freedom presents an obstacle to calculating the future of mankind on the basis of cosmic experiment, which seems to be enough for the prediction of faraway physical events.

Even if we shared the theological position and referred to the consciousness of absolute mind, we could not go beyond the Biblical interpretation of world's genesis. The hypotheses and predictions of modern cosmology keep its existential sense (deriving from the senses of time and space) in the context of emergence of life and against the background of human consciousness, which is open toward the eternity of absolute mind. Therefore, the Biblical interpretation of the world's genesis and modern cosmology appear to be mutually complementary systems of knowledge grasping the sense of being in the life of the universe.

## Nietzsche's Religion of Life

Friedrich Nietzsche foresaw the future of mankind through artistic metaphors. He was aware that it was incredible to calculate the development of the human mind on the basis of philosophical-scientific thinking. His ideas seemed riddled with absurdity, but this absurdity turned into reality very soon. He rejected God, but he was not an atheist. Using a phenomenological conception of the emergence of the sense of being (we have mentioned above), we dare assert that the rejection of God by Nietzsche played the role of a counterpoint that provides the Christian-theological position with existential sense.

Nietzsche was the son of a Protestant minister and, while criticizing Christianity, he took into account, first of all, the Protestant Church. Reformed Christianity ultimately denied the freedom of human will and upheld the domination of destiny over human life. (Calvinism). This seemed to be a Reformation against Christianity. Choosing the cross by His own will, Christ defeated death and got the believer free of the slavery of sin. Human freedom appeared to be the essential value of Christianity, and Nietzsche, the great philosopher of life, rebelled against the slavery to destiny premised by the Protestant Church. He dared set off, against a dogmatic religious system, the values of omnipotent life: vengeance in contrast with forgiveness, striving for power in contrast with modesty and humility, superman in contrast with God. Yet, these oppositions appeared to have figural meaning. Our phenomenological idea of the sense of being makes it clear that Nietzsche, thanks to this opposition, established the existential sense of the religion of life through the counterpoint of the values of the philosophy of life. He emphasized the subjectivity of the religion of life, of Christianity, focusing on the freedom the human being has as a gift of Christ.

Thus, we would like to consider Nietzsche's philosophy of life to be in harmony with the religion of life that is orthodox Christianity, a dogmatic system in which (in contrast to Protestant theology) the saving acts of God coexist with human freedom.

Religion of life and philosophy of life both have one and the same cosmological meaning. A philosophical-poetical vision of mankind's future helped Nietzsche to predict the prospect of time in the light of human freedom, against the determinism of the scientific vision. Accordingly, Christ, in explaining the Bible, was in perpetual discussion with scribes and scholars. The freedom of life goes beyond the physical world, which was saturated by sin. The future of the world is not an object of science; it is either a reflection of a metaphor out of the poetical vision of a writer-philosopher or the goal of the mystical vision of a believer, since the universe not only presents objective-physical reality but also encompasses the sphere of subjective forces that are responsible for engendering the cosmological sense of being.

## Tymieniecka's Phenomenology of Life

A quantum-phenomenological approach to the cosmos takes into account subjective freedom beyond the causality of physical reality. The phenomenon of quantum probability destroyed the classical scientific picture of the world. The "Phenomenology of Life" of Anna-Teresa Tymieniecka provides interpretation in line with this new situation in its detailed description of the hidden working of life and vision encompassing all the universe. Professor Tymieniecka considered human life not separately, as in the traditional conceptual framework, but treated vital human consciousness in the whole context of the emergence of life within the process of the individualization of being (Tymieniecka 2009, 35–36). One might be amazed by the wide scale of her description of hidden acts of vital forces deriving from the human creative condition. She revealed the conflict between the vital necessities and the inventive powers of human mind but, at the same time, the all-embracing glance of the philosopher unveiled the phenomenon of fittedness, which makes an invisible bridge between an elusive agent of life and the conditions of becoming for a being. The whole of the process of the individualization of being is rooted in the Logos of life, which is treated as a logos of creative development.

Tymieniecka evaluated this complicated process as a sequence of internal conflicts and integrations arriving at the individualization of life – that is, the play of imaginative freedom with vital necessities. She drew a sweeping picture of these forces at play along the whole chain of the creative unfolding of the Logos of life, coming to the seemingly infinite expansion of human potential. All the steps of this chain – organic matter, animal life, the functions of the brain, the emergence of free will, the individual self, human consciousness, social life, inter-subjectivity, and striving for the infinite telos – all these manifestations of life are not derivable from each other. They have arisen through the freedom of play on the theatrical stage of the universe.

We are certainly in great sorrow that this famous philosopher of our time cannot develop further her ideas about the almighty of logos of life, yet, we should find new, hidden points in her system in the whole context of existential phenomenology. It becomes clear that the irreducible nature of the forms of life alluded to in speaking of "life in itself," which is independent of matter and which, thanks to just this independence, reveals a capacity for fitness is the context in which to treat the creative feature of imitation in the world. One way to explain and understand this inborn imitation is to admit that it is rooted in the openness of the cosmos to the Logos of life. If the vital word of God was responsible for summoning the world into being, that would mean that the phenomenon of freedom accorded matter with a genius for self-imitation and, thus, self-interpretation. These were the gifts of Logos of life, which hovered in the darkness at the world's creation.

Professor Tymieniecka extended creative imagination beyond human consciousness into the infinite realm of the cosmos. The confrontation of a special agent of life with vital necessities appears to drive the will of her ontopoiesis of life. She focused on the phenomenon of achievement as an existential basis of endless development of human consciousness. Achievement indicates not only the scientific-technical and technological progress of consciousness but, also, means the works of art that unfold the creative mind beyond itself.

## Partnering with the Playfulness of Life's Stream

I feel the importance of this vision in my writer's experience. When writing fiction, I have a sensation of involvement in the stream of life that appears to be independent of my own will and which, like Heraclitus' river, it is impossible to enter twice. My imagination seems to go beyond me so that I cannot control the creative process carrying me away. But, this is not a completely uncontrollable and unconscious act. Entering the poetical river of fiction, I am aware that it is I who celebrates the birth of new, artistic forms of life; I try to keep a balance between awareness of myself and oblivion in the stream of creativity. This balance helps me to keep subject-object integrity within the phenomenological act of becoming a being.

Tymieniecka's idea of the tension between imaginative freedom and vital necessities I accept as a sensation of balance between the opposite poles that are the subject-self and the artistic object within the integrity of creative process. I am not the omnipotent author of my story. I am the partner of my characters in developing the text.

This development has the sky as its limit and all my effort to summarize and finish the story seems to be an artificial violence inflicted on a supernatural stream of poetical visions where the imaginative freedom of life plays with infinite eternity. This playful force does not belong to me but comes down upon me from above, intimating the unknown mystery of the higher realm.

The cosmos as a sphere of sense-forming acts is comparable to the creative acts in which my muse stays in touch with eternity. Therefore, when modern cosmology interprets and predicts the future of celestial events, it should also take into account the subjective self of the observer, the endlessness of his psycho-emotional life. The error of physical cosmology consists, in fact, in the observer's treating the cosmos as an especially external reality; his glance is always directed outward. He forgets that he ponders the mega-sphere according to the models, forms, hypothesis, ideas, theories that have originated from human minds and are grasped by his own mind. Observing the geometrical structure of the heavens, he unconsciously ignores the geometrical-logical structure of his own consciousness. Eventually, he comes to the discovery of dark matter, that is, matter that cannot be objectified according to the models and ideas of the human mind. Yet, that does not mean that dark matter presents a cosmos that is absolutely closed for us and thus sets a theatrical stage for unchecked fantasy.

It is my conviction that dark matter is the sphere of the subjective forces of universal life, which I previously ignored in the undertaking of the external observation of cosmos, just as I ignored my own internal, subjective self. Dark matter is a sphere of cosmos and, at the same time, it would be the hidden area of my consciousness, both being elusive for objective knowledge. Dark matter appears to be a rebellion of the subjective phenomenon that is life against the objectification of being. In writing a story I have the chance to deal with this cosmic dark matter. Artistic words arise independently from my thought and carry me away, words rooted in the black mystery of cosmos, and I should always take into account this internal-external mystery if I wish to be an observer of the universe.

The ideas developed here mean that a human being is not an incredibly small particle of the world, since he or she has consciousness that is open toward the cosmos. The sky is not endless. It has limits, but its celestial gate is open toward the infinitude of eternity. If we consider our mind within the same perspective from as that of the infinite telos, then it could be comparable in scale to the scale of the cosmos, and the huge number of stars and the tremendous distances of space and time going beyond our imagination would be illusive. The Bible, phenomenology of life, the subjective freedom of consciousness, the quantum approach to the cosmos, these provide a new methodological basis for completing modern cosmology in investigation of the everlasting life of the universe.

## Note

1. In physical cosmology, according to current scientific theory, the diameter of the observable cosmos is thought to be about 93 billion light years. The diameter of the entire cosmos is unknown. However, according to Alan Guth's "inflation theory" the actual size of the cosmos is at least 15 orders of magnitude larger than observable universe, approximately $10^{26}$ light years.

## Works Cited

Dolidze, Mamuka. 2002. Phenomenology in Science and Literature. In *Phenomenology World-Wide. Foundations—Expanding Dynamics—Life-Engagements*, Analecta Husserliana LXXX, ed. Anna-Teresa Tymieniecka. Dordrecht: Kluwer Academic Publishers. Print.

———. 2013. The Meaning of Existence in the Context of Phenomenology. In *Developing Husserl's Ideas in the Contexts of Phenomenology of Life and Modern Georgian Philosophy*, 48–58. Tbilisi: Phenomenological Society and Centre of Interdisciplinary Sciences of Georgia. Print.

———. 2014. Intentionality of Time and Quantum – Phenomenological Sense of Space. In *Phenomenology of Space and Time*, Analecta Husserliana CXVII, ed. Anna-Teresa Tymieniecka, 67–87. New York: Springer. Print.

Guth, Alan. 1997. *The Inflationary Universe*. New York: Perseus Books. Print.

Hawking, Stephen. 1998. *A Brief History of Time*. London: Bantam Dell Publishing Group. Print.

Tymieniecka, Anna-Teresa. 2009. Human Development Between Imaginative Freedom and Vital Constraints. In *Culture and Philosophy*, 30–40. Phenomenological Society of Georgia: Council for Research in Values and Philosophy. Print.

# Part IV
# Eco-Cosmology

# *Welt*. At the Origins of Eco-Phenomenology: Heidegger's Concept of "World" in the Work of Anna-Teresa Tymieniecka

**Stefano Veluti**

**Abstract** Starting from a textual basis, this paper researches the role of Martin Heidegger's thought in Anna-Teresa Tymieniecka's appropriation of phenomenology and then analyzes the philosophical notion of "world" in Heidegger's thinking; following the Heideggerian evolution and development of this issue.

**Keywords** Tymieniecka · Heidegger · Eco-phenomenology · World · Man · Intentionality · The unity of everything-there-is-alive

## Tymieniecka's *Phenomenology and Science in Contemporary European Thought* (1961)

Anna-Teresa Tymieniecka's thought emerged as a direct and explicit pursuit of the path opened by the founder of phenomenology, Edmund Husserl. He had gathered a group of scholars who, after reading his *Logische Untersuchungen* (1900–1), identified themselves using the name "phenomenologists." They pursued Husserl's rigorous method of founding knowledge, which critiqued psychologism. However, this path aiming at a new foundation of experience turned out to be unclear and ambiguous. Phenomenology's history has always been, from its origins, full of proscriptions and criticism. This does not mean, however, that it was not a rich and fruitful school for – and not only for – Western philosophy.[1]

From its origins, the debate within this philosophy was peppered with disagreements. The publishing of the journal *Jahrbuch für Philosophie und phänomenologische Forschung* (1912), edited by Husserl himself, was meant to be the start of a fruitful exchange of opinions and research between scholars who recognized themselves as practicing this philosophy. This was partly so. But, the first publication,

S. Veluti (✉)
University of Macerata, Macerata, Italy
e-mail: stefano.veluti@protonmail.com

 
W. S. Smith et al. (eds.), *Eco-Phenomenology: Life, Human Life, Post-Human Life in the Harmony of the Cosmos*, Analecta Husserliana CXXI,
https://doi.org/10.1007/978-3-319-77516-6_12

containing Husserl's second most important text *Ideen zu einer reinen Phänomenologie und einer phänomenologischen Philosophie. Erste Buch: allgemeine Einführung in die reine Phänomenologie* (1912), saw the beginning of many controversies between Husserl and his disciples. We cannot dwell on that subject, but we would like to underline that, despite the failure of Husserl's research program – which would be carried forward only by his nearest assistants, Ludwig Landgrebe and Eugen Fink – phenomenology's path has gone on, broadly influencing present-day philosophies (Spiegelberg 1994, 4, 6ff.; Tymieniecka 2002).

The most important break in phenomenology's history was between Husserl and his "most beloved disciple" Martin Heidegger.[2] Taking into account the historical reconstructions made by H. Spiegelberg and by A.-T. Tymieniecka's team, it would be a mistake to confine phenomenology to the thought of Husserl and his close assistants Landgrebe and Fink. We must give value to all the departures from the Master's thinking, especially Heidegger's project. Despite his harsh criticism toward his teacher and despite his own affirming that "we would do better in the future to give the name of phenomenology only to that which Husserl himself has created and continues to produce," (Heidegger 1988, 29) Heidegger never completely forgot his phenomenological roots, as shown in his first important text *Sein und Zeit* and in his *Denkweg* generally. He will assert, in a later text, that only through the Husserlian "categorical intuition" did he have the possibility of grounding his thought (Heidegger 1986, 373ff).

Anna-Teresa Tymieniecka herself counted Heidegger as part of the phenomenological circle: in her encyclopedia *Phenomenology World-Wide,* an article on Heidegger affirms that, (Tymieniecka 2002, 240ff); he is covered in a section that includes articles on Max Scheler, Edith Stein, and Hedwig Conrad-Martius. She had previously affirmed Heidegger's importance in her very successful (especially in the USA) *Phenomenology and Science in Contemporary European Thought* (1962). In this publication, which We are here going to analyze Tymieniecka's presentation in this important book of the founding phenomenological thesis, especially focusing on the foundation of scientific knowledge.

From this text there emerges the vital and active sense characterizing the philosophic tradition that led her to consider Heidegger's heresy – "the gravest danger," in Husserl's words – as a developing part of phenomenological philosophy (Husserl 1994, 397). Furthermore, Tymieniecka prides herself in the eclectic assembly of her phenomenological philosophy:

> You were asking me about Husserl and Ingarden, but it was not simply these that I read. I was reading practically all of the phenomenological writers of the Göttingen School. And then when Heidegger's works were beginning to be more disputed after the war, I read all that appeared by him. And, as I said, I was involved on discussions with existentialists. Of course, I read everything of Sartre. I read everything of Merleau-Ponty when it appears. So I was developing philosophically in a rather large orbit, not continuing in Ingarden or Husserl. (Rainova 1993)[3]

In this article, we want to underline the importance of the explicit roots in Tymieniecka's eco-phenomenological philosophy because we think that it has not been sufficiently analyzed.

When she was awarded an *honoris causa* degree by the University of Bergen on the 29th August 2009, she explained – in an interview done that same day by scholars Lars Peter Torjussen, Johannes Servan, and Simen Andersen Øyen – that *Phenomenology and Science in Contemporary European Thought* was the product of a confrontation between the phenomenological philosophy and the logical-analytical thought originating from as expressed by Alfred Tarski. He had met Tymieniecka at the pier in New York when she arrived in the United States in 1955 and seen to her employment at the University of California at Berkeley; he had decided to read *Logische Untersuchungen* with Tymieniecka weekly, after hearing her talk about Husserl. Tarski, during the 1950's, was interested in the pragmatic impact of philosophy on science and everyday life; but, he was also fascinated by phenomenological theories: "Well, but these Husserlian analyses are all theoretical and abstract, but to what do they lead? There is no practical result from them. There are no plans and no technical innovations, no solutions to problems of the world" (Torjussen et al. 2008). Tymieniecka, in response to these his perplexities, wrote *Phenomenology and Science,* intending to showing the link between the two, and, eventually, the influence of phenomenology on reality and on the "problems of the world," to be achieved through science newly defined. The starting point is "the concept of a mind alienated from physical nature" (Tymieniecka 1962, xxii).[4] This concept is rooted in the modern philosophic and scientific tradition, conventionally dating back to Decartes: it is he who first utters the *vexata quaestio*, the problematic origin of contemporary philosophy.

*Phenomenology and Science* was quite successful and was also used, in the USA, as an introduction to phenomenology (Torjussen et al. 2008). The text's approach to this theme is different from the traditional structure of the confrontation between European philosophy and other cultural demonstrations: it was generally aimed at celebrating the "humanist spirit in Continental thought," and it radically opposed European philosophy to Anglo-American science's aridity (Tymieniecka 1962, xvi). The thesis is expressed in the work's introduction: "Contemporary European philosophy, reaffirming the mutual obligation of philosophy and science, arose at the beginning of the twentieth century in response to the challenge laid down by the nineteenth-century conception of modern science" (Tymieniecka 1962, xix).

The divide between philosophy and science was partly produced – and justified – by the "necessity for a distinction between the search for ultimate principles and the search for immediate causes," but it has two important consequences that marked nineteenth-century philosophy (and possibly even twentieth-century philosophy) (Tymieniecka 1962, xvii). The first consequence is the conception that the method derived from "natural science" is the only valid method for every field of knowledge. "First, the startling practical results made possible by the natural sciences led to the acceptance of their method as the only fit standard for truly scientific endeavour" (Tymieniecka 1962, xviii). This meant a "rush for science" in every discipline without evaluating whether "there are methods other than the 'scientific' which might satisfy the criteria of precision and verifiability by means of which the scientific method had gained its prestige" (Tymieniecka 1962, xviii). The second consequence is reductionism, a philosophical position that states that what is complex can be explained through the identification of its origin in simpler forms, as exemplified in "Darwin's theory of biologic evolution, Spencer's extension of it to social phenomena, and Hegel's historical development of the spirit" (Tymieniecka 1962, 19). This scheme could, perhaps, be useful for a scientific-naturalist analysis, but it cannot be so for the study of art, culture, social life and institutions (Tymieniecka 1962, 5). The problem with reductionism – mainly in Anglo-American philosophy – is that it is reflected in the philosophical thought. Tymieniecka wonders:

> Philosophy's claim to be a science, or at least scientific, had suddenly become problematic. If several natural sciences embrace all truly legitimate inquiries, and if philosophy is constructed in traditional manner, there seems to be no properly philosophical task; and indeed abstract thought, lacking empirical verification, appears to be positive hindrance to legitimate research. How could philosophy maintain its respectability in face to be acknowledged methodical, well confirmed and practically efficacious? The traditional philosophical aim of *founding* and *unifying* the sciences also seemed misguided. (Tymieniecka 1962, xix)

We can detect two problems: on the one hand, science has lost touch with the authentic human experience; on the other hand, it has become the model for all knowledge, misleading not only itself, but also every other field of knowledge.

The Phenomenological movement, born in Europe at the start of the century, could, in Tymieniecka's opinion, offer a convincing solution to the problem of the radical separation between science and all other knowledge: "On phenomenological grounds new dimensions of the human life are recognized as autonomous, many rejected factors of cognition are restrained, new data are taken into account, and a new basis of cognitive evaluation is established" (Tymieniecka 1962, xx).

Tymieniecka's interpretation of phenomenology has three main themes, each of them developed by an author. These themes are: first, the foundation of an absolutely certain knowledge; second, knowledge of others; and third, a non-physicalistic conception of the world. The three thinkers presented as addressing these themes are Edmund Husserl, Karl Jaspers, and Martin Heidegger.[5]

The most important gain achieved by Husserl's philosophy was, in Tymieniecka's opinion, the opening of "a universal inquiry that is founded on absolutely sure knowledge. The certainty of this foundation consists, firstly, in clearing the ground for knowledge from presuppositions; secondly, in elaborating the notions of cognitive media in correlation with the nature of the objects of cognition – "the principle of all principles"; thirdly, in adducting the specific type of cognitive insight and evidence thus prepared" (Tymieniecka 1962, 16). Husserl, through the use of the condensed program of applying "the principle of all principles" – affirming that "whatever present itself in 'intuition' in primordial form (as it were in its bodily reality) is simply to be accepted as it gives itself to be, though only within the limits in which then it presents itself" – gives every kind of science enough space to found themselves and guarantees the certainty and the emphasis requested by the scientific method, avoiding any reductionism (Husserl 1952, 95).

The direct application of the phenomenological method will be carried out by "post-Husserlian thinkers" thanks to whom "the highly technical potentiality of [phenomenological] analysis came to light" (Tymieniecka 1962, 18). It was especially Roman Ingarden who gave the two most important contributions to

Husserl's phenomenology: the many-layered structure of phenomena, and the intentional unity of the various strata of a phenomenon (Tymieniecka 1962, 22–23). This last one, linked to human ontological unity and to experience's inflexible variety, is solved by Tymieniecka use of the category of "meaning" which replaces the perpetuity assured by the link of efficient causality: "'Man's destiny' and the 'meaning of life' are expression of this continuity whose successive stages cannot be understood as effects of preceding ones, but which are "motivated" in the way meanings which cannot act on each other can yet have bearing on each other" (Tymieniecka 1962, 57–58, 59–60; Ryba 2002, 433).

Husserl's phenomenological method is the basis for addressing the remaining two themes by Jaspers and Heidegger. Karl Jaspers advances a holistic comprehension of the human being: he shows (with Ludwig Binswanger and other scholars) that "man cannot be conceived as confined to organic functions variously manifested," proving "the existence and the role of a spiritual 'existential' process in which man is basically involved, and which parallels the organic process" and admitting the necessity of a "purely metaphysical dimension toward which this spiritual process points" called by him "transcendence" (Tymieniecka 1962, 112). Tymieniecka interprets Jaspers's philosophy in an interesting way:

> In recognizing that individual humans are directed toward transcendence, Tymieniecka's phenomenology recognizes neither a dualistic strife between spirit and matter, nor super-nature and nature, nor even a triadic opposition between the natural, the existential or the purely metaphysical. Rather, the *existentialism* of her phenomenological approach construes every tension as mutually implicative. (Ryba 2002, 433)

In surpassing the materialistic and reductionist view of humans (in Jaspers' opinion, usefully gone beyond especially in psychology and psychoanalysis), it is possible to consider relationship with other people as not impossible, but, as a structural part of existential and spiritual human nature, by mediation of the communication necessary for the realization of the human being (Tymieniecka 1962, 75ff).

Tymieniecka's interpretation of Heidegger's philosophy is strongly influenced by the main question of her text: the correction of the scientific vision of the world through the vision of phenomenological thought. Just as she had Husserl and Jaspers, she analyzes Heidegger's philosophy in three phases: Firstly, Tymieniecka focuses her attention on its most relevant theoretical contents. Secondly, she links the new outline of reality given by Heidegger to some scientific issues, describing the results that this approach has produced in science's contemporary history.[6] Finally, she concludes, drawing on the findings of the two previous phases of her work.

The theme that she introduces is the separation between man and nature that is typical of modern philosophy (Tymieniecka 1962, 117). At first, there seems to be an incongruity: this separation between man and the so-called "external world" should have already been solved through Husserl's phenomenological method, encountered in the first part of the book. However, the argumentation here is on a deeper level. It aims at *ontologically* justifying every kind of experience, not only scientific study, without just *gnosiologically* founding those sciences in an apodictic way and so offering a sure basis to them.[7] To quote Heidegger himself, commenting on Husserl's voice in his *Encyclopaedia Britannica* article 'Phenomenology': "Gehört nicht eine Welt überhaupt zum Wesen des reinen *ego*? Vgl. Unser Todtnauberger Gespräch <1926> über das 'In-der-Welt-sein' (*Sein und Zeit* I, §12, §69) und den wesenhaften Unterschied zum Vorhandensein 'innerhalb' einer solcher Welt" (Husserl 1968, 274). This issue is addressed anthropologically, and it gives a hint about her point of view on the gist and reason of phenomenological philosophy (Tymieniecka 1962, 176).

It is possible to better understand the real value of this issue through reviewing an example taken from a concrete problem in a particular science (psychiatry): the disorder of schizophrenia. This is presented by Tymieniecka in an interpretation of psychoanalyst Eugène Minkowski's thought in *Le temps vécu: études phénoménologiques et psychopathologiques.* Schizophrenia would mean, in Minkowski's opinion, the patient's loss of contact with his "I-here-now," "an elementary and irreducible affirmation of the dynamism of life independent of all rational knowledge." The bottom line in this illness is that "a schizophrenic loses all vital contact with reality while becoming prisoner of a perfectly static, rational universe" (Tymieniecka 1962, 146–7). The sufferer's recovery will involve "helping the patient to rebuild his spatio-temporal perspective by an adequate stimulation of his imaginative activity" (Tymieniecka 1962, 155).

Tymieniecka links Minkowski's studies to Heidegger's phenomenological discoveries.[8] In fact, this separation between man and nature – his habitat, the surrounding world – emerging in the schizophrenic disorder, stablishes itself in an anthropological and ontological concept that contemporary thought inherited from a lasting philosophic tradition that has its roots in ancient philosophy. Hence, "in order to find a view of the world in which man and nature are conceived within one homogeneous system, it is necessary to go back as far as the pre-Socratics; the pre-Socratics did not differentiate the universe into heterogeneous realms but sought a single principle with references to which all aspects of the universe could be explained" (Tymieniecka 1962, 118).[9] After their first interpretations there emerged a separation between man and nature (here, probably, reference is made to the Aristotelian re-rendering of pre-Socratic philosophy); however, it is with Descartes that this disjunction becomes a permanent part of philosophy, through the "formulation of the essentially human in terms of consciousness, as contrasted with nonconscious extended bodies" (Tymieniecka 1962, 117). The next step will be made by Kant: he expands consciousness' domination to all possible knowledge (admitting, however, in the *Kritik vom Urteilskraft* that other possibilities of experience are possible) and "depreciates" the status of nature (Tymieniecka 1962, 118). However, the notion of the "intentionality of consciousness" introduced by Husserl allows us to evade this division. This issue is, nevertheless, central to modern philosophy: *knowledge* is the main human activity, and it ends up being "a subjectivistic conclusion," that is, in "Husserl's theory of the absolute status of intentional consciousness" (Tymieniecka 1962, 119). The ultimate step forward will be made by Martin Heidegger with his notion of the "world," introduced in his *Sein und Zeit.* Tymieniecka is mainly interested in showing how Heidegger's studies indicate a new way of understanding space and time by which they are grasped as strictly human characteristics; in advancing this outlook, Heidegger goes beyond the sclerosed modern and contemporary experience of scientific philosophy.

The most important phenomenological remark Tymieniecka makes in this work concerns our original *being-in-the-world*, presented as a peculiar form of intentionality. "Heidegger assumes that the intentionality of consciousness is prior to man's emergence as an understanding being and to the emergence of the world-for-man as a system of meanings." This priority is not consequent on the transcendental nature of consciousness but reflects the structure of *being-in-the-world* itself. The world is "the man's necessary counterpart" (Tymieniecka 1962, 119) and it is thusly described:

> Considered at the most fundamental level, that of intentionality, that realities of man's environment do not consist, indeed, in the objective view of things that surround us (by "objective" is meant "objectified by perception"). We know them, of course, as object, but at the fundamental level of our concern with them they are nothing more than utensils (*pragmata*). A utensil is not identical with a thing – it does not exist independently. It exists only in the context of a double reference: That toward other utensils and that toward a man (*Dasein*). … Each utensil refers to the whole system of utensils, and as each utensil contains a reference to a type of *Dasein*, the world appears as a system of systems of those referential relations. Man (*Dasein*) is the ultimate point of reference, which does not refer any further but exists in itself. However, he can exist only in this particular mode of being, within these referential relations. It is man who is the source of the possibilities which engender the system of relations and which, as a whole, constitute the world. These possibilities of man give meanings to things, posit the mass an intelligible totality which we call the "world." (Tymieniecka 1962, 129–130)

Tymieniecka clearly underlines the differences between Heidegger's point of view and that of Kant and of Husserl. As Kant says, everything in this system has a final reference, that seems to be the human being; but, Tymieniecka affirms that "in opposition to the Kantian conception, Heidegger, in joining man's creative prerogative to the powers that condition man, is not conceiving intentional consciousness in terms of a set of static rules and laws fixed once and for always for a recurring universe." And she continues: "He assumes at the start that man is involved within a perpetual self-creative process with respect to the world. In other words, consciousness is not constructed merely as creating an external world which recurs, but as a constantly re-creating the individual man, with reference to the world that emerges through the same operations" (Tymieniecka 1962, 119–120). Here, Tymieniecka is referring clearly to the structure of the "circle of understanding," formed by the "understanding" [*Verstehen*], developed in our "projecting" [*Entwerfen*], and our "interpreting" [*Auslegung*], the final step, which becomes the basic segment of the understanding process. "Understanding is the existential being of the ownmost potentiality of being of *Da-sein*," and so it is what constitutes the world, seen as the totality of *Dasein*'s possibilities (Heidegger 1996, 134). Heidegger names this opening of the world "project": "the project character of understanding constitutes being-in-the-world with regrets to the disclosedness of its there as the there of a potentiality of being" (Heidegger 1996, 136). Interpreting, eventually, is the "appropriation of understanding in being that understands" which modifies *Dasein* itself and creates a new Understanding (Heidegger 1996, 140).

Man's never-ending re-creation of himself, described as the hermeneutic circle, is explained in *Sein und Zeit* through the use of the notion of *Sorge*: this is the unitarian structure that supports the fundamental relation between *Dasein* and the world. This is, in Tymieniecka's opinion, the main difference between Husserl's phenomenology and Heidegger's philosophy. She translated the German term *Sorge* as "concern" (instead of the more common "care") to avoid a "psychological" misinterpretation of this concept (Tymieniecka 1962, 194). "Man's concern with the possibilities about which he himself must decide gives to him his most specific character. Effectively it is man's capacity to be concerned with which constitutes the most fundamental aspect of intentionality" (Tymieniecka 1962, 125). This concern is, in fact, defined by Heidegger as the "being-ahead-of-oneself-already-in (the world) as being-together-with (innerworldly beings encountered)" (Heidegger 1996, 180). This bidirectional definition marks the originality of the two directions themselves: the world and man (*Dasein*). Here, is found the solution to the phenomenological problem of assigning a prime role to intentionality such that will avoid any "idealistic" conclusion (Tymieniecka 1962, 119). As a matter of fact, concern pertains not only to its intentional forms conceived in a cognitive way: "whatever the form of the concern may be... and whatever the object, the reach of man's concern prescribes his world" (Tymieniecka 1962, 124).

These are Heidegger's 'corrections' to phenomenology, in Tymieniecka's opinion. Now she can present the "crucial achievement" of Heidegger's thought: a new conception of time and space (Tymieniecka 1962, 129). Space is founded on the structure of concern: "spatiality of the world, conceived as an intentional system, means the essential constitutive organization of man (conceived as a net of intentions and their source) toward other beings in the world totality." This idea of space does not replace the concept of geometric space, but, nonetheless, it becomes its basis (Tymieniecka 1962, 130).

Time, instead, in Heidegger's opinion, is not based on *Dasein*, which is derived from time itself: "in point of the fact the constitutive structure of man's concern... appears on a closer examination to be composed of forma or shape which take time, which underlie the intentional projection." The 'ek-static' characteristic of temporality in its three phases – past, present, and future – unifies and describes the movement of intentionality, which is rendered as concern and as capable of 'grasping the entire structure of man's "life"' or, in other terms, of his complete 'functioning,' man being specifically a dynamic (non-static, but ek-static) being" (Tymieniecka 1962, 132).

## Heidegger's Development of the "World" (*Welt*) Theme

The introduction of Heidegger's theme of "being-in-the-world," as proposed by Anna-Teresa Tymieniecka, is obviously limited by purpose her book itself, which aims at showing the advantages of the phenomenological method in the process of understanding reality, especially in relation to scientific work.

The main source text for treatment of this "world" is obviously *Sein und Zeit*. When it was written, many of Heidegger's texts describing this issue had not been published yet (for example, Heidegger's university lectures at Freiburg and Marburg given prior to the publication of *Being and Time* and now part of the nearly finished Heidegger-Gesamtausgabe).[10] The essays of Heidegger that we want to examine here ("On the Essence of Ground," "The Origin of the Work of Art," and "The Age of the World Picture"), were published in single editions before the publication of Tymieniecka's book (the first essay in 1929, and the remaining two in 1950),[11] so it is very likely that she had read them.

The notion of "being-in-the-world" indicates the original communality between *Dasein* and the world: human existence is not, at first, isolated and then added to by the world, but it is originally characterized by a "disclosedness" to the world. Heidegger wants to overcome the modern separation between *res cogitans* and *res extensa*, subject and object, mind and external reality.

This structure is the origin of the world and of man (*Dasein*), but this does not mean that these "polarities" are liquefied in what generates them; on the contrary, they preserve their characterization even in the unfolding of this structure. The misunderstanding of this preservation has led to the crystallization of those two polarities and resulted in the oblivion of the dimension in which those polarities can be conceived.

Heidegger identifies three factors pertaining to the notion of "being-in-the-world":

1. "*In-the-world*": In relation to this factor, we have the task of questioning the ontological structure of "world" and of defining the idea of *worldliness* as such.
2. The being which always is in the way of being-in-the-world. In it we are looking for what we are questioning when we ask about the "who?"
3. *Being in* as such: The ontological constitution of in-ness itself is to be analyzed. (Heidegger 1996, 50)

The first factor is addressed through a phenomenological analysis of the formation process of the "world" and through a *critica* of the conception of the world formulated by Descartes. Adressing the second factor involves the idea of *Mitdasein*, of "They" [*das Man*] and of "Self"; those ideas are crucial to the underlying question of "who is" the entity living in the world. Addressing the third factor means describing the relationship between *Dasein* and the world, the presentation of a fundamental analysis of "Attunement" [*Befindlichkeit*], of the "circle of understanding" [*Zirkel des Verstehens*], of the "everyday being of the there" [*Das alltägliche Sein des Da*], and of "falling prey" [*Verfallen*].

We want to dwell here on the first factor in relation to the formation process of the world, since this seems the topic most relevant to Tymieniecka's assigned task in her *Phenomenology and Science in Contemporary European Thought.* First of all, it is important that we do not take for granted something that is otherwise important: the "world," as described in *Being and Time,* is an aspect of the structure of being-in-the-world and is the main constituent of *Dasein*. We must consider the world "not as beings essentially unlike Da-sein that can be encountered within the world; but, rather, as that '*in which*' a factical Da-sein 'lives.' Here world has a pre-ontological, existentiell meaning" (Heidegger 1996, 61). This means that the "world" is not to be intended in an actual way, but as part of the disclosedness as such.

Heidegger's thought on the idea of the world sums up two different but co-present meanings of the world itself (Figal 2012, 500): the meaning of the ancient Greek word κόσμος, central to the pre-Socratic philosophers, and the New Testament conception of the term "world."[12] In "On the Essence of Ground" [*Vom Wesen Des Grundes*], an essay Heidegger wrote in 1929, he sums up the dynamics of those two meanings by saying:

> World refers to a *"how" of being* of beings, rather than to these beings themselves. (2) This "how" determines beings *as a whole*. In its ground it is the possibility of every "how" in general as limit and measure. (3) This "how" as a whole is in certain manner *prior*. (4) This prior "how" as a whole is itself *relative* to human *Dasein*. The world thus belongs precisely to human *Dasein*, even though it embraces in its whole all beings, including *Dasein*. (Heidegger 1998, 112)

Citing the New Testament conception of the cosmos, he states:

> Κόσμος0 οὗτος in Saint Paul (cf. I Corinthians and Galatians) means not only and not primarily the state of the "cosmic," but the state or the situation of the human being, the kind of stance he takes *toward* the cosmos, his esteem for things. (Heidegger 1998, 112)

These two passages are interesting for our thesis not for their semantic analysis of terms, but because they underline the actual position of the world (the translation of the term κόσμος) in the "phenomenological overview" in Heidegger's philosophy. The world is prior to every intention directed toward a single entity that is always filtered by the "prior how as a whole" spoken of above. Furthermore, as emerged in the second remark, the relationship between man and world is not primarily cognitive but is "practical," namely, related to our life.

It is this prior-ness of the "practical" that leads Heidegger to create the world through the mediation of "Relevance" [*Bewandtnis*] (Heidegger 1996, 77ff). As a matter of fact, things are intentioned originally as "useful things"; they are characterized by their "serviceability." This prior-ness is always characterized by "what-for" and "wherefore" features, making the entity accessible only in a system of "references." This system constitutes the entities' condition of *ontological* possibility and is not a secondary characteristic posterior to their existence as "objectively present" (Heidegger 1996, 78). This is the first important point of Heidegger's thesis. Philosophic thought, from Aristotle onward, has explained "things" taking into account their mere presence, applying the categories of "matter" and "form." Heidegger states, instead, in "The Origin of Work of Art":

> Serviceability is the basic trait from out of which this kind of beings look at us – that is, flash at us and thereby presence and so be the beings they are. Both the design and the choice of material predetermined by that design – and therefore the dominance of the matter-form structure – are grounded in such serviceability. … Accordingly, matter and form are determinations of beings which find their true home in the essential nature of the equipment. (Heidegger 2002, 10)

Serviceability, being originally aimed at, is the prime ontological element of things. But, it is not only that. "Relevance" – the entity's necessity of being linked to any other entity – leads the entity itself to be anticipated by a totality: "*Which* relevance things at hand have is prefigured in terms of the total relevance. The total relevance … Is 'earlier' than any single useful thing." In *Being and Time*, Heidegger gives the example of the hammer: "a hammer has to do with hammering, the hammering has to do with fastening something, fastening something has to do with protection against bad weather. This protection 'is' for the sake of providing shelter for *Da-sein*, that is, for the sake of a possibility of his being" (Heidegger 1996, 78). The *Dasein* is always the last "what-for," namely that for the sake of which serviceability operates. These phenomenological observations do not aim at supporting a kind of 'anthropologizing' of the world, a humanization of it, but at underlining the unity of the world's priority in contraposition to the emerging of entities. In order to meet the entity, a preliminary disclosedness is necessary; through this disclosedness the entity can emerge. "That within which *Dasein* understands itself beforehand in the mode of self-reference is that for which it lets beings be encountered beforehand. *As that for which one lets beings be encountered in the kind of being of relevance, the wherein of self-referential understanding is the phenomenon of the world*" (Heidegger 1996, 80). The world 'in itself' is constituted as "significance," namely, the totality of the encounters preliminarily given to the man as *Dasein* (Heidegger 1996, 81). The entities' being is characterized by significance: it manifests itself in a non-thematic but necessary way.

However, the world as represented by Heidegger is limited to "a world of tools and works" [*Zeug- und Werkwelt*] (Figal 2012, 500–1). Even though he clarifies – in *Being and Time* – that the priority of the functionality aspect of things does not exclude things that are not characterized by a "what-for" aspect (considered 'unuseful things', such as broken tools or objects with an unknown usefulness), this idea of the world has too many limitations.[13] There are entities, such as works of art, storms, deserts, that cannot be understood through the appreciating the aspect of serviceability. They demand a new way of conceiving the *existential* of world, a wider one.[14]

Heidegger's first attempt at framing such a broader view is to be found in "On the Essence of Ground," a text published a little later than *Being and Time* (Figal 2012, 502–3). This attempt is based on a more ontological interpretation of the "*Hermeneutik Zirkel.*" Already, in *Being and Time,* the understanding that is the project of the world is a general understanding of beings: "in the projectedness of its being upon the for-the-sake-of-which together with that upon of significance (world) lies the disclosedness of being in general" (Heidegger 1996, 138). In the *Ground* essay, it is no longer the investigation of "daily-life" that is important: the starting point here is one the aim of studying *Dasein* as "*the distinction between being and beings* (ontological difference)": there is a truth about "beings in their being," and a different one about the "being of beings" (Heidegger 1998, 105).[15] This difference is possible because of an essential characteristic of disclosedness that constitutes *Dasein*, one called by Heidegger "transcendence" (Heidegger 1998, 106). This notion has a radically different meaning: it is neither about a subject that transcends vis-à-vis an object (or vis-à-vis other subjects), nor about God's transcendence. This term indicates the preliminary "surpassing" of beings vis-à-vis the world (Heidegger 1998, 108), on whose basis intentionality is founded (Heidegger 1998, 106): "the positive component of transcendence is the project of World" (Richardson 2003, 166).[16] The meaningful innovation of this essay, compared to the *Hauptwerk* (at least in relation to the notion of "world") is the fact that "being-in-the-world" is understood starting from transcendence. This indicates – in a less 'utilitarian' way than that in *Being and Time* – the original reference between *Dasein* and the world, namely the being-in-the-world phenomenon, and it allows a determination of the innerworldly beings that is wider than "a world of tools and works."

Heidegger makes a radical extension in the notion of world in his essay "The Origin of the Work of Art" [*Der Ursprung des Kunstwerkes*], framed as a lecture during the 1930s and published in its latest version in *Holzwege* in 1950. In this essay, the world does not appear anymore to proceed from a useful thing. To exemplify: apparently Heidegger describes a useful thing, a pair of peasant shoes, but describes these using a work of art, the famous depiction of a pair of peasant shoes made by Vincent van Gogh (Heidegger 1998, 13).

What is revealed during the contemplation and the description of the work of art is not the simple useful thing that is represented – the peasant shoes – but also the "peasant world" through which, alone, it is possible to understand the object taken into account. Art is the means through which a world proposes and imposes itself (Heidegger 1998, 14ff).

The world has here a wider notion. The following quotation is the entire passage in which Heidegger presents the notion of the world; in it, we can find many similarities with the notion developed in *Being and Time,* but, also, an attempt to not restrict that notion to the aspect of serviceability:

> World is not a mere collection of the things – countable and uncountable, known and unknown – that are present at hand. Neither is world a merely imaginary framework added by our representation to the sum of things that are present. World worlds [*weltet*], and is more fully in being than all those tangible and perceptible things in the midst of which we take ourselves to be at home. World is never an object that stands before us and can be looked at. World is that always-nonobjectual to which we are subject as long as the paths of birth and death, blessing and curse, keep us transported into being. Wherever the essential decisions of our history are made, wherever we take them over or abandon them, wherever they go unrecognized or are brought once more into question, there the world worlds. The stone is world-less. Similarly, plants and animals have no world; they belong, rather, to the hidden throng of an environment into which they have been put. The peasant woman, by contrast, possesses a world, since she stays in the openness of beings. In its reliability,

> equipment imparts to this world a necessity and proximity of its own. By the opening of a world, all things gain their lingering and hastening, their distance and proximity, their breadth and their limits. In worlding there gathers that spaciousness from out of which the protective grace of the gods is gifted or is refused. Even the doom of the absence of the god is a way in which world worlds. (Heidegger 1998, 22–3)

This 'description' of the world partly corrects and partly studies in depth the position exposed in *Being and Time:* it studies the notion in depth because the starting point of the event of the world is the openness, namely, an open place in which the entities can manifest themselves, and it is founded on the historical truth of being; it corrects the notion because the openness in *Being and Time* was described in a way too similar to the human. "*Dasein*" is not the man-subject, but the disclosedness itself, the openness (the "*da*" of the "*Sein*") in which only man can reach his truth. This new formulation solves the issue of the openness of the world as more 'vague' only apparently; Heidegger purifies the world from any subjectivistic traces (deriving from Kant's and Husserl's philosophies), stating that "the world worlds" [*das Welt weltet*],which underscores the impersonality of totality's openness (Von Herrmann 1964, 170ff).

A further central extension of the concept of the world, which we, here, merely outline, affects the structural relationship between the world itself and the "Earth": the finite scope of possibility not yet realized in the given *viable* possibilities of the world. "The earth is the unforced coming forth of the continually self-closing, and in that way, self-sheltering. World and earth are essentially different and yet never separated from one another. World is grounded on earth, and earth rises up through world" (Heidegger 2002, 26). These ideas, hard to understand without an explanation, will be carried out in other essays ("Building Dwelling Thinking" [*Bauen Wohnen Denken*], "The Thing" [*Das Ding*], and "...poetically man dwells..." ["*...dichterisch wohnet der Mensch...*"]), all of which deal with Heidegger's concept of the *Geviert,* which expresses the four essential preliminary relevancies (sky and earth, mortals and divinity) of every intention of things.

The question we have now to ask ourselves is the following: why have we lost memory of this phenomenon? As a matter of fact, although it be explicit in Heidegger's discourse, the world that makes up every intention, seen as a system of meaning and references, seems alien. What determines this our current perception of the phenomenon of the world?

Heidegger's answer to this question sinks into his profound and sibylline ontology of history, but it can be appreciated leaving behind a deep understanding of it for a moment.

In an essay written in 1938 and collected in *Off the Beaten Track,* Heidegger names the modern age "the age of the world picture" [*Die Zeit des Weltbildes*]. In this age, the world is not denied, but is grasped as a mere "picture":

> Understood in an essential way, 'world picture' does not mean 'picture of the world' but, rather, the world grasped as picture. Beings as a whole are now taken in such a way that a being is first and only in being insofar as it is set in place by representing-producing [*vorstellend-herstellenden*] humanity. Whenever we have a world picture, an essential

> decision occurs concerning beings as a whole. The being of beings is sought and found in the representedness of beings. (Heidegger 2002, 67)[17]

In modern thought, the world exists because it is represented. This "representedness" becomes the core of an epochal turning point. To represent means, in Heidegger's philosophy, "to bring the present-at-hand before one as something standing over-and-against, to relate it to oneself, the representor, and, in this relation, to force it back to oneself as the norm-giving domain" (Heidegger 2002, 69). In the representation there is always an inherent non-themed adaptation of what is presented to the requirements of usability. These worlds are deeply linked to other later well-known reflections of Heidegger dealing on the essence of technique, or, to the remarks collected in the well-known 'esoteric' text *Contributions to Philosophy* about "*Machenschaft*," "Machination," namely, the principle that "dominates the history of being in Western philosophy up to now, from Plato to Nietzsche" (Heidegger 1999, 89). Heidegger's explanation of the phenomenon of technique is metaphysical: throughout Western philosophy, man seen as a 'rational subject' has taken the place of the "subjectum," namely, the "ὑποκεὶμενον," "that-which-lies-before, that which, as ground, gathers everything onto itself" (Heidegger 2002, 66). And so is lost the transcendence of the world – a significance already-given to every human intention and action.

Thus – going back to our initial remarks on the development of phenomenology – our claim is that in all these formulations and revisions of the concept of the "world," what is constant is an aspect of the concept itself that characterizes the true innovation of Heidegger's philosophy compared to Husserl's thought and to the rest of the phenomenological movement.

As Vincenzo Costa states, the real gain in Heidegger's philosophy is the exteriorization of meaning from the subject to the world. "Heidegger allows us to say goodbye to the Cartesian tradition: because *he puts meanings within the world*; *those meanings are originally exteriorized* and, by being so, subjectivity itself is exteriorized. Subjectivity is not linked to the meanings by an action of reflection – namely, directing its attention to what happens within itself – but by an action of relation to the possibilities in the world" (Costa 2003, 224).[18] Thus, meaning is not given to the world by the subject, but is 'found' in the world by the subject who acknowledges the network of references before the single object. This network of references – the significance – constitutes the object itself and is called the "world." In all the variations and developments here outlined – and they are not the only ones – even if there is a modulation in the characterization of the unitarian aspect of the world (considered as "*Zeug-und Werkwelt*," and then as "the positive component of transcendence" or as "the rising up of earth"), there is something that they all have in common. What is constant is the priority of the givenness of the totality of the meanings compared to the givenness of a single entity. We can therefore, conclude, with a nod to Tymieniecka's early book, that "Heidegger shows us a non-anthropological consideration of the concept of the world, since it is not man who gives shape to the world, but it is the world itself that 'unlocks' man to himself, being a 'thrown' possibility" (Costa 2003, 249).[19]

## Conclusions. Kindredness Between the World and the "Unity of Everything-There-Is-Alive"

It is not possible to trace a textual account of the influence of Heidegger's remarks on the world on the eco-phenomenological thought of Anna-Teresa Tymieniecka. After her 1961 essay, she did not thematically treat Heidegger's philosophy any more, quoting it only sporadically – but, by doing so, underlining its importance as a reference, as emerges in the interview previously cited.

We believe that it is possible to link Heidegger's description of the peasant world made in the essay "The Origin of the Work of Art" to Tymieniecka's concept of "unity of everything-there-is-alive" presented in Tymieniecka's *Tractatus Brevis* (1986). The former describes the opening of the peasant world starting from Van Gogh's representation of a pair of shoes:

> From out of the dark opening of the well-worn insides of the shoes the toil of the worker's tread stares forth. In the crudely solid heaviness of the shoes accumulates the tenacity of the slow trudge through the far-stretching and ever-uniform furrows of the field swept by a raw wind. On the leather lies the dampness and richness of the soil. Under the soles slides the loneliness of the field-path as evening falls. The shoes vibrate with the silent call of the earth, its silent gift of the ripening grain, its unexplained self-refusal in the wintry field. This equipment is pervaded by uncomplaining worry as to the certainty of bread, wordless joy at having once more withstood want, trembling before the impending birth, and shivering at the surrounding menace of death. This equipment belongs *to the earth* and finds protection in the *world* of the peasant woman. From out of this protected belonging the equipment itself rises to its resting-within-itself.
>
> Whenever in the late evening she [the peasant woman] takes off the shoes, in deep but healthy tiredness, and in the still dark dawn reaches for them once again, or passes them by on the holiday, she knows all this without observation or reflection.... In virtue of this reliability the peasant woman is admitted into the silent call of the earth; in virtue of the reliability of the equipment she is certain of her world. (Heidegger 2002, 14)

And the latter, instead, concerns the essential link between the environment and the humanity that inhabits it:

> The gardener times his life by his patience and wakefulness in following the advance of sprout to a leaf, then to a blossom, and finally, a fruit. The pulsation of his veins fall in rhythm with those of juices of the fruit while it turns slowly from the green to the red, ripens and weighs heavily from the branch. He feels the thirst of the roots drinking with them every drop of the rain. The forester lives day by day with his trees maturing with their getting of age to be lumbered; he starts a 'new' life-season planting a new forest.
>
> The farmer lives with the season of the crops and rendering of the harvest. Beginning over again the sowing and germinating and irrigating and harvesting, he rhythms his existence by the cycles of vegetation. A sailor lives by the whims of the weather and the majestic unpredictability of the sea-Element. (Tymieniecka 1986, 16)

It is surely necessary to establish whether this "assonance" is owing to Tymieniecka's reading of Heidegger's works or to the Husserlian mediation of the ism of the world, especially presented in *Die Krisis der europäischen Wissenschaften und die transzendentale Phänomenologie* (it is Tymieniecka herself who affirms the dependence that links the late Husserl and Heidegger's hermeneutic phenomenology), or to Max Scheler's thought.[20]

Nevertheless, it appears that the theme of the constituting of the "world" is shared by both eco-phenomenology and hermeneutic phenomenology. The phenomenon of being-in-the-world has been more explicitly acknowledged by Tymieniecka as a fundamental part in the correct understanding of human experience (Tymieniecka 1962, 129). Tymieniecka then bases her phenomenological path on different grounds than those of Husserl and Heidegger, but this does not mean that she has refuted the phenomenon of the world as previously theorized. On the contrary, here is a starting point for a more radical foundation of experience. The *human condition*, intended in a "cosmological" sense, must be able to base and justify, because of its originality, the experience of being-in-the-world (Verducci 2012, 93).

## Notes

1. Spiegelberg defines phenomenology as a movement, in order to stress its pluralism. He affirms: "(1) Phenomenology is a moving, in contrast to a stationary, philosophy with a dynamic momentum, whose development is determined by its intrinsic principles as well as by the "things," the structure of the territory which it encounters. (2) Like a stream it comprises several parallel currents, which are related but by no means homogeneous, and may move at different speeds. (3) They have a common point of departure, but need not have a definite and predictable joint destination; it is compatible with the character of a movement that its components branch out in different directions." (Spiegelberg 1994, 1–2).
2. On this event, see the introductory note in C. Sinigaglia, "Saggio introduttivo," in Edmund Husserl,. *Glosse a Heidegger*, ed. Corrado Sinigaglia. Milano: Jaca Books, 1996. Translation of Roland Breeur. "Randbemerkungen Husserls zu Heideggers *Sein und Zeit* Kant und das Problem der Metaphysik *Husserl Studies* 11 (1994): 3–63; and also E. Husserl, R. Ingarden (ed.), *Briefe an Roman Ingarden. Mit Erläuterungen und Erinnerungen an Husserl*, Phaenomenologica 25, The Hague: Nijhoff, 1968.
3. To quote *Phenomenology and Science in Contemporary European Thought* itself: "if phenomenology consisted only of the application of a strict method, perfect determinate in scope and interpretation, then its philosophical significance, like that of the program of the Unified Sciences, would soon become sterile. Only new facts could be added. But phenomenology conforms to no program – although unified in its general orientation, it is intrinsically divergent in doctrinal interpretations. The new insight of those doctrine enrich in turn the basic phenomenological orientation and become incorporated into a deeper philosophical view." (Tymieniecka, 1962, 64–65).

4. Further readings on the relations between phenomenology and Anglo-American thought, see Analecta Husserliana XXVI – *American Phenomenology, Origins and Developments*, ed. Eugene F. Kaelin and Calvin O. Schrag.
5. The heterodoxy found in those three representatives of phenomenological philosophy chosen by Tymieniecka is clear.
6. The term "science" in Tymieniecka's philosophy indicates not only the natural and mathematical sciences, but also all the disciplines identifiable as the "human sciences," such as psychology and sociology.
7. This passage presents as the main issue the fundamental theme of Tymieniecka's interpretation of the relation between Husserl's phenomenology and Heidegger's ontology.
8. The real relation between Minkowski and Heidegger is one mediated by Ludwig Binswanger's psychoanalytic thinking.
9. This link between the pre-Socratic thinkers and Heidegger's thinking is very important because it will be acknowledged in later texts. For example, in the fourth volume of her *Logos and Life*, Tymieniecka states: "Heraclitus insistently affirmed that whereas common knowledge is "in the open" and shared by all, knowledge of true reality, "wisdom," is hidden. Yet we have access to it. His idea of the "uncovering of the hidden," of *aletheia*, has attracted particular attention in our times. It inspired the metaphysical speculation on the hidden that in Heidegger is identified with the notion of Being. Although Heraclitus' insight into the "hiddenness" of the roots of reality appears in the phenomenology of life as a very pertinent feature of double-faceted reality, it finds on that ground a non-speculative, concrete crystallization. Not a hidden being, but a primogenital origination and genesis of beings as *onta* comes out of hiding, principles and vehicles of becoming, of living beings as such. What is revealed are the "inner workings" of life's unfolding, generating, progressing, which simultaneously bring in the differentiation of beings and their existential interrelations, temporality, space" (Tymieniecka 2000, 295).
10. Just now the well-known *Schwarzen Hefte*, the "Black Notebooks" are being published: these are collection of Heidegger's philosophic and personal notes written between 1930 and at least 1948, and they make up the last volumes of the *Gesamtausgabe*, which will have a total of 102 volumes.
11. The essay "The Essence of Ground" was first published as a paper in *Jahrbuch für Philosophie und phänomenologische Forschung* (Halle a.S.; Max Niemeyer, 1929) written in honor of Edmund Husserl's seventieth birthday and, at the same time, also published as a single edition. From 1949 onward it was published by Vittorio Klostermann Verlag, with an introduction. The essay "The Origin of the Work of Art" was first published in its current form in the collection *Off the Beaten Tracks* (in 1950), and then published again by Reclam in 1960. The last essay we have quoted, "The Age of the World Picture," was first published *Off the Beaten Tracks*.
12. Günter Figal's text reads as follows: "Mit diesem Verständnis der Welt, kombiniert Heidegger die beiden Weltbegriffe, die als Vorgänger der hier unterschiedenen gelten können. Indem er die Gesamtheit dessen, was ist, von ihren

im Dasein eröffneten "Bedeutsamkeit," also von ihrer Bedeutung und Wichtigkeit für das Dasein her versteht, bringt er das Verständnis der Welt als eines geordneten Zustandes wieder ins Spiel, der mit dem griechische Wort für Welt, nämlich κόσμος, gemeint war. Und indem Heidegger diesen Zusammenhang von *menschlichen* Dasein her konzipiert, geht er auf des neutestamentliche Verständnis der κόσμος zurück, also auf der Verständnis der Welt als "Zustand" und "Lage" der Menschen" (Figal 2012, 500).

13. Vincenzo Costa thinks the same, as stated in *La Verità del mondo* (Costa 2003, 235ff). Friederich-Wilhelm von Herrmann faces this problem differently: in his main text *Die Selbstinterpretation Martin Heideggers* he affirms that in *Being and Time* Heidegger had considered just a "portion" of the world (von Herrman 1964, 53).
14. It is important to remember that the "world," in Heidegger's opinion, is not thought of as something objectively present, in relation to the modern subject-object duality: it is a fundamental component of *Dasein*, an existential part of it, and the lack of it would cause the failure of the opening the *Dasein* represents. (Heidegger 1996, 49ff).
15. About this theme, see the work of Massimo Marassi *Ermeneutica della differenza Saggio su Heidegger* (Hermeneutics of Difference. Essay on Heidegger).
16. The theme of the relation between the project of the world and the understanding of being is problematic, and would need a broader discussion. Richardson emphasizes a meaningful difference between the two terms in the *Kantbuch* del 1929 (Richardson 2003, 147).
17. Heidegger's position is clearly against the theory of *Weltanschauung*, part of the idealistic and historicist debate, especially in Dilthey's philosophy. On this theme Heidegger states: "The fundamental event of modernity is the conquest of the world as picture. From now on the word "picture" means: the collective image of representing production [*das Gebild des vorstellenden Herstellens*]. Within this, man fights for the position in which he can be that being who gives to every being the measure and draws up the guidelines. Because this position secures, organizes, and articulates itself as world view, the decisive unfolding of the modern relationship to beings becomes a confrontation of world views; not, indeed, any old set of world views, but only those which have already taken hold of man's most fundamental stance with the utmost decisiveness. For the sake of this battle of world views, and according to its meaning, humanity sets in motion, with respect to everything, the unlimited process of calculation, planning, and breeding" (Heidegger 2002, 71).
18. The original quotation, in Italian, is: "Heidegger ci offre gli strumenti per prendere congedo dalla tradizione cartesiana: perché *colloca nel mondo i significati, che sono dunque esteriorizzati all'origine*, e con essi è esteriorizzata la soggettività che non si rapporta ai significati riflettendo, cioè dirigendo la propria attenzione su ciò che accade al suo interno, ma rapportandosi a possibilità che sono nel mondo."
19. Another theme related to this issue, is that of human freedom, in relationship with the totality of meanings, also called later by Heidegger "system." I have

partially faced this question in my recent paper "Dinamica dell'evento. Lettura dei Contributi alla filosofia di Martin Heidegger a partire dalla questione della libertà umana" [Dynamics of Enowning. A reading of Martin Heidegger's Contributions to Philosophy Starting from the Issue of Human Freedom], *Rivista di Filosofia Neo-Scolastica* 4: 831–859.

20. See Tymieniecka 1962, 119.

## Works Cited


Costa, Vincenzo. 2003. *La verità del mondo. Giudizio e teoria del significato in Heidegger*. Milano: Vita e Pensiero.

Figal, Günter. 2012. *Oggettualità. Esperienza ermeneutica e filosofia*. Milano: Bompiani. Trans. Antonio Cimino. Translation of *Gegenständlichkeit. Das Hermeneutsiche und die Philosophie*. Tübingen: Mohr Siebeck Verlag, 2006.

Heidegger, Martin. 1986. In *Seminare*, ed. Friederich-Wilhelm Von Herrmann. Frankfurt a.M: Vittorio Klostermann Verlag.

———. 1988. *Hegel's Phenomenology of Spirit*. Bloomington: Indiana University Press. Trans. Parvis Emad and Kenneth Maly. Translation of *Hegels Phänomenologie des Geistes*. Frankfurt a.M.: Vittorio Klostermann Verlag, 1980.

———. 1996. *Being and Time*. New York: SUNY Press. Trans. Joan Stambaugh. Translation of *Sein und Zeit*. Tübingen: Max Niemeyer Verlag, 1953.

———. 1998. *Pathmarks*. Cambridge, UK: Cambridge University Press. Ed. William McNeill. Translation of *Wegmarken*. Frankfurt a.M.: Vittorio Klostermann Verlag, 1967.

———. 1999. *Contributions to Philosophy (From Enowning)*. Bloomington: Indiana University Press. Trans. Parvis Emad and Kenneth Maly. Translation of *Beiträge zur Philosophie. Vom Ereignis*. Frankfurt a.M.: Vittorio Klostermann Verlag, 1989.

———. 2002. *Off the Beaten Track*. Ed. Julian Young and Kenneth Haynes. Cambridge, UK: Cambridge University Press. Translation of *Holzwege*. Frankfurt a.M.: Vittorio Klostermann Verlag, 1950.

Herrmann, Von, and Friedrich-Wilhelm. 1964. *Die Selbstinterpretation Martin Heideggers*. Meisenheim a.G: Hain.

Husserl, Edmund. 1952. *Ideas: General Introduction to Pure Phenomenology*. Trans. W. R. Boyce Gibson. New York: Macmillan. Translation of *Ideen zu einer reinen Phänomenologie und einer phänomenologischen Philosophie. Erste Buch: allgemeine Einführung in die reine Phänomenologie*. Halle a.S.: Max Niemeyer Verlag, 1912.

———. 1968. In *Phänomenologische Psychologie. Vorlesungen Sommersemester 1925*, ed. Walter Biemel. The Hague: Nijhoff.

———. 1976. In *Die Krisis der europäischen Wissenschaften und die transzendentale Phänomenologie. Eine Einleitung in die phänomenologische Philosophie*, ed. Walter Biemel. The Hague: Nijhoff.

———. 1993. *Logische Untersuchungen*. 7th ed. Tübingen: Niemeyer.

———. 1994. In *Briefwechsel*, ed. Karl Schumann, vol. VII. The Hague: Kluwer.

———. 1996. *Glosse a Heidegger*. Milano: Jaca Book. Ed. Corrado Sinigaglia. Translation by Roland Breeur, "*Randbemerkungen Husserls zu Heideggers* Sein und Zeit *und* Kant und das Problem der Metaphysik". *Husserl Studies* 11(1994): 3–63.

Kaelin, Eugene F., and Calvin O. Schrag, eds. 1986. *American Phenomenology. Origins and Developments*, Analecta Husserliana XXVI. Dordrecht: Kluwer.

Marassi, Massimo. 1990. *Ermeneutica della differenza. Saggio su Heidegger*. Milano: Vita e Pensiero.

Rainova, Ivanka. 1993. *Interview with Anna-Teresa Tymieniecka*. Web. 3 September 2014. http://www.phenomenology.org/archive/.
Richardson, William J. 2003. *Heidegger. Through Phenomenology to Thought*. New York: Fordham University Press.
Ryba, Thomas. 2002. Anna-Teresa Tymieniecka's Phenomenology of Life. In *In Phenomenology World-Wide, Foundations – Expanding Dynamics – Life-Engagements*, Analecta Husserliana LXXX, ed. Anna-Teresa Tymieniecka. Dordrecht: Kluwer.
Sallis, John, ed. 1970. *Heidegger and the Path of Thinking*. Pittsburgh: Duquesne University Press.
Spiegelberg, Herbert. 1994. *The Phenomenological Movement. A Historical Introduction*. 3rd ed. Dordrecht: Kluwer.
Torjussen, Petter, Johannes Servan, and Simen Andersen Øyen. 2008. *Interview with Anna-Teresa Tymieniecka*. Web. 3 September 2014. http://www.phenomenology.org/archive/.
Tymieniecka, Anna-Teresa. 1962. *Phenomenology and Science in Contemporary European Thought*. New York: Noonday Press.
———. 1980. *Heideggers Philosophie der Kunst: eine systematische Interpretation der Holzwege-Abhandlung der Ursprung des Kunstwerkes*. Vittorio Klostermann Verlag: Frankfurt a.M.
———. 1986. Tractatus Brevis. First Principles of the Metaphysics of Life Charting the Human Condition. In *Phenomenology of Man and the Human Condition. Part II, The Meeting of Occidental and Oriental Philosophies*, Analecta Husserliana XXI, ed. Anna-Teresa Tymieniecka. Dordrecht: Kluwer.
———.2000. *Impetus and Equipoise in Life-Strategies of Reason, Logos and Life Book 4*, Analecta Husserliana LXX. Dordrecht: Kluwer.
———. 2002. In *Phenomenology World-Wide, Foundations – Expanding Dynamics – Life-Engagements*, Analecta Husserliana LXXX, ed. Anna-Teresa Tymieniecka. Dordrecht: Kluwer.
Veluti, Stefano. 2014. Dinamica dell'evento. Lettura dei Contributi alla filosofia di Martin Heidegger a partire dalla questione della libertà umana. *Rivista di Filosofia Neo-Scolastica* 4: 831–859.
Verducci, Daniela. 2012. *La fenomenologia della vita di Anna-Teresa Tymieniecka. Prova di sistema*. Roma: Aracne editrice.
World Phenomenology Institute. Web. 3 September 2014. http://www.phenomenology.org.

# Life and Human Life in the System of World Coordinates on the Basis of Extreme Dynamic Equilibriums

**Nikolay N. Kozhevnikov and Vera S. Danilova**

**Abstract** Life is a complex of natural formations, and Human Life is even more a complex of cultural formations. Their geneses connect with the principles, rhythms, and laws of self-organization, deterministic chaos, and the coordinate system of the world on the basis of a marginal dynamic equilibrium. All natural and cultural formations strive to achieve three fundamental equilibriums: identification, a communication-network, the full-time existence of its formation. However, they never reach these limits. The coordinate systems are created by parts of natural and cultural formations, at the expense of the energy that is balanced within them. It creates a general "cell dynamic equilibrium" for its formations and coordinate systems.

The surrounding world consists of two unequal parts. On the one hand, there are the chains and structures of interrelated limit dynamic equilibriums. In the formations at various levels of organization of the world, they are the same. On the other hand, there is the rest of the world, covering nonequilibrium processes and phenomena. All the natural and cultural formations and their structures can be related to the ultimate three types of fundamental equilibriums. These limits are mapped to the coordinate axis, which is associated with these fundamental limits. All the processes on the unknown levels in the world are balanced by means of the "marginal boundary surfaces" such as "inertia," "quasi-static processes," "spirituality." They are simpler than the others and are associated with the natural coordinate system of the world.

**Keywords** Dynamic equilibriums · Limit equilibriums · Identification · Communication-network · The full-time existence of a formation · Natural and cultural formations · Marginal boundary surfaces · Inertia · Quasi-static process · Spirituality

---

N. N. Kozhevnikov (✉)
Philosophical Society of the Sakha Republic (Yakutia), Sakha (Yakutia), Russia

V. S. Danilova
North-Eastern Federal University, Yakutsk, Russia



W. S. Smith et al. (eds.), *Eco-Phenomenology: Life, Human Life, Post-Human Life in the Harmony of the Cosmos*, Analecta Husserliana CXXI,
https://doi.org/10.1007/978-3-319-77516-6_13

Many aspects of "the world coordinate system on the basis of limit dynamic equilibriums" would better be investigated with the help of a phenomenological approach. The ontological approach makes it possible to identify the grounds for a coordinate system in reality; the metaphysical approach outlines the notions of the coordinate system's basics and its universalism; and the phenomenological approach makes it possible to "see and catch on to" the coordinate system as a whole, getting into its very existence with the help of sequences of limit relative equilibriums, which ascend to it.

## Laying the Ground

A natural coordinate system of the world, that is, one based on the ultimate dynamic equilibriums generated therein, must exist because, despite the complexity of the world, it is perfectly organized, reasonable, optimal, and stable. All its levels have associated with them the cycling of matter, energy, information. On the other hand, complexity and chaos are widespread in the world and, in future interactions with humanity in all spheres of its activity, will be intensified, as will be the need for understanding these processes as well. That is why it is logical that self-organization and deterministic chaos are accepted as grounds for a natural coordinate system of the world.

Ultimate fundamental equilibriums are defined by different types of deterministic chaos, emerging by means of self-organization and encompassing all levels of the hierarchy of the world. The world is extremely complex; its processes are unbalanced and nonlinear. However, the natural coordinate system based on these ultimate fundamental equilibria, which it creates in the process of its development, is simple, and available to be explored in all areas of human knowledge and activity. Many philosophers and scientists (E. Rutherford, R. Feynman) either emphasized that linear, simple representations of the world may describe it quite reliably, or presumed their existence (I. Kant). Such a coordinate system is especially needed in the humanities, where the criteria of knowledge are quite vague but there is an objective part of knowledge which can be explored on the basis of cognitive and digital approaches.

The natural coordinate system was something of which various thinkers throughout human history were aware. Some of its aspects were identified with God, Eidos, Spirit. Of special importance for this is the experience of ascetics, of hesychists, and apophatic theology, the concept of *epoché*. The most complete definition of the coordinate system is given in mathematics, astronomy, and geography. Examples of coordinate systems include inertial systems in mechanics, quasi-static processes in thermodynamics, and the coordinate system framed on the basis of fundamental physical constants by M. Planck.

The main idea applied in developing our approach is formulated in the following way. On the one hand, the chains and structures of interrelated limit dynamic equilibriums in the formations at various levels of organization of the world are the same. On the other hand, all the rest of the world covers nonequilibrium processes and phenomena (Kozhevnikov and Danilova 1993, 19). The main idea of the natural coordinate system of the world is that all the natural and cultural formations and their structures can be related to the ultimate fundamental equilibriums of three types. These limits are mapped to the coordinate axes, which are associated with the fundamental limits.

## The Three Fundamental Equilibrium-Limits

All natural and cultural formations may be associated with three types of ultimate fundamental equilibriums. According to our proposed concept, the natural coordinate system encompasses the whole universe. This approach opens up opportunities for studying the unity of processes and phenomena at all levels of the hierarchy of the world and makes the case for the integration of world cognition. Finding such a universal natural coordinate system should be the main task of philosophy, which is associated with the ultimate foundations of science and culture nowadays.

All the natural and cultural formations strive toward three extreme fundamental equilibrium-limits by means of self-organization processes. Firstly, any natural and cultural formations from the spheres of non-living, living, and spiritual (elementary particles, molecules, gaseous nebulae, living organisms, individuals, personality) strive toward a self-identification limit (the *I-limit*). Secondly, all these formations strive toward a communication-network limit, that is, they tend toward completeness and a stable equilibrium with their environment. An individual is defined by his or her communications. Cultures exist through dialogues within and outside themselves (the *C-limit*). Thirdly, the existence of all natural and cultural formations is limited to the full time or term of their existence (the *K-limit*), which can be determined only from the point of view of the neighboring structural levels and is unattainable from within these formations. All these limits cause and yet remain unachievable in real processes, which are caused by conflicting trends, and therefore the natural and cultural formations reach only some intermediate dynamic equilibrium states.

Any natural or cultural formations can be connected with these limits by the three coordinate axes. Already in the first concepts of philosophy, an orientation to the identification limit was decisive. The dialectics of Plato sought to distill essences, bypassing all that is transient, ancillary, and incidental. In later philosophical systems, special approaches and methods for the identification of entities according to the properties of things were created through consistently discarding all psychological, personal properties, and so on. For example, Kant closely correlated human identity with what he called the initial synthetic unity of apperception.

The communication-network limit involves a very wide variety of mechanisms that ensure the completeness of the communication system, which is unattainable for us, as well as knowledge of all the peculiarities of its formation. For example, the process of cognition comes about through the development of several networks:

rational, cognitive, those connected with the subconscious. The separate processes of self-organization are in the area of transcendental and include all possible mechanisms ensuring the pursuit of that limit. Communication within has contributed to the emergence of numerous interdisciplinary areas. Identification and communication within are most clearly expressed in the spheres of culture, philosophy, and science.

Investigation of the full terms of natural or cultural formations is connected with jumps and radical changes in our ideas about them. The view from without changes estimates of the activity of individuals. Those who were considered outstanding are, with time, perceived more modestly, and numbers of neglected personalities are later seen to actually be very important figures. This view discards all minor details, leaving only what is most important; the work of a scientist is much clearer to his successors than to his contemporaries, but the details of his personal life are more hidden from us. In the case of support deriving mainly from this limit, it gives vast opportunities for myths, as takes place in modern pseudo-histories, for example. The limit represented by the full term of the existence of natural or cultural formations becomes the axis around which mythology and religion "spin" for many millennia. Approaching the limit, the clarification of the essence of this limit is to be a gradual and lengthy process. The innermost core of myths, religions and metaphysics is rooted in the unknown, the transcendental.

Many philosophers, such as I. Kant, G.W.F. Hegel, said that it took them almost all their lives to clarify the ideas that appeared at the beginning of their philosophical way. All of this can be seen as stages of interaction with the coordinate system of the world, first a "touch" grazing it, then the gradual formation of thought on it related to a process of reflection.

## The Natural Coordinate System and the Rest

The coordinate system is based on the equilibriums of dynamical chaos, created by specific natural or cultural formations, through that portion of energy that can be balanced. The result is a cell of dynamic equilibrium ("cell interconnection"), the system of coordinates, and this formation occurs simultaneously. The coordinate system has no location or any spatial-temporal constraints; it exists in every part of the world, at all levels of its organization. The coordinate system interacts only with open natural or cultural formations, with a tendency towards self-development or dialogue.

The formal split of the surrounding world into a natural coordinate system and the rest of the world resembles the "natura naturans" and "natura naturata" distinction made by B. Spinoza, but its meaning is quite the opposite. Spinoza selects the most active part of nature, that which ensures its self-development. We highlight the passive part, the aggregate limits (attractors), in relation to which the rest of nature are the processes of self-organization. Searching the coordinate system becomes a "coordinate method" for philosophy's researches, one which has universal flexibility and allows researchers to interact with almost any natural phenomenon and process. Any intermediate dynamic equilibrium, in which the fixed nature of a formation has not yet met the limits of self-organized identification and systematic communication limits will remain stable only if they will be gathered in "calibration" (space-time) nodes, which are separated from each other by intervals corresponding to the frequency oscillations of these fundamental limits, which can be considered the fundamental rhythms of the world. Calibration originating from the full-time existence of natural and cultural formations reveals the optimality and stability of these "steps" and rhythms. The combination of all three types of "steps" and rhythms that conform to these limits, build sustainable natural and cultural formations, and these "steps" and rhythms define the parameters of the spatial-temporal cells' coordinate system on the basis of "deterministic chaos" (Prigogine and Stengers 2000, 77). In their development, these "steps," rhythms, and cells are sent into coordinate system and can exist in the world for millions and billions of years as, for example, galaxies, planetary systems, the atmosphere and the hydrosphere of the Earth and their elements, and philosophical and religious vistas (hundred and thousands of years). In the case of deviation from these rhythms, all natural or cultural formations are destroyed.

Calibration of the above-mentioned limits means that they can be used as the basis of scales, and the units for these scales are obtained by multiplying the values of these bases by the corresponding coefficients. The cell made up of these fundamental limits will be the largest of all. For the corresponding coefficients for these limits (identification, systematic communication, the full-time existence of natural or cultural formation), let these be $k_1$, $k_2$, $k_3$.Then the units of scale will be respectively $k_1 I$, $k_2 C$, $k_3 K$. The cell formed by the fundamental limits can be defined as the "*ICK*," and a large-scale cell at a certain stage of development of natural or cultural formations as "$\kappa_1 I \kappa_2 C \kappa_3 K$." These cells are unusually stable. There are many examples demonstrating that their existence can last extremely long: billions of years in inanimate nature; millions of years in living nature; thousands of years in the world of ideas and knowledge. They have a minimum of energy, information, spirituality, and other similar settings, and their main characteristic is their becoming a "connected substance" and its specific values of connected energy, information, spirituality.

We should first of all pay attention to the size of the cell, because of their dimensions; corresponding to certain types of quality, they are huge. The first axis of the coordinate system, that of identification, is suited to describing the essence of the research object. The second axis, that of the communication system, characterizes the most stable and optimal state of existence of natural and cultural formations. The third axis encounters the whole life of the research object. However, calibration splits off cells of smaller scale; in the end, the same person considered as such a formation is now presenting quite foreseeable time intervals, limits, confining its identity to a specific time period and its availability for using its communications system. All of this must match a spatio-temporal coherence, to comply with the unitary natural rhythms.

## Calibration and Destiny

Calibration in that cell proceeds as follows. It is advisable to start with a "full-time existence" and choose several points (marks) determining its development process in the future, as well as the most important point that influences this process from the past. From these points we define those marks most important for the investigational phase of this process. Similarly, you can choose point-labels that characterize the essence of natural or cultural entities under study and that characterize the system-communication of each entity. It may occur in the intermediate version of the cell, but once one has covered all fundamental limits, all excess marks will disappear; that is, the above limits will filter out all that is unnecessary in this cell. The more specifically we carry out research, the less the cell itself is considered. In the limit of a specific interval duration, the entity is required at this stage of the process to use the option of a system of communication. For example, when the lifetime of a human individual has passed for a certain period, the time of his or her existence begins not from the day of birth but from the day of death because the mechanism of this reference is changed. Nobody knows the date of their death; it is hidden from everybody, and man is afraid of death. However, after a certain point, one begins to appreciate one's life in accord with that term; one starts to feel differently. This is a kind of "calibration" of the process of life – a signal sent somewhere unknown, transcendentally, comes back and lets you know that there's still time for some things, within specific creative boundaries, for learning something new, a remake of the old self. One begins to ruthlessly drop all unnecessaries associated with vanity and one's life line is cleaner; following the calibration signal clarifies this even more.

In a natural coordinate system, a person's life proceeds from the representation of destiny developed in this paper. It is enough to engage once in sustained interaction and life just expands and deepens, it being expressed, for example, in existential prayer, where man brings life to sacrifice its purpose so that the person can be defined as "willingness to sacrifice." Destiny (or calling) is to be obeyed and borne as something absolute, because it is directly connected with an absolute being. We are to follow it, even when the aim is not feasible. The ancient Chinese believed, therefore, that a philosophical concept is true only when it is embodied in the life of its Creator. This corresponds to the pursuit of a life in accord with the rhythms of the natural coordinate system. In addition, many thinkers and prophets have emphasized that the right to a destiny is not given to everyone, but only to the elect.

## Leveraging Limit Boundary Surfaces

The originality of the approach developed here is that through fundamental relative equilibrium, we "close" all prior and unknown to us levels of the world, leveraging the existing process's "limit boundary surfaces" such as "vacuum," "inertia," "quasi-static process," "spirituality." "On top" of them there remains only fundamental equilibrium value of the available research that is responsive to the processes of the unknown parts of the world. "A limit boundary surface" is composed of multiple equilibrium "cells," the existence of which is provided by processes seeking identification and hidden within this surface. Sustained interaction by the "cells" of this "surface" is provided by their desire for communication-network limits. The equilibrium parameters of a vacuum, inertial systems, quasi-static processes, and spirituality have been identified well enough. The full term, the whole of the times of the studied natural and cultural formations, defines the basic rhythm of oscillation-related "limit on boundary surfaces." The equilibrium concepts remaining outside of these "surfaces" are simple and associated with the natural coordinate system of the world.

"A limit boundary surface" corresponds to a vacuum (a fixed state of quantum fields with minimum energy, zero-momentum, angular momentum, electric charge, and other quantum numbers) "closing" the whole unknown to us part of the world, providing an equilibrium for the sustainable existence of the vacuum. The properties of a "vacuum" define the properties of all the major states of the world, leading to the emergence of quantum fields, which in modern science are the most fundamental and universal form of matter and the basis for all of its concrete manifestations. All elementary particles are the quanta of certain physical fields that continuously interact with each other. Thus, the vacuum can be considered as the simplest system of reference for the level of organization of a world defined by elementary particles.

The next level of this "cover" deals with forming established sustainable matter and gravitational interaction. The parameters reference system for remaining in this "boundary surface-age" of the world is inertia, and its measure is the mass that has allowed the formation of ideas about inertial systems as a basis for all other more complex systems of reference on the level of organization of the world. Another ultimate boundary surface distinguishes a complex object's macrocosm, each of which consists of a huge number of particles and is characterized by a "thermodynamic equilibrium." At this level of the hierarchy of the world, views of equilibrium are associated with the concept of "dynamic chaos." The "limit boundary surface" of life leads to the formation of complex self-organizing systems, molecular chains, and preliminary "pre-life" structures. On the level of the soul, the individual has a matching level of identification and all social formations: a clan, tribe, ethnic group communication system through which people can survive. At the level of a spiritual person, there are two ways to God: internal (existential) and external. Modern philosophy, moving from the subject of classical philosophy, to knowledge of the inside of the investigated process, focuses on personal-dimensional complexes that are the person in an ethical context. Here, human-dimensional personality corresponds to the "*I*-limit," and the human-dimensional complex, to the "*C*-limit." The time-life of a human-dimensional personality expands in comparison with the ordinary individual, the existence of which ends with his death. This person engaged with the planetary problems of humanity becomes only a lacuna as these issues endure. The existence of this person may be continued in the cells of the noosphere, or in a more general form in the cells of the coordinate system of the world.

Next, the "limit boundary surface" is connected with the spiritual, which is a complementary balance between all the "*I*" subsystems, above all, the four pillars of the body, the mind, the subconscious, and super-consciousness. In the spiritual man, all these subsystems are in harmony: the mind and the body are healthy, the subconscious mind is well-organized and controlled by consciousness, and super-consciousness (cultural codes, religion, ideology, traditions, ethos) is humanistic. These subsystems' equilibrium interact with each other, all within in a person: his mind, the subconscious that is closed within this "limit boundary surface," over which stand the main parameters of spirituality: "free will" and "cultural secular asceticism." This "cultural and secular asceticism" is a bound state of all the intellectual, social, individual manifestations of personality and is able to be the foundation of a universal synthetic culture.

Limit boundary surfaces ("veils") correlate with Kant's "noumena" ("things in themselves") and "phenomena" ("things for us"), although having a somewhat different meaning (for us, Kant's "noumenon" is a limit concept, limiting the claims of sensuality). What is hidden from us under the limit boundary surface is unavailable to research based on the equilibriums corresponding to this surface, such as spirituality, for example. Research and its related theories, concepts, deal exclusively with phenomena.

## Distortions to Be Avoided

Ideas about the natural coordinate system of the world and its basic concepts – cells, layers, the fundamental limits – can be greatly distorted. Such distortions yield many approaches to classical philosophy and science. But the absolutizing of only one of the above-mentioned limits leads to an unstable equilibrium that sooner or later will cease to be, will be unable to sustain long-term spatial, temporal, and spatio-temporal coherence with natural rhythms and the communities and the processes inherent to the various levels of organization of the world. The cells within "equilibrium-covers" can be regarded as static and dynamic. For identification (identity) of the "equilibrium-covers" fairly static cells are based on two of the limits, the *I* limit and the *C* limit. One can register views about cells of the following types: a marginal cell, cells with two limits, cells with three limits. An isolated marginal cell lets us just "catch" the natural coordinate system. Cells with two limits can be considered by making a static cross-section. Marginal cells having three limits allow the use of the natural coordinate system in research of dynamic processes. The environment of the University in the Middle Age can serve as an example of the formation of artificial cells, a kind of simulacrum of the natural coordinate system. On the one hand it displays a process of self-identification, but one that was not the result of the self-organization of an individual (personal identification). In this process everything was set to be like God (the principle of Absolute personality). On the other hand, there was a clear focus on the "*C*-limit": a common language (Latin), public theses (which lasted sometimes for several days), rapid crowded debates. However, all of this social formation and communication system also corresponded to certain predefined rules. As a result, the cell was created artificially; self-organization is excluded on purpose.

The communication-network limit is widely used in the social sphere when a particular community of people gather together around a characteristic exclusive claim to be the most correct, for example, most close to God, or most closely following true teaching. This is a very powerful temptation, for exclusivity makes life easy and comfortable. The most prominent examples are Orthodox directions in religions of primitive ideology. Modern universalism may be linked with the natural coordinate system on the basis of planetary and human limit states based on the equilibria of interconnected terrestrial environment.

## Accessing the Natural Coordinate System

The natural coordinate system appears to us as a "tabula rasa," although, in fact, its organization is rather complicated. This coordinate system is pure being, the perfect design of all possible limiting conditions, which may not blend with natural and cultural formations, though natural processes periodically interact with it in accordance with its internal rhythms. The coordinate system should be opened directly and be available to everyone. The interconnected system of dynamic equilibrium can be accessed by developing sensual submission or by using theoretical models drawing on the "emptiness" of Taoism or the religious experience of ascetics. Every individual may interact with different levels of the coordinate system: personal, the cultural inheritance of their ethnic group, the planetary. For this purpose, these should be stacked, summing up the cells of all these levels. Finding the relationship with the coordinate system starts with the ability to listen to one's own quietness, finding the fundamental mood for beginning the process of cognition, touching various kinds of "emptiness." In this way, the individual forms the original cell of the natural coordinate system – which "meets" the person, an individual, involving them in their self-organization and development.

The search for the natural coordinate system does not mean aspirations to identify a universal philosophical category or concept. It is about the study of the unity of the benchmarking processes and phenomena at different levels of the structural organization of the world. This matches one of the main principles of synergy, the claim that the processes of the creation (of growing complexity and order) has a single algorithm, regardless of the nature of the systems in which they are carried out. However, our approach considers the unity of the algorithms of these processes in their interactions with the natural coordinate system. A natural coordinate system cannot be comprehended within a single principle or one idea; it gradually becomes clearer as one uses it. This awareness can start with any phenomenon: the respective religious, artistic, scientific, everyday processes engaged at any time.

All this correlates well with the concept of autopoiesis in modern philosophy (Maturana and Varela 1980, 78). Autopoiesis is a process of generation of the elements of the system itself as something different from the surrounding world, the natural "seeding" of this system. The item is a certain operation, which requires energy and inclusion in the relevant cause-effect relationship. Autopoietic interactions can be identified everywhere: in inanimate and animate nature, in the human and social spheres. Modern philosophy has moved its focus to the knowledge inside the investigated process, without which the existence of a natural coordinate system would be impossible, and is developing ideas about the human-dimensional complex. That is to say, instead of a subject of classical philosophy, it focuses on human-dimensional complexes. This last defined has many facets, for here is man in the context of ethics. A human-dimensional personality corresponds to the *I* – limit, and the human-dimensional complex to the *C*-limit. The time-life of a human-dimensional personality is an expanded one entering into the planetary problems of humanity, compared with the ordinary individual, the existence of whom ends with his or her death. Again, the existence of this person may be continued in the cells of the noosphere, or in a more general form in cells in the coordinate system of the world.

## Approach to the Coordinate System in Stages

A coordinate system is not initially a given. It is developed with the participation of natural and cultural formations. The sphere of the living world around us and all its levels (genetic, cellular, ontogenetic, population-specific, biogeocenosis, biospheric, celestial) possess specific mechanisms for interaction with the coordinate system. On the one hand, the quasi-steady-state processes of a living being's development at all these levels participates in the development of the world coordinate system. On the other hand, these very processes occur on the basis of the coordinate system.

A phenomenological approach makes it possible to sequence the phenomena of spirit, consciousness, presence, being. Our approach assumes the sequencing of limit-relative equilibria, including those based on deterministic chaos, which are quite easily organized in the sphere of life, are obvious, and are strongly linked with each other.

A living being is developed through relative equilibria that were first formed accidentally but, then, as a result of a self-organization process, steady states of these equilibria were realized, and after that their sequence was realized, the ascent of which helped to make the notion of a coordinate system more and more clear. The process of coordinate system development as "the thing itself" is endless, but at each stage the particular type of interaction of natural or cultural formations with this coordinate system is realized.

Regarding human beings, individuality, and personality, these can experience interaction with the coordinate system directly on the basis of limit dynamic equilibria. According to the ascent from human being to personality, these experiences increase in depth and variety. A personality's life, from a certain level of its development, may flow directly into the coordinate system, like the quasi-equilibrium sequence of his direct experiences. Individual and personality interaction with the coordinate system has two main stages to be singled out.

The first stage follows from the fact that the individual's ways interact with the specific elements of the environment. If he manages to enter into the sequence of the equilibrium amid changing conditions, set up for the immanent configuration of these interactions, then, with their help, he will be able to directly find the main point of the coordinate system. The coordinate system may be seen at once, in its entirety, but just for a moment. This apprehension as a whole inevitably contains the components of the natural attitude and can be realized through any natural or cultural formation at any stage of their development. Random interaction with the coordinate system surprises, making it possible to see its harmony and beauty, and this feeling turns out to be so unusual and mind-bending that one wants to come back to it again and again and looks for ways to induce such returns.

At the second stage, for these interactions to become stable and efficient, a certain order of actions is required to maintain the functioning of the sequential equilibria. An individual may rely on great number of equilibria and rhythms (space, time, space-time). A human possesses about one hundred circadian rhythms, and the individual may build the sequence of equilibria, using an endless variety of combinations among sequential relative equilibria and quasi-steady-state processes. For instance, first, one can approach the coordinate system with the help of unconsciousness, wherein one can use the equilibria and stable rhythms of body, and then shift over to equilibrium state of consciousness, and come back to the equilibrium state of unconsciousness and the body again. This cycle can repeat again and again, involving new types of dynamic equilibria and quasi-steady-state processes, forming them from elements of different world organization levels. There, quests for equilibrium states become more stable with time, ensuring a progression to awareness of the deep levels of the coordinate system. In each of these subsystems of living being, unconsciousness, and consciousness, there is the formation of elements that create a unit interacting with the coordinate system, and this unit is common for it and the individual.

## The Coordinate System's Approach to the Person

Personality, especially a creative personality, has more chances to interact with the coordinate system since, in this case, relative equilibria and quasi-steady-state processes may be used in science, art, literature, education, and so on. Equilibrium interrelation through particular cultural elements may be realized more variously than can be realized by the system at a living being level (individual, population, etc.). Accordingly, a sequence of elements and their structures is built, which forms the sequence of relative equilibria, one connected with the human, individual, and personality and focused on fundamental limits, to which the personality moves closer and closer in the process of his or her activities. The main self-organizing processes here are people's internal dialogues with themselves and with others, but in each elementary act of these dialogues there shall be reference to the coordinate system. For a dialogue to be real, it should be carried out from the individual to the coordinate system and from thence to the other individual. The coordinate system should be present between dialogue participants in the same way. Cultural formations in these processes become free from everything intermediate, transient, external, sensual-objective. For example, they tend to the principles of secular asceticism.

The approach, based on sequences of relative equilibria ascending to the three fundamental limits here emphasized, is considered to be a middle path between the traditional ontological and the phenomenological approaches. It correlates very well to the concept of the "third dimension" of human reality, developed by M. Merleau-Ponty and represented as the scope of life relations, which unite nature and consciousness. According to this concept, the human being is not opposed to the world, but is a part of the "world's flesh," in which he lives as in initial integrity.

The closest connection with the coordinate system leads to its constant "sounding," like Hesychast prayer. The coordinate system "knocks itself" on the individual and personality. This "knock" is very quiet, like God's "knock," but then it touches upon far more subsystems, connected with the body, unconsciousness, and so on. It seems as if the coordinate system "grows" in the individual and even more in the personality. Under the conditions of the modern world, the rivalry between different cultures and their diversification make it more complicated and difficult to find the coordinate system on the basis of limit dynamic equilibria. But the need for this coordinate system in the modern world also becomes more significant, as well as the need for the development of notions of it.

Modern phenomenological approaches have many aspects. "Phenomenology is against the unrestrained relativism of our times, when the criteria of significance, truth, validity are rejected and the measure of utility for human existence is accepted instead of them. It is understood as the ontopoiesis of life; phenomenology persistently searches the final grounds of all things" (Tymieniecka 2005, 152).

## Works Cited

Kozhevnikov, N.N., and V.S. Danilova. 1993. *The Stage of the Self-Development of Nature and the re-Romny Stage of the Development of Mankind.* In Summary XIX World Congress of Philosophy. Vol. 1. Moscow. Print.

Maturana, H.R., and F.J. Varela. 1980. *Autopoiesis and Cognition: The Realization of the Living*, Boston Studies in the Philosophy of Science. Vol. 42. 2nd ed. Dordrecht: Springer. Print.

Prigogine, I., and I. Stengers. 2000. *Time, Chaos, Quantum: Towards Solving the Paradox of Time.* Moscow: Editorial URSS [In Russian] Print.

Tymieniecka, A.-T. 2005. "Fenomenologiia otvechaet na vyzov pragmaticheskoi proverki" Russian. *Voprosy filosofii* 11: 145–162. Print.

# Eco-Phenomenological Vision: Balancing the Harmony of the Earth

**Debika Saha**

**Abstract** Phenomenology, a movement which started its journey through the writings of Edmund Husserl, has now occupied a special place not only to the philosophers but also to those who are really eager to save the earth from environmental degradation. But phenomenology deals with certain key concepts like its resistance with naturalistic attitude and keeping engaged within the 'intentional' realm, which forces one to question: how could there be a phenomenology of nature? Though it may seem outwardly that phenomenology deals with certain concepts which stand in quite opposite directions to those who talk about environmental wellbeing of the earth, if one goes deeper the picture presents a different view.

The return to "things themselves" and the critique of scientific naturalism both point in the direction of much contemporary environmental thought. In fact, phenomenology offers a space for the interdisciplinary examination of our relation with nature. The nature-culture vision provides an open horizon for the exploration of all possibilities regarding our relation with the cosmos as a whole. The present paper will try to unveil how, within the phenomenological realm, eco-phenomenological vision is possible.

**Keywords** Eco-phenomenology · Naturalistic attitude · Phenomenology of nature · Nature-culture vision · Intentional realm

Anna-Teresa Tymieniecka, in her interview on 27 August, 2008, in Bergen, Norway, shared her vision of eco-phenomenology in the following way. According to Tymieniecka's philosophy, human beings should be considered as a human condition within the unity of everything that is alive. That means human being unfolds and generates in a mutual contributive relation to all the other living beings. This is the spirit of eco-phenomenology, which demands a kind of openness to enter more deeply into the 'sensorial present' and to recover the moral sense of our humanity by uncovering the moral sense of nature. The eco-phenomenologists argue that it is

D. Saha (✉)
Department of Philosophy, University of North Bengal, Darjeeling, West Bengal, India

 
W. S. Smith et al. (eds.), *Eco-Phenomenology: Life, Human Life, Post-Human Life in the Harmony of the Cosmos*, Analecta Husserliana CXXI,
https://doi.org/10.1007/978-3-319-77516-6_14

possible to save the earth from environmental degradation with the help of phenomenological vision.

The present paper is an attempt to trace back the phenomenological vision as initiated by Husserl and later developed by Heidegger which helps to formulate the eco-phenomenological vision to regain the balance of the earth.

Phenomenology is a science of the essences of consciousness and of the ideal essence of the objective correlates of conscious acts. To reach these essences, it is necessary to bracket the naturalistic attitude. Consciousness should not be viewed naturalistically as part of the world at all, as it is due to consciousness that the world exist for us. Husserl is of the opinion that it is not that consciousness creates the world in any ontological sense. This view is formed under the naturalizing tendency whereby consciousness is viewed as cause and the world its effect. But, for Husserl, the world is opened up or disclosed though consciousness. It is not possible to know the world without consciousness. But, if one treats consciousness as part of the world, then the foundational open role of consciousness get ignored. It is in this sense that Husserl is against the naturalistic attitude. As consciousness plays the most vital role in all aspects of knowledge, so the proper approach to the study of consciousness itself must be a transcendental one.

Now here, a question arises: if Husserl is against the naturalistic attitude, how could there be a phenomenology of nature? In answer to this question, it may be pointed out that Husserl is not against the 'idea of nature'. *Ideas II*, one of the most important of Husserl's works, begins with the discussion of the 'idea of nature' in general and then discusses about material, animal, and human nature and concludes in the realm of personhood and spirit. Here, Husserl discusses the way through which we relate to our bodies and the surrounding world.

Along with the work on social constitution and human personal world, Husserl discusses at the same time the work on transcendental subjectivity. This attitude proves that Husserl did not treat the two approaches as one of conflict. On the contrary, he shows that both were necessary to the full understanding of the constitution of the objective world, including the aspect of nature and culture.

This relation is also visible in Heidegger's later writings, wherein he offers a deep, insightful analysis of the encompassing nature of the global technological framework which now threatens to engulf genuinely human modes of existence. Despite his orientation towards the question of being, Heidegger's thought is deeply phenomenological. In one of his essays, "My way to phenomenology," Heidegger claims that what he gained from phenomenology was the practice of "phenomenological seeing" (Heidegger 1972, 78). As he comments in the introduction to *Being and Time*:

> The following investigation would not have been possible if the ground had not been prepared by Edmund Husserl, with whose Logical Investigations phenomenology first emerged. Our comments on the preliminary conception of phenomenology have shown that what is essential in it does not lie in its actuality as a philosophical movement. Higher than actuality stands possibility. We can understand phenomenology only by seizing upon it as a possibility. (Heidegger 1962, 62–63)

In fact, the concepts of 'world' and 'environment' Heidegger owes to Husserl's *Ideas II*. But, as a phenomenologist, Heidegger's position is quite different from Husserl. According to Heidegger, phenomenology is the attempt to make manifest the matters as they manifest themselves. In *Basic problems of phenomenology*, Heidegger denies that phenomenology is a method in any specialized sense. Phenomenology is a new way of seeing rather than a set of philosophical propositions. He claimed that it was Husserl's vision of 'seeing' which is important for him. This 'seeing', for Heidegger, meant doing away with all philosophical theories, whether idealist or realist, and cultivating a 'pure vision'. Though he identifies with Husserl's slogan "Back to the things themselves," he had also claimed that the vision of phenomenology should really be *'Set Dasein free'*. Phenomenological vision must be able to understand *Dasein* from within the concrete particularity of a lived life.

Coming back to eco-phenomenology, let us try to set up a crucial step in making room for eco-phenomenological perspectives in the environmental discourse. In *Eco-Phenomenology: Back to the Earth Itself*, Charles Brown and Ted Toadvine observe:

> The intersection of ecological thinking with phenomenology, the momentum that drives each toward the other, begets a new cross disciplinary inquiry; eco-phenomenology. Eco-phenomenology is based on a double claim: first, that an adequate account of our ecological situation require the methods and insights of phenomenology; and second, that phenomenology, led by its own momentum, becomes a philosophical ecology, that is a study of the interrelationship between organism and world in its metaphysical and axiological dimensions. (Brown and Toadvine 2003, ix–xxi)

This interdisciplinary perspective grew out of a gap between culture and nature, between action and thought, between techne and a new ethos.

Following David Wood's article 'What is Eco-Phenomenology', we may point out some of the important aspects that phenomenology shares with eco-phenomenology. Wood begins his essay with the issue of naturalism. Naturalism is concerned primarily with the laws of causality. But, the phenomenological standpoint views causality as but one dimension that structures the possibility of perception. It is due to our unique embodiment that we are confined to understand perception on the grounds of causality and, for this reason, naturalism views causality as constituting, rather than qualifying, perception. But, embodiment is a purely phenomenological structure; that is, it qualifies perception on its basis. What Wood contends here is not to find a space for phenomenology within naturalism, but to demonstrate that naturalism is but one way by which to approach the world.

Here we may mention Erazim Kohak's definition of naturalism. Kohak, a leading eco-phenomenologist, describes naturalism as "reflect[ing] the late medieval division of reality into two realms, conceived of as almost two distinct natures, one 'natural' the other 'supernatural' ... Thus naturalism came to mean a philosophy which accepted as normative of 'reality' the reality construct of the science favored by a given 'naturalistic' thinker" (Kohak 1984, 7).

David Wood also shows that "phenomenology was born but of resistance to the threat of naturalism and elucidates 'the need to rescue nature from naturalism'" (Brown and Toadvine 2003, 211).

Wood shows that the grounds by which we understand the world can be assessed either by means of intentionality or by causality. As he comments: "if we accept that connection to practical agency is central to intentional meaning, it does locate intentionality within an inter active nexus from which causal powers cannot be separated … If this is so intentionality is firmly lodged within my bodily existence, within the natural world" (Brown and Toadvine 2003, 222).

But, Heidegger explains the above point in *Being and Time* in the following way:

> If one understands Nature ontologico-categorically, one finds that nature is a limiting case of the Being of possible entities within the world. Only in some definite mode of its being in-the-world can *Dasein* discover entities as nature. This manner of knowing them has the character of depriving the world of its world hood in a definite way. 'Nature' as the categorical aggregate of these structures of being which a definite entity encountered within the world may possess can never make world hood intelligible. But even the phenomenon of 'Nature' can be grasped ontologically only in terms of the concept of the world, that is to say in terms of the analytic of *Dasein*. (Heidegger 1962, 93–94)

So, an eco-phenomenological enquiry must address the question of the meaning of being. Thus, our direct experience of the world is grounded on our "Being-there."

In fact, the ecological crisis is a crisis of meaning. As Charles S. Brown shows, it is ultimately the meaning of nature and humanity that faces a big problem nowadays. Alienation is the worst enemy that creates a big gap between humanity and nature.

This tendency of alienation, along with the population explosion and technological power, now dismantles the Earth's bio-web. Phenomenology's specific contribution to eco-phenomenology rests in the attitude of respect for experience. This respect for experience is one of the main pillars on which the concept of eco-phenomenology stands (Brown and Toadvine 2003, 3–5). It is through phenomenological reflection that it is possible to discover the oneness of the human person and the rest of the world.

## Works Cited

Brown, Charles S., and Ted Toadvine, eds. 2003. *An Introduction. In Eco-Phenomenology: Back to the Earth Itself.* Albany: State University of New York Press. Print.

Heidegger, Martin. 1962. *Being and Time*. Trans. John MacQuarrie and Edward Robinson. Oxford: Basil and Blackwell. Print.

———. 1972. Being and Time. *On Time and Being*, Trans. Joan Stambaugh. New York: Harper and Row. Print.

Kohak, Erazim. 1984. *The Embers and the Stars*. Chicago/London: University of Chicago Press. Print.

# From Anna-Teresa Tymieniecka's Eco-Phenomenology to Paul Ricoeur's Hermeneutics. The Role of the Human Being in the Global Context of Cosmos, Chaos and Evil

**María Avelina Cecilia Lafuente**

**Abstract** The recently widespread term 'eco-phenomenology', following from the concept of Ecology, leads science and philosophy to reformulate the link between the human being and the rest of beings and the cosmos in general. Nevertheless, a clear opposition arises between scientific-objective procedure, centered on causal laws, on the one hand, and the phenomenological approach, focused on intentionality and consciousness, on the other. Some authors suggest that the field of eco-phenomenology could help to elaborate a middle ground between intentionality and causality. Anna-Teresa Tymieniecka develops this middle ground. Always interested in scientific fulfillment, she centered her Philosophy of Life and the Cosmos on her concept of "Ontopoiesis", which is a fundamental key to understanding her perspective on eco-phenomenology. "Imaginatio Creatrix" points to the creativity implied in all processes of life and especially in the work of human imagination in inventing "possible worlds". However, it is Paul Ricoeur's task to explore the content of mythical narratives in order to go from the products of symbolic imagination to an interpretative hermeneutics that takes into account the genesis of evil – and good – its possible end, and the human role in the entire global process.

**Keywords** Eco-phenomenology · Anna-Teresa Tymieniecka · Paul Ricoeur · Hermeneutics · Evil · Myths · Meaning

---

M. A. Cecilia Lafuente (✉)
University of Seville, Seville, Spain
e-mail: cecilia@us.es



W. S. Smith et al. (eds.), *Eco-Phenomenology: Life, Human Life, Post-Human Life in the Harmony of the Cosmos*, Analecta Husserliana CXXI,
https://doi.org/10.1007/978-3-319-77516-6_15

## Introduction

The human being cannot be properly understood as an isolated being, without first having been situated in the complex context of the whole reality of the Cosmos – "order" in the etymological sense. However, that "whole reality" can also present unpredictable, chaotic, harmful aspects to human experience which can be recognized as Chaos.

Nowadays, humans, belonging to the Great Chain of Being and, more specifically, included in the Unity-of-everything-there-is-alive (to use Anna-Teresa Tymieniecka's expression), are perhaps more aware than ever of their cosmic origin and of their role, if not in the Cosmos, at least on Planet Earth. That awareness could be a first step, made at the ontological level, following the guidelines of the Phenomenology of Life; it could also guide us down the path towards the approach recently called eco-phenomenology, which tends to develop its thinking not only at a theoretical level but also at a practical one. The latter reminds us of the responsibility we have, as conscious beings, towards other living beings and, ultimately, in the fragile equipoise of our Planet, hitherto the only one habitable by humans.

Consequently, our effective involvement in the world – natural as well as cultural – has to be understood in the integral framework of life in order to take into account to what extent we humans actually have the capacity and the responsibility to contribute to consolidating either the state of the cosmos and its equilibrium or to a state of chaos and disorder.

Having said that, we cannot forget that the search for meaning and the task of giving sense to our own existence is a crucial dimension of humanity. In fact, Tymieniecka also gives a fundamental role to meaning in her phenomenology of life. In doing so, she sheds light on the central aim of this paper: to move from the ontological eco-phenomenological approach, to an interpretative hermeneutical field wherein meaning is the focal point to be established. For this reason, at the end of my reflection here, I shall highlight Paul Ricoeur's study of human symbolic expression. It is obvious that humans present by means of the latter, an implicit awareness of chaos, disorder, and evil, something that really worries them independently of the culture and the historical moment in which they are integrated. However, at the same time, humans do not stop at that stage; on the contrary, all of these negative phenomena lead them to search for a positive world (consciously or unconsciously, and by means of reason or of symbols), for a *Lebenswelt* where they can take root guided by a concrete *Weltanschauung*.

In effect, human beings are fragile creatures, always at risk: harm, disasters, misfortune, illness, and evil can all be stalking. Immediately, several questions arise. Among them: Why? Where is the origin of all of those negative events? Do we humans have some responsibility in the introduction of chaos and evil into the cosmos? Can we do something about it, to eliminate it, to restore the "cosmos-order-positivity" (if there was one) in our chaotic Cosmos-World? In fact, are we ultimately able to interfere in such a huge global phenomenon?

Tymieniecka insists on situating the human being in the Unity of Life by means of conceptual language. In contrast, Paul Ricoeur departs the strictly rational path, which not valid for him in this case. He takes a detour, doing an accurate analysis of four important groups of myths that represent four respective human expressions concerning the origin of evil and the role and responsibility (or lack of them) of the human being in negative – for humans – manifestations in the world.

In short, this study will be specifically focused, in its final point, on Ricoeur's analysis of symbolic-mythical expressions, on the realm of the human being's relationship with the whole of reality. We shall discover a wide field: the place and impact of evil in the human world and vice versa; the active or passive role of humans in the introduction-perpetuation of evil, on the one hand, and the possibility of reaching – or reconquering – a positive ordered global world, on the other, at least as far as our capacities can be developed.

## Ecology – Eco-Phenomenology

The gnoseological-epistemological meaning of Ecology is the "study of the house/environment" in which we live; it is the science that studies the relationships of organisms among themselves and with their environment or, looking at the facts, the set of relationships between organisms and their environment. Today, on the practical level, we can consider it as action or policy enhancing the protection of the air, water, and other natural resources from pollution or its effects.

As is well known, Ernst Haeckel introduced the term in 1866, but it has only recently come into widespread use, after the United Nations conference on the environment, held in Stockholm in 1972. The reason why these obvious facts are related here is that, in doing so, some other fundamental questions can be posed.

Does this mean that humans only recently realized their unavoidable connection with other living beings, along with their dependence on them, on the one hand, as well as their impact on the conditions of the Earth and the consequences for themselves, on the other? Should we look at different cultures, we can verify that this is not the case. It seems evident that humans have always been aware of their intimate connection to the whole chain of living beings and to the environment in general. However, attitudes taken have always been different – even opposing – not only between cultures but also throughout history. Let us consider, for instance, prehistoric paintings that apparently imply a strong human link with animals within the environment.

I would say that it is possible to find two main and opposing views. The first view is that the human being is installed at a superior level vis-à-vis all other beings and has the right – sometimes the duty – of spreading throughout the world and dominating all living beings. The second view is that the Earth – perhaps also the Cosmos – is the household of all living beings, including humans, who must respect it and take from it only what they really need. In that case, the Earth would be our domestic realm (*domus* in Latin); it would be our *οἶκος* in the original Greek sense, from which derives "*eco*", the prefix widely used nowadays, as in ecology and eco-phenomenology.

I consider the explicit message at the beginning of the *Bible* to be a good example of the first attitude. After having created all the diverse elements of Nature and living beings such as terrestrial, aquatic, flying, God decided to create humans giving many privileges to them:

> 26 And God said, Let us make man in our image, after our likeness: and let them have dominion over the fish of the sea, and over the fowl of the air, and over the cattle, and over all the earth, and over every creeping thing that creepeth upon the earth.
> 27 So God created man in his own image, in the image of God created he him; male and female created he them.
> 28 And God blessed them, and God said unto them, Be fruitful, and multiply, and replenish the earth, and subdue it: and have dominion over the fish of the sea, and over the fowl of the air, and over every living thing that moveth upon the earth.
> 29 And God said, Behold, I have given you every herb bearing seed, which is upon the face of all the earth, and every tree, in the which is the fruit of a tree yielding seed; to you it shall be for meat.
> 30 And to every beast of the earth, and to every fowl of the air, and to every thing that creepeth upon the earth, wherein there is life, I have given every green herb for meat: and it was so.
> 31 And God saw every thing that he had made, and, behold, it was very good. And the evening and the morning were the sixth day. (*Standard King James Version Bible*, Genesis 1. 26–31).

With the opposing perspective, we can find several modes of thinking, of conceiving religion – directly connected to man's role in the world – corresponding to the second attitude: a respectful feeling concerning every living being. Regarding this approach, we can mention several cultural groups still having guidelines even now aimed at living in peace with nature (this is at least their theoretical worldview; the level of practice is different). For instance, some such groups include Asian cultures like Hinduism and Buddhism and, on another continent, the so-called "American Indian Poems and Prayers", among them a well-known one to the "GREAT SPIRIT" – even if some researchers have doubts concerning the real source of the poem:

> Great Spirit,
> give us hearts to understand;
> Never to take from creation beauty more than we give;
> Never to destroy for wantonly for the furtherance of greed;
> Never to deny to give our hands for the building of Earth beauty;
> Never to take from her what we cannot use.
> Give us hearts to understand.
> That to destroy Earth music is to create confusion;
> That to wreck her appearance is to blind us to beauty;
> That to callously pollute her fragrance is to make a house of stench;
> That as we care for her she will care for us.
> We have forgotten who we are.
> We have sought only our own security.
> We have exploited simply for our own ends.

We have distorted our knowledge.
We have abused our power.
Great Spirit whose dry lands thirst,
help us to refresh your lands.
Great Spirit whose waters are choked with debris and pollution,
help us to cleanse your waters.
Great Spirit whose beautiful Earth grows ugly with mis-use,
help us to restore beauty to your handiwork.
Great Spirit whose creatures are being destroyed,
help us to find a way to replenish them.
Great Spirit, whose gifts to us are being lost.
in selfishness and corruption,
Help us to find the way to restore our humanity. ("First People". Web. 18 August 2014).

This poem, together with the "Ten American Commandments", represent the attitude of respect regarding the Earth and everything existing on it – especially humans – and, on a practical level, taking responsibility for our actions:

1. Treat the Earth and all that dwell thereon with respect.
2. Remain close to the Great Spirit, in all that you do.
3. Show great respect for your fellow beings.
   (Especially Respect yourself)
4. Work together for the benefit of all Mankind.
5. Give assistance and kindness wherever needed.
6. Do what you know to be right. (But be careful not to fall into self-righteousness).
7. Look after the well being of mind and body.
8. Dedicate a share of your efforts to the greater good.
9. Be truthful and honest at all times.
   (Especially be truthful and honest with yourself.)
10. Take full responsibility for your actions. ("Native village" Web. 15 August 2014).

It is not my aim here to evaluate whether some of these "Native American poems, prayers, and commandments" did or did not originate from indigenous sources. In my opinion, the most important question is not whether the considered origin – the Native Americans – is real or fictitious (there have been some serious debates about it) but, rather, their deeper meaning. Furthermore, a similar feeling of belonging to the Earth, of sharing it with "everything-there-is-alive" (in Anna-Teresa Tymieniecka's expression) together with an attitude of respect is still manifested in a few cultures. In fact, some cultural groups existed until at least a few decades ago – even nowadays, if the "global and predominant culture" has not already assimilated them – hunters and gatherers living in the Kalahari Dessert, or Eskimos of the North Pole, among them (Harris 1997, Ch. 12 and 15; Kottak 2013, Ch. 8).

In short, the human being considers himself either as a privileged being in the cosmos – created in the image of a Supreme Being – and having the right to occupy and dominate all the corners of the Earth or as a living being among others, sharing the same *οἶκος*. With the latter, he would have – as supposedly the most intelligent living being – the responsibility of looking out for the well-being and order – the *cosmos*. Humans should look for the equipoise of everything-there-is-alive, including themselves and the environment – natural and cultural – of the world in which they live.

Nowadays, ecologists – both scientists and political movements – could say (even without considering the previously given religious, mythical, or literary quotes) that the first viewpoint could have in the past and, even now, continues to lead us to risks that would have been unimaginable only a couple of centuries ago. In fact, scientific studies are starting to alert us to the serious human impact working all kinds of changes, on water, soil, forests, climate, air, and all domains of our world. Thus, we are becoming more and more aware of the problem, analyzed from diverse perspectives: ethical, economic, political and philosophical.[1]

Nevertheless, we could say that most of these ecological studies and perspectives, which are serious and necessary in my view, deal more with naturalism and causal laws than with our "consciousness of", or with intentionality, which is the main concept to be applied in phenomenology – one of the fields wherein the present reflection is located, with the other being hermeneutics. In the mere naturalistic approach, the human being's search for meaning is left aside. In actuality, we face two divergent – but not necessarily, incompatible – approaches, corresponding respectively to the domain of causality, on the one hand, and that of intentionality, on the other. These two approaches have been separated for a long time and even considered to be irreconcilable. Are they really so? Must philosophy or, more concretely, phenomenology renounce approaching such problems so fundamental not only for the basic survival of human beings but also for satisfying our need for sense? From a reductionist point of view natural science, on the one hand, and philosophy, on the other, seems to be in irreducible conflict.

However, two approaches – post-naturalism, and post-Husserlian phenomenology – have tried recently to go beyond that conflict, finding place for a rapprochement that some writers call eco-phenomenology. Having presented the opposition to be overcome, let us grasp the meaning of "eco-phenomenology." As an example, here is the account of it given by author David Wood:

> What then is eco-phenomenology? I have argued that eco-phenomenology, in which are folded both an ecological phenomenology and a phenomenological ecology, offers us a way of developing a middle ground between phenomenology and naturalism, between intentionality and causality. I argue that our grasp of Nature is significantly altered by thinking through four strands of time's plexity – the invisibility of time, the celebration of finitude, the coordination of rhythms, and the interruption and breakdown of temporal horizons. Also by a meditation on the role of boundaries in constituting the varieties of thinghood. Eco-phenomenology takes up in a tentative and exploratory way the traditional phenomenological claim to be able to legislate for the sciences, or at least to think across the boundaries that seem to divide them. In this way, it opens up and develops an access to the Nature and the natural that is both independent of the conceptuality of the natural sciences, and of traditional metaphysics. (Wood 2003, 234–235, Conclusion).

In this context, *eco-phenomenology* could be described as the pursuit of the relationalities of engagement with the world, both with humans and with other creatures. This engagement is situated in a middle position of relationality, a space that is neither purely objective, since it is reciprocally constituted by a diversity of life experiences connected to the movements of other organisms, nor purely subjective, because it is a field of relationships between bodies, which are always material. Consequently, we cannot say that it is governed exclusively by either causality or intentionality. Assuming such a viewpoint, Phenomenology, situated in this area of in-betweenness, could be able to transcend its initial antagonism to naturalism.

Thus, the split between "objective" nature, on the one hand, and intentionality and consciousness, on the other, fades away. The new phenomenological path would be centered on the mode for which humans are always searching, and generating meaning in the world. This is actually a human distinctive mode of being, given that it refers to the space and time that has unfolded where and when our consciousness revels, it is a characteristic feature that is quite different from a causal relationship: humans are conscious about/of something. Nevertheless, at the same time, we are not pure consciousness; we cannot forget that we are "embodied beings", something that has already been pointed out by Merleau-Ponty and by most post-Husserlian phenomenologists. Merleau-Ponty understands Nature as Flesh regarding the body, as well as the world. Supported by his analysis of the phenomenology of perception, the opposition between, on the one hand, Nature and perception in terms of passivity and, on the other hand, Culture, which is considered to be active, is diluted. Furthermore, as can be felt in our *lived* experience, there is a disconnection between neither subject and object, nor between mind and body. Consciousness, being natural and perceptual, points to nature; in fact, there is a meeting point where interior and exterior "natures" converge. In other words, we are, before *the activity of* a subject-object, *a being-in-the-world*. Merleau-Ponty was right when he underlined the fact that intentional consciousness is embodied and develops its activity *in* the world, so that knowledge has its basis in the body and *lived* perception. Thus, Merleau-Ponty situates the body in the core of his phenomenological analysis of perception and, reciprocally, the mind – mainly the phenomenon of perception – is actually rooted in the body and the world, as appears in his article "*The Primacy of Perception*".

For Merleau-Ponty, the lived body, conscious of itself, is reciprocally subject and object; the body itself simultaneously sees and is seen. It sees itself seeing; it touches itself touching (Merleau-Ponty 1964, 162). The body is the human conscious insertion in the world, the presence of the world within the human being. I naturally communicate with the world – previously discovered and objectified by my conscious reflection – with a world I inhabit precisely because I am an embodied being; I am a being-in-the-world in coexistence with other beings and this is a phenomenon revealed by my perceptual mode of being, by my perception. My mind – perceiving mind – is, above all, an incarnated mind (3–4).

My view on this is that similar points – concerning the interaction between our material senses and the outside world – can also be found in contemporary neuroscience: an approach that we could consider as being parallel to or even as a scientific support for the kind of phenomenology that underlines our fundamental feature of being necessarily embodied beings living in a material world. In this sense, the Spanish neuroscientist Francisco Mora has in one of his books, *Cómo funciona el cerebro* (How the Brain Works), a chapter titled "Does the world we see actually exist outside the brain?" (Mora 2011, 87–112).[2] Mora insists on the fact that we are only able to perceive everything around us thanks to our "windows open to the world", to our senses that "translate" the world's events so that they are accessible to our brain. What happens in the world is communicated to the brain in a symbolic language only understood by each individual brain. Mora specifically comments on the case of sight and takes the view of an orange as an example.[3] Initially our retina does not perceive the orange as an object pure and simple. On the contrary, the retina analyses and separates the orange into its diverse constituents: color, movement, depth, form, and relationship with other objects placed in its space. All of these diverse elements are sent to the brain one by one through different and parallel routes. Furthermore, some of the mentioned constituents of an object (the orange in this example) are previously divided into smaller elements, as happens in the case of the shape. Mora calls those smaller elements "atoms of perception" that follow a complex process of integration and convergence that takes place later in diverse areas of the brain. A new analysis of the ingredients that arrive at the brain happens again before being "kept" somewhere in it (Mora 2011, 87–94). There is no knowledge, no objective element if it is not first "elaborated" or "translated" by the brain after the complex process that has been undertaken by the corporeal senses; they transport to the brain, through diverse paths, perceptions, and sensations produced by external stimuli. That being so, it is possible to find a parallel approach – to a certain extent – to philosophical phenomenology: no object without subject, and vice versa.

Another good example, similar to the phenomenological idea of humans as intentional embodied consciousnesses developing their knowledge based on bodies and on *lived* perception, is the case of color, which Mora describes as having a strategic survival value for living beings. He presents a good overview of the process. When he wonders if outside-world colors actually exist, the answer is clear: the brain, based on the retina receptors, the neuronal structure, and the consequential working program, actively constructs the color that we see. Nevertheless, wavelengths coming from an object as well as from the objects near it also play a central role. Therefore, brains (subject) have an active role in the perception of the world – proof of this being achromatopsia, the impossibility of seeing colors following on a lesion in a certain part of the brain – though wavelengths (an object) are also indispensable in perceiving them (Mora 2011, 94–103).

Besides, Mora is clearly convinced (based on recent research only possible thanks to the rapid progress of medical diagnostic technology) that emotions and sentiments are the origin, the main "on switch" of human action.[4] Namely, our brain – as well as other living being's brains – codifies positive and negative functions: either pleasure and reward, or fear, punishment and pain, all linked to the feeling of willing to be alive. Our survival, our personal stability, our reasoning and coherence, our relationships with other humans, our values, as well as our ultimate meaning of existence depend on this work of our limbic, emotional brain (Mora 2011, 115). Consequently, a novel method for understanding the human being and its place in nature as much as in the human life-world emerges, coming from either philosophy-phenomenology or from neuroscience, in both projects going beyond classical dualism.

As we can see, the new eco-phenomenology places the "natural human body" in the world by discovering our perceptual and intersubjective interactions as well as our relationships with our sheltering-surrounding world. In this respect, it opens us up to dimensions of nature's meaning and value. At the same time, it can provide us with methods of approach different from those that, following only the conceptual-ideal approaches or the constitutive function of Husserlian transcendental consciousness, find it difficult to reach those dimensions. In short, eco-phenomenology plays a leading role in gaining a better understanding of nature, our place in the world, and our ethical responsibility towards it.

To prove this, I shall first summarize Anna-Teresa Tymieniecka's global philosophical approach to life, cosmos, and the human being's place in that context. As will be shown, imagination and creativity play a basic role and establish a bridge allowing us to continue our work with a survey of Paul Ricoeur's hermeneutical comprehension of the same question, this time taking into account the message given (and at times hidden) by means of symbolic narratives and, more specifically, through myths. Indeed, Ricoeur focuses his attention on Mediterranean myths that express the origin of evil; but this problem implies, as we shall see, a general cosmo-vision. A further-reaching area of cosmological and anthropological problems underlies Ricoeur's *The Symbolism of Evil*: for instance: the origin of everything, evil included; the possibility of its end; the way it affects the human being; the human place in the global reality; the possible human responsibility concerning evil and negativity, and so on.

## Anna-Teresa Tymieniecka's Eco-Phenomenology

Anna-Teresa Tymieniecka centres her philosophical work on both a theoretical and a practical approach to life and the cosmos. *Ontopoiesis* is the key to properly understanding her perspective on eco-phenomenology. After a long process of philosophical elaboration concerning of all the levels of metaphysical reality and the interwoven phenomenon of "logos and life", her view on the subject of eco-phenomenology is set forth clearly in the following statement:

> Actually, my account of ontopoiesis is an eco-phenomenology. Ontopoiesis reaches to the very germs of ecology: development and genesis. I have published several essays related to this. *In* Passions of the Earth [Analecta Husserliana, Vol. LXXI], *I show how the human being is an ecological fruit and how the human being is formed by the earth and sucks the juices of the earth. I have also written things about the cosmos and the cosmic dependencies of the human mind and human development. You see, the self-individualization of life, which is the basic instrument of ontopoiesis draws upon the laws of the cosmos and the earth.* This is the most fundamental ecology that can be done. So, we have just touched the essence of my philosophy, the base—our relationship to the earth and to the cosmos. ("An Interview with Anna-Teresa Tymieniecka", Web. n.p. 20 August 2014).

It is evident that Anna-Teresa Tymieniecka's meaning of the term eco-phenomenology is much wider than the previous meanings we have covered here.

Addressed here is not merely a gnoseological question of "developing a middle ground between phenomenology and naturalism", which is already a significant step; neither is it only an overcoming of the dualism of nature/culture, body/consciousness and soul, and so on, given that our embodied intentional consciousness develops its activity in a world that is both natural and cultural. Anna-Teresa Tymieniecka assumes, of course, all these premises, as well as the necessary aim of understanding the place of humans in nature, their ethical responsibilities towards it, and the decisive role of contributing to 'meaning-bestowing', as she puts it, upon human existence.

Actually, she adds and insists on some relevant points that can already be recognized in the summary contents of her phenomenology of life:

> (1) the self-individualization of life circumscribing the context of phenomenological investigation, (2) the creative act of the human being which brings us into the center from which the human mind draws all the rays of order, and (3) the human condition which grounds the creative act as man's foothold within the unchartable schema of life. (Tymieniecka 1990, 5)

I shall try to summarize the more relevant points and features I find in her eco-phenomenology.

## *First*

Anna-Teresa Tymieniecka's approach is global; she faces reality as a whole; "entirety" and "complexity" are, accordingly, two cornerstones of her eco-phenomenology. Every element on Earth, even in the Cosmos, is linked with each and all of the others. This is especially important in order to understand the chain of living beings and the development of their features – including, of course, the human being.

## *Second*

Following the first point, she expands the field of human relationships with nature beyond the Earth, reaching the Cosmos; thus, human faculties, including rationality, have their roots in and are results of the evolutionary progress of the universe itself:

> Indeed. To bring out from the recesses of the hidden, unmanifested operation subtending life and beingness the *entelechial design* which carries the existential profile of the ontopoietic route and which personifies the autonomous selfhood of the self-individualizing being-in-progress, we have to take all into consideration. It is the entelechial design which, through the crystallization of its innermost virtualities delineating the ontopoietic route of the living being, precipitates its sufficient reason, establishes its ontic ground. And in

> performing this function it is suspended upon all cosmic forces and laws, the innumerable factors conditioning bios, zoe, psyche, the human spirit, society and culture, which are differentiated into innumerable circuits. In other words, the ontopoietic nucleus –the ground of beingness of the living individual- reposes in itself while it spreads and maintains tentacles involving the circuits of all. (Tymieniecka 1998, 64–65).[5]

This is why Tymieniecka prefers the concept of "the human condition" and leaves to the side that of "anthropology", understood as a knowledge centered on humans considered in isolation from other beings – especially living beings. The human condition dwells within the unity-of-everything-alive that depends in turn on the laws of the Earth and Cosmos.

## *Third*

As a counterpoint to the previous feature, individualization, self-individualization is actually linked to the origin of life and continues being a decisive dynamic element all along the process of the emergence of diverse living beings at different stages, but this process does not imply that any living beings are situated at a superior or inferior level. If that were the case, the life process would open the way towards the human being and its knowledge: Anthropology. However, this is not the case here.

## *Fourth*

Consequently, Tymieniecka refuses to develop an Anthropology as such, one separate from her general conception of the unity of life. "Yes, that is what I am saying. But here there is one thing you have missed completely, and which is completely essential to my philosophy. There is no anthropology in my philosophy. I have dissolved the notion of anthropology in a [sic] essay which I published three years ago."[6] It is only possible to understand properly the human being if we consider the human as an element inserted in the chain of living beings, or as a condition within the unity of everything alive, to employ Tymieniecka's term:

> *I arrived at the very central point, namely that human being can not be considered in itself as such, that there can be no anthropology that considers human being as such, in the middle of other things almost by chance. On the contrary, human being should be considered as a human condition within the unity of everything there is alive.* That means the human being unfolds and generates in a mutual contributive relation to all the other living beings. (Torjussen et al. 2008. "An interview with Anna-Teresa Tymieniecka)

Nevertheless, the Archimedean point for the phenomenology of life is, for her, the creative-inventive dimension of the human being.

## *Fifth*

Creativity – *Imaginatio Creatrix* – is primarily the crucial basis of Tymieniecka's phenomenology of life; more precisely, human creative activity is the bridge between reason and life, between theoretical and practical perspectives. It is also a decisive crux for understanding her eco-phenomenology. The critique of reason was always for her a primary task, an analysis unfolded especially in the first book of her work *Logos and Life,* in opposition to the description of creative experience, to the point that *ontopoiesis* is actually the basis and origin of the evolving of reason.

Thus, creativity is the essential human feature for her phenomenology of life and the guideline along which everything finds its proper place; it is also essential for her eco-phenomenology.

## *Sixth*

With the human creative act as the axis of phenomenology of life, Tymieniecka opens the door to what she calls the "phenomenological realism of possible worlds", setting up the creative against the constitutive function of the human being and unfolding human reality by means of the consideration of multiple and open possible worlds (Tymieniecka 1989, 36–39). In fact, intentional-ideal structures have their origin and are rooted in the creative process. At the same time, the creative act now has a crucial role and priority, one ignored by Husserl, as well as by other phenomenologists (Scheler, Merleau-Ponty), according to Tymieniecka. The creative act, specific to the human being, appears as the prototype of all action, theoretical or practical; at the same time, it is the sense- giving factor.

## *Seventh*

Searching for meaning and awareness is undoubtedly a human task. Having said that, the philosophical analysis of creativity, understood as the essential human feature, constitutes for Tymieniecka "the Archimedean point for the phenomenology of life" (Tymieniecka 1989, 342). It is important to underline that the human faculties – Imagination/Memory, Will, and Intellect – emerge from a "source experience", discovered by creativity, giving life a new orientation. From this point of view, the true human action – characterized by awareness, the need and the capacity for searching for a meaning of life and self-interpretation – finds its own path (Tymieniecka 1989, 13–16).[7] In the same work, Tymieniecka clarifies the subject of meaning and praxis:

> Thus, the individual's life-course assumes a transmuted interpretative modality and, subsequently, a *specifically human significance* due to the works of the *creative orchestration*

> of his interpretative system. On the one side it has to be emphasized that the creative orchestration of human functions with its differentiation into three constructive faculties: *imagination, intellect,* and *will* – as well as with its four sources of meaningfulness: the *vital, poetic, intelligible*/structural, and *moral* senses – is geared in the spontaneous unfolding of its activity to the progress of the specifically human life. It is the vehicle of culture, civilization, and human history.
>
> On the other side... it is the *entire system of individual functioning which in the case of the human being takes the form of creative orchestration.* (Tymieniecka 1989, 164)

To sum up, the concept of creativity not only plays an important theoretical role in Anna-Teresa Tymieniecka's thought process, but it also constitutes the bridge towards practical life, as it shows us that human creative activity tends to surpass the actual mode of being in the world (the object of constitutive phenomenology). Thus, it opens new paths (by means of creativity) for exploring other possible worlds. Imagination comes into play and, in this respect, our discourse is ready to connect with the role of narratives, myths, and literary stories dealing with the human being's life meaning and its interpretation; thus, we are at hermeneutics' doors.

## Paul Ricoeur: Narratives-Myths and Hermeneutics

The human being living in the *Life-World* often has the paradoxical impression, the subjective experience, that outside as well as inside oneself there is a double, valuable possibility. On the one hand, it is a positive possibility, close to *cosmos*-order, with preexistent rules that can, to a certain extent, be known and on which one can count without significant contrarieties; on the other hand, chaos, dysfunction, and evil are always menacing.

If eco-phenomenology shows humans to be beings linked to everything around, the problem of the relationship between *cosmos* and *chaos,* their origin – especially evil's origin, so disturbing for the human being – must be included in its outlook, which problem is directly related to the subject of meaning (the sense of Cosmos and the meaning of the human being). All cultures have always tried to understand and give a solution to the cosmological and existential problems intertwined with this disturbing phenomenon of chaos and evil. Nevertheless, according to Paul Ricoeur, the rational path gets lost in the comprehension of the problem. Narratives and, more specifically, Myths, by using an ambiguous and multivocal language, are able to explore the different and opposing aspects of reality simultaneously, so that humans reach directly a more global perspective of the problem. According to the French thinker, narratives provide linguistic beings – that is humans – with a crucial tool for establishing meaning and a coherent guideline in a world that often appears incoherent. In regards to the same question, Ricoeur states that literature is an indispensable implement for finding the cohesion of life in a paradoxical context of "incoherent coherence" (Ricoeur 1985, 3: 200). This cohesion becomes an organized development of life experiences (107–108).

On the whole, narrative discourses and, in general, collective representations like myths are thus privileged fields for the exploration of diverse possibilities in order to apprehend human existence, for they relate to the creativity-imagination and the openness to possible worlds that Tymieniecka highlighted. They point out the need, as well as the capacity, to unfold the virtual creativity of the human being, who is constantly searching for meaning in order to understand his role in the cosmos. Nevertheless, human creativity is limited; we are limited beings. We can only develop our creative imagination in the context of our existing circumstances. First, we are embodied beings, situated in a perspective given by our body and our senses, something emphasized by Merleau-Ponty and by neuroscience, as we saw above; second, we are also limited, shaped, to a certain extent, by our cultural world; but at the same time, our linguistic capacity (*Logos*) tends towards infinitude. Therefore, we deal with a finite being that speaks of its finitude and is aware of being-in-perspective; that being is, by this fact, on the way to transcending finitude by means of the dialectics of perceiving/meaning:

> I am not only a located way of looking, but a longing to say something in a mode of intentional transgression of the situation; as soon I speak, I speak of things concerning their faces not perceived, as well as in their absence. Thus, the completed perceptive intention, that gives me the presence perceived in the living present, which is the present of presence, is never alone and naked; it is precisely because it is full, that it is always taken in a relationship of fulfillment, more or less complete regarding another aim, that completely traverses the aforementioned perceptive intention, that literally transits through it, and to which the speech is originally linked. This aim is the longing-of-saying of the saying. At birth, I enter the world of language that precedes and envelops me. The silent gaze is retaken in the discourse that articulates its sense; this characteristic of saying of the sense is a continuous overrunning, at least in intention, of the perspective aspect of the perceived here and now. (Ricoeur 1960b, 45)[8]

The problem of imagination and creativity, only secondary in Ricoeur's *Fallible Man* (*L'homme faillible*), is central in other works of his. In *The Symbolism of Evil,* as the hermeneutic problem is outlined in the conclusion, the idea is advanced that "symbols *offer* something to be *thought about*"; "The Symbol Gives Rise to Thought" ("Le symbole donne à penser") is precisely the title of the Conclusion of *The Symbolism of Evil* (Ricoeur 1960a, 347–357). This statement can lead us toward the idea of the dialectics between construction and reconstruction, production and reproduction, for anthropological comprehension.

Nonetheless, we have to consider works like *La métaphore vive,*[9] *Temps et récit,* and *Du texte à l'action* in order to discover all of the theoretical and practical strength of Ricoeur's theory on imagination, to be assumed in any symbolic, poetic, or narrative work. 'Poetics' (*la Poétique*) is now the new reflection, that promised in Ricoeur's analyses of will and freedom; this reflection is centered on the cultural recreation of human existence, by means of human imagination and creativity, thanks to the 'poetic' capacity of the human being.

It is mainly in *La métaphore vive* where Ricoeur deals copiously with the subject of imagination, elaborating a theory centered on the notion of semantic innovation. In this context, 'to imagine' means to restructure semantic fields, to "see as" being imagination that, in this work, is linked to a certain use of language. Imagination really plays a basic role in human action, understood as an open task—"L'action sensée considérée comme un texte" (Ricoeur 1986, 197). It enables the exploration of new possibilities always open both to the individual and to the cultural group. Regarding that explorer dimension, we can notice that it offers a series of guidelines in its passing from theory to praxis (a move searched for by Ricoeur in *Temps et Récit* and *Du texte à l'action*). This process is clear in the domain of narrative discourse. Thus, the 'variations of imagination' in fiction locate us in a 'vast laboratory', in which humans search for possible solutions when they face their own fundamental problems: for example, the enigmas of our temporal condition, as shown in *Temps et récit III*. A possible transition from meaning to reference is, then, facilitated within the field of language by means of fiction—"L'imagination dans le discours et dans l'action" (Ricoeur 1986, 220–228). Nevertheless, the fundamental anthropological analysis must go beyond the merely theoretical or linguistic point of view and reach the practical level, both at the individual and at the collective levels—"L'imagination dans le discours et dans l'action," par. "L'imaginaire social" (Ricoeur 1986, 228–236).[10]

Here I shall leave aside the practical important question of imagination as carried out in historical human time: whether ideology-tradition, on the one hand, or utopia, on the other. From now on, I shall focus on Ricoeur's description and interpretation of four mythical types centred on evil, its origin, its possible end, and the human role in the global, cosmic process. For that, it is necessary to clarify the idea of Myth proposed by Paul Ricoeur:

> I shall regard myths as a species of symbols, as symbols developed in the form of narrations and articulated in a time and space that cannot be co-ordinated with the time and space of history and geography according to the critical method. For example, exile is a primary symbol and not a myth, but the history of the expulsion of Adam and Eve from Paradise is a mythical narration of the second degree, bringing into play fabulous personages, places, times, and episodes. Exile is a primary symbol and not a myth, because it is a historical event made to signify human alienation analogically; but the same alienation creates for itself a fanciful history, the exile from Eden, which, as history that happened *in illo tempore*, is myth. (Ricoeur 1960a, 18).

Thus, myth is primarily presented – in a verbal or written language – through a narrative form as if it were a novel or a drama; but, in myth, we deal with a kind of space and time far beyond any existent comparison. In fiction we can always find, at least by means of our imagination, a possible space and time. In contrast, in myth, the 'time' of *in illo tempore* is completely 'outside' any space and time imagined; it is the very origin of everything. Moreover, myth is made of symbols and, so, we have to ask of what a symbol consists.

A symbol is, for the French thinker, a type of sign, an expression communicating meaning; but it is a very singular sign, given that it is far from being transparent in its intention of significance and communication. The symbol is ambiguous. It "conceals in its aim a double intentionality": one literal, and another one to which the first leads; "symbolic signs are opaque, because the first, literal, obvious meaning itself points analogically to a second meaning which is not given otherwise than in it… This opacity constitutes the depth of the symbol, which, it will be said, is inexhaustible" (Ricoeur 1960a, 15).

However, myth's symbols must not be confused with analogy, as it is impossible to objectify the analogical relation connecting the second meaning with the first one: "It is by living in the first meaning that I am led by it beyond itself" (Ricoeur 1960a, 15). Finally, mythical symbols (full of contents that "ask" for interpretation) are nearly the opposite of logical symbols (merely formal).

After having analyzed in the first part of *The Symbolism of Evil* several important symbols implied in myths of evil – defilement, sin, and guilt – Ricoeur insists on the fact that myth must not be confused with history; and it has nothing in common with etiological purpose, with any causes suited to rationality. Thus, myth is neither history nor explanation, but it takes the form of narration adding a new level of meaning to mere symbols. Myth is an expressed language that becomes a kind of narrative, made of symbols.

Besides this, for Ricoeur, the first step is to face the interpretation of mythical consciousness:

> What is essential for us here is to understand why that consciousness structured lower than any narration, any fable or legend, nevertheless breaks out into language under the form of narration. If the phenomenologists of religion have been more concerned to go back from the narration to the prenarrative root of the myth, we shall follow the opposite course from the pre-narrative consciousness to the mythical narration. It is in this transition that the whole enigma of the symbolic function of myths is centered. (Ricoeur 1960a, 166).[11]

The French thinker agrees with several points of the phenomenology of religion, but he insists on two characteristics of myth: it is an expression in language and the symbol assumes in a myth the form of language.

For the Phenomenology of Religion, the myth in taking the form of narration is only a verbal aspect of an action, a rite. In fact, both myth and rite point beyond themselves to an archetype, which must be imitated and repeated over and again and which signifies, somehow, a plenitude, an entirety, a cosmic whole:

> The myth-narration is only the verbal envelope of a form of life, felt and lived before being formulated; this form of life expresses itself first in an inclusive mode of behavior relative to the whole of things; it is in the rite rather than in the narration that this behavior is expressed most completely, and the language of the myth is only the verbal segment of this total action. (Ricoeur 1960a, 166–167).

In contrast to the phenomenologists of religion, for Ricoeur, that wholeness is expressed in the total action of the rite more than in the verbal form of myths. Thus, both myth and rite point to an all-inclusive totality of beings (cosmos or chaos, for the moment) that can be understood in eco-phenomenological terms (regarding the global reality in which the human being is immersed), even if Ricoeur does not employ the term 'eco-phenomenology'.

Besides, there is a common structure at the base of all myths, one that ramifies in a diversity of myths depending on the content and the culture. In order to clarify the meaning of this common structure, Ricoeur remarks that in addition to this common structure, he assigns three clear functions to myths, which functions are easily recognizable in the myths of evil (Ricoeur 1960a, 161–163):

(a) *Concrete universality*. Myths present man as a concrete universal. "To embrace mankind as a whole in one ideal history", 'time' represents all times and "'man' is manifested as concrete universal; for instance, in the case of the Adamic myth, Adam signifies man" (Ricoeur 1960a, 162). According to Saint Paul, we have all sinned with Adam, so that we deal not with an individual experience but with an archetype by which humanity is presented. The figure of the hero plays the role of this archetype – the ancestor, the demigod, and the like – depending on the culture. Adam is evidently the figure of the 'first man' in the Bible, in our culture, but this is not the only archetype on Earth.
(b) *Temporal orientation*. Myths manifest the universality of the human being but, at the same time, they display its concrete aspect thanks to the movement introduced in human experience by narration. In this sense, they hold a concrete function by introducing time in a dramatic narration: a beginning, an end, a tension, some characters, and other such narrative aspects. It is not just a question of a present experience.
(c) *Ontological exploration*. From my point of view, there is actually an ontological-anthropological function that myth serves. In Ricoeur's words:

> Still more fundamentally, the myth tries to get at the enigma of human existence, namely the discordance between the fundamental reality – state of innocence, status of a creature, essential being – and the actual modality of man, as defiled, sinful, and guilty. The myth accounts for this transition by means of a narration; but, it is a narration precisely because there is no deduction, no logical transition, between the fundamental reality of man and his present existence, between his ontological status as a being created good and destined for happiness and his existential or historical status, experienced under the sign of alienation. Thus, the myth has an ontological bearing: it points to the relation – that is to say, both the leap and the passage, the cut and the suture – between the essential being of man and his historical existence. (Ricoeur 1960a, 163).

Consequently, myth – as is the case with literary-fictitious texts too – has a basic role in anthropological apprehension, especially when we face the most relevant problems of our existence. It can reveal to us directly, even if it is through intuition, many aspects that escape conceptual or logical language. However, symbolic narratives and, more specifically, myths do not give us a complete comprehension of anthropological or cosmic problems: the mythical consciousness does not experience the plenitude; it only points to it, and to the beginning and the end of a basic global History. As Ricoeur put it, "the plenitude that the myth points to symbolically is established, lost, and re-established dangerously, painfully. Thus, it is not given, not only because it is signified and not experienced, but because it is signified through a combat" (Ricoeur 1960a, 169–70). Thus, there is, at the basis of Ricoeur's exploration of myths, a search for a plenitude, a global conception of the human being and its place in the world, an eco-phenomenology to be dynamically discovered and rediscovered by means of an interpretation, which also could be in process.

Consequently, Ricoeur's approach could be called an "eco-phenomenological hermeneutics".

As already mentioned, in his book *The Symbolism of Evil,* Ricoeur focuses his exploration on some groups of Mediterranean myths concerning evil and its context,. There is no doubt that the question of the role of evil in the world, chiefly in human existence, was never completely solved and is one of those decisive anthropological problems often considered rationally by philosophy – with not much success, as I see it – that are also present in every culture precisely expressed by means of religion and myth. Nevertheless, before focusing on myth (a language that reveals the factual presence of the chaos and evil), Ricoeur had already carried out reflections about crucial questions concerning human action. In *Le volontaire et l'involontaire*, he described the structures of human will from an eidetic phenomenological perspective; the question of evil was left aside, because it had no place at an eidetic level. Then, in the first part of *Finitude et culpabilité: L'homme faillible*, he goes a step further, this time changing the method to an empiric one and trying to understand the human mode of being in the middle of the dialectics between finitude and infinitude. In this work, Ricoeur finds a fragile creature, a limited human being characterized by "disproportion", by non-coincidence with himself. This condition, together with fallibility, could open the path to chaos and evil. The origin, the occasion, and the capacity for misfortune, adversity, and so on, could be within the human being. Nevertheless, even if this is the case, this does not imply that humans are actually the subjects who introduce evil into the world. There is always a gap (*un saut*) between the capacity for evil's actualization and the negative fact itself.

In this context, and taking into account the title of Ricoeur's main work on myths (The Symbolism of Evil), one could think that the French thinker is only interested in a partial facet of the human being and the world. If this were the case, it would not be appropriate to insert that analysis of myths into the global approach of eco-phenomenology. However, even if Ricoeur centers his investigation on evil, he never forgets the global context. We can find in that kind of myth subjects such as cosmogenesis, how evil can disappear – if at all possible – from the human world, the possible return from Chaos to Cosmos, and the destiny of the human being, among other subjects.

Thus, the aim of Ricoeur's analysis goes much further than the problem of evil, pointing implicitly to cosmology and explicitly to anthropology and preparing the path from phenomenology to hermeneutics. In this context, it could be meaningful to employ the term eco-phenomenology as the apprehension of human being's role in the Cosmos, as our *οἶκος* and of meaning more generally. But, this time, the process is carried out starting from symbols and myths' multivocal language as the basis for a philosophical interpretation. In *The Symbolism of Evil,* Ricoeur approaches the problem by means of extra-rational paths (given the incapacity of reason in that field), trying to reach the meaning of the human being and that of the totality of things, both of which are necessary to human life.

Apart from the incapacity of reason when it deals with serious anthropological problems – evil here – the human yearning for comprehending the gap between the capacity and the actualization of evil leads us to myths, to listening to them, to what the human being intuitively apprehends and presents in collective narrations concerning human experience –through diverse modes depending on different cultures. Then, proceeding from symbols and myths, philosophy has the task of interpreting their complex contents by means of a hermeneutical method (a complex question that would take me in far away from the main subject of the present article).[12]

In fact, symbolic-mythical language points to human existential experience as such. It is a conflict between finitude and infinitude:

> The chaotic and arbitrary aspect of the world of myths is thus the exact counterpart of the discrepancy between the purely symbolic plenitude and the finiteness of the experience that furnishes man with "analogues" of that which is signified. Narration and myths, then, are needed to consecrate the contour of the signs of the sacred: holy places and sacred objects, epochs and feats, are other aspects of the contingency that we find in the narration. If the plenitude were experienced, it would be everywhere in space and time; but because it is only aimed at symbolically, it requires special signs and a discourse on the signs; their heterogeneity bears witness to the significant whole by its contingent outcroppings. Hence, the myth has the function of guarding the finite contours of the signs, which, in their turn, refer to the plenitude that man aims at, rather than experiences. (Ricoeur 1960a, 169).[13]

Plenitude is consequently only aimed at, or glimpsed, and signified dramatically, usually by means of an arduous symbolic conflict between cosmos and chaos, good and evil, in a cosmological context, where the mythical narration discloses "the hidden meaning of human experience" (Ricoeur 1960a, 170). Ricoeur even agrees with some thinkers who find a biological role in myth: to protect humans against anxiety: "If myth-making is an antidote to distress, that is because the man of myths is already an unhappy consciousness; ... for him, unity, conciliation, and reconciliation are things to be *spoken of* and *acted out*, precisely because they are not *given*. Myth-making is primordial, contemporaneous with the mythical structure, since participation is signified rather than experienced" (Ricoeur 1960a, 167–168). As Ricoeur maintains, it would be an arduous task to take into account all the myths on Earth that treat the conflict of cosmos and chaos, good and evil. That is the reason why he decides to begin with a "typology" *a priori,* always remaining open to the possibility of amending it *a posteriori*, once we are in contact with the experience.

He highlights four types of myths – all from the Mediterranean area – concerning the problem of cosmos/chaos, as well as good/evil, centering on the origin and the end of evil. I will comment briefly on them.

## *The First Type: The Drama of Creation*

In this group of myths, the origin of evil coincides with the origin of everything: "it is the *'chaos' with which the creative act of the god struggles.* The counterpart of this view of things is that *salvation is identical with creation itself*'. In this type of myth, evil is actually chaos and salvation is equal to creation. What is the role of humans in this case? Apparently, this third type of myth is linked only to cults and rituals that reproduce the conflict situated at the beginning of creation. Rituals are, thus, essential for achieving a renewal of the drama by active human participation, but humans are not originally implicated in the production of chaos and evil.

The myths of Babylon, Homer, and Hesiod have similar content; a recessive form is the myth of the Hebrew king. We can also consider a mutant version, the stories of the Hellenic figures of the Titans.

## *The Second Type: The Tragic Myth*

Ricoeur denominates this class of myths tragic because he finds its complete manifestation in Greek tragedy, viewing it not only as a literary expression but as a narration related to a theology: "the tragic theology of the god, who tempts, blinds, leads astray" (Ricoeur 1960a, 173). Paradoxically, the hero appears to be guilty, but has not committed any fault. However, the hero cannot avoid it; it is implicit in his actual existence. How is it possible to reach any kind of salvation? It seems that it cannot come either from the gods – in fact, they provoke chaos in human life – or from humans, trapped as they are unwillingly in a tragic situation. The only possible salvation is to understand and accept fate and necessity; in fact, this is a form of freedom. However, there is another essential aspect: a tragic salvation consisting of an aesthetic deliverance, produced by the spectacle, that in turn produces pity for oneself. The human being is not guilty; on the contrary, in the second case, the human is a victim, a kind of "toy in the hands of gods".

## *The Third Type: The Adamic Myth*

Now, the human being appears to be the focus of the origin of evil and he can participate in ending it. In Ricoeur's words, "there is a change of type with the idea of a 'fall' of man that arises as an irrational event in a *creation already completed*" (Ricoeur 1960a, 172). Salvation is, in this type of myth, a kind of great deed related to the primary creation that is already closed. However, a "new" history, a human history but one with divine intervention – the coming of Christ, for instance, within the Christian perspective – starts to develop so that salvation can overcome the original fault.

Myths belonging to this third type are the opposite of those of the drama of creation type. The "fall", human guilt or responsibility for chaos and evil in the world, is an idea excluded from the field of those myths; on the contrary, in the "Adamic" or "Eschatological vision of history" type of myths, the burden of responsibility concerning the origin of evil falls back onto the human being's shoulders:

> Thus the cleavage effected, with the second type, between the irrational event of the fall and the ancient drama of creation provokes a parallel cleavage between the theme of salvation, which becomes eminently historical, and the theme of creation, which recedes to the position of 'cosmological' background for the *temporal* drama played in the foreground of

> the world. Salvation, understood as the sum of the initiatives of the divinity and of the believer tending toward the elimination of evil, aims henceforth at a specific end distinct from the end of creation. (Ricoeur 1960a, 173).

In this third genre of myths, creation is perfect, given that a perfect divinity accomplished it; thus, perfection is its main feature. It is the fault of humans that then destroys order, which stands in contrast with the "drama of creation" type of myths, wherein violence and chaos are present at the very beginning.

## *The Fourth Type: Myths of Exiled Soul*

Ricoeur describes this type of myth as "marginal" or "solitary" because it introduces a new characteristic: there are two elements in the human being: body (material) and soul (immaterial or spiritual). Hence, dualism enters into the new apprehension of the human being. Even if the French thinker depicts this kind of myth as "marginal", because it is very different from the three others, he must recognize that it has always played a crucial role in Western culture, both in religion and in the development of Greek philosophy.

Plato (Orphism before him) and Christian religion, for instance, focus on the final soul's future (salvation or condemnation). The soul comes from somewhere else, from a superior level and is, in fact, lost in the world as we see it, the common, everyday experience:

> This myth is the only one [comparing it with the three other types] which is, in the proper sense of the word, a myth of the 'soul' and a myth of the 'body'. It tell how the 'soul', divine in its origin, became human – how the 'body,' a stranger to the soul and bad in many ways, falls to the lot of the soul – how the mixture of the soul and the body is the event that inaugurates the humanity of man and makes man the place of forgetting, the place where the primordial difference between soul and body is abolished. Divine as to his soul, earthly as to his body, man is the forgetting of the difference; and the myth tells how that happened. (Ricoeur 1960a, 280).

Salvation and deliverance arrive in these myths through knowledge: "If now, turning toward deliverance, we ask what type of 'salvation' goes with this type of 'evil', one answer forces itself upon us: while the not-to-be-avowed theology of the wicked god excludes philosophy and finds fulfillment in the spectacle, the myth of the exiled soul is *par excellence* the principle and promise of 'knowledge', of 'gnosis'" (Ricoeur 1960a, 300).

Nevertheless, this typology, constructed by Ricoeur to establish guidelines for the vast and culturally diverse production of myths, is not enough to yield understanding of their symbolic "message". The diverse types of myths are not completely different narrations, totally apart from each other. Full interpretation passes from the Static – typology – to the Dynamics of the myths, in other words, the relationships among them, the passage, absorption, complementarity, and affinities of elements, symbols of one type of myth to those of another type in the dynamic historical process. Ricoeur gives us a good example of this dynamic:

> One test for our typology … will be to understand why the myth of the exiled soul and the myth of the fault of a primeval man could sometimes merge and blend their influences in an indistinct myth of the fall, although these two myths are profoundly heterogeneous. In addition, the secret affinities of the Biblical myth of the fall carry it toward the myth of chaos and the tragic myth rather than toward the myth of the exiled soul. (Ricoeur 1960a, 174).[14]

Notwithstanding, almost at the end of *The Symbolism of Evil,* Ricoeur surprises us by speaking of the preeminence of one myth over all others. The Adamic myth is privileged because, according to Ricoeur, it allows much complexity and can reaffirm in one way or another "the essential truths of the other myths" without abolishing them; on the contrary, they receive a new form of life when they are appropriated by the Adamic myth. The author tries to justify his choice by analyzing how a dynamic process of circularity reaffirms each one of the three other myths. I shall not discuss this point here, considering that my aim is now centered on the image that the human being envisages for or constructs of himself, his place in the world, and his role regarding chaos and evil – by means of a symbolic collective imaginary. I think that we could call this apprehension of the whole and of human beings an "eco-phenomenology" – a term that is not used by Ricoeur himself – focused on the symbolic language that opens the way to interpretation and hermeneutics: "Le symbole donne à penser".

## Conclusion

We can deduce from the entire path followed in this paper, considering diverse perspectives on eco-phenomenology, that however the term is understood, in a wider or more restricted sense, we deal with an approach we cannot avoid nowadays. There are several reasons for this. We are more worried than ever when we consider the impact of human technology on the future of our planet. At the same time, we are also recovering – at least according to some contemporary thinkers with whom I agree – the idea of belonging to the wholeness of the Earth, of the ensemble of living beings and, it seems, to the Universe, to address our last remote origin. In this sense, diverse approaches suggest that we are "the dust of stars" to use figurative speech, but also perhaps in a literal sense, given the interconnectedness of every element in the Universe.

In Anna-Teresa Tymieniecka's work, the meaning of eco-phenomenology is certainly broad, for she relates the human being to the Universe as a whole.

Other thinkers like David Woods formulate the question mainly in gnoseological-epistemological terms, and situate eco-phenomenology halfway "between phenomenology and naturalism, between intentionality and causality", as we have seen, opening up a bridge toward Nature and the natural, "independent both of the conceptuality of the natural sciences and of traditional metaphysics".

From the neuroscience point of view, philosophical eco-phenomenology – Merleau-Ponty's especially – is corroborated by demonstrating the ineludible belonging to nature of humans as embodied, somatic beings that are necessarily linked to "exterior" nature through the "windows" of their senses.

In Paul Ricoeur's thought, an eco-phenomenological hermeneutics could be derived especially in the field of symbols and myths – as exemplified in particular in the myths of the origin and the end of evil that he pondered – from their symbolic language and their meaning. These myths do display our own pre-consciousness of belonging to a whole and, at the same time, of our own responsibility, which consciousness coincides in this point with all the other approaches. Ricoeur's perspective encompasses many of the aspects already seen in the previous perspectives: particularly, the necessary impulse we humans have to attain an apprehension of ourselves. Philosophy, science, religion, and myth follow diverse approaches in this task. Ricoeur chooses means beyond those of rationality and causal knowledge of nature; he includes, as does Anna-Teresa Tymieniecka, imagination and human creativity. However, if for her creativity seems directly linked to awareness, in Ricoeur's perspective the messages of creative imagination are more or less veiled. Humans "listening" to symbolic narrations only become conscious of them and are able to integrate their content into conceptual philosophy by means of the detour of interpretation. Given that, perhaps it would not be impertinent to call Ricoeur's approach eco-phenomenological hermeneutics.

To sum up, the unavoidable theoretical perspective, eco-phenomenological in this case, should encourage us to go further and to emphasize, on a practical level, the relevance and the need to assume our human, always problematic, responsibility. It is necessary to underline our role, even if it is partial, given that, on the one hand, we do not control most natural negative events and, on the other, we arrive to life in an already established (déjà-là) cultural world. Despite that, we humans "feel" that we are able to participate to a certain extent in the whole process of Nature and History or, at least, we wonder if we can do something about the situations we confront therein. Specifically, are we decided to respect nature and other humans? We wonder if we are set to participate in either the development of order, cosmos, and kindness when we deal with other humans or, on the contrary, in general disorder, chaos, and hatred, usually linked to violence. We still have a choice.

## Notes

1. We can mention a few of them: "Ecosystems and human well-being", *International Health Regulations*, World Health Organization 2005. Print; Evans, Michael, "Environmental Disasters". Thu, 12 May 2011 10:50:01 GMT, www.earthtimes.org. Web. 18/09/2014; Golemann, Daniel. *Ecological Intelligence*. New York: Broadway Books, 2009. Print. Küng, Hans, *Global Responsibility: In Search of a New World Ethic*. New York: Crossroad Publishing Company, 1991. Print; McWhorter, Ladelle and Stenstad, Gail, editors. *Heidegger and the Earth: Essays in Environmental Philosophy.* Toronto: University of Toronto Press, 2009. Print. "Philosophy, Humanity and Ecology:

Philosophy of Nature and Environmental Ethics", edited by Oruka, Odera. *Proceedings of the Nairobi World Conference of Philosophy, July 21–25, 1991*. Print; Stracher, Glenn B., "Coal fires burning out of control around the world: a Global Catastrophe." *International Journal of Coal Geology*, 59, Issues 1–2, 12 July 2004: 7–17. Print. Zimmerman, Michael E. et al. *Environmental philosophy: from animal rights to radical ecology*, 4th ed. Upper Saddle River: Prentice Hall, 2004. Print.

2. Francisco Mora has doctorates in Medicine from the University of Granada, (Spain), and in Neuroscience from Oxford University. He is Professor at the Complutense University (Madrid); he is also an adjunct professor at the University of Iowa.
3. It is important to put into context the aforementioned similarities between Merleau-Ponty's approach and that of neuroscience: these coincide when they underline the role of the body and the material world both for knowledge and for human life. Nevertheless, for the French thinker there are two fundamental human levels not to be forgotten: the human being is both rooted and transcendent; there is a dialectic of archeology and teleology. This dialectic also appears as a fundamental theme in Paul Ricoeur's works: *De l'interprétation. Essai sur Freud* (Paris: Seuil, 1965) and in other hermeneutical essays. This implies an essential ambiguity of human life. While Merleau-Ponty also takes up the example of color, in his comments on the primacy of perception, his approach is different from that of Mora, as it is evident in *The Primacy of Perception* 15.
4. A subject developed in his book *El reloj de la sabiduría* [The Clock of Wisdom], Madrid: Alianza, 2005. Print.
5. The underscoring is mine.
6. She refers to "The Human Condition Within the Unity-of-Everything-There-is-Alive and Its Logoic Network," Analecta Husserliana LXXXIX, Dordrecht: Springer, 2006. Quoted by herself in "An interview with Anna-Teresa Tymieniecka".
7. For the crucial role of creativity, imagination in exploring diverse possibilities, self-interpretation, and meaning, see also "The Moral Sense. A Discourse on the Phenomenological Foundation of the Social World and the Ethics," in Analecta Husserliana XV, 1983; "Poetica Nova, The creative crucibles of Human Existence and of Art", in Analecta Husserliana XII, 1982, 1–93; "Imaginatio Creatrix. The 'Creative' versus the 'Constitutive' Function of Man, and the 'Possible Worlds'", in Analecta Husserliana III, 1974, 3–41.
8. "Je ne suis donc pas seulement regard situé, mais vouloir dire et dire comme transgression intentionnelle de la situation; dès que je parle, je parle des choses dans leur faces non perçues et dans leur absence. Ainsi, l'intention perceptive finie, qui me donne la présence perçue dans le présent vivant, qui est. le présent de la présence, n'est. jamais seule et nue; elle est. toujours, en tant que pleine, prise dans une relation de remplissement plus ou moins complet par rapport à une autre visée qui la traverse de part en part, qui la transit littéralement et à quoi la parole est. originairement liée; cette visée est. le vouloir-dire du dire. En naissant j'entre dans le monde du langage qui me précède et m'enveloppe. Le

regard muet est. repris dans le discours qui en articule le sens; et cette dicibilité du sens est. un continuel dépassement, au moins en intention, de l'aspect perspectif du perçu ici et maintenant." My translation. Ricœur, P., *L'homme faillible*. Paris: Aubier Montaigne, 1960. This book completes, together with *La symbolique du mal*, the ensemble *Finitude et Culpabilité*. In fact, the Spanish translation appears in a single volume that includes both works and has the title *Finitud y culpabilidad*, divided into Books 1 and 2. In English translations, these works have been brought out in two volumes: *Fallible Man*, and *The Symbolism of Evil*. From now on, quotes from *La symbolique du mal* will be taken from the English edition, Trans. Emerson Buchanan, Boston: Beacon, 1967.

9. *La métaphore vive*. Paris: Seuil, 1975.
10. A problem thoroughly developed by Ricoeur in his *Lectures on Ideology and Utopia*, New York: Columbia University Press, 1986.
11. The underscoring is mine.
12. Ricoeur wrote several books and articles on the question of hermeneutics; e.g. *De l'interprétation. Essai sur Freud*. Paris: Seuil, 1965; *Le conflit des interprétations* (*Essais d'herméneutique I*). Paris: Seuil, 1969; *La métaphore vive*. Paris: Seuil, 1975; The three volumes of *Temps et récit*; *Du texte à l'action*; and *Écrits et conférences. Tome II: Herméneutique* Paris: Seuil, 2010, among others.
13. The underscoring is mine.
14. Another good example: in the Adamic myth, man introduces evil in the world but, at the same time, there are other symbols that point to the fact that evil was already there: the serpent and, outside the myth but in the Bible, the "fallen angels".

## Works Cited


Bible. Standard King James Version. Web. 15 September 2014.

First People. American Indian Poems and Prayers. http://www.firstpeople.us/html/Great_Spirit.html. n.p. n.d. Web. 18 August 2014.

Harris, Marvin. 1997. *Culture, People, Nature: An Introduction to General Anthropology*. 7th ed. Boston: Pearson Higher Ed USA. Print.

Kottak, C.Ph. 2013. *Mirror for Humanity: A Concise Introduction to Cultural Anthropology*. 15th ed. New York: McGraw-Hill. Print.

Merleau-Ponty, Maurice. 1964. The Primacy of Perception and Its Philosophical Consequences. In *The Primacy of Perception and Its Philosophical Consequences and Other Essays on Phenomenological Psychology, the Philosophy of Art, History and Politics*, ed. and trans. James M. Edie. Evanston: Northwestern University Press. Print.

Mora, Francisco. 2011. *Cómo funciona el cerebro*. Madrid: Alianza. Print.

Ricoeur, Paul. 1960a. La symbolique du mal. Paris: Aubier Montaigne. Print. English edition: *The Symbolism of Evil*. New York: Harper & Row, 1967. Print.

———. 1960b. *Finitude et culpabilité: L'homme faillible*. Paris: Aubier Montaigne. Print. English ed. *Fallible Man*. Ed. and trans. Charles A. Kelbley, New York: Fordham University Press. 1986. Print.

———. 1985. *Temps et récit*. 3 volumes. Paris: Seuil, 1983, 1984, 1985. Print. English ed.: *Time and Narrative*, 3 vols. Trans. Kathleen Blamey and David Pellauer, Chicago: University of Chicago Press, 1984, 1985, 1988. Print.

———. 1986. *Du texte à l'action*. Paris: Seuil. Print. English ed.: *From Text to Action: Essays in Hermeneutics II*, trans. Kathleen Blamey and John B. Thompson, Evanston: Northwestern University Press, 1991. Print.

Ten Native Commandments. http://www.nativevillage.org/Inspiration-/ten_native_american_commandments.htm. n.p. n.d. Web. 16 August 2014.

Torjussen, L. P., J. Servan, and S. Andersen Øyen. 2008. "An interview with Anna-Teresa Tymieniecka", in Bergen, Norway, 2008. http://www.phenomenology.org/Bergen-Interview.htm. n.p. Web. 20 August 2014.

Tymieniecka, Anna-Teresa. 1989. *Logos and Life*, Book 1: *Creative Experience and the Critique of Reason*. Dordrecht: Kluwer Academic Publishers. Print.

———. 1990. Phenomenology of Life and the New Critique of Reason: From Husserl's Philosophy to the Phenomenology of Life and of the Human Condition. In *Man's Self-Interpretation in Existence: Phenomenology and Philosophy of Life, Introducing the Spanish Perspective*, Analecta Husserliana XXIX, ed. Anna-Teresa Tymieniecka, 3–15. Dordrecht: Kluwer Academic Publishers. Print.

———. 1998. Phenomenology of Life (Integral and Scientific) as the Starting Point of Philosophy. *Phenomenological Inquiry* 22. Print.

Wood, David. 2003. [The same article] in: Charles Brown and Ted Toadvine, eds. *Eco-Phenomenology: Back to the Earth Itself*. Albany, New York: State University of New York Press. Ch. 12: 211–235. Print.

# Ego: The Cross Point of Divine Illumination and Social Reality

**Konul Bunyadzade**

**Abstract** Ego is here considered from two perspectives: as the manifestation of divine wisdom and as the tipping point of external influences. Ego is the matrix of the human being. It looks like a stable core, protecting the being of the person. However, every internal and external influence is an impulse toward its next manifestation. Thus, in the first instance, Ego is the end of the divine line connecting God and human beings. It is the first to receive a divine message and interpret it. Ego is a door to the world of ideas and the divine sources of Truth. In the second instance, Ego is considered in the context of society, as personality is the manifestation or reaction of the Ego appropriating its environment.

**Keywords** Ego · Soul · Contradiction · Harmony · Ishragism

## Introduction

The human being is a unity of spirit and body, both a symbol of the divine world and a manifestation of matter. Ego is the harmonic and dynamic essence of this unity. The human being is a resident of the divine and corporeal worlds at the same time, and the Ego shares in the creativity and ability of the Creator. Therefore, a human being who can cognize his or her essence and mission is useful not only to society but also to all humanity. Anna-Teresa Tymieniecka called this dual trajectory of human cognition eco-phenomenology. Other thinkers and philosophers have tried to discover the human being's world – the initial and final point, the source and motivation of knowledge – through, in one instance, the Sufi concept of homeland (Bunyadzade 2014, 135–149).

K. Bunyadzade (✉)
Azerbaijan National Academy of Sciences, Baku, Azerbaijan
e-mail: bkonul885@hotmail.com

 
W. S. Smith et al. (eds.), *Eco-Phenomenology: Life, Human Life, Post-Human Life in the Harmony of the Cosmos*, Analecta Husserliana CXXI,
https://doi.org/10.1007/978-3-319-77516-6_16

## Who Is the Ego?

Many thinkers have indicated that the spirit is the good and gentle part of the human being while the body leads to low desires. Plato wrote:

> Of the nature of the soul, though her true form be ever a theme of large and more than mortal discourse, let me speak briefly, and in a figure. And let the figure be composite—a pair of winged horses and a charioteer. Now the winged horses and the charioteers of the gods are all of them noble and of noble descent, but those of other races are mixed; the human charioteer drives his in a pair; and one of them is noble and of noble breed, and the other is ignoble and of ignoble breed; and the driving of them of necessity gives a great deal of trouble to him. (Plato 1871)

Plato wanted to present the force of knowledge that could take the human being to heaven, to the level of gods, but also show how the power of earthly desires humiliates the human being. Both features of the human being, divine and earthly senses, are included in a single soul, which guides or should guide one's self, and nobody else. Although the human being has divine features and the chance to ascend to a god's level, he has a material body and should not ignore that. This theme of the dual character of the human being has often been the object of research and discussion. As it seems from this thesis, Plato's soul and Ego are the same. Consequently, the features attributed to the soul belong to Ego too. In other words, Ego drives the chariot and holds the reins.

However, if human beings are created by God with great love, how can part of each be ignoble and negative? According to Plato, the "two horses," the divine and earthly features of a human being, should be together. If one part operates alone, it can take the human being to an extreme pole. Therefore, the perfection of Ego is in the harmony and unity of one's divine and material sides. Each side should complete the other.

Ego must also have wisdom and know the names (essence) of all creatures. As Plato says: "The wing is the corporeal element which is most akin to the divine, and which by nature tends to soar aloft and carry that which gravitates downwards into the upper region, which is the habitation of the gods. The divine is beauty, wisdom, goodness."

Several ages later, a Quranic verse stated: "And He taught Adam the names – all of them" (*al-Baqarah* 31). Indeed, the human being can grow to the level of God only through knowledge: "Then he approached and descended. And was at a distance of two bow lengths or nearer" (*an-Najm* 8–9). The human being has a material body, but wisdom can raise him to the highest level, where there is an open door to the divine world.

Further, Ego must have the will to direct his knowledge to this aim. The Ego must unite the forces of both "horses" to penetrate both worlds. Schopenhauer pointed to the Will, which "alone gives him the key to his own phenomenon, reveals to him the significance and shows him the inner mechanism of his being, his actions, his movements" (Schopenhauer 1969, 100). Indeed, for Schopenhauer, the will is "the innermost essence, the kernel, of every particular thing and also of the whole. It appears in every blindly acting force of nature, and also in the deliberate conduct of man, and the great difference between the two concerns only the degree of the manifestation, not the inner nature of what is manifested" (Schopenhauer 1969, 110).

However, the philosopher associates Will with the body, which is detached from the divine world and divine knowledge – the wings of Ego. Consequently, the essence of Ego is imperfect and incomplete. Thus, Ego is the inner force of the human being by which to perfect his or her personality and to connect the divine world with the material world.

## The Conflicting Moments of Ego

In Ancient Chinese philosophy, wisdom should serve society. Human beings can reach perfection only in society; outside, the individual is unnecessary and can be destroyed. Guan Yin Tzu wrote: "The wise men don't leave society. For this reason other creatures cannot capture them" (Gurevich 1991, 40). The Sufi thinker Junaid Baghdadi sees the value of the *arif*, the wise man, in his good deeds for society and humanity: "They are the signs of the true path of God. They do more good deeds than anybody else. They protect the nation from all troubles" (Junaid 1970, 75). If we imagine the human being as a circle, one half-circle receives knowledge from the material and divine worlds, and the other half-circle realizes this knowledge. The process of a person's completion and perfection, therefore, depends on society. Thus, society is simultaneously a field test for knowledge, a fertile ground for creativity, and a springboard for new ideas. When society is full of instability and illness, it is impossible to stand on it and begin a new circle of cognition. It is no coincidence that thinkers turn to society to find the roots of the decline of morality and ethics. However, just how great is society's responsibility?

Society is a complete structure that includes individual Egos. Some Egos are native and congruent with society. Their personalities do not suffer because they have their own place. Whether in a palace or a slum, this place is enough for the manifestation of Ego. If the thought of the human being adheres to the frame of society and his Ego or superpersonality is on the level of his residence, his place satisfies him. What a paradoxical moment: a brick on a grand balcony and a crumbling brick in a tenement are the same to one who considers a place his best fate! Indeed, these thoughts are on different levels but equally bounded, involving fully realized and manifested essence, full harmony between the Ego and society. However, the one living under the crumbling brick understands his limited position and tries to improve; it is required by his position in society.

Indeed, the fine building is more perfect than the slum. The brick in the grand building has completed its mission and cannot do anything beyond its immediate surroundings. To do anything out of its system is meaningless. Khalilov Salahaddin writes:

> In a highly organized society the life of every human being depends not on himself but on forces having an external character. The natural, social, economic, and cultural environment of the human being was formed before his birth. Owing to the fact that this environment involves processes on a higher level than the level of the individual, it is more inert and conservative. Even small changes demand great inner energy. (Khalilov 2007, 9–10)

Meanwhile, a brick in a cheaper building can compare his position with a higher one. It wants to improve its position and thinks, proposing projects. This desire causes crises of mind and morality in developed countries and motivates reconstruction in third world nations.

Thus, a paradox takes shape: an imperfect society is necessary to improve and perfect the Ego. However, every Ego should also form his or her own environment. The existing environment can quickly impair and unravel a person. As Nikolai Berdyaev says: "Contradiction is richer than similarity" (Berdyaev 2003, 57). The definition of light is in darkness, and the life stories of famous thinkers and philosophical masters serve as examples of that. However, imperfect society affects the Ego differently: the Ego of a human being can improve, and the inner world can rise to the perfect level, but in real life—imperfect society—that human being may be alone and unhappy, even perceived as crazy. Abu Turkhan writes, "Indeed, the tragedy of a perfect man is born from his desire to increase the highest moment of happiness. This moment is manifested in his creativity and mission of savior" (Abu Turkhan 2012, 51–52).

The question arises, where is the root of the contradiction between Ego and society? Historically, social discontent is associated with social problems. The formation of some philosophical, religious, or mystical trends can be related to social circumstances that are important and leading contributors to their development. Examples appear in both Western and Eastern philosophy: the Cynics from the Ancient period, Protestantism, Wahhabism, the Franciscans and Dominicans, or Hurufism. However, the common features of these trends make it clear that social circumstances are only the visible side of the problem. The leader of each these trends held the Ego of a single, defined person. In other words, these movements are the result of the desire of an Ego to escape his inner contradictions and voice his own truth, to realize a new society.

The group or small society gathered around the thought of such an Ego will eventually experience discontent and shifts, collecting additional forces according to time and circumstance. The truth of the founding Ego gets transformed to placate contradictions and conform to new circumstances. This is a long process, and after some period the real truth is forgotten. The best example is Sufism. The true essence of Sufism is the love and worship of God, perfecting a person's thought to serve society. However, after five ages of formation and growth, Sufism took on different essences: each Sufi tarikah (order) had its own system and could influence society. Instead of its true aim, Sufism shifted to represent the political views and tactics of its leader, the murshid of the tarikah. Moreover, the thoughts of the followers only developed within the framework defined by the murshid. The same changes can be observed in other philosophical trends, such as how the challenges of Nietzsche to humanity and true belief in the Nineteenth Century became a denial of holiness and humanity in the Twentieth Century. Consequently, if an Ego is in contradiction with the small, local society, he creates and looks to external circumstances for self-realization.

The contradictions between Ego and society in its first stage are not connected with social problems; then, by coincidence, it is justified by social concerns and can be directed to a different perspective. Moreover, frequently, an Ego who could not find common language (harmony) with society and who stood in contradiction to his or her environment, did not find new support from society's downtrodden. On the contrary, those who joined the new movement were frequently from among the higher ranks and financially secure. The contradiction of the Ego with society is in itself invisible and incomprehensible. When it is visible, it becomes a contradiction setting the person against society.

## Conclusion

The Ego, who inhabits two worlds – divine and material – unites in his essence the main beliefs of both. Ego feels sad at home because he is always separated from half of himself, and each part prevents him from completely surrendering to the other. According to the level of his thought, this home can be divine or material, indicating a relationship between the home and the Ego: the Ego receives knowledge from its source and feels passion for this source. It looks to the love or passion of the lower light or to the higher light held out in the Ishragism of Suhrawardi. Ego is a broader concept than light because it includes both light and darkness, yet it balances them and allows for the full realization of their potential.

## Works Cited

Abu Turkhan. 2012. [Abu Turkhan's world of wisdom. In *The Representation of Prof. Dr. Salahaddin Khalilov*]. Baku: Chashiogli. Print.

Berdyaev, Nikolai. 2003. *Dukh i realnost* [Spirit and Reality]. Moscow: OOO ACT Press. Print.

Bunyadzade, Konul. 2014. Wahdat al-wujud and Logos of Life: The Philosophical Comparison. In *Phenomenology of Space and Time. The Forces of the Cosmos and the Ontopoietic Genesis of Life*, Book Two, Analecta Husserliana 117, ed. Anna-Teresa Tymieniecka, 135–149. Dordrecht: Springer. Print.

Gurevich, Pavel Semyonovich. 1991. *Chelovek. Misliteli proshlogo i nastoyashego o ego jizni, smerti i bessmertii. Drevniy mir – epokha prosvesheniya* [A Man. Thinkers of the Past and Present About His Life, Death and Immortality. Ancient World – The Enlightenment]. Moscow: Political Literature Press. Print.

Junaid, Baghdadi. 1970. *Rasail al-Juneyd* [Epistles of Junaid]. Ed. Dr. Ali Hasan Abd al-Qadir. Istanbul: Cağaloğlu. Print.

Khalilov, Salahaddin. 2007. *Meneviyyat felsefesi* [Philosophy of Morality]. Baku: Azerbaycan Universiteti Press. Print.

Plato. 1871. *Phaedrus*. Trans. Benjamin Jowett. New York: C. Scribner's Sons. Sacred-texts.com. Web.

Schopenhauer, Arthur. 1969. *The World As Will and Representation,* vol. 1. Trans. E. F. J. Payne. New York: Dover Publications, Inc. Digitalseance.files.wordpress.com. Web.

# Part V
# Eco-Ethics and Environmental Theories

# Phenomenology as Ecology: Movement from Ego- to Geo- and Eco-Thinking

**Ella Buceniece**

> *The fact that a technology, which by definition alters the limits of nature, is required in order to violate them, means that the limits of nature exist.*
>
> – Umberto Eco

**Abstract** A paradigmatic change of thinking is taking place at present – away from the transcendental themes of mind, ego, and language, towards the world, reality, immanence, and the realms of the Earth and the cosmos. This change marks the ecological turn of philosophy. Awareness of the new ecological situation has produced a whole range of new directions of research – e.g., eco-philosophy, environmental philosophy, philosophy of nature, deep ecology, ecocriticism, ecofeminism, and ecoscepticism. This article is therefore concerned with the relations among eco-philosophy, environmental philosophy, and phenomenology. Phenomenology occupies an important place in the apprehension of the present-day ecological situation. This holds both for the classical version of phenomenology with its well-known concepts of the world – Umwelt and Lebenswelt from Husserl – and for the present-day phenomenology of life and its ontopoiesis – the approach of Tymieniecka. It also holds for the more modern modes of ecological conceptualization: geophilosophy and the new realism of Deleuze, Guattari, Embree, and Eco, among others.

**Keywords** Tymieniecka · Phenomenology as ecology · Eco-philosophy and environmentalism · World · New realism · Umwelt · Phenomenon of the cosmos

---

E. Buceniece (✉)
Institute of Philosophy and Sociology, University of Latvia, Riga, Latvia



W. S. Smith et al. (eds.), *Eco-Phenomenology: Life, Human Life, Post-Human Life in the Harmony of the Cosmos*, Analecta Husserliana CXXI,
https://doi.org/10.1007/978-3-319-77516-6_17

It is well known that ecology as a discipline was thematized in the nineteenth century (1866) by German biologist and philosopher Ernst Haeckel, who coined the term. Initially it was spelled as *oecology*, coming from the Greek *oikos*, meaning the family household and its daily operation and maintenance.

At present the problematics of ecology have transcended the narrow traditional meaning of the term, for the essence of the "family" and its environment or the ecosphere have profoundly changed in such a way that Anna-Teresa Tymieniecka called it "the human condition." It encompasses not only the surrounding nature, but the whole biological, cultural and social set-up, including the cosmic dimension, united as it is by the logos – the self-individuating principle of life and of everything-that-is-alive.

Awareness of the new ecological situation has produced a whole range of new directions of research, including eco-philosophy, environmental philosophy, philosophy of nature, deep ecology, ecocriticism, ecofeminism, and ecoscepticism. Therefore, I suggest that we may speak of a new turn – an ecological turn – as we speak with regard to the previously experienced anthropological, linguistic, and ontological turns. Of special importance for the demarcation of the ecological turn may be the fundamental phenomenological ontopoietic approach of Anna-Teresa Tymieniecka. But more on that later.

The present ecological situation, justly referred to as the ecological crisis, is actually an estrangement of humanity and of individual human beings from the essential bases of existence: from involvement with nature and natural wholeness. This follows from the fact that the aims of development are looked for beyond human beings as living beings (or beings that have been born, as Husserl put it) and without awareness that humans as living beings are part of a whole ecosystem, so that the consumerist attitude towards the surrounding environment is tantamount to the self-destruction of mankind and civilization. I hold that the solutions of our ecological problems are looked for in two directions: technocentric and ecocentric. The first of these approaches is concerned with solutions pertaining to consequences. The chief concern of the second approach is negotiating the relations existing between humans and the world; it requires reinterpretation of our understanding of the world and of reality, of the place of humans within the system of nature; it requires a new system of values, one designed to "encounter life" (a term coined by Lester Embree) so as to include therein also the ecosphere.

## Eco-Philosophical Trends

Thus, philosophy becomes involved in the apprehension of the ecological situation. It concerns both well-established philosophical trends and those branches of eco-philosophy that have acquired popularity recently. Fox Warwick, a notable representative of modern eco-philosophy, has listed about nine major approaches to our relationship with the world around us that have been mapped out by environmental philosophers in recent years. Warwick has delineated three thematic groups within these nine approaches:

> First there are those approaches that ascribe only a *use value* or *instrumental value* to the non-human world. This covers the approaches of *unrestrained exploitation and expansionism, resource conservation* and *development, and resource preservation.* Second are those approaches that argue for a criterion of moral consideration that would attribute an *intrinsic* value to at least some, and perhaps many, members or aspects of the non-human world. This covers the *sentience based* approach (that is, the animal welfare approach), the *life* based approach, and the *holistic integrity* approach. Third are the co-called "radical ecologies" of *deep ecology, ecofeminism,* and *social ecology.* (Warwick 1995, 70)

What is the place of phenomenology within the whole process of devising of a new strategy for apprehension of the ecological situation? If eco-philosophical thinking in the sense of "vigorously moving away from human-centeredness" took place during the mid to late 1970s and if we know that phenomenology originated at the beginning of the twentieth century and that it had always possessed a characteristic of "human-centeredness," the question arises as to how is it to be connected with eco-philosophical and environmental approaches (Warwick 8). I hold that phenomenology is to be counted, in a certain sense, as belonging to both classical phenomenology, with its concepts of Umwelt and Lebenswelt, and the new French phenomenology, which seeks conceptual substantiation in the new materialism, and to as well Anna-Teresa Tymieniecka's phenomenology of life as well (Mullarkey 2006, 56).

## Eco-Philosophy's Interface with Environmentalism

As is usually the case with new spheres of knowledge, there is a wide range of definitions of eco-philosophy and environmentalism to be found in the works of various scholars. There is a variety of typologies arising out of widely differing presuppositional bases. There are scholars attempting to unite the two disciplines, and those who insist on considering them as separate entities. A very systematic overview of these approaches represented by various authors is given by Warwick. Thus, ecopolitical theorist Murray Bookchin discusses the relations of the two disciplines: Environmentalism refers to "a mechanistic, instrumental outlook that sees nature as a passive habitat composed of 'objects' such as animals, plants, minerals, and the like, that must merely be rendered more serviceable for human use," while "ecology or social ecology refers to an approach that rests on the ecological principles of unity in diversity, spontaneity, and the nonhierarchical nature of ecological communities" and which sees "that attempts have be made to overcome the splits between society and nature, mind and body, thought and reality that mark Western images of the world" (Bookchin 1982. 86; Warwick 31). I hold that both approaches are not intrinsically different; they are actually complementary, for the formation of ecological awareness has taken place by way of making use of the notion of the environment and vice versa.

## Classical Phenomenology and the Environment

It is well known that Husserl's phenomenology set out to counter naturalism by returning to the *Sache selbst,* including the *Sache* that is nature. Thus, nature as a phenomenon is no longer an object, and a human being is no longer a mere subject; the appearance of both of them is simultaneous within the initial ego-constituting processes. What is more, pure reason in phenomenology does not create a hierarchy of things (entities), in contrast to traditional metaphysics, where some entities – that is, spiritual ones – occupy a higher position than their material and natural counterparts. That conclusion follows from the latter's being accessible to us (I hold that naturalism is a cannibalism of nature), for those things then have being only as a correlate of mind, as they have been apprehended by us in accordance with the qualities of mind: as percepted, as remembered, as expected, as visualized, as imagined, as discriminated, as accepted by faith, as possible, and as evaluated. Mind in phenomenology does not permit one to use, to consume things while the "nature of things" has not been reflected upon, and has not become clear. Even in his early work "Philosophy as Rigorous Science" Husserl called the phenomenological approach a method leading to clarity – to exact pure apprehension, in other words – the ability to ecologically see the world as a precondition for the existence of humanity and civilization. Owing to the discovery and description of mind, the world (in the first place) enters the mind in a pure manner; besides, mind is an essential, if not the most essential, element of the structure of the world. Mind does not exist outside the world. According to Latvian phenomenologist Teodors Celms, a pupil of Husserl, mind, in general, is that very point in the development of the world where the world not only exists but becomes aware of itself. Not seeing the world in a clear fashion is one of the ways leading towards ecological-anthropological catastrophe. Merab Mamardashvily, a Georgian-Russian phenomenologist, says that this happens because the laws of the human mind are broken, and the structure called "civilization" goes down with it.

The next glaring deficiency of egoecology or humanocentricity is bound up with its unawareness that the natural sciences cannot tell us all about the natural world: they refuse to consider the pre-scientific experience of humans; they ignore the close intertwining and overlapping of things that influences the understanding of reality, providing it with perspective (Llewelyn 2003, 56). John Compton, in his article "Phenomenology and the Philosophy of Nature," holds that in reality what we experience "is always this perspectival unity; that which is more than we see, which has another side, an inside, and as yet unexpressed capacities; that which relates to and interacts with other things" (Compton 1988, 79). Classical phenomenology has tagged this a-theoretical world, that in which humans fundamentally live their lives, with the terms of Lebenswelt and the surrounding reality or Umwelt. The role of phenomenology in apprehension of the present-day ecological situation is directly related to the concepts of Umwelt and Lebenswelt, which are intended to thematize the world (*Welt*) in its communal reality as a phenomenon of the world-cum-human being, not just as a sum total of all objects, not just as a sum (All-Nature), but as a process and as the experiences of the subject (for the concepts of the World and of the environment are not identical). In this subjective experience, the Umwelt enters as nature and also as the world of objects in the totality of culture. Husserl announces the theme of the Umwelt in his *Ideas I* in 1913, and treats it extensively in his later "Crisis" works. Here, Husserl is not so much concerned with the crises of nature, but rather with the crises of a science severed from the human Umwelt.

Today the Umwelt theme has become topical within the context of the ecological crisis and calls for a rethinking of the phenomena of the world and nature. This theme has attracted scholars' attention anew. Jung-Mi Lee has discussed it in a dissertation "Lebenswelt und Einstellung in der Phenomenologie Husserls," that accentuates the significance of the phenomenological approach: "In day-to-day life we are not observing the world, but only objects; in the scientific approach we observe the world, not the being of the world (Sein der Welt). Only a phenomenological approach makes it possible to thematize the world" (Lee 1999, 125). The logical connection between phenomenology and environmental philosophy is attested also by Don E. Marietta: "Even though environmental philosophy that stresses the involvement of humans in the system of nature and produces an ethic of responsibility to nature was not well developed in their time, the existential phenomenologists found in Edmund Husserl's work an approach to human consciousness and experience that lays a groundwork for environmental thought" (Marietta 2003, 121). Thus, Husserl's phenomenology fits naturally within the project of modern environmental philosophy by way of adjusting its terminology for new uses, For example, alongside the "surrounding world" the term "invironing" is being used.

This theme has also been developed today by phenomenologist Lester Embree in his article "The Possibility of Constitutive Phenomenology of the Environment." He describes the stance of classical constitutive phenomenology concerning the world as the "encountering of objects" and "objects as encountering" – in a word, as environmental encounters – and he concludes that the notion of "encountering" (to offer a preferred term) is to be related to both mind and experience. Therefore, the surrounding world is that sphere which phenomenologists refer to as "lifeworldly nature" and which may further be specified as "lifeworldly environment" (Embree 2003, 39). This latter sphere is, according to Embree, cultural in a concrete way. The author objects to culture being understood as an aggregate of artefacts of culture to be modified by human activities. There are objects of culture that are beyond the reach of the modifying activities of humans: the Sun in the morning and during daytime, and in the evening the stars as part of the surrounding world, or encountering life (Embree 2003, 40). This author seems to accord with the idea that the environment is part of the cultural world for he states that "the environment can be approached not only naturalistically but also culturally, and environmentalism can be political, as well as, in a rather broad signification, aesthetic" (Embree 2003, 38).

## Post-modern Recognition of the Bounds Nature Places on Thought

In my opinion, the thinking of post-modern interpreters also has ecological consciousness as a backdrop the by way of their acknowledging the inescapability of reality for the substantiation of interpretations. Thus, Umberto Eco (in a discourse given at the 2013 World Congress of Philosophy in Athens) concedes that there is some sort of a boundary for interpretations determined by reality, which – in a stance distinct from the old type of realism – he says is susceptible to changes owing to habit. Eco calls this position Minimal or Negative Realism: "Language does not construct being *ex novo*: it questions it, in some way always finding something *already given*" (Eco 2015, 8). Yet, this *already given* is not a thing in itself; it is reality as lines of resistance to our interpretations. "This idea of lines of resistance, by which something which does not depend on our interpretations challenges them, can represent a form of Minimal or Negative Realism according to which facts, if scarcely tell me that I am right, frequently tell me that I am wrong" (Eco 2015, 9). Thus, the ideas of Minimal Realism offer us an opportunity to increase the certitude of our interpretation and, at the same time, to foster modifications of reality – not in any direction whatever, but in the direction of minimal safety (and this may sometimes even offer a glimpse of the truth): "However, by producing these interpretants we grow a habit, that is, an ability to act upon the surrounding reality, and the proof that a given series of interpretants works is given by the fact that through our habit we can modify the reality. If this does not succeed, it means that our interpretations were dominated by the continuous risk of fallibilism. The possibility of a failure means that there is something controlling and limiting our interpretations" (Eco 2015, 12).

One can notice the turn in the direction of eco-phenomenology also in the geophilosophical approach discernable in the thought strategies of Gilles Deleuze, Felix Guattari, Paul Virilo, and others. Deleuze and Guattari also reject the subject *cum* object approach as a means of grasping a thought; they see the field of thinking in the conjunction of territory and the earth. The earth is not only a single element; the earth is a movement with regard to a place. Thus the earth encompasses all elements so as to deterritorialize the territories. Deterritorialization and reterritorialization are two intertwined movements of the earth; they have manifested themselves – and continue to do so – in the form of "imperial spatium" and "political extensio," and the structure of this movement has provided for different manners of thinking in various territories of the earth. Thinking is and is not exterritorial; it is a form of compatibility of the incompatible forms of the earth and various territories; it is both relative and absolute: "… deterritorialization is relative insofar as it concerns the historical relationship of the earth with the territories that take shape and pass away on it, … with the cosmos and the stellar system of which it is a part. But deterritorialization is *absolute* when the earth passes into the pure plane of immanence of a Being-thought, of a Nature-thought of infinite diagrammatic movements. Thinking consists in stretching out a plane of immanence that absorbs the earth (or rather, adsorbs it)" (Deleuze 1994, 88). Such a type of absolute thinking, being immanent to the earth may, possibly, take place within a certain territory, and yet it means the origination of a "future new earth" (Deleuze 1994, 88). So it seems possible to preserve the earth as oikos, and Heimlichkeit, if we saturate ourselves with it, in order to create it anew, or if we create a new reality within a specific territory.

> Philosophy is a geophilosophy in precisely the same way that history is a geohistory from Braudel's point of view. … Geography is not confined to providing historical form with a substance and variable places. It is not merely physical and human but mental, like the landscape. Geography wrests history from the cult of necessity in order to stress the irreducibility of contingency. It wrests it from the cult of origins in order to affirm the power of a 'milieu' (what philosophy finds in the Greeks, said Nietzsche, is not an origin but a milieu, an ambience, an ambient atmosphere). (Deleuze 1994, 95–96)

This ecological thinking also reterritorializes in different ways in various countries: "the expression 'ecology' is often extended to encompass much more, particularly in Europe; in the United States it chiefly designates a biological discipline concerned with organisms and their environments. This discipline also includes work in chemistry, geology, meteorology, and other naturalistic sciences" (Embree 2003, 38).

## Tymieniecka's Phenomenology of Life

Anna-Teresa Tymieniecka's phenomenology of life has from the start reterritorialized itself as eco-phenomenology, and her work is of fundamental importance for the present-day ecological turn, even if she formulated this position in the terms of eco-phenomenology only in the interview she gave in Bergen, Norway in 2010.[1] She says in this interview:

> Actually, my account of *ontopoiesis* is an eco-phenomenology. Ontopoiesis reaches to the very germs of ecology: development and genesis. I have published several essays related to this. In *The Passions of the Earth* (Analecta Husserliana LXXI), I show how the human being is an ecological fruit and how the human being is formed by the earth and sucks the juices of the earth. I have also written things about the cosmic dependencies of the human mind and human development. You sees the self-individualization of life, which is the basic instrument of ontopoiesis draws upon the laws of the cosmos and the earth. This is the most fundamental ecology that can be done. (Tymieniecka 2015)

According to Francesco Alfieri, Tymieniecka is also the originator of the third phase of phenomenology. Thus, one may even assert that Tymieniecka's phenomenology of life, by way of being the beginning of the third wave of phenomenology, largely conflates with the manner of thought that I designate as the ecological turn. The basic principles of both approaches are similar: to consider the processes of life, of subjectivity in an inextricable unity with the regularities of the Universe, and vice versa:" Proceeding from our generic roots in the earth's soil, the logos of life upon which the entirety of ordering and vital sense is suspended traverses intentional-creative-ontopoietic becoming *within* and expands into an unconfined horizon of becoming *without*, and so the ontopoietic becoming of life finds a completion" (Tymieniecka 2011, 11).

A similar balancing act between humans and the Universe, between "the same" and "the other," and the importance of respecting of such a stance for the present ecological situation, is pointed out also by Thomas M. Robinson in his analysis of Plato's dialogue *Timaeus*:

> The respect will take two forms. On the one hand it will be respect for environment and its creatures as *other*, but nonetheless creations of divine. On the other hand it will be respect for the environment as *the same*, sharing with us as does the all important feature of rational soul. Which leads us directly to the argument from utility and self-interest. If my own rational soul is composed of the same stuff as the soul of the world itself, to be damage to the world is in effect to be damage to myself. (Robinson 1999, 159)

Which is to say that humans are responsible for their immediate environment and for the world at large; this encompasses the cosmic dimensions as well. Bence Peter Marosan offers the approach of transcendental morphology as a phenomenological discipline by the very fact of its acknowledging that "at the core of world-phenomenon is the phenomenon of cosmos. Ultimately, it is the phenomenon of the cosmos that could shed light on any phenomenon in the world, be it environment, encircling ring (Umring), home-world or any other" (Marosan 2014, 88).

It is noteworthy that Anna-Teresa Tymieniecka proposed a similar idea in the 1960s. Gary Backhaus in "Introduction to Phenomenological Inquiry" accentuates the "cosmological attitude" when discussing the most essential aspects of Tymieniecka's open system of phenomenology of life. He deals, in particular, with Tymieniecka's work *Why is there something rather than nothing? prolegomena to the phenomenology of cosmic creation,* which shows that the insights of Tymieniecka as life- and eco-phenomenologist were in advance of the times, or – to use the words of Nietzsche – non-contemporaneous.[2] Eco-philosophical thinking makes use of such notions as the human condition, environmental imagination, place, dwelling, rhythm, earth, and cosmic law. These are basic concepts of Tymieniecka's work; however, they are rarely used today outside the publications of phenomenologists of life. I hold that this is to be explained not only by the originality of Tymieniecka's thought, but by what could be called the "cosmic autonomy" of her views. I fully agree with Francesco Alfieri to the effect that: "The originality of her thought and the full autonomy of her investigations are immediately clear from the difficulty experienced by every researcher every time they seek to fit Tymieniecka's thought into established conceptual schemes inherited from philosophy" (Alfieri 2014, 22).

Of course, the participants in the phenomenological congresses of the World Phenomenology Institute and authors in the Analecta Husserliana series, most especially Daniela Verducci, in respective articles and monographs have done an excellent job in providing explanations and interpretations of the conceptions of Anna-Teresa Tymieniecka. However, much work still has to be done in the contextualizing of the autonomous and non-contemporaneous monumental heritage that is hers, by way of making the "established conceptual schemes" ever more comprehensible and significant. Within the context of my present theme, the investigation of the relationship between eco-philosophy and environmentalism, the work of

Anna-Teresa Tymieniecka may be said to synthesize – in an original fashion, that is, and using a different type of vocabulary – both approaches.

I am deeply grateful to Professor Tymieniecka – a prominent thinker of the twentieth and twenty first centuries and equally a magnanimous humanist – for her philosophically stimulating friendship over several decades. If one would have to attempt to characterize Professor Tymieniecka, I would say: a Cosmic Woman – a Woman of the Universe.

## Notes

1. This was on the occasion of the World Phenomenology Institute's 60th International Congress of Phenomenology, organized by Anna-Teresa Tymieniecka.
2. A reminder: the movement of eco-philosophical thought originated only in the mid-1970s.

## Works Cited


Alfieri, Francesco. 2014. "The Inseparable Link Between "Cosmology" and the "World of Life" in the Philosophy of Anna-Teresa Tymieniecka: The Originality of a New Perspective on the "Real Individual and Autonomous Being" and a Comparison with the "Phenomenological Realism" of Hedwig Conrad-Martius." In *Phenomenology of Space and Time: The Forces of the Cosmos and the Ontopoietic Genesis of Life,* Book 1, Analecta Husserliana 116, ed. Anna-Teresa Tymieniecka, 21–36. Dordrecht: Springer.

Bookchin, Murray. 1982. *The Ecology of Freedom: The Emergence and Dissolution of Hierarchy.* Palo Alto: Cheshire Books.

Compton, John. 1988. Phenomenology and the Philosophy of Nature. *Man and World* 21: 65–89.

Deleuze, Gilles and Felix Guattari. 1994. *What Is Philosophy?* Trans. H. Tomlinson and G. Burchell. New York: Columbia University Press.

Eco, Umberto. 2015. *Some Remarks on a New Realism.* Web, March 13, 2015. www.wcp2013.gr/files/items/6/649/eco_wcp.pdf.rnd.

Embree, Lester. 2003. The Possibility of a Constitutive Phenomenology of the Environment. In *Eco-Phenomenology: Back to the Earth Itself,* ed. Charles S. Brown and Ted Toadvine. Albany: State University of New York Press.

Lee, Jung-Mi. 1999. *Lebenswelt und Einstellung in der Phänomenologie Husserls.* Diss., Wuppertal.

Llewelyn, John. 2003. Prolegomena to Any Future Phenomenological Ecology. In *Eco-Phenomenology: Back to the Earth Itself,* ed. Charles S. Brown and Ted Toadvine. Albany: State University of New York Press.

Marietta, Don E. 2003. Back to the Earth with Reflection and Ecology. In *Eco-Phenomenology: Back to the Earth Itself,* ed. Charles S. Brown and Ted Toadvine. Albany: State University of New York Press.

Marosan, Bence Peter. 2014. Transcendental Morphology: A Phenomenological Interpretation of Human and Non-Human Cosmos. In *Phenomenology of Space and Time: The Forces of the Cosmos and the Ontopoietic Genesis of Life,* Book 1, Analecta Husserliana 116, ed. Anna-Teresa Tymieniecka, 285–325, Dordrecht: Springer.

Mullarkey, John. 2006. *Post-continental philosophy: An outline*. London/New York: Continuum.
Robinson, Thomas M. 1999. The Cosmic Environment: Some Implications of Plato's *Timaeus*. In *Philosophy and Ecology. Greek Philosophy and the Environment*, ed. Boudouris Konstantine and Kalimtzis Kostas, vol. I. Athens: International Center for Greek Philosophy and Culture.
Tymieniecka, Anna-Teresa. 2011. Inspirations of Heraclitus from Ephesus Fulfilled in Our New Enlightenment. In *Phenomenology/Ontopoiesis Retrieving Geo-Cosmic Horizons of Antiquity*, Analecta Husserliana 110, ed. Anna-Teresa Tymieniecka, 3–12. Dordrecht: Springer.
———. 2015. An Interview with Anna-Teresa Tymieniecka. Lars Petter Torjussen, Johannes Servan, and Simen Andersen Øyen. 2010. Web.13 March 2015. http://www.phenomenology.org/Bergen-Interview.htm.
Warwick, Fox. 1995. *Toward a Transpersonal Ecology: Developing New Foundations for Environmentalism*. Albany: State University of New York Press.

# "Song of The Earth": An Eco-Phenomenology

**Kimiyo Murata-Soraci**

*The splendor of the simple.*

—Martin Heidegger

**Abstract** This paper probes into a dimension of "απειρον" in Anna-Teresa Tymieniecka's conceptualization of "eco-phenomenology" as seen through the lens of an egalitarian ethic and the holistic vision of deep ecologists such as Aldo Leopold, Val Plumwood, Dawne McCance, Max Oelschlaeger, and James Lovelock, who strive to create a free biosphere and eco-system of all the living beings beyond the bounds of humanism and speciesism. We will explore Tymieniecka's and deep ecologists' shared question concerning how to dwell with others on the earth via a backtracking through Plato's χώρα, Husserl's "Ur-Arche," and Heidegger's "region," so as to illumine not only the depth of Tymieniecka's philosophical roots but also the fecund potency of her "eco-phenomenology" for fostering a new conception of justice (*dikē*) by which a symphony of earthy "life, human life, and post-human life" will be conducted.

**Keywords** Eco-phenomenology · Ethics · Critical animal studies · Deep ecology · biosphere · Earth · χώρα · Ur-Arche · Region · Dwelling · Plato · Husserl · Heidegger

K. Murata-Soraci (✉)
Chuo University, Tokyo, Japan



W. S. Smith et al. (eds.), *Eco-Phenomenology: Life, Human Life, Post-Human Life in the Harmony of the Cosmos*, Analecta Husserliana CXXI,
https://doi.org/10.1007/978-3-319-77516-6_18

Let us open our meditation on eco-phenomenology and the issue of gathering (*Versammlung*) by reciting the ancient Homeric Hymn:

> Gaia, mother of all, I sing, oldest of gods,
> Firm of foundation, who feeds all creatures living on earth,
> As many as move on the radiant land and swim in the sea
> And fly through the airs—all these does she feed with her bounty. (Sargent 1975, 79)

The earth on and in which we stand and find always already standing, just as the Homeric poet did, is the "indestructible" matrix of all natural phenomena, historical events, and cultured worlds (Heidegger 1959, 147). It enables, supports, and shelters the destinying (*Geschick*) of the historical happenstances of all life-forms in their finite luminosity. We all are of the earth.

According to the ancient Greek myth of Cura (Care), which Heidegger cites in *Being and Time* as the springboard of his existential-ontological interpretation of Care (*Sorge*) as the Being of Dasein, we humans received the name "homo" because we "are made out of "humus" (earth) (Heidegger 1959, 242). The Cura myth tells us of the three parentages (Jupiter/Care/Earth) that humans receive, and Earth remains closest to us of these three. By sheltering and holding the dead, Earth's prevalence over man and his fabricated senses of life and the world surpasses the reign of Care, who watches over man during his life. Does not this hint that Earth surpasses the anxiety of *Jemeinigkeit* and the time of Care cleared in *Being and Time*? Care's ecstatic (*ekstasis*) temporality "shook off" (*abbauen*) the ontotheological reign of Jupiter, who formerly held man's knowledge of self, life, and the world in reference to a transcendental spiritual substance that remains fixed outside of history and thereby kept man oblivious to the questions of who he is and where "we" really come from (Heidegger 1984, 22; Derrida 1989).

Earth, being man's material and native origin, outlasts the time of not only the "pre-" but also the "post-metaphysical" ground of life and man's spiritual (*Gemütes*) character. Is not the earth a *third* ground to which we must turn so as to meditate on the "humanitas of homo humanus" after the stages of philosophical wonder and anxiety (Heidegger 1993, 248)? Is not the earth a gold mine of possibilities from which we may summon a common ground of kinship, solidarity, and mutual responsibility among earthly creatures of all kinds who posses not only a non-historical but also a historical place of origination? Would not the earth be the just and necessary ground for us today to rethink the "**in**" of "being-**in**-the-world" from, and on the basis of the "**in**" of **in**habitant-on-the-earth so as to break through current impasses of ecological thinking?

Today, ironically, the earth seems to be the one who bestirs us and calls to us from underneath a time-honored footing of objectivism and instrumental thinking in order that we may take a good look at her pale appearance as either a rootless planetary "disk" or a swollen (*sunya*) ball of bloodless objects that are a "standing-reserve [*Bestand*]" for use and service for human consumption, aesthetic taste, and efficient lifestyle (Heidegger 1993, 322). We humans reign heedlessly on her massive wasteland. The earth solicits us to rethink what it means for us to share with innumerable life-forms her native soil (*heimatlicher Grund*) and provokes us to find new ways of feeling, thinking, and speaking about the kinship among earthly denizens of all sorts beyond the fence of anthropocentrism and speciesism. In our response to the uncanny call of m(O)ther earth, whom as we shall see shortly Plato called "Χώρα," and Husserl "Ur-earth," and Heidegger "region," let us step earthward so as to "map out"[1] Tymieniecka's "eco-phenomenology" on the basis of m(O)ther earth and to illumine several issues to be attended to for making peace with both Mother Earth and our earthly kindred.

Today, we fulsomely hear and read about interventions and studies on the environment and nonhuman animals conducted by advocates, ethicists, and governments in order to save our habitat from eco-cide. Peter Singer's *Practical Ethics* denominates environmental issues on a vast scale—global warming, the melting of glaciers, flooding, droughts, the sinking of land, deforestation, loss of wilderness, nuclear waste, the methane emissions of the livestock; the list goes on and on (Singer 2011, Chaps. 9 & 10). As we confront and ingest various other species at meal time, pollution caused by factory farms' manure wastes rises to the top of health and environmental concern for living beings in general. Dawne McCance directs our attention in her *Critical Animal Studies: An Introduction* to the Mallin and Cahoon study on how animal manures consisting of microbes survive the normal round of disinfection treatments; these pollutants then enter and contaminate water systems and poison aquatic animals while imperiling the health of humans and livestock and intensifying global warming (McCance 2013, 20–21).[2]

In her book, McCance gives detailed accounts of the denaturalization and maltreatment of and loss of biodiversity among animals by industrialized farming. Both McCance and Singer agree with Jeremy Bentham's advocacy that the capacity for suffering be the criterion for equal consideration of the interests of nonhuman animals' well-being. However, she raises interesting and vital questions as to whether the ontotheological nature of law can appropriately speak for the rights of nonhuman animals since a mode of apophantic discourse undergirds, unbeknownst to the advocates and ethicists sensitive to the plight of animals, the conventional stance of the rational animal's (*zōon logon echon*) sovereignty over nonhuman animal lives and their domains (McCance 2013).[3]

McCance also questions whether Singer's utilitarian ethical position does not fall short of equality and justice for the nonhuman animals' and their rights to their different ways of being because Singer posits unquestioningly the "mental capacities" of "normal human adult" to be the benchmark for the capacity for suffering and for an equal consideration of interests not only among humans but, also, among human and nonhuman animals.[4] Thus, she wonders if Singer still stands on the fence of modern subjectivity and speciesism. Her questions unveil, if anything, the recalcitrant grips of the metaphysics of presence, the discourse of substance and intentionality, and dichotomous thinking, and thus, indicate the present impasse over holistically regathering all life-forms in peace, justice, and hospitality with an eye to their differences in relation to the earth.

In the field of environmental ethics, I will here name three spokesmen for the rights of nature because of the relevancy of their outlooks to my topic of eco-phenomenology. They are Aldo Leopold, Max Oelschlaeger, and James Lovelock.

Their approaches to environments are holistic, egalitarian, polycentric, respectful of the symbiotic relation between particular lands and their denizens for nurturance and freedom, in accord with their ways of being and not being, inclusive of inoperable spaces and gaps and of atmospheres in virtue of their giving priority to the lands themselves. A terrain, a non-sentient being, is elevated to the holder of prime value among all living and non-living creatures not for the reason of aesthetic beauty, or as potential property and resource locale for humans, but because of the land's enabling power to situate any life taking place and play itself out along with others. Leopold pioneered such a holistic "land ethic" (Singer 2011, 251). Oelschaeger proposes an "ecocentric" and not "egocentric" system because it is only in and through a global codependence of natural systems that "even life itself must be set in a larger evolutionary frame of reference that contains inorganic constituents" (Oelschlaeger 1991, 293). Lovelock reiterates the deep ecology of "biosphere" shared with Leopold and Oelschaeger and says in *Gaia: A New Look at Life on Earth*, "The Earth's living matter, air, oceans, and land surface form a complex system which can be seen as a single organism and which has the capacity to keep our planet a fit place for life" (Lovelock 1987, x).

Through our brief encounter with these ethicists of critical animal studies and environments, we recognize the need for another discourse that would enable not only the rights of animals and of nature of which these ethicists speak but also the intelligent presentation of their accounts as a unified system of knowledge. According to Heidegger, logos originally means "to gather, collect, read" (*lesen*) in a sense of placing one thing with another in a better arrangement, and the German "lesen" is related etymologically to the Latin "legere," the root of which the word intelligence bear (Heidegger 1959, 125, 122). To remain faithful to the earth, as Nietzsche urges, and to think intelligently with *physis* and the earth, as Heidegger advocates, deep ecologists must become aware of the inadequacy of the conventional views on rationality, apophantic discourse, the unified system of knowledge (*Wissenschaft*), which is what the "-logy" of ecology carries. For their method of gathering on equal footing vast and diverse biotopoi and the unique inhabitants therein on the common ground of the earth tends to drift laterally, instead of moving teleologically to an ultimate ground of unity, thereby resisting a familiar pattern of producing a coherent system of knowledge by placing parts into a hierarchical order on a ground of common identity among present beings. Perhaps, it is this lack of unity or a difference from the common practice of systematizing, as such, that motivates Singer's critical appraisal of deep ecology, for example, Lovelock's system of the "biosphere" (Singer 2011, 253).

Be that as it may, in response to the inadequacy of the term "ecology," I deem that Tymieniecka's term "eco-phenomenology" rightly befits a renamed deep ecology, provided the "eco-" of eco-phenomenology hosts the wide range of usages in regards to "dwelling" derived from mortal Dasein's "ex-istential" spacing of "innan" and its linkage to the Latin "colo" (Heidegger 1962, 80). As Casey points out, Latin "colere," which means "to inhabit, care for, till, worship," and is related to the familiar word "culture," which means "placed tilled" in Middle English (Casey 2009, 336). Thus, the "eco-" of eco-phenomenology takes us back to the Greek οικονομια of the οικος,, the management of household economy in a sparing manner. The derivatives of *oikos* are *oikia* ("dwelling") and *oikesis* (administration); the same root "weik" found in oikos is borne by ecology and economy (*weik* + *nemein*, to allot, to manage) (*The American Heritage College Dictionary* 1997, 1606).

In light of this (r)"eco-"(llection) on the interconnection between dwelling and the ordering of all sorts of lives of things' right emplacement, how to keep our alertness up and open to and care for a living unity of diverse biotopoi, with their worldly relations both within and beyond of each sphere or biotopes of inhabitants, on the ground of this earth and in relation to it moves us to further question how to speak "intelligibly" **of** the earth. For, having and maintaining "intelligence" concerning the earth, as shown in the deep ecologists' ecosystem, is to be fully gathered only when the earth is let to manifest and speak in and of "herself" to the common everyday world of the "biosphere." This last question appears to be missing among deep ecologists. As this question appears to be crucial in finding another discourse for the deep ecosystem, we shall now re**co**llect the three ways of speaking of the earth used by Plato, Husserl, and Heidegger at this crucial juncture of the philosophical (de-/re-trans-) forming of the world.

In the *Timaeus*, Plato's divine architect Demiurge presupposes the choric space when he molds the cosmos (the world) in the design of geometrically-informed forms. Χώρα is called a "third kind" which is absolutely necessary (*Ananke*) for the generation of all forms to come within a grid of the non-sensible and the sensible realms. The Χώρα "in" which whatsoever becomes lands, first of all, and makes its contacts with is not a static "receptacle" (49b) (Plato 1989, 1176). As the verb χωρέω means "to go forward, to be in motion or flux," χώρα overflows into the eidos as the idea's necessary a priori element and into the stuff of things as their indirect matrix (Sallis 1994, 4–5).[5] It "nurses" all kinds of what becomes by permeating them thoroughly, thus, extending herself all the way along with them; χώρα receives back all the images and names of both the non-sensuous and the sensuous types and shelters them in their decomposition. This "matrix" (*ekmageon*, 50c) is like a "mother" (50d). χώpa makes room for whatever becomes so that they may take their first "seat" (*hestia*, hearth) (52a–b). The divine craftsman says:

> And there is a third nature, which is space and is eternal and admits not of destruction and provides a home for all created things, and is apprehended, when all sense is absent, by a kind of spurious reason, and is hardly real—which we, beholding as in a dream, say of all existence that it must of necessity be in some place and occupy a space, but that what is neither in heaven nor in earth has no existence. (52b)

As Sallis points out, the verb χωρέω has also the sense of "to make room for another, to give way or withdraw" (Sallis 1994, 4).

However, it is χώpa's intrinsic disposition and power to not only let others dwell by a spontaneous self-withdrawal but also to bring into the confluence of being a neutral place for the generation of all entities (having a topological role as the "seat"), this while becoming strewn into and with the imprinted forms (having a participatory role as the "nursing receptacle"), and this has been covered over beneath the Demiurge's intuition of place in terms of extension (space) when he creates the κόσμος. Thus, χώρα is unremembered in the orbits of metaphysics since Plato. In *An Introduction to Metaphysics*, Heidegger refers to a crucial event of loss in the *Timaeus* with respect to the Greeks' experience of place in terms of thing's upsurging out of concealment; thus, the Greeks' intuitive understanding of place as the intimate place of a thing's dwelling became displaced by the experience of the spatial in terms of extension after Plato (Heidegger 1959, 66). This oblivion of χώρα has played since a definitive role in the formation of man's basic disposition in the world and toward the self and the other.

Husserl raises critical questions about the modern man's objective attitude and regard for the earth in his essay "Foundational Investigations of the Phenomenological Origin of the Spatiality of Nature." He says that "we Copernicans, we moderns" have forgotten that earth is the "'primitive home-place,' as ark of the world" (Husserl 1981, 230). Our experience of corporeal nature, observations of celestial bodies, and experiential knowledge of them are all "relative to the earth basis ark and 'earthly globe' and to us, earthly human beings, and Objectivity is related to the All of humanity. What about the earth-ark itself? It is not itself already a body, not a star among other stars" (Husserl 1981, 228).

There is a good parallel between the χώρα and the earth as the "Ur-Arche." Just like χώρα, the earth functions both a topological and a participatory base for all possible experience of bodily knowledge. The earth enables a round of motion and rest; but, the earth cannot be reduced to either one. This older base (as a whole) can neither be captured by reference to any one body (part), nor can it be explained by a sum total of all bodies (parts). Just as the χώρα remains an unnamable abyss or pure void, the earth remains unidentifiable. The Demiurge says in the *Timaeus*: "[w]e may liken the receiving principle to a mother, … the matter in which the model is fashioned will not be duly prepared unless it is formless and free from the impress of any of those shapes which it is hereafter to receive from without" (50d). Husserl says, "the earth is the ark which makes possible in the first place the sense of all motion and all rest as mode of motion. But its rest is not a motion" (Husserl 1981, 230).

For Husserl, the earth (*Ur-Arche*) is originarily not only an *arche* of human experience of earthly knowledge but, also, "another primitive home" (*Urheimat*) of the "cosmopolis" to which "'we,' that is, humanity as a whole, is referred back, that is, borne as by an ark" (Sallis 1993, 98). Husserl, thus, underlines: "All development, all relative histories have to that extent a single primitive history of which they are episodes. In that connection it is indeed possible that this primitive history would be a togetherness of people living and evolving completely separated, except that they all exist for one another in open, undetermined horizons of earth-space" (Husserl 1981, 228).

Through their usage of "χώρα" and of "Ur-Arche," and, regardless of their opposite ways of unveiling, both Plato and Husserl indicate the precedence of a place for life prior to world-shaping. They hint at an impersonal sense of freedom embedded in the simultaneous counter-movement of unconcealing-in-withdrawal which the figures of χώρα and Ur-Arche seem to repeat purely and naturally. Then, the question we share with the deep ecologists becomes that of how we are to gather our daily common world of "cosmopolis" with the primordial movement (*Wesen*) of the placing the life, be it that of the χώρα or the Ur-Arche/Ur-heimat, in such a way that all living and nonliving inhabitants can embody and realize the primordial universals in the concrete, allotted place of living for each. In other words, how are we to reconstruct our impoverished world with the bivalent universals of the historical (deep ecology) and the non-historical (the third path of eco-phenomenology), for the common destiny of all terrestrial tenants is an issue for us and for eco-phenomenology. At this juncture, Heidegger's thought of *alētheia* and of "region" presents us a beacon of guiding light.

As Heidegger has shown in *The Origin of the Work of Art*, setting the world on the earth (which is of *a priori* necessity for life and for world-shaping) in an irreducible strife without a telos of finality would imitate (*mimēsis*) the primordial draw of unconcealment-in-withdrawal and its natural shade of topological and participatory co-turning (Heidegger 1993, 174). A non-dialectical strife between the contraries, the world and the earth, allows "the contraries to correspond to one another, and in that they let one another reciprocally come forth" (Heidegger 2012, 5). The strife of reciprocal co-bearing between the world and the earth gives us a chance to perceive the earth's sovereign power over all generations of life and worlds and, above all, presents us an occasion to be bestirred by the earth's immense capacity for watching over inhabitants of all sorts without a monocentric focus.

In the face of m(O)ther earth's boundless (ἄπειρον) sovereign power in bestowing the place of dwelling and watching over us all, we "humus" are allowed to see the delimitation of our will to ground and essence looming in the "-isms" of biologism, anthropocentrism, speciesism, and objectivism; above all else, we are allowed to see both our comical and tragic reign over the earth in building a homeless human world. As Heidegger says, the irreducible *polemos* between the world and the earth brings forth freely a measure of dimension to both the places and the inhabitants elsewhere in, on, and around the earth. Only by placing before and after the m(O)ther earth our stance of egocentric mono-vision, as well as our bodies of knowledge concerning the senses of life and the world tilled by philosophical wonder and mortal anxiety, can we be ready to "let world prevail" (Heidegger 1998, 127).

In that letting (*lassen*), we come to let the world watch over all spheres of life-forms in justice, hospitality, and harmony. By letting the earth and the world watch conjointly and coextensively over us all, we humus are able to hear other life-forms speak in and of themselves to us. For on the "place [*Ortschaft*] that encompasses all locales and-time-play-spaces, [w]hat bestirs in the showing of saying is owning" (Heidegger 1993, 414).[6] Thus, we can make peace (*Friede*) at once with Mother Earth and with the World-full of co-tenants. In bringing to close my overture to an eco-phenomenology, allow me to cite these words from Heidegger's "Building Dwelling Thinking":

> To dwell, to be set at peace, means to remain at peace within the free,
> the preserve, the free sphere that safeguards each thing in its essence.
> The fundamental character of dwelling is this sparing.
> It pervades dwelling in its whole range (351).

## Notes

1. Edward Casey makes an interesting remark about a deep connection between cartography, chorology (<chōra), and ecology in his *Getting Back Into Place*, Second Edition (Bloomington and Indianapolis: Indiana University Press, 2009), p. 439. Casey says that cartography was once called chorology, which meant a science of place. Ernst Haeckel, who first used the term ecology, systematically associated the term to chorology in *The History of Creation* (1873). Accordingly, I am using the verb "to map out" in a sense of journeying and going through an unexplored place.
2. According to Dawne McCance, the microbes that animal wastes contain include "pathogenic bacteria, such as Escherichia coli, salmonella, and streptococcus, pathogenic protozoans, and a number of viruses." (McCance 2013, 21)
3. In the text, McCance acknowledges her indebtedness to Derrida's studies on the animal in *The Animal That Therefore I Am* and *The Beast and the Sovereign,* Vols. I & II.
4. For Singer's and McCance's detailed discussions and their viewpoints on this issue, read *Practical Ethics*, Ch. 4 and *Critical Animal Studies*, Ch. 3, respectively.
5. Judith Butler in *Bodies that Matter* directs our attention to an etymological link between matter and matrix and points out that matter as the place of generation is intimately tied to the issues of cause and origin (p. 31). She further states that in the *Physics* (4: 209b) Aristotle identifies *hyle* with the notion of "receptacle" (*hypodokhe*) in the *Timaeus*, and just as the Platonic notion of "receptacle" is nonidentifiable and unthematizable, Aristotle's notion of "hyle" remains non-definable in the *Metaphysics* (1035b). For a detailed discussion, see Judith Butler, *Bodies that Matter* (New York: Routledge, 1993), p. 253.
6. I am indebted to David Farrell Krell for his discussion on the animal world, *Walten*, freedom, and language in Heidegger. See David Farrell Krell, *Derrida and Our Animal Others* (Bloomington and Indianapolis: Indiana University Press, 2013), p. 112.

## Works Cited

Casey, Edward. 2009. *Getting Back into Place*, 3rd ed. Bloomington: Indiana University Press. Print.

Derrida, Jacques. 1989. *Of Spirit*. Trans. Geoffrey Bennington and Rachel Bowlby. Chicago: University of Chicago Press. Print.

Heidegger, Martin. 1959. *An Introduction to Metaphysics*. Trans. Ralph Manheim. New Haven: Yale University Press. Print.

———. 1962. *Being and Time*. Trans. John Macquarrie and Edward Robinson. New York: Harper & Row. Print.

———. 1984. *Was Ist Das—Die Philosophie?* Pfullingen: Neske. Print.

———. 1993. In *Basic Writings*, ed. David Farrell Krell. San Francisco: HarperCollins. Print.
———. 1998. *Pathmarks*. Trans. William McNeil. Cambridge: Cambridge University Press. Print.
———. 2012. *Four Seminars*. Trans. Andrew Mitchell and François Raffoul. Bloomington: Indiana University Press. Print.
Husserl, Edmund. 1981. In *Husserl: Shorter Works*, ed. Peter McCormick and Frederick Elliston. Notre Dame: University of Notre Dame Press. Print.
Lovelock, James. 1987. *Gaia: A New Look at Life on Earth*. Oxford: Oxford University Press. Print.
McCance, Dawne. 2013. *Critical Animal Studies: An Introduction*. Albany: SUNY. Print.
Oelschlaeger, Max. 1991. *The Idea of Wilderness: From Prehistory to the Age of Ecology*. New Haven: Yale University Press. Print.
Plato. 1989. *The Collected Dialogues*. Trans. Edith Hamilton and Huntington Cairns. Princeton: Princeton University Press. Print.
Sallis, John. 1993. Question of Origin. ed. Leonard Lawlor. *Derrida's Interpretation of Husserl. The Southern Journal of Philosophy* 32(Supplement, 1993): 86–106. Print.
———. 1994. Of the χώρα. *Epoche: A Journal for the History of Philosophy* 2: 1–12. Print.
Sargent, Thelma Trans. 1975. *The Homeric Hymns: A Verse Translation*. New York: W. W. Norton & Company. Print.
Singer, Peter. 2011. *Practical Ethics*, 3rd ed. Cambridge: Cambridge University Press. Print.
*The American Heritage College Dictionary*, 3rd ed. 1997. Boston: Houghton Mifflin Company. Print.

# Cultural Sustainability: Lines of Reflection for a Human Life in the Harmony of the Cosmos

**Alessandra Lucaioli**

**Abstract** The main aim of the paper is to propose a more robust criterion for the concept of sustainability, one reflecting the importance of taking into account the cultural dimension as a fundamental pillar of authentic human and virtuous development. The theoretical assumption behind this new concept is that cultural processes are also to be considered structural elements within the socioeconomic system. Cultural sustainability concerns actions that affect the way in which a community expresses its identity, safeguards its traditions, and builds shared values; indeed, meanings, narratives, and constellations of symbols enable a society to recognize and identify itself in the image of world that it has developed. Therefore, starting from a brief reconstruction of the theoretical debate in literature about the definition of sustainability, and showing how, in the headquarters of national and European institutions, the concept of cultural sustainability is still absent, the paper is intended to clarify that it is not only for our physical dwelling – and to cushion the ecological crisis we are experiencing today – that we have to take into account chemical, physical, economic, or social parameters. What is needed, as the eco-phenomenological tradition already recognizes, is a re-conceptualization of human values and of our relationship with nature and place in the harmony of the cosmos.

**Keywords** Cultural sustainability · Identity · Place memory · Territory · Harmony of cosmos

## Introduction

Dwelling is decidedly an essential feature of humans' fundamental way of being in the world. This is testified to by the fertile anthropological reflections originating from the phenomenological-hermeneutic and metaphysical tradition that refers to dwelling, as in the various interpretations of the theme by Melchiorre,

A. Lucaioli (✉)
University of Macerata, Macerata, Italy

 
W. S. Smith et al. (eds.), *Eco-Phenomenology: Life, Human Life, Post-Human Life in the Harmony of the Cosmos*, Analecta Husserliana CXXI,
https://doi.org/10.1007/978-3-319-77516-6_19

Merleau-Ponty, and Heidegger. Human beings indeed inhere in space and time and find their constituent coordinates therein. By virtue of their transcendental allocation, human beings have with the world a familiarity more ancient than thought, as Merleau-Ponty declares. Yet, the constituent bond with a place and familiarity with the spatial dimension do not automatically result in a set of instructions telling us how to dwell. Therefore, inhabiting the Earth remains, especially today, a difficult art for accomplishing the task of taking care[1] of the world around us is becoming increasingly hard.

On one hand, there is the extreme challenge presented to Earth-dwelling man today by globalization. We are drawn to both enjoyment of its opportunities and resistance to its relocating corollaries, which are multiple in form: management of energy and the earth's resources in general; building methods; consumption awareness; protection of cultures and home production. On the other hand, there is a challenging multidimensional crisis that has underlying structural problems connected to each other: from the economic crisis to the financial one, from the crisis of the environment to that of happiness or the meaning of life. These are nothing other than the result of the logic and ideology of recklessly pursuing hypertrophic growth instead of promoting authentically human and virtuous development. This has encouraged the human cohort dwelling on the Earth to own and control it, justifying the irresponsible exploitation of all of its resources. However, the need to anchor the meaning of our existence and any possible future for us and other living systems in sure and shared values taking the Earth as an essential horizon of meaning leads to a radical rethinking of our ways of living and dwelling.

## What Is the Meaning of Sustainable Development?

The concept of *sustainability*[2] emerges within this horizon. Deriving from the Latin verb form *sustinere*, which is made up of the prefix *sub-* (up) and *stinere* (hold), the word refers both to the ability to hold something up over one's head and to an attitude of protecting, defending, nourishing, and preserving something or someone. The term became popular and officially settled in the global lexicon only when the international community, driven by the need to examine the regulatory relations between the natural world and human beings, defined development as being "sustainable" when it "meets the needs of the present without compromising the ability of future generations to meet their own needs" (World Commission on Environment and Development 1987, 43). In this sense, sustainability must be understood as a call to leave our children a legacy that is not worse than that we have received from previous generations. It means, in short, trying to pursue objectives that were deemed incompatible for a long time: protection of ecosystems and socioeconomic development. All this is based on the idea – one structural for the concept of sustainability – of a solidarity-based ethics: solidarity with other individuals existing here and now, with those who are living elsewhere at the moment, with those yet to come and – in certain variants of the idea of sustainability – with other living species.

The connection with the principle of responsibility, formulated for the first time by Hans Jonas in 1979 and conceived as a fundamental guide for human action, is evident. The necessity to "act in such a way that the effects of your action are compatible with the permanence of an authentically human life on the Earth" is the prerequisite for ethics based on the self-limitation of man's transformative powers where they may have a destructive impact on nature (Jonas 1990, 30). The notion of sustainability also involves the idea of quality. While analyzing the metaphorical value of graphic signs in the alphabets of ancient and modern languages, Austrian scholar Alfred Kallir attributes a strong guttural sound to the letter *q*, one capable of evoking hardness and endurance, as in the word *quality*, which evokes the ability to resist anything that might cause a deterioration of present things. It is no coincidence that we consider quality products or services to be those which endure over time.

At this stage, the definition of sustainability is limited to the environmental sphere. In 1992, on occasion of the Earth Summit in Rio de Janeiro, the concept was taken up again and partially integrated. This conference has the merit of focusing debate on the question of how to reconcile the objectives of environmental policy with those of development policy, recognizing that the global protection of the environment is possible only if economic and social outlooks are also considered.

At the same time, such a conciliation seems unreachable as the environment continues to be subsumed under economic rationality. As emphasized by many supporters of the degrowth movement, the concept of sustainable development is the most absurd oxymoron ever created. Actually, per its definition, no development can be sustainable even though at an etymological level both words can be located under the reassuring wing of *oikos*. Within this paradigm there is no place for the respect for the environment sought by ecologists, or the respect for humanity sought by humanists, but only the safeguarding of the egoistic interests at the center of classical economics.

Nevertheless, the hypothesis of a conflict between environmental concern and economic development can turn out to be contradictory. Actually, one cannot exist without the other, and, moreover, economic development can involve technological and scientific improvements that can be useful for environmental protection. However, it is only with the Copenhagen summit and the Amsterdam Treaty of 1997, that the European Union defined the three pillars of sustainability. This principle, better known as the three pillar model, establishes that sustainability is not limited to only the natural heritage that we pass on to future generations but also involves the economic achievements and the social institutions of our society, such as the democratic expression of our will or the pacific resolution of conflicts. In this sense, sustainable development is based on three pillars: ecological, economic, and social. Economic sustainability means the capacity to create, in a long-lasting way, work and revenues so as to support the population in conditions of the free market and transparency. In speaking of social sustainability we refer to a capacity to grant humans conditions of wellness and access to opportunities (safety, health, instruction) in an equal way across social classes, genders, ages, and, in particular, across present and future communities. If just one of these pillars fail, then sustainability inevitably fails.

By the way, I reckon that this theoretical paradigm still has a "narrow" point of view on sustainability, for it fails to contemplate a fourth pillar that should be added, namely, culture. The Amsterdam Treaty, signed by all of the member states of the European Union, makes no mention of cultural sustainability. This means not only that culture is not considered instrumental for human development, but also that the prescribed model of development is based on the precepts of classical economics. Thus, it cannot promote real human flourishing since it attributes the human behavior to the rational behavior of *homo oeconomicus*.

UNESCO is the only institution that has deepened the vision of sustainability by raising the socio-cultural pillar, highlighting the importance of cultural diversity and the need to maintain it. It recognizes human cultural diversity as being as necessary as biodiversity is for Nature. Cultural diversity is considered one of the roots of development and is to be taken into account not only as it contributes to economic growth but also as a means by which to realize a better life from an intellectual, emotional, spiritual, and moral point of view. This is already a small step forward, but it highlights only one dimension of culture, that of its diversity. This gives insufficient recognition to cultural processes as structural elements of the same weight as the others within a social and economic system.

## Toward a Definition of Cultural Sustainability

Given all the above, is it then actually a must to talk about cultural sustainability? And what does it mean?

Monica Amari has been the first to formulate this concept. Recently, Ms. Amari, an expert in cultural politics and processes, has written provocatively that maybe in the near future a minister of economics could be held responsible for violating cultural rights at the very moment he decides to reduce funding for culture. This paradoxical scenario is used by the author to highlight that many of the problems faced by Western society right now could be worked out if the European Union would recognize the concept of cultural sustainability, which she defines as "a right / a duty for the society to maintain the conditions necessary for reproducing cultural processes" (Amari 2012, 12). These processes, in a dynamic and recurrent way, lead to the recognition and creation of a new cultural heritage for the community.

The author considers a territory to be a living organism that breathes, eats, and communicates, and in this way she arrives at her formulation of cultural sustainability. This means granting a territory a subjectivity, and consequently recognizing it as a "legal person." Amari then wonders if a territory's rights would belong to the consolidated category of the preservation of material assets or to the rules that protect human rights. The Fribourg Declaration grounds cultural rights on the human rights that must belong to each individual in a society. Amari says that, "the cultural sustainability represents not only the opportunity but also the right, for the system of the territory to be able to create and re-create the essential conditions that would give it the opportunity to communicate its identity, values and strengths" (Amari 2012, 12). By the way, I reckon that the legitimacy of cultural sustainability rests not only on the fact that a territory is the subject of cultural rights – which actually belong to the category of human rights – but on a precise anthropological basis that includes the human being in his transcendental allocation, that is, his spatial nature and ability to create a spatial area.

If a human being is always positioned in some place, then it is clear that his allocation will be taken into account when talking about sustainability. This means that sustainability has to refer to the human being as a whole. In other words, if place and territory are really taken into account with no spatial fetishisms, that is, if they are considered as social products and productive factors of sociality, then a more "robust" concept of sustainability will be necessary, one that will transcend any reductionist one. Therefore, we do not have to bear in mind only physical, chemical, economic, or social parameters for a dwelling, or simply to stem the crisis we are facing right now. As recognized by the eco-phenomenological tradition, what we need is a re-conceptualization of human values and our relationship with nature,[3] which I deem to require even more value within the human transcendental allocation. I think that this relationship could express itself in the key of cultural sustainability, which relates to all those actions that condition the way a community expresses its identity, safeguards its traditions, and creates shared values; actually, significant tales and symbols help society to recognize itself, its identity, and the image it has created of itself outwardly.

The European Landscape Convention agrees with this concept. It recognizes people as a milestone of the landscape concept itself, highlighting in this way the nature of landscape as a social product that represents and talks to the society that every day constitutes and transforms it. The ELC interprets landscape as a product of the interrelations between a population and its ambience. In this way, a landscape is wherever such interrelations occur: "in urban areas and in the country side, in degraded areas as well as in areas of high quality, in areas recognized as being of outstanding beauty as well as everyday areas" (Council of Europe 2000, 18). But, the interrelation between humans and landscape cannot be limited just to this. Actually, the landscape is not simply the object of human perception and a mere background to its actions. A landscape is a living reality that is constantly modified by such actions, acquiring always new and different characteristics and meanings. In this way, a landscape can be considered the expression of local culture, it being that its construction is guided by economic mechanisms and socio-cultural values that rule the way a certain society acts and the meaning of its signs and symbols. At the same time, being a mirror for the interrelation between a population and the area in which it lives, a landscape becomes an element of cultural identity for its inhabitants.

## Forms of Cultural Sustainability: Identity, Symbol, and Place Memory

Following the epistemological revolution enacted by the new definition given by the European Landscape Convention, Laura Bonesio and Ettore Rocca pondered the often ambiguous significance of this concept. Actually, as well explained by Rocca, it can indicate three different realities being one: the landscape as *perceived* by humans; the landscape as a *reproduction* in a treaty, a movie, a tale, a picture, a painting; and finally the landscape as *part of the territory* modified by human action. These distinctions talks about the landscape and the territory as symbolic spaces, as cultural and common memory but, above all, even though this is not clearly expressed, about place identity and the way a community expresses it.

It is unquestionable that it is possible to recognize a place in the measure in which it has an identity. This declaration involves not only mere naturalistic data or a subjective/emotional concept of territory, even though it includes both. At the same time place identity cannot be construed as a fixed, stable, static entity aprioristically given. Identity does not mean stillness or stiffness but, rather, a unity gained through multiple aspects. Christian Norberg-Schulz would indicate it with the term *genius loci*, referring to all those sociocultural, architectural, linguistic, and habitual characteristics that "give character to a place." A place identity is rather the prerequisite and result of a cultural – and consequently common – interpretation that has configured it over time in a dynamic and coherent way. This is "the result of a long story of relations with other places and, consequently, necessarily an open, permeable, product of connections and links" (Massey and Jess 1995, 52).

This identity can be recognized from the outside and, consequently, can be represented through tales and representations. That is why we can say that people enact places, give life to the *representation of spaces*[4] through which a society gives form to itself though images, symbols, and rituals. This outlook is clearly taken by Cornelius Castoriadis when he writes that whatever is displayed in the historical-social world is indissolubly intertwined with symbols without, at the same time, being limited precisely to this. Both individual and collective real acts, as much as the material products without which no society could survive, are not always symbols. But both are not possible outside a symbolic net and its existential component: imagery. Castoriadis writes:

> Each society has tried to give an answer to some essential questions: Who are we as community? What do we represent one for the other? Where are we? What do we want, what do we desire, what are we missing? A society has to define its identity, its articulation, its world, its relation with the world and the object it contains, its own needs and wishes. If these queries be left with no answer, if there be no definition, then the human word would not exist, nor society or culture, because everything would be an undifferentiated chaos. The role of imaginary significations is to answer these questions. And it is clear that such an answer cannot be given by reality or rationality. (Castoriadis 1998, 72)

That is to say that each society elaborates an image of the natural world and its context, trying each time to create some kind of significance in which it can contextualize what is important not only for community life but also for the community itself. Each society has social imagery that is at the same time the consolidated result of the representations and the creeds that become in some way institutions, and there are moments in which it creates these significations. Therefore, the State, the enterprise, the supermarket are social imagery, socially consolidated, instituted, and rationalized owing to the permanence of things: buildings, ministries, factories, shops, symbol. Thus, to adopt a notion of cultural sustainability that recognizes the value of social imagery as a special way through which a society represents itself and has the roots of its identity means to safeguard the capacity to create links, interactions, and aggregation processes among different subjects that, even though very different from each other, recognize themselves as part of shared representations. Nevertheless, at the same time we have to say that this does not mean considering society and its imagery significations to be unchangeable or eternal, because such constructions can be modified by the society that has created them or even by the next generations. From such a mistake, one could derive the wrong idea that cultural processes are stable and fixed, while we know that they are dynamic and unstable. Rather, it means recognition that social imagery is fundamental if we want to reach a state of social entropy.

Place or territorial identity is, then, a symbolic construction that, to survive, needs to have its roots in memory too, in its cultural and collective dimension. Maurice Halbwachs was the first to systematically write about memory as a social phenomena. He highlights how the collective memory in the shape of memory of a shared past can exist only with fixed space-time coordinates, with a symbolic connection to the group with itself, and with a continuous reconstruction of the same memory. This means that the group elaborates a social memory, a memory of itself that is always placed in one space and one time; such points in space, if retraced either physically or with the imagination, can "sustain" and grant a continuum to a shared representation of the collective self. This social memory is the result of a selection process, carried out by the group, which stresses the elements that highlight its differences compared to other groups. Finally, the memory reconstruction the group undergoes ensures that new things also can be represented in a kind of reconstructed past. Then it comes to light that memory is not only a natural datum but also a cultural construction. All the "memory phenomenon" is the result of an activity that consists in taking up certain elements and charging them with precise symbolic significance, while removing others from the picture we want to represent.

Adopting a notion of cultural sustainability that recognizes the value of memory's role means to then preserve the faculty thanks to which human beings create a connection between past and present, a fundamental ingredient of identity. "It is easy to understand that identity is a matter concerning memory and recollection: as much as an individual can develop its personal identity and maintain it notwithstanding the days and years passing by, in the same way a group is able to reproduce its identity as a group just through its memory" (Assmann 1997, 61). But this also means to preserve, on one side, group coherence, as memory is a factor of cohesion and stability for social groups, and on the other side, to preserve the permutation, self-examination, self-transcendence, conjecture, prefiguring capacities that memory can trigger.

## Conclusions

Today, owing to the world's multidimensional crisis and globalization processes, a new way of living and dwelling are needed. An understanding of cultural sustainability is needed such that can satisfy the requisites of authentic and virtuous human dwelling and can engender human flourishing as it takes into consideration – in terms of rational geography –territorial heritage, cultural heritage, and material culture:

> Only a well-educated consciousness of space can sustain the right relation with the natural environment. The maintenance of a green economy will have a positive and long-lasting effect and will be a valuable alternative to the calculative economic mentality only if places are not considered as purely goods, instruments for attaining wealth, mere resources. The environmental crisis (Norberg-Schulz) is first of all a human crisis, and it is possible to face it only by understanding the concept of place, landscape, territory. (Danani 2013, 88)

Starting from the consideration of human transcendental allocation, we have reached a more complex idea of sustainability. It is a possible consideration that once we have transformed reference to place into anthropological content, the heuristic strength of the concept of sustainability can extend beyond the question of place. In this way, we initiate a new reflection on sustainability's cultural dimension as pertaining not only to territory but also to other spheres/contexts.

## Notes

1. The credit for having brought the theme of care back to the centre of philosophical reflection belongs to Heidegger, who described it as a fundamental existential quality.
2. The term sustainability has its origins in the field of forest economics. At the beginning of the eighteenth century, the economy of Saxony was based mainly on silver mining. However, timbering the mines along with charcoal-fired foundry furnaces led to the destruction of forests and a considerable shortage of wood. The then administrator of silver mining, Hans Carl von Carlowitz, strongly criticized the mentality of the time, which focused solely on short-term profit. Von Carlowitz witnessed how easy profit eventually compromised wealth. In his 1713 *Sylvicultura oeconomica,* the first comprehensive treatise on forestry, he advocated a "respectful" use of forests that felled only as many trees as would subsequently grow again.
3. On this matter, Claudio Napoleoni has written that nature is not the environment. Even though the environment exists, nature is another thing. To recognize that nature exists means that humans will come back to their humanity and stop considering themselves to be the ones predestined to rule the world.
4. Henri Lefevbre was the first to speak of the difference between spaces of representation and representations of space.

## Works Cited

Amari, Monica. 2012. *Manifesto per la sostenibilità cultural*. Milan: Franco Angeli. Print.

Assmann, Jan. 1997. *La memoria culturale. Scrittura, ricordo e identità politica nelle grandi civiltà antiche*. Trans. Francesco De Angelis. Turin: Einaudi. Print.

Castoriadis, Cornelius. 1998. *L'enigma del soggetto. L'immaginario e le istituzioni*. Trans. Riccardo Currado. Bari: Edizioni Dedalo. Print.

Council of Europe. 2000. *The European Landscape Convention (ECL)*. Strasbourg: Council of Europe Publishing. Print.

Danani, Carla. 2013. *Abitanti di passaggio*. Rome: Aracne. Print.

Jonas, Hans. 1990. *Il principio responsabilità. Un'etica per la civiltà tecnologica*. Trans. Pier Paolo Portinaro. Turin: Einaudi. Print.

Massey, Doreen, and Pat Jess. 1995. *A Place in the World? Places, Cultures and Globalization*. Oxford: Oxford University Press. Print.

World Commission on Environment and Development. 1987. *Our Common Future*. Oxford: Oxford University Press. Print.

# The Geology of Movement. The Earth and the Dynamic of Phenomenalisation in Merleau-Ponty and Patočka

**Renato Boccali**

**Abstract** The overexploitation of the Earth is not merely a question of natural resources management. According to an eco-phenomenological approach, the intertwining between man, environment, and habitat can be conceived in new forms. This paper discusses motivations for such a change of perspective focusing on eco-phenomenology, paying specific attention to the critical reception of Husserl's short manuscript on the Earth (*Overthrow of the Copernican Theory in the Usual Interpretation of a World View*) by Merleau-Ponty and Patočka. Merleau-Ponty outlines a living geology while Patočka defines the dynamic of the manifestation of appearance as a real "science of movement". In both cases, and thanks to the Husserlian original intuition, we have the opportunity to question the ecosystem of the primordial and topological space that lays the foundations for a phenomenological cosmology.

**Keywords** Earth · Husserl · Merleau-Ponty · Patočka · Movement · Cosmology

## Eco-Phenomenology and the Geo-Cosmic Horizon

To what extent can phenomenology address current environmental problems and planetary-scale challenges? How might phenomenology bring us a new awareness for the human-induced environmental changes that are increasingly causing destruction and degradation of natural environments, ecosystems, biodiversity, and landscapes? What can phenomenology really add to ecological discourse?

According to Charles S. Brown and Ted Toadvine, "Having constituted ourselves in opposition to nature, we adopt values and purposes that threaten earth itself. Only a reconceptualization of our place and role in nature can work against the tragic disconnection from ourselves and from the wellspring of our being. To begin this

R. Boccali (✉)
IULM University, Milan, Italy
e-mail: renato.boccali@iulm.it

 
W. S. Smith et al. (eds.), *Eco-Phenomenology: Life, Human Life, Post-Human Life in the Harmony of the Cosmos*, Analecta Husserliana CXXI,
https://doi.org/10.1007/978-3-319-77516-6_20

task by reconnecting us with our most basic and primordial experiences of the natural world – such is the power and promise of eco-phenomenology" (Brown, Toadvine 2003, xx). Indeed, the task of phenomenology is to provide an analysis of how we are embedded in a network of relations and interactions in the natural environment through which our lives continually unfold. The reciprocal, interactive, dialogical nature of our earthly embeddedness calls for a reconceptualization of our human condition in the natural world of Earth and in the cosmos. In Tymieniecka's words, the ontopoiesis of life or eco-phenomenology shows "the entire schema of specific elementary conditions that allow planet Earth to become the site of germination and sustenance" (Tymieniecka 1995, 149).

In order to investigate this site of germination and sustenance, I propose to question Merleau-Ponty and Patočka's notion of the Earth in the light of Husserl's short manuscript on the Earth (*Overthrow of the Copernican Theory in the Usual Interpretation of a World View*). Questioning the ontopoietic ground of the geo-cosmic horizon, I will try to address understanding the Earth and our place in it at a fundamental, ground-up level, in order to open up new issues, reassessments or overviews of the ecosystem.

## Merleau-Ponty and the Earth-Ark: Latency, Differentiation and Intertwining

During Merleau-Ponty's short stay at the new-born Husserl Archive in Louvain, he was able to read an unpublished manuscript with this label on its envelope: *Overthrow of the Copernican theory in the usual interpretation of a world view. The original ark, the earth, does not move. Foundational investigations of the phenomenological origin of corporeality, spatiality, nature in the primary sense of the natural sciences* (Husserl). The main the argument made therein was the proposal of an overthrow of the current interpretation of the world with the purpose of demonstrating Earth's proto-spatiality and immobility as the original ark and source of movement. This original phenomenological investigation on the spatiality of nature aims to neutralize the scientific knowledge based on the idealized geometrical operations and constructions of astronomy so as to open up a survey of the bodily experience of space. The Earth is thus the zero point of every measurement, the kinesthetic absolute zero, the pivotal and fixed point of orientation, a total body or *Boden-Körper* (body ground) and not only a mere *Körper* (object body). The Earth as originary ark is the root of our spatiality. It is neither a simple object in front of us nor a simple place. As natal soil and shared homeland, or *Urheimat*, the unmoving Ur-earth is placelessness, founding an *Urhistorie*, an originary history of primordial existences.

It is clear that Earth as soil or *Erfahrungsboden* (soil of experience), in its primordiality will never be an objectivized body. Only after a gnoseological process of idealization will Earth be reduced to a mere object-body as a measurable member of the solar system, thus establishing the validity of Copernican science. But, in any case, the phenomenological reality of the Earth as a pre-object is not revoked, because it represents the general framework of our perception. As Merleau-Ponty noted, "It is the massive body beyond which my *Unwelt* structure is opened – On the contrary, the world of science is an infinite world that looks over everything without being planted here. This is the difference between *Unendlichkeit* and *Offenheit*" (Merleau-Ponty 2000, 230–231).

It is now time to question the ultra-physical ground – which is not a metaphysical one – of the physical world of bodies. I said that Earth is not an object-body but a soil of experience, that it is the origin of the movement and rest, and that it contains object bodies and living bodies. But which is the difference between them? And how can one differentiate living bodies? As a living body, my body is not a sheer object but has several intentional objects. It is the center of several bodies in movement or rest against the background of a motionless soil. Between these bodies there are some living bodies intentioned as mere body-objects. At the same time it could be said that I am a mere body-object for another body, a living body.

This reveals the depth of the origination of the life-world that implies a generative differentiation of life in distinctive structure of living beings throughout a singularizing process. In any case, differentiation and individualizing life processes entail interrelation and interacting in the unity of the primordial Earth as germinating seeds in a common soil. As ark of the world and primordial dwelling, Earth is a structural *a priori* founding a fleshly and intercorporeal unity of the sentient and the sensible. Merleau-Ponty speaks of a *parenthood* or filiation between Earth, my living body, and other bodies (persons, animals, stones, etc.) based on a common flesh.

Thanks to Husserl, Merleau-Ponty moves forward an "ontological rehabilitation of the sensible" suggesting a dynamic structure of manifestation through what he calls *écarts,* or gaps (Merleau-Ponty 1964, 167). In fact, in his late works Merleau-Ponty speaks of the "flesh of the world" to suggest the idea of a contexture of bodies. The expression is not exempt from ambiguities; therefore, the ways in which Merleau-Ponty refers to it should be taken into serious account. He speaks of the "thickness of the flesh between us and the 'hard core' of Being" (Merleau-Ponty 2001, 169). This thickness is a real gap in the hard core of Being that introduces a redoubling of non-Being, producing an irreducible distantiation. The flesh is thus the texture of the world. But in the permanent weaving of life there are always plenty of irreducible fissures, splits, folds (*plis*), so that at the origin of the primordial and massive unity of Being, before "segregation" or differentiation in multiple dimensions, there are always several folds or facets. Merleau-Ponty speaks about a "latency" that produces differentiations, according to a lateral relation of *Ineinander* or intertwining, which makes it possible to conceive of the non-differentiation of an ambiguous, transcendent, and non-totally positive Being. Depth implies the carnality of the sensible on the strength of the dimensionality of Being in a lateral unity that unfolds the inherence of a physical, vital and spiritual order.

As Merleau-Ponty remarks, with his phenomenological reflection on Earth, Husserl "looks for an 'aesthetic' ground and rehabilitates the philosophy of Nature, the framework of the perceived world" (Merleau-Ponty 2000, 232). Therefore,

Earth is the ontological cradle of the originary openness (*Offenheit*) that yields the totality of the living beings as primordial soil. “This ‘Earth’ is an ark: it brings the possibility for every living being to be above the void, the deluge – seed of the menaced world where everything blooms again. It is ‘nature’ in the sense of the perceptive cosmogony, neither in itself nor for God, but our horizon” (Merleau-Ponty 2003, 174).

## Patočka and the Phenomenalisation of the World: The Appearing of the Earth and the Sky

During the same years in which Merleau-Ponty’s thought developed, Patočka was working on the complex legacy of the Husserlian concept of the *Lebenswelt* as developed in *The Crisis of European Sciences and Transcendental Philosophy*. The question was to work up the notion of natural world beyond the objective and geometric idealities of the natural sciences so as to open up the way to reconquer the concrete life-world underneath the mathematical deductions. The “constitutive and genetic” work of elucidation was done in the light of “Husserl’s innovative investigations” to gain the transcendental subjective ground that makes it possible to understand the constitution of such transcendental idealities as body, things, time, and world (Patočka 1976, 65, 67). Very quickly, Patočka focuses his attention on the Husserlian concept of *Ineinander*, to the very idea of the world as a preliminary and interactive totality of overlapping systems that serves as the original matrix of every experience.

This brings him progressively to an independent position and to a general reconsideration of phenomenology as “the study of the movement of the appearing of everything that is” (Patocka 1976, 179). He gives priority to this appearing, despite transgressing the Husserlian principle of the universal a priori correlation between the world-phenomenon and pure ego-centered experience. What Patočka is looking for is an asubjective phenomenology.

Indeed, Patočka recognizes a double level of manifestation. The first one is the movement of the individualization of entities by means of the world. The second one is the appearing of entities by means of one of these entities, someone to whom they appear thanks to the multi-layered motion of his existence in the background of the world – that is to say, the embodied consciousness of a person. The individualization of humans and of the world as a whole are two coextensive manifestations of what Renaud Barbaras calls “the *anthropocosmic* ground of phenomenalization” (Barbaras 2002, 6). Therefore, there is not correlation but cooriginality between them because the world is a non-objectifiable and ever-present whole, an horizontal givenness in which entities individualize and show themselves to someone to whom they appear on the background of the world.

By virtue of our motion, the appearing of the world unfolds itself. The manifesting of the world in its whole is possible only with the presence of human beings, but it cannot be reduced to them. Indeed, human beings are the condition of phenomenality. They give a form to world, transforming it into the background for the appearing of the entire realm of phenomena. But, as a whole, the world cannot be constituted from pure immanence. It is something essentially "containing" from which humans emerge in the self-individualization process of life. This process is a fundamental movement and, as Patočka wrote in an article devoted to Aristotle and movement, "movement is what gives things the being that they are; movement is a fundamental ontological factor" (Patocka 1976, 129).

In order to better understand the nature of movement, Patočka suggests a genetic phenomenological investigation to gain the primeval *referent* of a movement as such, referring explicitly to Husserl's manuscript on the Earth. In the primordial world as a whole movement is the basis of all manifestation and generative differentiation of life. But to have movement the world has to singularize itself in a natural world through an ontopoietic self-individualization movement that generates a "solid" referent for any kind of movement – that is to say, a massive ground that allows physical movement and rest. For Patočka this universal pillar is the Earth as "the bearer and the referent of all relations", the immobile substrate upon which every movement is possible (Patocka 1989, 255). "Earth is the prototype of everything massive, corporeal, material; it is the 'universal body' of which all things are in some sense a part" (Patocka 1989, 255). For this reason it is defined as "power" in an Aristotelian sense, because it acts permanently (and not occasionally as force) as a specific "domination" of elements and things, non-living beings and living beings. Domination by the Earth is manifest in every vital movement as a power, particularly for living beings. Therefore the Earth is not only the constant substrate, the firm ground underfoot, but also a mother's lap sustaining, providing and preparing everything. In a word, it makes possible life itself. Its mode of being is accessibility and nearness and it gives to everything primarily its "where".

But Earth is not the only referent, because there is also distance. Here Patočka goes beyond Husserl's text, maybe with a nod to Heidegger. Indeed, distance is possible only in the light of an intangible and inaccessible referent, one not to be managed through bodily touch: "all that encloses our horizon without closing it in, all that constitutes the outside as something that constantly encloses us in an interior" (Patočka 1989, 256). To it pertains the sky, the light, the darkness, the heavenly lights and bodies, and thanks to the alternation of night and day, the sky gives everything a "when", by way of their coming and passing.

The movement of life in the natural world is thus articulated between Earth and sky as the fundamental coordinates of vital and existential orientation. This living dynamism unfolds body-centered, self-individualizing life processes that in human beings produce the self-constitution of human existence. The realization of human existence through the appropriation of possibilities is a fundamental existential movement that can take place in three different forms and three different existential positions (Patočka 1988).

The first of these forms is the movement of the rootedness that makes it possible for a human being to be received in life by means of the protection and acceptance of others. In this way an original dwelling – an *Urheimat,* to use Husserl's word – is set up. The existential rootedness thus presupposes a past, something that is preliminarily opened, a contingency where vital movements can take place. This foundational horizon of givenness reveals that human movement is always a co-movement, as a consequence of reciprocity with others.

The second movement of life is functioning and integration in the context of the natural world. The others are present not only through their bodies but also through their works, projects, and activities. This movement reveals the necessity of subsistence and the presence of others as a community of work. Here, the natural world is disclosed as exteriority and, at the same time, the domain of objectivity and manipulation. Life is progressively fragmented in a movement of repetition and reproduction, excluding any kind of reflection on purposes and goals. This is a movement of life-extension by means of the destitution of self in the stark realm of indifference and self-evidence. The corresponding existential attitude is the necessity of defense and the struggle for procuring what is needed.

But there is a third movement: the self-discovery of one's own possibilities by the act of self-dedication and the assumption of one's own finitude – that is to say, becoming conscious of the alienation produced by the second movement and of the question of the meaning of such a finitude. This is the real movement of truth generated by the self-sacrifice that poses the situatedeness of existence. To express this unexpected insight, Patočka uses the metaphor of an earthquake. It is a tectonic movement producing the possibility of the assumption of freedom for humans, shaking the fixity of everyday life and the base finitude of repetition. The existential earthquake shakes the rootedness of existence and reveals the life in its nakedness. This vertigo of freedom implies the desubstantialization of Earth and sky and the openness of a mystery in which humans are actively involved.

As a result of an a-subjective phenomenology, Patočka realizes that the existential movement opens up the possibility for humans to question the appearing of Earth and sky so that they could manifest a primary ground, that is, the natural world as a world of movement, a prehistoric world appearing on the basis of the original world as a whole.

The world as a whole can never appear as something particular but as something containing all individual realities and all living domains in its framework. Thanks to the existential movement of humans, the world appears as an *Umwelt*, a surrounding life-world, so that space gains a center. The fundamental movement of life consolidates life in the appearing thanks to the three vital movements: rootedness and acceptance (becoming a center), functioning and integration in the natural world (the first-person experience of reaching the center) and the care on the part of the soul (an explicit relationship with the whole).

## Phenomenological Cosmology

In conclusion, it seems that the Husserlian phenomenological investigation of Earth enables us to question the ecosystem of primordial and topological space as a "total voluminosity". On the one hand, Merleau-Ponty outlines a living geology based on the living dialectics of intercorporeality; on the other hand, Patočka defines the dynamic of appearing as a "science of movement". Merleau-Ponty stresses the filiation between Earth and body (*Leib*), while Patočka considers Earth and sky as the horizon structure of human motricity given by an ultimate and unshakable world as a whole. In both cases, a cosmology arises: a cosmology of the visible for Merleau-Ponty and a cosmology of movement for Patočka, revealing a pre-objective and irreducible ground, the *oïkos* of the *logos* as the source of every possible eco-logy.

## Works Cited

Barbaras, Renaud. 2002. L'individuation de l'homme. *Fogli Campostrini*, 2.2. Print.

Brown, Charles S., and Ted Toadvine, eds. 2003. *Eco-Phenomenology. Back to the Earth Itself*. New York: SUNY Press. Print.

Merleau-Ponty, Maurice. 1964. *Signs*. Trans. Richard C. McCleary. Evanston: Northwestern University Press. Print.

———. 2001. *The Visible and the Invisible followed by working notes*, ed. Claude Lefort. Trans. Alphonso Lingis. Evanston: Northwestern University Press, 1969. Print.

———. 2000. *Parcours deux 1951–1961*. Lagrasse: Verdier. Print.

———. 2003. *L'institution. La passivité. Notes du cours au Collège de France (1954–1995)*, ed. Dominique Darmaillacq, Claude Lefort, and Stéphanie Ménasé. Paris: Belin. Print.

Patočka, Jan. 1976. *Le monde naturel comme un problème philosophique*. Trans. Jaromír Danek, Henri Declève. The Hague: Martinus Nijhoff. Print.

———. 1988. *Le monde naturel et le mouvement de l'existence humaine*, ed. and Trans. Erika Abrams. Dordrecht: Kluwer. Print.

———. 1989. "The 'Natural' World and Phenomenology." *Jan Patočka. Philosophy and Selected Writings*. Ed. Erazim Koan. Chicago: The University of Chicago Press. Print.

Tymieniecka, Anna-Teresa. 1995. Nature in the Ontopoiesis of life: From cosmic dissemination to the human cultivation of the logos. In *Heaven, Earth, and In-Between in the Harmony of Life. Phenomenology in the Continuing Oriental/Occidental Dialogue*. ed, Anna-Teresa Tymieniecka. Analecta Husserliana 47. Dordrecht: Kluwer. Print.

# On Two Versions of Phenomenological Transgression – Anna-Teresa Tymieniecka and Jean-Paul Sartre

**Piotr Mróz**

**Abstract** The present paper focuses on the problem of transgression, to wit, broadening upon, as well as the overcoming of the fundamental tenets of Husserlian phenomenology: *epoché* and the problem of the transcendental Ego. The author juxtaposes the concepts of Jean-Paul Sartre's existential phenomenology and Anna-Teresa Tymieniecka's phenomenology and eco-phenomenology following the path of the logos of life. Apart from some substantial differences in their approaches to this challenge, the French and the Polish philosophers share the same opinion that in view of reaching the unmediated source of all being and its cognition one must give up all remnants of idealism and belief in the inner life of a personal Ego.

According to Anna-Teresa Tymieniecka, diverse forms of the crisis characterizing the age of modernity, as well as postmodernity have had a definite metaphysical dimension. Viewed from the phenomenological (logos of phenomenology, resp. eco-phenomenology) existentialist angle, this condition, inimical to human development—our free, spontaneous, and creative aspect demands an overall and fundamental reconceptualization of human values, along with recognition of man's *eidos*. The latter, within the frame of reference of the philosophical currents in question, is closely related to human consciousness, along with its intentional objects.

The following paper attempts to present the problem of consciousness (*subjectivity, ego*, or *selfhood*) as being part and parcel of this definite metaphysical dimension of which Tymieniecka spoke. The two ways of the transgressing Husserlian "orthodoxy" will be the main topic of our presentation.

**Keywords** Tymieniecka · Sartre · Transgression · Phenomenology · Husserl · Ego · Consciousness

P. Mróz (✉)
Institute of Philosophy, Jagiellonian University, Kraków, Poland
e-mail: p.mroz@iphils.uj.edu.pl



W. S. Smith et al. (eds.), *Eco-Phenomenology: Life, Human Life, Post-Human Life in the Harmony of the Cosmos*, Analecta Husserliana CXXI,
https://doi.org/10.1007/978-3-319-77516-6_21

Both Sartre (chronologically first, being of the second generation of the phenomenological movement) and Tymieniecka (of the third generation), in their respective existential phenomenology and *phenomenology of life* and *eco-phenomenology* credit Husserl with having provided the right tool for overcoming most of the antinomies, contradictions, misconceptions, and aporias blocking access to Reality-as-it-is—to Being in its multifarious aspects and expressions (phenomena). Moreover, it was owing to a reinterpretation of the subjective pole that a lost relation with the transcendent reality was to be found anew and reinstalled (Merleau-Ponty 1966, iii–xi). Thus, the creator of *Logische Untersuchungen*—one of the most seminal works of twentieth-century philosophy, soon followed by his *Ideen*—blazed the trail, pointing to a new direction toward which all philosophical investigations should orient themselves, in order—as Kołakowski has it—to reach the domain of absolute certitude (Kołakowski 2001, 1–30). In other words, phenomenology, as an original project (that is, in the sense of source), should, as it were, *bracket out* most of the philosophical (cultural) tradition—a long history of continual failure insofar as indubitable cognition was concerned. With what, then, did Husserl find fault in this history of human consciousness' unrealized (if attainable, realizable at all) encounter with Reality?

It stands to reason that such a vast area of discourse can not be covered here with the slightest claim to even a sketchy exposition, let alone provision of a definitive interpretation. However, let it be said that what could be regarded (with certain stipulations) in Lyotardian terms as "*grand narrations*" (the way we used to narrate the story of both the world and our endless efforts to grasp its different aspects) fall under two main paradigms. Let us characterize the first one as being idealist, egological, spiritual, subjective, interior, and dualistic, as well as rational and logocentric; and the other one as being realistic, empirical, scientific, materialistic, exterior, and non-egological. The idealist tradition harks back to the celebrated Platonic myth of an immortal soul imprisoned within a body (matter) to Plotinus, Boethius, St. Augustine, St. Bonaventure, the Renaissance Platonists, and Montaigne. With the arrival of Descartes (and that was the culmination of the idealist-subjectivist paradigm), we have learnt that human soul has an exclusive abode—namely, consciousness endowed with inborn dispositions to ideas—with the idea of the Supreme Being, God, having the uttermost importance. Hence, cognition was the only access to (and certainly the indubitable proof of) the existence of God, myself, and the world. This almost axiomatic formula formed a legacy, one passed on to such influential thinkers of this vast persuasion as Leibniz (an important thinker for Tymieniecka), Spinoza, Fichte, Schelling, Kant, Hegel (the latter's philosophy treating mind, or consciousness as identical to reality, and *vice versa*), down to Bergson, Brunsvig, Lavelle (points of negative reference for Sartre).

The first narration treated consciousness (often equated with the soul, mind, subjectivity, selfhood) as an immaterial, more often than not immortal, entity (of God's provenance). It was a pure translucent force with constitutive and cognitive faculties at its disposal. What was the most critical quality of consciousness seems to have been its almost limitless capability to impose laws and rules on the transcendent world, which lost—in the process of such constitution—its intrinsic character of an ontologically independent structure. Moreover, consciousness was thought of as a pregiven element, essential to rational human nature—one making us be what we are (unique persons). To sum up, the idealistic paradigm, extolling the role and position of conscious, rational human beings, degraded the Reality "out there," thus, misinterpreting the true character of the relation between consciousness and the world. This approach—based on the insurmountable dualism of *res cogitans* and *res extensa*—produced another, more dangerous dualism: that of consciousness and the world (Ayer 1979, 317–320).

The other main narration—initiated by Aristotle, continued by Roger Bacon, St. Thomas Aquinas, the British Empiricists, certain French philosophers (d'Holbach, La Mettrie), Comte and the nineteenth-century positivists, down to the Vienna Circle, Ryle or Koestler—vehemently opposed all forms of idealism, trying to get rid of the prevalent vestiges of this unjustly usurped privileged position cherished by consciousness (spirit). The myth of the *ghost in the machine* had to be demystified (in the Weberian sense), and the cogito, resorting to the mysterious and elusive matrix—consciousness itself—empirically explained away. In the crudest version of this tradition (Avenarius, Engels, Lenin), consciousness was nothing but a function (a set of functions) of matter. Hence, what was regarded as "a domain of spirit" turned out to pertain to nature (*physis*), becoming an object, bearing all liabilities of a mere element of the material world (Baumer 1977, 400–405).

It appears that these two paradigms committed high treason, finally paying a high price for it. The notion of consciousness (in the idealist tradition), rendering reality dependent upon conscious acts, lost irretrievably the world's richness, its qualities, mystery, all those factors which were so fascinating, enticing creative responses on the part of men. To paraphrase Kant's famous dictum: Ego without the transcendent world becomes a formal, empty, aseptic entity, magically—as it were—imprisoned in its own interior. The second narration/*episteme*, acting under various guises and names, turned its critical attention to consciousness. Not only was the latter dethroned and deprived of its privileged status of an omnipotent, vital, and constitutive power, it was reduced (like its own acts in the infamous illusion of immanence) to a "thing," an "object" found in the natural, or cultural world.

But the turn of the nineteenth and twentieth centuries witnessed another, a more serious crisis in this respect. Dramatically enough, having lost all solid ground and metaphysical point of orientation, man turned out to be a seat of dark, uncontrollable forces, drives, instincts, which if unleashed, were most likely (as history was soon to prove) to lead us to mass self-destruction, suffering, constant tension—in a word, to a total uncertainty (Tymieniecka 1990, 3–14; 2008).

This condition of epistemological despair, combined with man's hectic quest for his/her identity in the world of social, cultural, economic crisis of the first decades of the twentieth century, has been reflected in philosophy just as in most human expressions: in visual art, literature, and poetry. As Tymieniecka remarked on one occasion, the time was ripe for change, and demanded the thorough reconceptualization of the human being and his/her position—to resort to Scheler—in Cosmos (Mróz 1997, 5–9). The world of ideas was anxiously waiting for new proposals and (possibly) solutions, radical responses to this ever-deepening crisis (Kołakowski 2001, 10–15).

It cannot be denied that the Husserlian project of *Strenge Wissenschaft* perfectly exemplified (and in a way not limited to the history of ideas only) and constituted a most radical, seminal, and creative attempt at a total alteration of outlook in Kuhnian terms, one affecting almost all areas of philosophical discourse of the way we thought about and perceived the world we happened to find ourselves in. What is understood—and not only by the present author—by the very term *phenomenology* (along with its modifications—the natural issue of the Husserlian matrix, as it were) is not a unified body of doctrines, but more or less identifiable presumptions, premises, presuppositions, ideas, and finally, solutions. Viewed from the perspective of several decades, phenomenology seems primarily to have been, and still is, an inspirational movement of great impact.

In his inquisitive essay on the quest for certainty, Kołakowski points out that the German master, throughout his long academic career, reflected, in one of the most profuse of philosophical outputs, and had been realizing this project, time and again changing his points of view, alternatively emphasizing different aspects of the investigated discourses (domains), as he put it himself: "starting it all anew" (Tymieniecka 1990, 3–14). His own stance— *never accept anything in dogmatic intransigence, get rid of the presumably established truths the moment they prove inadequate*—was to provide a perplexed generation exposed to all aporias and contradictions (as was the case at the beginning of the twentieth century) with a reliable tool, a remedy to dispose of all misconceptions inimical to human development, our spontaneity and creativity [Tymieniecka], or our free access to the real, true world [Sartre] (Dobson 1993, 9–20; Tymieniecka 2008).

And so there arrived a philosophy that so strongly appealed to young, promising, but disillusioned thinkers, who—like the first generation of the Husserlian provenance—immediately perceived the wealth of possibilities inherent to Sartre's phenomenology and in the third generation (disciples of disciples), continued the Husserlian project, as Tymieniecka did. As Anna-Teresa Tymieniecka retrospectively stated after the passing of so many years: "Looking back today, we see that phenomenology has entered all sectors of knowledge and in praxis, as much as in theory. Phenomenology is discussed in physics, embryology, and elsewhere—not by everyone, of course, but by some. So its relevance is a simple matter of fact nowadays" (Spiegelberg 1994, 470–479).

However, one point should be made here before we proceed any further. The Husserlian critical attitude (turned into a unique method) taught his fellow travelers never to take anything for granted (to avoid the natural stance, as the thought idiom has it). This deconstructive—if we may use this term—stance acted, as it were, as a double-edged sword. Soon, the internal force of transcending the given (be it some part of reality, or ideas concerning the latter) inherent to phenomenology, led to some vital, approved of or rejected by the Master himself, transgressions (Lyotard 2011, 15–36).

We have chosen as instances of this transgression the Existential Phenomenology represented by young Jean-Paul Sartre (the French ramification of the phenomenological movement), and by Anna-Teresa Tymieniecka (her logos of phenomenology, ontopoiesis, and eco-phenomenology), both of whom can be considered to provide substantial evidence of theoretical and methodological fidelity to the old Master but, at the same time, to have initiated a transgression carried out in the interest of an unhampered and complete return to the things themselves.

As is well known, the Husserlian Project—based on the idea of a return to the things themselves—employed *epoché* (the "bracketing" of all knowledge, presuppositions, pregiven ideas being of special significance) and had an overwhelming belief in the existence of "real" things to be then arrived at, which operation involved a series of reductions—phenomenal, eidetic, and transcendental. It was this aim and the rigor of the means deployed that made phenomenology so attractive. The celebrated principle of the intentionality of consciousness and its acts, revealed by description of what appears in the process of the constitution of an "object," which was to be regarded as absolutely given, certain—in a word—indubitable datum, an intuition, and insight, was widely acclaimed, and Husserl was credited with boldness and great intellectual acumen (Sartre, *L'idée* 4–10). Thus, in these two versions of transgression of phenomenology, we find a mixture of admiration, approval, and fascination, along with severe criticism, especially in the domain of the three reductions which, according to Sartre and Tymieniecka, the Master did not carry out "to the very end."

Let us begin, then, our more detailed presentation with Sartre's existential phenomenology. Like Tymieniecka, Sartre makes the main tenets of the Husserlian method his own, as he states time and again in his early studies. Of these, the Transcendence of the Ego seems to have acquired a special significance for his later, full-fledged existentialism (Spiegelberg 1994, 473). The very term, phenomenology, appears in almost all of the titles (or subtitles) of Sartre's major works, and especially those belonging to the first and second stages of his philosophy. The presence of the phenomenological approach with its meticulous but never boring descriptions (superseding formal theorem-like propositions) and its intuitive insight into diverse phenomena is vividly felt in his theoretical as well as artistic work (such as his masterpiece, *La Nausée*), and is present in his critical essays (*Situations*) as well. Nevertheless, one cannot avoid posing the important question of the extent and nature of the relationship between Sartre's philosophical endeavors and the Husserlian method. According to Herbert Spiegelberg, a definite answer cannot be given (Spiegelberg 1994, 470–533). Although the Frenchman credits the Husserlian reconceptualization of philosophical investigation with having set us free from the fetters of the aforementioned antinomies concerning the human nature and our cognitive faculties, in a word, of having overcome psychologism and naturalism, thus making possible the restitution of the lost relation between consciousness and the world, the Sartrean conception of phenomenology is not to be uncritically identified with that of Husserl's project (Mróz 1997, 24). As is widely known, Sartre's phenomenological studies fall within the general framework of the Husserlian paradigm. They are descriptive, conducive to eidetic insight; they esteem the intentional structure of consciousness which, as he admits in an almost poetic cant, was Husserl's greatest contribution to modern philosophy. He [Husserl] has reinstated horror and charm in objects. He has restored to us the world of the artists and prophets, terrifying, hostile, and dangerous, with harbors of grace and of love. Here, we are liberated from Proust and liberated, at the same time, from the inward life (Mróz 1998, 31–40). This kind of philosophic reflection enabled one—young Sartre seems to have admitted—to grasp true, adequate relations with the transcendent world. In other words, it makes it possible to take into philosophical consideration "two partners," as it were, in grasping the sense, meaning of the sensory and the intellectual richness of that which directly appears to our awareness. Hence, the celebrated trio—*Ego, Cogito, Cogitatum*—turns out to be a true and unrivalled starting point of a new phase in modern philosophy.

In one of his earliest studies we find the author of *L'Imagination* openly saying that "we know today that we must start again from zero, and disregard the whole phenomenological literature" (Sartre 1957, 60–62). Thus, while the project of a thoroughgoing rebuilding of the philosophical heritage seems to have been adequately realized by means of *epoché*—the three types of reduction—it is the very concept of reduction that brought about the transgression of Husserlian orthodoxy both in Sartre's and Tymieniecka's work.

Having acquainted himself with the idea of the phenomenological *epoché*, Sartre readily takes advantage of this salient remedy, allowing him to chase out all misconceptions and illusions concerning the most vital relation: that of all conscious acts (intentions) and the "objects," the "things" at which those acts are directed.

The main tenet of all phenomenological thinking—consciousness is always consciousness of something—opened up for Sartre the path to inquisitive analyses of three modes (three intentional acts) constituting three different intentional objects. These modes were perception, imagination, and emotion. Not only did this kind of description (carried out in such studies as *L'Imagination, L'Imaginaire,* and *La Transcendence de l'Ego*) reveal the workings of three different modes of intentionality (hence three aspects of a given fragment of reality), but it also yielded much data concerning the essence of consciousness (Ego, the subjective pole) itself.

Out of the basic, fundamental modes of activities of consciousness displaying Ego in its pre-reflective and reflective stages, Sartre has chosen the three aforementioned levels or modes of constitution and yielded objects for his analyses: levels aimed at overcoming both the psychological and "objective" (mechanistic) senses given to objects in the pre-phenomenological literature (Sartre 1971, 25). The overriding assumption (based on the intuition presumably leading to the sought after domain of the phenomenologically certain) is that they all pertain to one integral consciousness, but each operate a different existential thesis "embracing" the ontological status of the intended object. Putting it differently: they assume radically separate stances in their own unique way, hence, constituting respectively the world-as-it-is-perceptually-given, as it is imagined, or as it is emotionally affected.

The first mode grasps—as Sartre has it—the surrounding reality (later on, Sartre would underline the dramatic aspect of my "thrown-ness into" it, *my facticity*) in an ordinary, "normal," standard way. The world is presented as something real, existing, "graspable," within the range of my, and others', consciousness (Sartre 1971, 25). The pen I perceive at this very moment (while writing, for I do not use a computer) always appears against a real (present at all times) backdrop, which is "composed" of my study, in my apartment, the street I live on, my hometown (Kraków), my country, and, to recall the words of Joycean hero Stephan Dedalus, the whole cosmos (Universe). Thus, the perceived qualities (namely, those constituted by perceptual consciousness), are part and parcel of the infinite whole—Being as Phenomenon, presenting itself in what shall be referred to in the next stage of Sartre's philosophy as the *hodological space.* (A word of caution should be conveyed here: this being is always what appears to us—such as this perceived Waterman fountain pen—as there is no "something" hidden behind the phenomenon here). It seems obvious that it would be a mere impossibility to grasp all its traits, qualities, attributes, and interconnections; that is, to embrace the richness of its very substance. Thus, to bestow a definite, completely "fulfilled," and realized once and for all meaning upon this ever open-ended item (both from the ontological and epistemological points of view) would require infinite procedures (Sartre the writer and essayist uses the apt metaphor of a Godlike eternal look, which human beings are denied). We are, then, fully justified in stating that perception is a mode of active consciousness that, step by step, act by act, intention by intention, painstakingly tries to reveal the true, actual face of a partly given, and partly hidden (from perception) reality (Caws 1993, 293–317). No matter how long the act of perception (perceptual intention) lasts, the perceptually constituted objects will tend to conceal, to keep secret some of their properties while revealing others. It is only through eidetic analysis that we acquire the knowledge which, to some extent, leaves out all the "perceptual" (*hyletic*) qualities, concentrating on the essence of given intentional objects. Being what they are, the objects of perception "belong to" me; they constitute my world, playing various functions in my existence. They are always "seen as," "regarded as useful or useless," thought to be "valuable, or not," "taken into consideration," "chosen for some purpose in mind," or simply "discarded." Briefly: it is perceptual consciousness—says Sartre—that "builds" the world (the human world, that is) around us. What is of a vital significance, is the fact that perception, like the rest of consciousness, is intentional; but, unlike the two remaining modes, it is endowed with more intersubjective value, by being attached to a world we are supposed to share with others (Sartre 1971, 336–340).

In contradistinction to perception, imagination (*l'imaginaire*) changes the ontological status of its intentional object. The latter, owing to specific procedures of imagination (consciousness in its own right), like the celebrated *recul* (retreat) from the surrounding reality, and the negation of certain elements of the latter, is not constituted through a continuous process of "reaching" the substance of the perceived entity. According to Sartre, the imagined object is "grasped" at once (as no perceived experience is required) but, at the same time, it stands alone, on its own, isolated—it presents itself as impoverished because all *hodological* relations have been cut off. The imagined object appears schematic and highly formalized since perceptual richness does not come into its constitution. As to its ontology, one may add that these objects may be carriers of contradictory qualities, making them paradoxical beings pertaining to the sphere of the anti-world—the world which has been made unreal (*irrealisé*), or deprived of real existence (*néante*).

What this analysis shows, however (and the young French philosopher avidly seizes the yield of this description), is the internal, inherent-to-consciousness power of negation. This, in turn, refers to one of the most precious, invaluable traits of consciousness: "its freedom from." The imaginative mode of constitution, like the third mode, that of the emotions, is a kind of imposition of our laws on the given facticity we find ourselves in. Moreover, all such constitutions are an indelible sign of consciousness' freedom, lucidity, transparency, and spontaneity, combined with creativity (the traits which Tymieniecka finds to be so essential to the *Logos of life*). It stands to reason that, for existential phenomenologist Sartre, the celebrated principle of intentionality (consciousness' always being consciousness of something, no matter what "kind" of object it might be) must preclude interpretation and understanding of our awareness in any substantial terms. Strictly speaking, under no circumstances can consciousness be reduced to anything beyond itself. It is so radically different from the object it intends, that any attempt at bringing them together would immediately deprive the human being of its above- mentioned most significant qualities. The very description (based on phenomenological insight) of all conscious acts was, in Sartre's case, soon to break through the limits that the German Master imposed on phenomenology. In other words, the intentional principle created some new possibilities, that were unforeseen, as it were, by the Husserlian paradigm. (These were to open up a vista for a full-fledged existential philosophy). Sartre maintains that eidetic analysis yields incontrovertible evidence. Combining the Cartesian formula "I think, therefore I am" with the concept of intentionality, and applying the phenomenological reduction (the latter will bring about the rift between Husserlian transcendental idealism and the Sartrean version of phenomenology), Sartre discovers the following indisputable fact having seminal consequences. The *I* is both an object of my thinking, as well as the subject carrying out this act. While the first *I* is endowed with essence (it is what it is) like any object (in this particular case, it is however more intimate), the *I*-subject is not; it does exist but it is not, will add Sartre the existentialist. As one Sartrean specialist underlines, the Frenchman presents consciousness (soon to be identified with human experience, the realities of man—*réalité humaine*) as a presuppositionless, absolute given, to which not only do phenomena appear, but also by which they are creatively constituted (Thody 1990, Chaps. 1, 2). The idea (based on the evidence of *le vecú*—the phenomenologically given and experienced) of admitting the creative, constitutive role of free, spontaneous acts of consciousness turned out to manifest a revolutionary force. In order to re-establish true, unhampered relations with transcendent reality we must be consistent with the propositions put forward by the Master. But, what should be done if he himself was not consistent? Sartre admits on one occasion that it was necessary to have taken an anti-Husserlian position. There is no phenomenological evidence on either the pre-reflective or reflective level (consciousness conscious of itself) for the existence of the so-called transcendental Ego (the inhabitant of consciousness); the postulating of such an entity would be inimical to consciousness itself (Sartre 1957, 31). Husserl had spared it, in the process of applying the reductions, in order to procure the unity of our acts. But, there is no such an exigency, underlines Sartre. It is through the "internal" (not psychological) structures of retention and protention that such a desired unity is safeguarded by the ontological structure of consciousness of the world. We find Sartre categorically declaring,

> Now it is certain that phenomenology does not need to appeal to any such unifying and individualizing *I*. Indeed, consciousness is defined by intentionality. By intentionality consciousness transcends itself. ... It will be said that a principle of unity within duration is nonetheless needed if the continual flux of consciousness is to be capable of positing transcendental objects outside the flux. Consciousness must be perpetual syntheses of past consciousness and present consciousness. (Sartre 1957, 38–39)

But, the brunt of the Sartrean argument points in another direction. Consciousness must be purified of all ego-logical structure in order to recover its primary transparency—a concept so dear to Sartre. This is a domain of absolute existence (sharply opposed to being). Thus, any introduction of a passive *hyle* (Ego) into this pure transparency, *lucid force*—as Sartre dramatically states—means the death of consciousness (Sartre 1957, 38–41). The latter is being born as if *ex nihilo*—nothing determines its doing so, nothing lies behind it. It is neither caused, nor generated. This primordial presence of and to the world attains unity through intentionality itself—still better, through the object it intends. This everlasting synthesis (union) with the transcendent reality underlies the character of impersonal spontaneity, and has nothing to do with the activity of the *Je du cogito* (the *I* of thinking), which is an object itself attaining the status of a subject after the act of reflection (Sartre, *Transcendence,* 38–41). In this way, the troublesome subject-object duality (dichotomy) disappears from philosophical preoccupations. "The world has not created me, the *me* has not created the world. There are two objects for absolute, impersonal consciousness" (Sartre 1957, 38–39). For the early Sartre, then, consciousness, freed and purified from all idealistic and materialistic remnants, cannot be treated as either an immaterial spirit or a substantial entity.

It seems that this transcendent force is co-present with the world—the world it constitutes, giving, imparting the sense to its diverse aspects. Such series of constitutions leave a trail of concrete intentional objects. Thus, is manifested consciousness' freedom, both freedom to do and freedom from, spontaneity, and ontological difference from the being-in-itself (a thing). These conclusions were to prove themselves of seminal consequence.

It was Sartre, the existential phenomenologist, who, for the first time in the twentieth century, turned our attention to a non-egological structure of conscious life. In order to fully embrace the surrounding world, intentional acts did not need to belong to an individualized Ego—the *I* of the cogito. Quite the contrary. The prevalent myth of the so-called inner life, led by a mysterious inhabitant of the deep recesses of consciousness, has been shattered to pieces. Both ontologically independent and radically different parties in the intentional move on the part of pure, transparent force came, nonetheless, together, yielding an intentional object—something completely unknown and unthought of before the triumphant arrival of Husserlian phenomenology. This was an intentional object encountered in diverse spheres of human activity. Owing to spontaneity and creativity (the *I* part), the transcendent world received new sets—so to speak—of qualities and traits. But, most importantly, it became the indelible sign of consciousness re-building the cognitive and spiritual ordering of something, what used to be regarded as inaccessible and distant from the sphere of consciousness. Similar, if not analogous, motifs will be discerned in the work of the Polish phenomenologist Tymieniecka, who, more often than not, enriched the Husserlian *universum* of phenomenology.

Anna-Teresa Tymieniecka has often been credited for the important position she occupied in North America, Europe, Asia, all points really, in so far as the dissemination of phenomenological ideas and methods were concerned. And like Sartre, the author of *Logos and Life* proved to be a critical and, what is more, an important creative reader of the profuse work of Husserl. But, as is well known by now, phenomenology, owing to the intellectual creativity of its many critical and innovative followers (disciples), inspires—as it were—its own transgressions. Such was the case of Sartre's and Tymieniecka's redefinition and reconceptualization of the Husserlian project of a "strict science." It is through meticulous analyses of the Husserlian texts that Tymieniecka assumes exactly the very Husserlian angle (position) of an everlasting beginner. This eventually led her to the sought-after sphere of primordiality—to all those areas and domains of research which, after the liberating impact of *epoché* and the reductions, could legitimately be regarded as cleansed from all forms of presupposition and postulates—in brief, ideas, which, instead of opening us up to the world-as-it-is, block the path to things themselves. It stands to reason that phenomenology (in almost all of its versions) accentuates the fundamental significance of the indubitable: the area of these experiences, which can stand for the certain, unmediated by any forms of representation. To a greater extent than can be discerned in Sartre's reading of Husserl, it is Tymieniecka who stresses the unquestionable datum of the phenomenological method: the principle of all principles, enabling us to describe all that appears to consciousness in its self-evident presence. Hence, her great stress on the intuition (*Wesenschau*) treated by Husserl as the location of all certain cognizance. The latter motif is, according to Tymieniecka, a leading-thread, a kind of *point de répere*—the main track of Husserl's investigations into the nature (*eidos*) of the very matter of what appears to consciousness, as well as the manner in which it appears (Tymieniecka 2005, xvii). As phenomenology puts forward the unquestionable priority of unhampered, spontaneous intuition (over explanatory hypotheses or axiomatic-like theorems), this type of direct insight requires a series of activities, to wit, the reductions. Tymieniecka has a point in stating that it is for this essential reason that phenomenology has been identified with its method, *epoché*, or reduction (but that, at the same time, its strongest point, if not properly executed, will turn out to be its Achilles' heel).

In a fashion hitherto unknown in the Western philosophy—says Tymieniecka—Husserl, from the time of his so-called transcendental turn, devoted his reflection to a quest for the systematic uncovering and exfoliation of essential, indubitable cognition (Tymieniecka 2005, xvii). Only on the condition of a thoroughly carried out reduction can the overall goal of all kinds of phenomenological investigation be accomplished. On the epistemological level, the reductions aim at disposing all the remnants of the so-called natural attitude—the naïve, but deeply-rooted conviction fueled by all the empirical sciences that we deal with objectively, hence, independent beings (natural entities), facts, "established truths," and our tendency on the ontological level to endow them with "real" existence. The two parties, if we are allowed to use this phrase, are poles apart. As with the Sartrean attitude, Tymieniecka is also of the opinion that the main, vital change proposed by the Husserlian program consisted of a restitution of the necessary relation of our consciousness (empirical and transcendental) to the world out there. It can only be upheld that Husserl is indisputably an initiator, one who, "in the footsteps of Descartes," opened up the ego-cogito within the dynamic framework of the tripartite schema: Ego (the I on the diverse plane of ontic levels); its acts (intentions), that is, Cogito; and finally its objects, the results of the constitution of the domain of genuine experience of the world-as-it-is-for-us. This celebrated triad (underlining, however, the unity of consciousness), along with such of its propensities as intentionality, led Husserl to a previously unthought of discourse of research—a multitude of intentional objects pertaining to the world of philosophy (ethics, aesthetics, sociology, politics), the area of exclusively human activity. But, one must constantly bear in mind—stresses Tymieniecka—that, along with the transformation of the "traditional' antinomian paradigm, geared up with the radicalization of the starting point (*epoché*), the "end product" is supposed to undergo a transformation as well. The discovery of the relation between the conscious I and the transcendent world had a series of consequences not to be underestimated. "Turning then to subjectivity"—states Tymieniecka—"Husserl proposed to unfold first philosophy, philosophy's absolutely necessary beginning, which will unfold out of its own inner necessity" (Tymieniecka 2005, xix).

What is this inner necessity then? It is, as Sartre conceived of it, a kind of indubitable certainty revealed or, still better, showing itself to the searching consciousness. Being a consistent phenomenologist, Tymieniecka is, at the same time, an independent thinker, presenting (unfolding) her own understanding of the *Logos of life*—hence, the logos of phenomenology—of how the multiple diverse ways our creative, constitutive nature come into being in relation to reality. No wonder that Tymieniecka's reading and re-reading of phenomenology is regarded, along with particular theoretical proposals, as a method no less critical than creative (like Sartre, she takes the Husserlian project itself up to its extreme possibilities, which Husserl himself was loath to see realized). It seems only natural that, on numerous occasions, Tymieniecka is ever ready to point to all those shortcomings, bordering on evident errors, committed by Husserl, and with the effect that the vital ideas that could have come to fruition never did. In other words, one might say that Tymieniecka is endowed with an enormous capability for transcending the given, on both the level of "reality" and the theoretical level. In an attempt characteristic of many critical philosophers of the phenomenological persuasion to reach the primordial, precognitive, the original (presenting itself at the very beginning), or still better, the absolute origin by turning to the subjective pole itself, Husserl—according to Tymieniecka—presented the principle of all principles, which she re-formulated in the following manner: "I am—the world is" (Tymieniecka 2005, xix). Although the move made on the part of Husserl was, in itself, unquestionable (aiming at overcoming the antimony of subjectivity and objectivity—for it is the very intentionality of consciousness, which directs our attention at reality),—Husserl (claims Tymieniecka) failed to account for an absolute (original in the sense of being a kind of an uncontaminated source) cognitive ground. To a great extent, that had hampered the disclosure of the unique relations of subjectivity and objectivity. Moreover, the temporal structure (protentions and retentions—as the Husserlian guarantees of respectively future and past experiences with constituted objects) proved to be of little help in revealing of the true nature of the temporality of our experiences with the world.

Tymieniecka's quest for the "ultimate grounding" belongs to this type of question which demarcates her version of phenomenology from other transgressions within the phenomenological movement. If such a grounding is to be found, pointed to, or unfolded in the constant process of phenomenological research, then the cognizing subject—free, spontaneous, creative, and as with the Sartrean interpretation, not egological in the personalistic sense—would be a legitimate co-factor of the preconstituted reality. Only in this case could one rightly declare that we have found the unquestionable "launching pad" as it were for all future (subsequent) moves and forays into the multifarious realms of investigation. But, Tymieniecka would never be satisfied with any ready, established, taken-for-granted proposals and solutions. In her never-ending quest for the true Logos of phenomenology, she focused her critical, inquisitive attention on still more original, more pristine modalities of Reason itself. As can be surmised, her project consists of going beyond, in overcoming, hence, in re-defining the method proposed by Husserl and continued uncritically by Eugen Fink.

We have reached a very sensitive point here. Both Sartre and Tymieniecka object to the way Husserl carried out the essential part of his methodological program, namely, the reductions. In sharp contrast to the Husserlian stance, Tymieniecka proposes to lift the veil from the last phase of our query, thus directly pointing to the ontopoietic plane of the Logos. This might be achieved by resorting to all three reductions: phenomenological, eidetic, and transcendental, and respectively the reduced levels of reality; it is a type of query similar to that we find in Sartre—the particular realms of investigation correspond to particular types of reduction. The latter operation, in turn, facilitates carrying out the cleansing of the field of research, hence, proper, adequate constitution. However, for both Tymieniecka and Sartre the basic tenets of the Husserlian method were not consistently applied, had not been taken advantage of, to the "very end." What then could the reductions achieve, provided they were applied as Sartre and Tymieniecka propose?

Tymieniecka speaks of spheres that border on the "givenness of life itself," while Sartre objects to the very notion of *epoché* as a hampering consideration that presents an evident obstacle in the process of our return to the transcendent world from the reduced one. All in all, both thinkers, although taking different points of view (Tymieniecka's *Logos* of life, and Sartre's ontological status of impersonal consciousness along with its fundamental intentions) do care about the idea of an existence neutralized by Husserl's science. Like Sartre, Tymieniecka gives short shrift to the celebrated switching off of the existential thesis Husserl had postulated and obstinately tried to carry out. The overwhelming experience, the phenomenon of existence, does not interfere with the phenomenological quest: in the case of Sartre, the goal would be the certain, and in Tymieniecka's case, the sphere of *Logos* (Reason and Rationality itself). But, as Tymieniecka "categorically states:" within the very nucleolus of the latter there "lurk difficult, sphinx-like questions" (Tymieniecka 2005, xix).

Such questions will lead Sartre's version of existential phenomenology to investigations into the very essence (still better—the lack of essence) of human reality. The *certain* demands its counterpart: the sphere of its negation in which we desperately search for something absolute, as Tymieniecka has it (especially in an age of global crises, which is as well a time in which the *Logos of Life* comes to the fore). The latter could have constituted the true beginning if only Husserl had been more consistent in realizing his initial project. The very beingness of Tymieniecka's reality (Sartre's *En-soi*) is expressed in the oft-dramatic search for true origins. For Sartre and Tymieniecka, this sphere should be sought in the impersonal (not egological) cogito, its existence in the world, while the third element of the triad's cogitations may yield—so emphasizes Tymieniecka—the object of cognition. This is the domain of First Philosophy, which concentrates on the "inner necessity" of self-presenting phenomena. In a superb and lively analysis, Tymieniecka re-interprets the well-known Husserlian formula of the principle of all principles. This indubitable experience, according to Tymieniecka, is conducive to the "everlasting process" of interrogation, opening up—as it were—the sphere of ideally-possible meanings. These are no less than the end results of the activity of the mind. Let us recall at this moment the apt Sartrean metaphor of *egoless* consciousness acting on the pre-reflective level, which constitutes its world (hence receiving its own integrity through the constitutions it is engaged in).

Throughout her whole career, Tymieniecka had been dealing with many aspects of consciousness, trying to relate them to the nature of the *Logos of Phenomenology*. It cannot be denied that, in the second step of the reductions, Husserl "seeks within the objectivity given in conscious acts the platform of the phenomena as the true nature of givenness" (Tymieniecka 2005, xix; Haney 2008, 45). Purifying and cleaning up operations allows First Philosophy to constitute itself as the ultimate grounding of reality. What is more important is the human being at its very center. All efforts in our phenomenological quest enable us to see, to intuitively grasp the things in themselves, which are constituted in absolute consciousness. It is only in this way that the emergence of the sphere of the absolute givenness is presented to us. It is noteworthy that Tymieniecka, much like the young Sartre, makes a move towards the very relation of consciousness and Being. This is an initial stage of the oncoming eco-phenomenology, whose project is—in a way—to bring to total fruition the Husserlian ideas. The fundamental difference with Husserl's thought would be the question of both the ontological and epistemological status of the transcendental Ego.

In her critical and relevant analyses of the Husserlian methodology, Tymieniecka pays special attention to the very process, to the internal *dynamis* of numerous attempts at reaching the true, authentic, and "uncontaminated" origins of the Logos of Phenomenology itself. After the substantial critique of the objects of experience has been made, there comes the time for questioning—the interrogation of experience itself. This procedure, says Tymieniecka, can bring us closer to the *eidos* of constitutive consciousness. As we will have remembered, the idea of constitution is geared up with the intentional structure of all conscious acts. With this "static and genetic" nature of objectivity (in which phenomena are related to consciousness), we come nearer to the basis of givenness. At this moment, here is the inevitable question of the world's ontic, ontological, epistemic, and epistemological status. It might seem that phenomenology, as the subjective (in the intentional sense) science of the world, the science whose essence points to its operational formula *ego cogito cogitatum*, may be said to have established the indisputably valid correlation between transcendental consciousness (the transcendental Ego) and the World, or if we recall Tymieniecka's phrase: the outlining of the life-world. For the majority of this phenomenologist's followers, this point appears to have brought to completion the initial project. But, that is not the case given Tymieniecka's constant re-interpretation.

As the philosopher herself will explain this complex problem in the best way, let us quote Tymieniecka's words:

> "All the horizons of consciousness and of the world in flux of the living present is grasped at its absolutely valid (i.e., cognitively purified of all natural *naïveté*) structural level opening its pattern of *passio* and *actio* for inspection as the seemingly ultimate transcendental level—ontological level—grounding empirical and the positive sciences of reality. With all that now being revealed, it would seem that Husserl had indeed reached the goal of his quest for the pure phenomenon of the givenness as such." (Tymieniecka 2005, xx)

But, this end result proposed by Husserl does not meet the requirements of Tymieniecka's *Logos* of Phenomenology. In her constant reinterpretation of this vital problem, she goes even further than the final point of Husserl, as we can see another essential, pivotal problem emerging. It is related to the legitimacy of consciousness itself and the reductive procedures. What Husserl took for granted, she submits to an arduous and meticulous process of questioning. Although the last reduction allows us to reach the realm of transcendental consciousness itself, establishing a universal First Philosophy cannot be identified with the *Logos* of interrogation itself. In other words, the ceaseless search must be continued, as any closer look at the interrogation of the transcendental quest reveals its incompleteness. Being most revealing, acutely perceptive—to paraphrase her words—the Husserlian thrust of intuition (bringing us nearer and nearer to the certain and necessary foundation of reality) did not, however, reveal this foundation, the origin of reality, thus, rendering almost all analyses and descriptions insufficient. Time and again, Tymieniecka accentuates all these vast intermediary spaces between the phases of both reductions and constitutions, making the true and dedicated phenomenologist interrogate Logos—the Logos that Humanity pursued "for centuries" (Tymieniecka 2005, xx). The most important value is here at stake: the true origin of Life and Cognizance. The author of *Creative Experience and the Critique of Reason* is fully justified in crediting herself for this realization, in so far as the relentless quest for the Logos is concerned. Not downplaying the role and the function played by the intentionality and the series of reductions (the transcendental reduction being of the utmost importance), Tymieniecka directs our attention towards creativity itself—capable of exposing "these further areas." We read in one of her texts that "true critique of phenomenological reason would thus expand to territory beyond the direct role of consciousness in the intellective, specifically human constitutive function. We have such a fresh approach to the origination of human being within the Human Creative Condition" (Tymieniecka 1966, 30). Thus, this creative, imaginative force (so visible in the meta-poietical approach) is capable of accounting for all manifestations of living beingness in its free, unhampered, and uncontaminated human expression.

For Sartre, the adequate expression of such an activity would be the sphere of art. It is the imaginary mode of consciousness (the celebrated *recul* present in all valuable works of art) that creatively transcends the given reality; it negates it, and through having been anchored in the *hyletic* strata of the world (sounds, colors), these works are endowed with an internal force to overcome the one-sidedness of the materiality of the world that consciousness is in.

The very moment we willingly negate the thingness of shapes, sounds, and colors, along with the sensations they induce in us, we are likely to enter into a different, higher (Tymieniecka might have said "closer to the very beingness") level of consciousness, by which the *eidos* of art reveals itself to us. In a word, our perception of the human reality is thus made richer and fuller.

But, there is yet another fundamental similarity between Tymieniecka and Sartre. Like the French existentialist, She is far from forsaking, "putting aside" the factual and actual, the perceptual sphere of reality, of the world we live in. This return to the factual, the rich sphere of our being-in-the-world, univocally points to the creative ontopoiesis of life itself. According to Tymieniecka, the creative and the imaginative are all expressive forces of the *Logos of Life* itself. Thus, only with such notions and ideas as creativity, spontaneity, creative imagination, and ontopoiesis can one reformulate *Logos* itself.

It is a widely known fact that Husserl had left as his legacy many pertinent and relevant ideas and postulates concerning the nature of consciousness and its relations with the transcendent world. In his project of reaching the true origin of both Cognizance and Life, he did not—in many cases—go to the very end, thus, falling short of successfully concluding his, more often than not, sharp intuitions and insights. Among those who took up the task of the constant, arduous, and creative effort of interpretation, reformulation of the core tenets of Husserlian phenomenology and saw what others could not see and perceive are both Jean-Paul Sartre, and Anna-Teresa Tymieniecka. The role they played in our expanding on the understanding of life and human existence must be taken into account in all phenomenological investigations.

## Works Cited

Ayer, A.J. 1979. Construction of our Theory of the Physical World. In *Philosophy as It Is*, ed. Ted Honderich and Myles Burnyeat. London: Pelican. Print.

Baumer, Franklin L. 1977. *Modern European Thought: Continuity and Change in Ideas 1600–1950*. London: Collier Mac. Print.

Caws, Peter. 1993. Sartrean Structuralism? In *The Cambridge Companion to Sartre*, ed. Christina Howells, 293–317. Cambridge: Cambridge University Press. Print.

Dobson, Andrew. 1993. *Jean-Paul Sartre and the Politics of Reason: A Theory of History*. Cambridge: Cambridge University Press. Print.

Haney, Kathleen. 2008. Noting Some of Tymieniecka's Gifts to Phenomenology. *Phenomenological Inquiry* XXXII: 45–47. Print.

Kołakowski, Leszek. 2001. *Husserl and the Search for Certitude*. South Bend: St. Augustine's Press. Print.

Lyotard, Jean-François. 2011. *La Phénomenologie*. Paris: PUF. Print.

Merleau-Ponty, Maurice. 1966. *Phénomenologie de la perception*. Paris: NRF Gallimard. Print.

Mróz, Piotr. 1997. *Four Essays in Existentialism*. Kraków: Aureus. Print.

———. trans. 1998. *Wyobraźnia*. Kraków: Aureus. Print. Translation of Jean-Paul Sartre, *Imagination*. Paris: Gallimard, 1986. Print.

Sartre, Jean-Paul. 1957. *The Transcendence of the Ego*. New York: Noonday. Print.

———. 1971. *l'Imaginaire*. Paris: Gallimard. Print.

Spiegelberg, Herbert. 1994. *The Phenomenological Movement*. Dordrecht: Kluwer. Print.

Thody, Philip. 1990. *Jean-Paul Sartre*. London: MacMillan. Print.

Tymieniecka, Anna-Teresa. 1966. *Why Is There Something Rather than Nothing? Prolegomena to the Phenomenology of Cosmic Creation*. Assen: Van Gorcum. Print.

———. 1990. *The Origins of Life. The Origins of the Existential Sharing-in-Life*. Dordrecht: Springer. Print. Analecta Husserliana XXVII.

———. 2005. The Logos of Phenomenology and Phenomenology of the Logos. In *Logos of Phenomenology and Phenomenology of Logos, Book 1: Phenomenology as the Critique of Reason in Contemporary Criticism and Interpretation*, Analecta Husserliana LXXXVIII, ed. Anna-Teresa Tymieniecka, xii–xxl. Dordrecht: Springer. Print.

———. *Bergen Interview 27* August 2008, Bergen, Norway. Web. 06 January 2015. http://www.phenomenology.org/archive/.

# Part VI
# Eco-Phenomenological Readings

# Sowing "A Quilt of Harmony": An Eco-Phenomenological Reading of Ben Okri's "Lines *in Potentis*" from *Wild* (2012)

**Rosemary Gray**

**Abstract** This article begins with a brief discussion of the terminology used, followed by an explication of its title, words taken from "A Wedding Prayer," a poem in Ben Okri's (Wild. Random House, London. Print, 2012) anthology, *Wild*, that is to say, the significance of sowing/sewing "a quilt of harmony" (20), that in relation to the broad yet symbiotic theme of Life, Human Life, Post-Human Life in the Harmony of the Cosmos. A close reading of Okri's 'Lines *in Potentis*,' a poem commissioned by the then Lord Mayor of London in 2002 in commemoration of the bombing of the City of London and featured in Okri's most recent anthology of poetry, is next. Both title and argument are predicated on the former poem, which is not analyzed in any detail. Axiomatic to the interpretation is the poet's own conception of *wild* as a defamiliarization of the familiar to regain a sense of equipoise through our link with the "stars." As I attempt to show in my reading of the focal poem, this is not aesthetic posturing; it has to do with mystical unrest viewed from an epistemic eco-phenomenological "enjoyment of literature, of beauty, of the sublime, the elevated, as well as our compassion for the miseries of humankind, [and] generosity towards others … inspired by the subliminal passions of the human soul," to invoke the words of the late Anna-Teresa Tymieniecka (Tymieniecka, The theme/The esoteric passion for place. In: Tymieniecka A-T (ed) Passion for place, Book II. Analecta Husserliana LI. ix–xiv. Kluwer Academic Publishers, Dordrecht/Boston/London. Print. 1997).

**Keywords** *A way of being free* · "A wedding prayer" · Eco-phenomenology · "Lines *in Potentis*" · Ben Okri · Ontopoiesis · Anna-Teresa Tymieniecka · *Wild* (2012)

---

R. Gray (✉)
Department of English, University of Pretoria, Pretoria, South Africa



W. S. Smith et al. (eds.), *Eco-Phenomenology: Life, Human Life, Post-Human Life in the Harmony of the Cosmos*, Analecta Husserliana CXXI,
https://doi.org/10.1007/978-3-319-77516-6_22

## Physis and Poiesis

I will begin with a brief discussion of the poet's Ben Okri's point of view, ontopoiesis, and eco-phenomenology. Rooting his argument firmly in ontopoiesis or phenomenology of life, Okri asserts in *A Way of Being Free* that "the poet is the widener of consciousness" (Okri 1997, 3). Implicitly invoking an eco-phenomenological standpoint, he explains how this raising of consciousness, or the process of ontopoiesis, occurs poetically: "[Poets] speak to us. Creation speaks to them. They listen. They remake the world in words, from dreams" (Okri 1997, 3). He muses about this mystical dialectic: "Intuitions which could only come from the secret mouths of gods whisper to them through all of life, of nature, of visible and invisible agencies" (Okri 1997, 3). Underlining the relation of eco-phenomenology to the fluid nature of reality in the same text, Okri explains, "The poet turns the earth into mother, the sky becomes a shelter, the sun the inscrutable god" (Okri 1997, 2).

Okri's view of the role of the poet and the source of poetic inspiration accords with the late Anna-Teresa Tymieniecka's concept of eco-phenomenology, elucidated in her introduction to her 1997 *Passion of Place* as the philosophy of "our relationship to the earth" (Tymieniecka 1997, 2). In defense of her unique brand of phenomenology, she explains not only its theoretical base but, also, its application. Drawing an analogy to climate change and the current ecological crisis, she later explains in the 2008 *The Passions of the Earth* that people generally see these as physical problems, resolvable through technological innovation. By contrast, eco-phenomenologists approach these problems from a metaphysical perspective, thus requiring "a fundamental re-conceptualization of human values and our relationship to nature" (Tymieniecka 2008, 2).

For Aristotle, too, natural generation and artistic creation are purpose-driven. And, as with Okri's "invisible agencies" and Tymieniecka's "metaphysical perspective" on climate change, there is no impassable gulf between *physis* and *poiesis*. Consider, for example, Aristotle's statement in his *Physics*: "It is absurd to suppose that purpose is not present because we do not observe the agent deliberating. Art does not deliberate." In other words, this is what constitutes the natural generation just referred to. Aristotle's illustrations serve to elucidate: "If the shipbuilding art were in the wood, it would produce the same results *by nature*. If, therefore, purpose is present in art, it is present also in nature. The best illustration is a doctor doctoring himself: nature is like that. It is plain then that nature is a cause, a cause that operates for a purpose."[1]

## Okri's Poiesis

And so I turn to illustrating Okri's re-conceptualization of human values in relation to Nature in his poems "A Wedding Prayer" and "Lines *in Potentis*" (Okri 2012, 20–22; 26–27). It is precisely this natural or purposive generation that gave rise to the title of this article, inspired by a line from Okri's "A Wedding Prayer." Evincing an eco-phenomenological viewpoint, Okri writes:

> … Love has brought two
> Rivers into one way, one dream;
> Has *sown a quilt of harmony.*
> And scattered some magic
> Fragrance upon the sea. (Stanza 1, ll. 6–10; emphasis added)

The metaphysical conceit of the love of two people having "sown a quilt of harmony" leading to their commitment to marriage is transcendently eco-phenomenological; the pun on SOW (connoting regenerative husbandry as in the nurturing of plant life as well as the procreation of life customarily anticipated by newlyweds) and SEW (as in the unifying bringing together of stitch craft) encapsulates the *double entendre* of the conscious acts inherent in spending a life together, as well as the responsibilities of wedded life. Symbolically, too, SOW invokes the Biblical axiom of "As you sow, so shall you reap," while SEW recalls Joseph's coat of many colours and all that it signifies.

This occasional poem was commissioned to celebrate the nuptials of a bride and groom and is, at once, profound and idyllic. It is dedicated to Ieva and Ivor,[2] to life and to the procreation of life as expressed in the injunctions to the bridal pair: "May you never lose your/Laughter, your playfulness,/And your music" is juxtaposed with "Be fruitful in enchanting deeds/And in futures" (Stanza 6, ll.65–67 & ll. 69–70).

Explicit in the second occasional poem, "Lines *in Potentis*," are the same epistemic eco-phenomenological motifs of ecology, love, music, dreams, nature, harmony and magic as the opening lines attest:

> One of the magic centres of the world;
> One of the world's dreaming places.
> Ought to point the way to the world:
> For here lives the great music of humanity. (Stanza 1, ll. 1–4)

In this focal poem, the canvas is broader and addresses a wider audience in its intersection of the diachronic with the synchronic, that is, of history and contemporary society. Addressed to the cosmopolitan society of Londoners, the "Workers of the world," the poem appeals for "[t]he harmonisation of/Different histories, cultures, geniuses and dreams" (Stanza 4, l. 42; Stanza 1, ll. 5–6). This is not simply a utopian vision for, as Eckhart Tolle explains, "At the core of all utopian visions lies one of the main structural dysfunctions of the old consciousness: looking to the future for salvation" (Tolle 2005, 308). Commissioned in 2002 for the London Assembly and inscribed around the curving structure of the Greater London Authority Building [the City Hall], "Lines *in Potentis*" was read by the poet in Trafalgar Square on 14 July 2005.

## Okri and Contraries

Whereas the first poem celebrates the conjoining of human life, the second commemorates post-human life after the bombing of London.[3] Okri's prayerful dream, as he recalls the horrors of the Blitz from a twenty-first century perspective, is akin to that of Revelations – his is a vision of "a new heaven and a new earth,"[4] the foundation of both consciousness being awakened or ontopoiesis. The recollection of historical facts is part of a purposive mythic pattern the poet employs in order to capture a series of moral crises that he not only portrays but also seeks to embody in this poem. This is borne out in the injunction to "tell/Everyone that history, though unjust,/Can yield wiser outcomes" (Stanza 1, ll.7–9). An enumeration of a list of contraries that follows articulates just what these "wiser outcomes" may be:

> And out of bloodiness *can* come love;
> And out of slave-trading
> *Can* come a dance of souls;
> Out of division, unity;
> Out of chaos, fiestas. (Stanza 1, ll.10–14; emphasis added)

The repetition of the modal verb "can" highlights the transformative metamorphosis envisaged. Explicit in Okri's Africanist cosmogony is the Blakean aphorism that "Without Contraries is no Progression" (Keynes 1966, 149); see "Blake's 'The Marriage of Heaven and Hell'" therein. As Okri, too, avers in *A Way of Being Free*, "There can be no absolutes: no absolute good or evil, no absolute way of living. No absolute truth. All truths are mediated and tempered by the fact of living. Being alive qualifies all things" (Okri 1997, 54). "Poets," Okri muses in the same text, "seem to be set against the world because we need them to show us the falseness of our limitations, the true extent of our kingdom" (Okri 1997, 2).

The epistemological implications are that, without paradox, without the various trials and tribulations in one's encounter with what religions call good and evil (but what Blake refers to as Innocence and Experience), one cannot attain perfection of Being.

Thus, the vision of London, "City of tradition, conquests and variety;/City of commerce and the famous river," as a sacred *axis mundi* – implied in the "magic centre" quoted earlier, alludes to an awakening of consciousness – that is, to an awakening to the realization of Presence and its power (*potens*) to effect change (Stanza 2, ll. 15–16). The flow of the river [of life] into the sea in "The Wedding Prayer" becomes the inscrutable flow of the cycle of the seasons, "Awaiting an astonishing command/From the all-seeing eye of Ra," in "Lines …" (Stanza 2, ll. 29–30). Not only does this aesthetic evocation of the Egyptian sun god transliterate John Keats's "Hyperion" – "One moon with alteration slow had shed/Her silver seasons four upon the night" – it also embodies the awakened consciousness. "When we look out on the world with all its multiplicity of astonishing phenomena," Okri asks rhetorically in *A Way of Being Free*, "do we see that only one philosophy can contain, explain, and absorb everything"? (Okri 1997, 19)? He elaborates, "I think not. The universe will always be greater than us" (Okri 1997, 19). As mere microcosms within the macrocosm, Okri suggests that "Our mind[s] therefore should be like Keats's thoroughfare, through which all thoughts can wander" (Okri 1997, 19). Okri surmises, therefore, that our minds "should also be a great cunning net that can catch the fishes of possibility" (Okri 1997, 9). In an interview given on the occasion of the conferring upon her of *Laurea Honoris Causa* in Philosophy (27 August 2008), Tymieniecka asserted that the very essence of her philosophy – and, it is evident, that of Ben Okri, too – "is our relationship to the earth and to the cosmos" (Tymieniecka 2008, 2).

## Disaster and Potential in Okri's Vision

The appeal in "Lines *in Potentis*" is for this "magic centre" of the global village to use its regenerative power to "Re-make the world/Under the guidance of inspiration/And of wise laws" (Stanza 4, ll.43–45). In a reworking of the Marxist tenet, "Workers of the work unite," the injunction here is for the workers to unite, by implication, not because they have nothing to lose but because the proletariat has everything to lose should they fail to heed the poet's plea. A history of natural and man-made disasters (The Great Fire, the Plague, and the Blitz) enables London's workers to comprehend their own potency and potential, to "Tell everyone that the future/Is yet unmade" (Stanza 2, ll.17–18). Thus, the poem seeks to address mystical unrest, not bourgeois capital. The closing lines, in particular, resound with this mystique, this eco-phenomenological sensibility:

> I want you to tell everyone
> Through trumpets played
> With the fragrance of roses, that
> A mysterious reason has brought us
> *All together*,
> Here, now, under the all-seeing
> Eye of the sun. (Stanza 5, ll. 57–63; emphasis added)

The two pivotal motifs that run throughout world mythologies – wonder and self-salvation – implicitly coalesce in these lines. The mystical is, as Joseph Campbell explains, in a chapter tellingly entitled "Cities of God" in his *The Masks of God*, "redemption or release from a world exhausted of its glow" (Campbell 2011, 35).

## Okri's Sense of the Wild and Tymieniecka's Sentience of the Logos

"Lines *in Potentis*" is featured in Okri's most recent anthology of poetry, *Wild,* as is "The Wedding Prayer." Axiomatic to my interpretation is the poet's own conception of the *wild,* detailed on the dustcover to *Wild* as "an alternative to the familiar, where energy meets freedom, where art meets the elemental, where chaos can be honed." More precisely, for this London-loving Nigerian poet, "the wild is our link with the stars …." In an interview in *The New Statesman* (29 March 2012), Okri stated, "I was interested in the wild, not as in wildlife or outside civilization but as a raw, formative energy that artists notice when they look at objects." This is not aesthetic posturing; it has to do with that which lies beyond reason and rationality. It is predicated on the mystery of eco-phenomenology, on "enjoyment of literature, of beauty, of the sublime, the elevated, as well as our compassion for the miseries of humankind, generosity towards others … inspired by the subliminal passions of the human soul," to quote Anna-Teresa Tymieniecka (Tymieniecka 1997, ix–xiv). Okri's concept of "wild" as that which becomes ontopoietically familiar when creative energy is released to contain cosmic chaos vicariously resonates with Tymieniecka's 2008 explication of the Sentience of the Logos which "is carried in various guises through all the individualizations of life" and "which from this first germinal coming forth of life, appearance of life, to the fashioning of the individual, which in the case of the human being, with its highest sentient spiritual unfolding … is really carrying the divine" (Tymieniecka 2008, 6). In "The Passions of the Earth," Tymieniecka asserts that "the human being is an ecological fruit … formed by the earth," whose life is sustained by "the juices of the earth" (Tymieniecka 2008, 6). This, in turn, explains "the cosmic dependencies of the human mind and human development" (Tymieniecka 2008, 6). Having posited the notion that this has nothing to do with theology but is, rather, just a metaphysical tendency of life itself, she added: "the self-individualization of life, which is the basic instrument of ontopoiesis, draws on the laws of the cosmos and the earth" (Tymieniecka 2008, 8).

One of these cosmic laws manifests in poetry, according to Okri in *A Time for New Dreams* (2011:3): "We are, at birth, born into a condition of poetry and breathing. Birth is a poetic condition: it is spirit becoming flesh. Death is also a poetic condition: it is flesh becoming spirit again. It is the miracle of a circle completed, the unheard melody of a like returning to unmeasured silence"(Okri 2012, 3). The cycle of the seasons and in the Cardinal Points of the compass, all of which feature in "Lines *in Potentis*" are likewise cosmic laws. Significantly, the first season mentioned in "Lines …" is Spring, a time of rebirth. In the poem, an anthropomorphized Spring (invoking its correlative – human birth) "waits/By the lakes, listening/To the unfurling daffodils" (Stanza 2, ll. 25–27). This, perhaps, is an allusive conflation of William Wordsworth's definitive nature poems, "To Daffodils" and "Tintern Abbey." The depiction of Summer shares the same esemplastic imagination: it "lingers with the hyperborean worms," perhaps an allusion to the dragons of yore (Old English "wyrms") conscientized to nature's purposive way. The choice of the epithet "hyperborean" endorses the indissoluble link between past and present, between man and nature, and highlights cosmic unity. The adjective (hyperborean) denotes or relates "to the extreme north," while as applied to the noun, "worms," it could signify inhabitants "of the extreme north" in accordance with the *New Oxford Dictionary of English* (900). Furthermore, embedded in the signifier is a tacit allusion to Greek mythology; the Hyperboreans were members of a race worshipping Apollo (the Greek god of the sun) and living in a land of sunshine and plenty beyond the north wind, coincidentally typifying an African setting (*New Oxford Dictionary of English* 900).

What then of the other two seasons? Both Autumn and Winter are, likewise, anthropomorphized. In an intertextual braiding of Keats's "Ode to Autumn," "Ode to the West Wind" and "Ode to a Nightingale," shot through with Negative Capability,[5] Okri's Autumn "dallies/With the West wind/And the weeping nightingales" (Stanza 5, ll. 50–52). Winter is depicted as clearing "its sonorous throat/At the Antipodean banquets, preparing/For a speech of hoarfrost/And icicles conjured from living breath" (Stanza 5, ll. 53–56).

Implicit in the evocation of the cycle of the seasons is not only a correlation with human life – *a priori* Being, birth, life and death and life-after-life or post-human life – but, also, a veiled bringing together of the four Cardinal points, namely North (hyperborean, already discussed), South (Antipodean; the Antipodes is an appellation for Australia and New Zealand), East (the land of the rising sun/Ra/Apollo) and West (the autumnal West wind). The seasons and points of the compass conjoin to unify the cosmos reflected in Okri's already quoted vision of the "harmonisation of different/Histories, cultures, geniuses, and dreams" (Stanza 1, ll. 5–6).

## Okri's Human Ecology

Okri's treatment of humankind in both poems discussed in this article is, likewise, rooted in eco-phenomenology. On the brink of a new life together in London, Viscount Ivor Guest and his bride, Ieva (a Latvian beauty), are enjoined to:

> … Travel
> Into one another, as into
> A country you have long admired,
> And read many fables about,
> And now find yourself
> Before its famed rivers
> Its inspiring mountains. (Stanza 3, ll.20–26)

In contrast, "Lines *in Potentis*" is addressed to the London populace, possibly the most cosmopolitan in the world, who are enjoined to attune to the "wild," to the Elizabethan Music of the Spheres; to "Create the beautiful/Music our innermost/Happiness suggests delight the future./Create happy outcomes" (Stanza 4, ll. 46–49). Londoners, Okri implies, are those who have awakened "to their essential true nature as consciousness and recognize that essence in all 'other,' all life forms" (Tolle 2005, 309). They feel a oneness with the ecological epistemology of the whole, as Okri suggests in "Lines …:"

> Tomorrow's music sleeps
> In our fingers, in our awakening
> Souls, the blossom of our spirit
> The suggestive buds of our hearts. (Stanza 3, ll. 31–34)

This single poetic quatrain foreshadows the closing lines of this poem with its "trumpets played/With the fragrance of roses" and reflects back to the opening line's "one of the magic centres of the earth," where heightened sense impressions prevail, where human sensibilities are awakened and optimized.

Ultimately, "Lines *in Potentis*" can be seen as an eco-phenomenological blueprint for "the way" to transform "nightmare" into "illumination", transliterating St Paul's Letter to the Philippians,[6] and encapsulating both Revelations' prophetic new earth or Blake and Parry's hymn to the New Jerusalem[7] (Stanza 2, l. 20). All four of these intertexts embody the awakened consciousness that Eckhart Tolle asserts "is changing all aspects of life on our planet, including nature, because life on earth is inseparable from human consciousness that perceives and interacts with it" (Tolle 2005, 308). This assertion, in turn, underlines the Aristotelian insight quoted at the outset of this article that there is no impassable gulf between *physis* and *poiesis*, and accords as well with an eco-phenomenological epistemology. "Ours is a condition," asserts Tymieniecka, "within the unity of everything alive, which depends on earthly and cosmic laws" (Tymieniecka 2001, 3).

## Conclusion

In an article entitled, "African Modes of Self-Writing," fellow African phenomenologist Achille Mbembe (2005: 252–253) debates the contention of this unity in the more prevalent discourse of emancipation and autonomy (e.g., Franz Fanon in his *Black Skins, White Masks*), which discourse Mbembe terms a "paranoid reading of history" (Mdembe 2002, 252–53; 253). He points to "a tension between a universalizing move that claims shared membership within the human condition (*sameness*) and an opposing, particularistic move that "emphasizes difference and specificity" by accenting, not originality as such, but the principle of repetition (*tradition*) and "the values of autochthony" (Mdembe 2002, 253). This latter invokes a fabricated mask of Africanity[8] – race, culture, and the like – whereas my argument is premised on *Ubuntu,* or the brotherhood of humankind in the context of cosmic harmony. It is inspired by the subliminal passions of a contemporary poetic soul.

In the introduction to her *Passion for Place*, Tymieniecka elaborates within an eco-phenomenological logic: "The soul's creatively orchestrated swings, together with the generative propulsions of imagination, distil the primitive strivings endowed with *specifically human life-significance*" (Tymieniecka 1997, ix). As an endorsement of an eco-phenomenological passion for place in the hands of a true poet, she evokes the resulting Affective Fallacy:

> On the wings of the creative imagination our subliminal passions carry us to unattainable realms, and we exhale beyond our frames with joy or enchantment as our dreams have us roam through shifting dimensions of sublunary reality. All passions 'fly,' charting nebulous spheres, as human strivings, dreams, forebodings, desires – in following their particular bents – seek fulfilment by crystallizing in a place … that they imaginatively fashion and qualify with the significance of expectant strivings. (Tymieniecka 1997, xii–xiii)

## Notes

1. Aristotle 1941. *Physica* II 8 199B27-31. In *The Basic Works of Aristotle*, ed. Richard McKeon. New York: Random House, p. 251.
2. The poem was commissioned by Ivor Guest, the 4th Viscount Wimborne, a friend of Ben Okri, for his marriage to Latvian beauty Ieva Imsa. Their daughter, Greta, was born in 2011.
3. "Lines ..." was commissioned by the then Lord Mayor of London, Ken Livingstone. It was read by the poet at the memorial of the bombing of London (Rosemary Clunie, personal communication).
4. Revelations 21: 1.
5. Douglas Bush (1966: 58) explains Keats's doctrine of Negative Capability or intuitive empathy in the poet's own words: "The setting sun always sets me to rights – or if a Sparrow come before my Window I take part in its existence and pick about the Gravel."
6. Philippians 4: 8: "whatsoever things *are* true, ... whatsoever things are pure, whatsoever things *are* lovely ... Think on these things."
7. The hymn "Jerusalem" by William Blake and Hubert Parry obviously informs Okri's vision; his second anthology of poetry is entitled, *Mental Fight*. Blake's well-known verse reads: "I shall not cease from mental fight/Nor shall my sword sleep in my hand/Till we have built Jerusalem/On England's green and pleasant land." Blake's epic poem, "Jerusalem," seems, in turn, to have been inspired by Revelations 21: "I saw the holy city, the new Jerusalem, coming down from God, prepared as a bride adorned for her husband." For Blake, Jerusalem (the True church) was the Bride and Emanation of Albion (the Eternal Man) and of Jesus in Eternity.
8. See Archie Mafeje, 2000. "Africanity: A Combative Ontology." *CODESRIA Bulletin* 2000, nos.1–2, 66–71.

## Works Cited


Aristotle. 1941. Physica ll 8 199B27-3. In *The Basic Works of Aristotle*, ed. Richard McKeon. New York: Random House.

Bush, Douglas. 1966. *John Keats: His Life and Writings*. New York: Macmillan. Print.

Campbell, Joseph. 2011. *The Masks of God: Oriental Mythology*. London: Souvenir Press. Print.

Fanon, Franz. 1972. *Black Skins. White Masks*. New York: Grove Press.

Keynes, Geoffrey. 1966. *Blake: Complete Writings*. Oxford: Oxford University Press. Print.

Mafeje, Archie. 2000. Africanity: A Combative Ontology. *CODESRIA Bulletin* 2000, nos. 1–2; 66–71. Print.

Mbembe, J.-A. 2002. African modes of self-writing. *Public Culture*. 14 (1): 239–273. Print.

Okri, Ben. 1997. *A Way of Being Free*. London: Phoenix. Print.

———. 2011. *A Time for New Dreams*. London: Random House.

———. 2012. *Wild*. London: Random House. Print.

Tolle, Eckhart. 2005. *A New Earth*. London: Penguin. Print.

Tymieniecka, Anna-Teresa. 1997. The theme/The esoteric passion for place. In *Passion for Place*, Book II, ed. A.-T. Tymieniecka. Analecta Husserliana LI. ix–xiv. Dordrecht/Boston/London: Kluwer Academic Publishers. Print.

———. 2001. The passions of the earth. In *The Passions of the Earth in Human Existence, Creativity, and Literature*, ed. Anna-Teresa Tymieniecka. Analecta Husserliana LXXI. 1–12. Boston/Dordrecht/London: Kluwer Academic Publishers. Print.

———. 2008. *Interview with Anna-Teresa Tymieniecka. The Pragmatic Test of Phenomenology*. Conducted by L.P. Torjussen, J. Servan, and S. Andersen Øyen. http://www.phenomenology.org/images/Interview-A-T-Tymieniecka-27-August-2008.pdf. Accessed on 26 Mar 2014.

# Eco-Phenomenology of Scientific Activity As Non-Routinized Routine: Stefan Banach's Café Method of Research and Its Contemporary Continuation

**Bronisław Bombała**

*There are two kinds of intelligence: one acquired,*
*as a child in school memorizes facts and concepts*
*from books and from what the teacher says,*
*collecting information from the traditional sciences*
*as well as from the new sciences.*
*With such intelligence you rise in the world.*
*You get ranked ahead or behind others*
*in regard to your competence in retaining*
*information. You stroll with this intelligence*
*in and out of fields of knowledge, getting always more*
*marks on your preserving tablets.*
*There is another kind of tablet, one*
*already completed and preserved inside you.*
*A spring overflowing its springbox. A freshness*
*in the center of the chest. This other intelligence*
*does not turn yellow or stagnate. It's fluid,*
*and it doesn't move from outside to inside*
*through conduits of plumbing-learning.*
*This second knowing is a fountainhead*
*from within you, moving out.*

*[Jalaluddin Rumi, Two kinds of intelligence]*

**Abstract** Fascination with mathematical formula, i.e., instrumental rationality, is the core of the crisis that plagues the contemporary world. Absolutizing the role of mathematics in building business strategies has become one of the main determinants of economic and ecological crises. The crisis experienced nowadays is perceived by most people as a physical problem to be resolved through technological innovation. However, phenomenologists notice also a metaphysical dimension of this crisis, calling for reflection and re-conceptualization of familiar business

B. Bombała (✉)
Uniwersytet Warmińsko-Mazurski w Olsztynie, Olsztyn, Poland



W. S. Smith et al. (eds.), *Eco-Phenomenology: Life, Human Life, Post-Human Life in the Harmony of the Cosmos*, Analecta Husserliana CXXI,
https://doi.org/10.1007/978-3-319-77516-6_23

strategies and the relationship between human beings and nature. Phenomenology, especially Anna-Teresa Tymieniecka's *phenomenology of life*, is able not only to challenge the basic conceptualization of utility in today's modern, industrialized civilization, but also to indicate the way in which the existing problematic situation can be remedied.

Anna-Teresa Tymieniecka's phenomenology, centered as it is around the creative experience, provides a good basis for an analysis of scientific activity. According to her, phenomenological research should be focused on will, imagination, and creativity – i.e., cognition that crosses the boundaries of reason. Stefan Banach's *café method* is an example of a phenomenological research system that fulfils these criteria. Banach's *café method* of research has developed from addressing a paradoxical situation in which a new tool of scientific research, one originally invented as a form of resistance against routinized methods of scientific work, has become a routinized process itself; still, it is an unusual one, since Banach's *café method* is characterized by an "initial spontaneity".

Over time, Banach's *café method* of research became a source for developing methods of spontaneous inference that are used in contemporary management science (e.g., *Open Space Technology*, *The World Café* and the *"3i" formula*), which leads to the conclusion that this non-routinized method of research has evolved into a routine of management, even though that form of management remains non-routinized.

**Keywords** Creative Experience · Self-Organization · Banach's Café Method · Owen's Open Space Technology · Phenomenological Praxeology · Bombała's 3i Heuristics

## Introduction

Phenomenologists sense a metaphysical dimension of today's economic, ecological crises, which calls for re-conceptualizing business strategies and the relationship between human beings and nature (Brown and Toadvine 2003; Abram 1997). Ecological phenomenology, or eco-phenomenology, is a sub-discipline developed by such researchers as Hans Jonas, Michael Zimmerman, Frank Schalow, David Abram, and Erazim Kohák, who have made substantial contributions. Bence Peter Marosan states that "The task of eco-phenomenology is to provide a phenomenological analysis of this strained relationship between the industrially and technologically determined human existence and nature. … The theoretical foundation of eco-phenomenology is the phenomenology of the human condition in the natural world of Earth and in the cosmos in general" (Marosan 2013, 119). The purpose of this analysis is to reach an understanding of this complicated relationship between man as such and nature. Marosan recognizes three main levels of eco-phenomenological analysis: the theoretical, the axiological, and the practical. The theoretical level clarifies the reasons why the human being is rooted in and/or alienated from the world of nature. The axiological level (ecological ethics) focuses on the ethical aspect of being rooted in or alienated from the world of nature on the basis of the human being's relation to ethical values (if human beings incorporate ethical values in their lives, they are rooted in the world of nature; accordingly, if ethical values are not incorporated in human beings' lives, they are alienated from the world of nature).[1] The practical level formulates directives and strategies for keeping human beings' political, economic, and social activity in harmony with the world of nature (Bombała 2013).

Phenomenology, especially Anna-Teresa Tymieniecka's "phenomenology of life," is able not only to challenge the basic conceptualization of utility in today's modern, industrialized civilization (instrumental rationality), but also to indicate the way the existing problematic situation can be remedied. Tymieniecka (1988) proposed two new concepts, namely, the *creative experience* and the *creative act of man*. Both concepts present a new anthropological quality: the relevant factors of consciousness and, simultaneously, the creative work enacted by the human person. Thus, in the context of Tymieniecka's phenomenology, philosophy goes beyond its previous limits, evolving from "just" a science into a manifestation of the creative experience (Barral 1991). Philosophy in this understanding is like a stream that constantly undergoes transformation by continually discovering new areas of phenomenological research. Hence, philosophy by its nature becomes an act of creative activity that provides a basis for other forms of human action. Tymieniecka's phenomenology, being centered on the creative experience, provides also a good basis for an analysis of scientific activity.

## Stefan Banach's Café Method of Research As Eco-Phenomenology of Scientific Activity

According to Tymieniecka (2011, 15), phenomenological research should be focused on will, imagination, and creativity – that is, on cognition that crosses the boundaries of reason. Stefan Banach's café method is an example of a phenomenological research system that fulfils these criteria. Banach's café method of research has developed from a paradoxical situation in which a new tool of scientific research, originally invented as a form of resistance against routinized methods of scientific work, has become a routinized process itself; still, it is an unusual one, since Banach's café method is characterized by an initial spontaneity.

Polish mathematician Stefan Banach strongly disagreed with the routinization and bureaucratization characteristic of European science. Therefore in his own research work he tended to use original working methods. He did not care about verbal perfection and, for the whole of his life, his way of expression was shaped in accord with the common man's manner of speaking. Instead of staying in the building of Lviv University, where he taught, he conducted classes and discussions with students primarily in a café named Scottish Café. Tables in this café were covered with slabs of marble, on which one could write in pencil and, what was even more important, from which one could quickly wipe off one's notes (Ciesielski 2007, 3).

Often the whole discussion consisted of only a few spoken words, with long periods of meditation interspersed with sudden bursts of conversation, a few lines written on the table, and, occasionally, even an outburst of laughter from one of the panelists, usually followed by long periods of silence during which one was only drinking coffee or cognac (Ulam 1969, 52). In this way Banach, whose preferred working style was based on taking part in discussions whose participants inspired each other, developed together with his colleagues and students a method of work that stood in opposition to traditional methods of scientific work. Nevertheless, this method has become an important element of mathematical research. Hence, Stefan Banach's café method of research was born.

The main problem with practicing this research style lay in the fact that many mathematical proofs are gone forever, having been wiped off the table top by a waiter. That is why, in order to save Banach's cafe achievements, his wife bought him a special notebook. It included both the mathematical problems and their solutions. This unusual diary, known as the "Scottish Book," survived World War II in the café locker room. In 1957 Stanislaw Ulam translated it into English and published it. A year later, this book was presented at the International Congress of Mathematics in Edinburgh.

Stanislaw Ulam, who had been a student of Banach, agreed to work with the Americans and moved to The United States of America. Later, he became a co-creator of the hydrogen bomb (a thermonuclear weapon) in the famous Los Alamos laboratories. Ulam claimed that working in Los Alamos on that bomb was the only time in his life after leaving Lviv when he participated in such intense brainstorming as he used to engage in at the Scottish Café. He recalled that there had been neither envy nor competition among the scientists in Lviv, since their common and only desire had been to experience a great mathematical adventure: "This type of session with Banach, and often with Banach and Mazur, made the atmosphere of Lviv something unique. Such an intimate cooperation was probably something completely new in the mathematical life, at least in terms of its having such a large scale and involving such great intensity" (Ulam 1969, 53).

The story of Banach's café method of research became a source of inspiration for developing the methods of the spontaneous inference used in contemporary management (methods such as *Open Space Technology*, *The World Café* and the *3i formula*). This method of management is used, for example, by the Brazilian company Semco, whose owner, Ricardo Semler, formulated a "golden rule" for management in his business in saying that its basic rule was that there were no rules. Another example is the American company W. L. Gore & Associates, which is described as a *flat lattice organization*; because of its organic-amorphous structure, the "routine" of amorphousness (as an element of management) remains a non-routine that cannot be routinized. Likewise, the Japanese Gemba Kaizen method also contains elements typical for such non-routinized routine.

## The Phenomenon of Self-Organization As a Remedy for the Problems of the Contemporary World: Open Space Technology

Open Space Technology is an approach to organizing scholarly meetings that has been around since 1985. This method was developed by Harrison Owen, the American organizer of The Annual International Symposium on Organization Transformation. The first two of these events were constructed in a most traditional manner. Although Owen had put much effort in preparing both of them, after the second one he decided to structure the future conferences differently, for according to many of the conference participants the most effective part of the meetings was not the presentations and panel discussions, but the coffee breaks. Thus, he asked himself how during future events he could create conditions in which it would be possible to combine the level of synergy and excitement present during the coffee breaks with the substantive activity and outcomes characteristic of a good meeting (Owen 1985, 2).

As a result, the Third International Symposium was organized in a most untraditional manner: at the point of arrival, the participants knew only when the meeting would start, when it would conclude, and generally what might be its theme, for there was no conference agenda available and neither planning committee nor management committee was present. In the meeting, all of its 85 participants were asked to sit in one large circle in order to schedule the symposium together. Each of the persons who wished to pursue a particular area of exploration in a debate during the conference should write a brief description of the proposed issue on a small placard, announce the topic to the assembled group, post the placard on the wall, and sit down. When no further topics were posted, the original proposers determined the time and place for meeting, and anybody interested in a particular topic then signed up. Much to the amazement of everybody in the room, two and a half hours later a 3-day agenda was thoroughly designed, of conference sessions in the form of multiple workshops, each having its time, place, participants and assigned convener (Owen 1985, 3).

Owen "opened the space" to the self-organization of the conference participants assigning much more responsibility to them in the third symposium than in the previous two symposia in order to maximize the level of their productive learning and contribution, minimizing, ipso facto, the level of grunt work for him to undertake as the main organizer of the conference. Thereby, Owen initiated the Open Space Technology that has become extremely popular throughout the world, a development that Owen himself describes as follows:

> For a number of years, Open Space was generally viewed, certainly by myself, as a pleasant but mildly aberrant phenomenon to be enjoyed only at our annual Symposia. The thought that it might have general utility in the world of commerce, government, and industry was never seriously entertained. Despite best efforts at nonchalance, Open Space crept into the world of work. Recently, the passage of Open Space has become something of a rush. It has now been experienced on every continent (with the exception of Antarctica) by groups of

> five to seven hundred and fifty. Major government agencies, large international corporations, small community groups, main line religious bodies, and more have all had the experience of creating intelligent and productive gatherings in a minimum of time with maximum enjoyment. Yes, it is true, Open Space is not only efficient, effective and productive, it is also fun. The original two and a half hours necessary for organization has now been reduced to less than an hour, even with groups of 500. And best of all, my status as solo Open Space practitioner has ended. Presently there are hundreds of people all over the world who regularly demonstrate that Open Space is a global phenomena and definitely not the private magic of Harrison Owen. (Owen 1985, 3–4)

The expression "doing the job" means here that with help of Open Space Technology, diverse, often conflicted groups up to 1000 people have managed immensely complex issues in a minimal amount of time during meetings in which self-organization played the key role, for self-managed work groups became the general mode of operation and distributed leadership turned out to be the norm, there being no advance agenda preparation and little to no overt facilitation. Important characteristics of these meetings have been: a created atmosphere of high energy (often experienced as playful); the phenomenon of conflict becoming a situation with the creative potential to eventually or inevitably produce truly fruitful outcomes; and the fact that the participants treated each other with respect perceiving diversity as a rich resource to be cherished as opposed to a problem to be managed, which respect is the core of Open Space Technology.

The technology is based on four principles and one law, which, however, function more descriptively than prescriptively; in effect they describe rather than control the process of the meeting, for both the principles and the law only acknowledge what the meeting participants would do anyway.[2] The four principles are:

1. "Whoever comes is the right people," which reminds meeting participants, especially the ones in small groups, that accomplishing a task does not require a large number of persons with the chairman of the board present, but only the presence of persons who care to accomplish the task; in an Open Space meeting this essential care is demonstrated solely by one's showing up for it, as therein the direction or control exerted in a meeting organized in a traditional way are absent.
2. "Whatever happens is the only thing that could have," which keeps meeting participants focused on the here and now, and eliminates all of the could-have-beens, should-have-beens or might-have-beens, for what happens in the meeting is the only thing that this group, in this space, at this time could do; hence, once something has happened, it is done, and nothing can change it, thus fretting and/or complaining about it is pointless.
3. "Whenever it starts is the right time," which sensitizes meeting participants to the fact that inspired performance and genuine creativity "do not run on the clock"–they happen (or not) when they happen, and one cannot force their occurrence "on cue."
4. "When it's over, it's over," which reminds meeting participants not to waste time, for once a task is accomplished, the participants should move on to another issue,

instead of rehashing the discussion about the problem just because there is still some time left in the session–as in the motto "Do the work, not the time."[3]

The law is called the "Law of Two Feet" (or the "Law of Mobility"), which states that meeting participants can enter or leave an Open Space session as they choose, for whenever during a session a participant finds him/herself in a situation where she/he is neither learning nor contributing, he/she should use her/his two feet and go someplace else. Such a place might be another group, or even outside into the sunshine. What is important is not to stay in an environment that makes the participant feel he/she is spending his/her time unproductively.

One of the most significant charges of the law is to make it exquisitely clear that participants themselves are independently responsible for their own experiences in Open Space and for the quality of their learning, because if any situation does not meet their needs for either contributing or learning effectively, it is incumbent upon the individual participant to make it so. In this way, all participants are given both the right and the responsibility to maximize their own input and outcome whilst attending the Open Space event. Their choices and power to change meeting circumstances are not subordinated to any authority figure, but depend entirely on their judgment, thanks to which the passion and spirit of the group is able to circulate and generate the needed results.

However, Owen himself emphasizes the fact that neither the principles nor the law are the cornerstone of the Open Space Technology, but rather the phenomenon of self-organization. The main function of the principles and the law is to eliminate all the possible guilt coming from activities that meeting participants would do anyhow by naming and officially allowing these, in order to enhance the functioning of the group through the elimination of major chunks of guilt and blame. Considering, for instance, that every meeting starts when it starts regardless of what the clock says, and that bored meeting participants exercise the "Law of Two Feet" mentally, if not physically, without even knowing this law, it is fruitless to reproach oneself or anyone else for these mechanisms since they occur automatically. Thanks to the principles and the law, participants in an Open Space event invest their resources constructively in the process of self-organization, instead of wasting time and energy on focusing on any feelings of compunction or regret towards other meeting participants. Furthermore, the process of self-organization starts spontaneously, for the Open Space Technology is based on the premise that groups having a clear purpose and a willingness to engage the new will automatically self-organize to achieve their goals.

Harrison Owen believes that in the natural world "there is no such thing as a non-self-organizing system" (Owen *Opening Space*). Human beings as part of this natural world should therefore not resist self-organizing, for self-organization as the most natural phenomenon seems to have been in existence within the realm of human systems in form of everyday occurrence from the very first moment when these systems came into being. This leads to the conclusion that the significance of Open Space Technology as a technique solely based on the phenomenon of self-organization is considerably more than improving the quality of organized meetings, as it can become a key to overcoming the feeling of being lost in today's world. As Owen (2015) states: "It is not about having better meetings, although that certainly takes place. It is about experiencing the mystery and power of self-organization to the end that we might learn to be at home in this rather strange, possibly new, universe."

## Hospitable Space in the Service of Science: The World Café

The World Café is a participatory method discovered in 1995 by a group of scientists and entrepreneurs while they were meeting at home of Juanita Brown and David Isaacs in Mill Valley, California (Brown 2002, 4–7). The 24 participants in the meeting were holding a large-circle dialogue which, however, was disrupted by rain. At that, they spontaneously formed into smaller groups to lead intimate table conversations about the questions initially intended to be discussed in plenum; whilst talking with each other they were noting their reflections on makeshift paper "tablecloths." Periodically, they changed discussion partners by switching the tables in order to secure a "flow" of important ideas and insights between all of the meeting participants. Thereby, they launched an innovative conversational process: the World Café, which soon was enthusiastically incorporated into the structures of thousands of organizations all over the world, as disparate as large multinational corporations, small nonprofits (community-based organizations included), government offices, and educational institutions. The key to this success lies in the simplicity of the World Café process and the principles that underlie it, which makes this method applicable to various situations and problems, regardless of the region of the globe in which they occur.

Within the framework of the World Café participatory method, a café ambiance is created while meeting for discussion in order to facilitate collaborative dialogue among the discussion participants by creating a living network of conversation and action, one conducive to the sharing of knowledge and new ideas. The meeting participants discuss a topic in small groups at several café tables, switching the tables at regular intervals and, at each of their new tables, being introduced to the discussion previously held at this table by a table host who remains at his posting (Brown 2002, 4–7). This allows for the enrichment of the proceeding conversations with the ideas that other participants generated in former conversations. At the end of the discussion process, the main ideas are summarized at a plenary session and follow-up possibilities are analyzed.

The discussion proceeds during the Café event in several stages (rounds). At the beginning of the consecutive rounds, the table hosts welcome the new guests and briefly share the main ideas, themes and questions of the initial conversation. The guests should be encouraged by the hosts to listen carefully and build the conversation on the contributions of the other participants. This is why participants' listening skills are crucial in the World Café method. Because the participants switch among the tables in several rounds of conversation, the ideas coming from their previous and current table conversations begin to link and connect, thanks to which all of the conversations in the room become cross-pollinated with insights from the prior conversations already by the end of the second discussion round. In the third round, the participants either return to their original tables to synthesize their findings, or they continue travelling to new tables, leaving the same or a new host at the table. Sometimes a new question to deliberate is formulated for this round in order to deepen the ensuing exploration. After several discussion rounds, a plenary conversation is initiated with the purpose of sharing discoveries and insights; with this discussion stage being a town-meeting-style conversation, not only do thinking patterns among the group become identified, but also collective knowledge grows and possibilities for action start to be elucidated.

Two factors of are key to creating a hospitable space necessary for the development of the café ambience: an appropriate social atmosphere and a comfortable physical environment, both of which contribute to creating a "safe" space, where everyone feels free to be themselves and to offer their most creative thinking, speaking, and listening. To develop the appropriate social atmosphere means to give each participant in the Café a feeling of representing an aspect of the whole system's diversity and to offer each of the participants a chance to contribute to the conversation in order to make more of the intelligence inherent in the group accessible while discussing. To develop the appropriate physical environment, understood as warm and inviting, means to provide a space with café tables and several accessories (such as, for example, small bud vases with flowers, candles, posters) in order to allow the participants a more intimate atmosphere while discussing than is usually present at business or scholarly meetings. The fact that creating a Café ambiance is easy and need not be expensive is a great advantage of this participatory method.

Since its discovery in 1995, the World Café as an innovative approach to overcoming problems has been used to help solve critical issues in such diverse fields as, for example, socially responsible business, health care, education, environmental protection, social welfare, conflict resolution, and sustainable development, which turns the discovery of the World Café and its subsequent world-wide development into one of the most pivotal moments in the evolving history of eco-phenomenology.

## Intuitions of Life and Creation: Roots of Phenomenological Praxeology

Stefan Banach's Café Method of Research has become for me personally an inspiration whilst searching for the essence of creative work, so much so that I have been led to invent what I call the "3i" formula as part of an educational strategy that aims to release students' creativity. I first limned this educational method during a train journey from Olsztyn to Warsaw in 2003. Over the following years, this initial draft was developed into the full "3i" formula, becoming one of the basic methods of phenomenological praxeology (Bombała 2008).

The "3i" formula consists of three metaphorical concepts: interpretation, inspiration, illumination. Interpretation is a version of the hermeneutical method that helps to expand students' knowledge resources through their critical analysis of the literature. Inspiration, being a variant of the phenomenological method in which students conduct thorough case study research, is used to define problems while searching for their essence and thus serves as a tool that enables a student's acquisition of diagnostic skills. Illumination (creativity) is a heuristic method that helps students to start creative designing; effectively, it is used to strengthen students' faith in their own abilities and develop their competencies (student expertise) (Bombała 2012, 51–60).

One can distinguish two stages of the teaching process: a cognitive stage and a creative stage. The first has the function of exploring students' preexisting knowledge, and the second serves as a trigger of their creative potential. The method that allows for a combination of both stages, enabling thereby the inventive creating of reality, is the aforementioned "3i" formula, which can be used to develop creativity not only among students, but also among employees during an empowering exercise. The didactic process within the framework of the "3i" formula starts from interpretation and inspiration and proceeds towards illumination:

- Interpretation means the analyzing of different schools of thought, concepts, methods and techniques;
- Inspiration means conducting thorough case studies research (a work of art or/and nature can also become the source of inspiration);
- Illumination means searching for one's own vision of an organization while using methods of creative thinking (meta-cognition and heuristic techniques).

When the "3i" formula is used for diagnosis within and improvement of an organization, its design is exactly the same as that of the "3i" formula described above when applied to the didactic process.

Moreover, in management science "3i" describes three ways to found an organization:

- Illumination (revelation) describes the creation and development of the organization according to the autonomous vision of its founder; for example, J. Robert Ouimet's vision of a company (Bombała 2014b);
- Inspiration means the process of creating a business under the influence of certain ideas, beliefs, or patterns (such as the servant leadership concept, which has a religious inspiration);
- Interpretation describes the most common pattern for the establishment and development of an organization, one that is based on the currently dominant and fashionable theories (for example, reengineering).

The "3i" formula in management is a method that is used not only to define a problem within an organization, but also to detect the causes behind it. This method allows also for creative experimentation, within the framework of which I use, inter alia, Clark Moustakas' heuristic method (Bombała 2010). Moustakas' heuristics is a process of inner exploration and discovery of the meaning of experience, as well as selecting the direction for further research and designing. It is also a way to discover oneself and build dialogue with others. This method requires a large commitment whilst looking into a question, until the moment a brainwave rolls or an answer is obtained. Moustakas outlines six phases of heuristic research: initial engagement, immersion, incubation, illumination, explication, and the culmination of research in creative synthesis. During the synthesis of data obtained while researching, one takes into consideration interviews, notes, works of art (poetry), and personal documents (autobiographies) (Moustakas 1990, 27–37).

The ontological and epistemological assumptions of phenomenological praxeology are based on two philosophical approaches (Bombała 2014c). The first of these is intuitionism, especially the intuitionism of John Henry Newman who proclaimed that concrete knowledge and direct conviction (*assent*) were more important than scientific knowledge. The basis of conviction is direct intuition of reality and a special "inference sense" (*illative sense*), being an individual reasoning ability differing from reasoning by general rules and methods (Newman 1992). Newman's approach has been enriched by the philosophy of Henri Bergson, who claimed that there were two existing types of knowledge: rational knowledge (that is, mathematical natural science), and intuitive knowledge. Bergson critiqued the rational understanding of the world, which he admittedly found necessary, as being incapable of adequate recognition of reality. Rational cognition, using abstraction and analysis, decomposes reality into artificially isolated fragments; it immobilizes what is in fact a variable; it interprets movement as a sum of static states; it reduces qualitative transformations of things to quantitative differences; thus, it distorts reality. Thereby, this type of knowledge results in the reification and hypostatization of characteristics into independent beings, which is why the knowledge of mathematics and the natural sciences must not claim to ultimately explain reality. The human being has been equipped with two kinds of cognition: rational cognition, thanks to which he turns into a laborer transforming the world (*homo faber*); and intuitive cognition, thanks to which he becomes a wise man (*homo sapiens*). Intuition is cognition in duration (*intus legere*), for data obtained through intuitive cognition are difficult to formulate linguistically (as part of a spoken/written language) but can be captured and communicated through the use of metaphors and by means of artistic expression (Bergson 1946).

The second foundation of the ontological and epistemological assumptions of phenomenological praxeology is phenomenology, especially Anna-Teresa Tymieniecka's phenomenology of life, an extension of Heidegger's and Bergson's philosophies that has become an important component of phenomenological praxeology (Bombała 2014a). Tymieniecka states that rationality is not the sole way of cognition since there are instinctive as well as intuitive signposts in the course of life (Tymieniecka 2011, 15). According to Tymieniecka, the "passive" nature of classical phenomenology was unable to elucidate what reasons stand behind the creative activity of human being. Therefore, she introduces a new category (function) while describing the phenomenon of a human being's activity, namely, the human being's creative function, which exists beside the human being's constitutive function. This approach of Tymieniecka's can inspire further research in various fields of the humanities broadly defined, for assumptions concerning the creative nature of the human being may provide the basis for new developments of classical problems of philosophy, as well as of the social sciences and art. Tymieniecka's distinction between "constitutive" and "creative" subjectivity became a source of inspiration for me while developing a theory of phenomenological praxeology that covers both the philosophy and praxis of management, one that, in contrast to the theory of classical praxeology, is based not solely on rational formulation but also and mainly on intuitive cognition.

## Eco-Phenomenology of "Learning-To": *To Be an Aficionado*

Eco-phenomenology of scientific activity also places the process of education into a new perspective. Daniela Verducci states that:

> It would seem that the task of education is no longer simply to articulate and develop a paradigm of humanity, known and shared in its essentiality, as in the case of the Greek *paideia* or of the *animal rationale* of the Latin tradition, but to promote man's capacity for transcending and surpassing himself ... This capacity for transcendence, however, does not anchor him to any stable idea of himself, but, rather, continually pushes him beyond himself, positioning him before the ever new challenges that history sets before him. (Verducci 2008, 23)

Nevertheless, in the current model of education, students' cognitive development, rather than their self-realization, is the preferred objective of the didactic process. Thus, a new model of education is needed in which one would aim for the full development of students, that is, with students' cognitive development occurring inseparably from their self-realization. Furthermore, during the self-realization process, students must feel free to search for the meaning of a phenomenon and to be a self-interpreter of their own experiences (Selvi 2014, 256). As noted by Carl Rogers, searching for meaning and relying on process rather than on static knowledge is the foundation of education in the modern world. This approach is called metacognition, which is defined as "thinking about thinking," and helps students while learning (Ridley et al. 1992, 293–306). Metacognition focuses on self-reflection. Reflective thinking allows students (or employees) to understand the undertaken actions and to develop effective patterns of action, i.e., to acquire new knowledge, skills and competencies. The most commonly used techniques are "learning-by-doing" and "experimental learning." By knowing how to learn, and what strategies are the best to use, learners acquire valuable skills that distinguish "student experts" from "student beginners."

Carl Rogers perfectly captured the essence of the analyzed problem by explaining that the initiation of such learning is not based only on a leader's teaching skills, scientific knowledge, program planning, the use of audio-video equipment, a strict program of teaching, lectures, or masses of books (although each of these can be used as an important resource). Such learning is based on certain qualities relating to the attitude that occurs in the personal relationship between the teacher and the learner (Rogers 1983, 105–106). The first of these qualities that facilitate learning is the authenticity of the teacher, which means that the teacher is able to manifest feelings and communicate them. This quality is the foundation of the teacher's veracity and credibility. As a result, the teacher is able to bring a genuine engagement with learners into the didactic process. The second quality is an attitude of valuing, acceptance and trust, which facilitates the learning process. This attitude allows for acceptance of the fear and hesitation that a student may feel when a problem arises, as well as acceptance of the satisfaction coming from an accomplished task. Regarding the third quality, Rogers states that trust creates a climate for self-initiation. Such creation of the proper climate is based on empathic understanding. This approach triggers enormous emotional and intellectual potential.

Moustakas approaches this issue similarly. He developed a profile of an "ideal" teacher. This teacher recognizes his student as a unique person, gives the student the feeling of being a special individual and therefore incomparable to others. Communicating with students in a direct and honest language, this teacher creates an atmosphere of freedom, openness, trust, willingness to disclose one's own thoughts and feelings. Such a teacher enters the student's world in order to understand, affirm, and encourage students to create space for their development – the teacher recognizes his/her students' right to make choices (Moustakas 2001, 54–55).

Max Scheler's theory that love and hate are primary acts in relation to the knowledge of an object is of significance in the eco-phenomenological context of "learning-to." A person reacts emotionally with the primary feelings toward the object of cognition, i.e., with love or hate (Scheler 1986, 228–318). Only on this basis is true cognition possible. Scheler states that the aficionado is always the forerunner of the researcher and therefore deserves special attention. This statement contains a deeper meaning, because it is not purely sensual feeling, but rather an emotional spiritual experience whose qualitatively ideal equivalent are values. Only then can values be defined in a clear and direct way. Acts of love and hate are the prerequisite and foundation of all other acts of emotion, on the basis of which values are defined (Scheler 1987, 272–275). One can consider Scheler's positing the act of love as the cause of knowledge and creativity to be the foundation of eco-phenomenology of "learning-to," for love always activates cognition and volition and is the mother of the spirit and the intellect.

I try to use an approach similar to those of Rogers, Moustakas, and Scheler whenever I work with students. I carry out common searching with them, in which we use the "3i" formula. During workshops, students form research teams and prepare panel discussions based on literature analysis and their own research. The aim of each team is to present the obtained knowledge in an interesting way to the other members of the group. However, the most successful project has been a discussion club and the student magazine "3i." It was an informal group, unlimited by any regulations, and its aim was to develop the scientific, social, and cultural interests of its members. Similar to Banach's group that met in the Scottish Café, the place of our meetings was the Irish Pub Carpenter Inn in the Old Town of Olsztyn, located next to the Castle which once was the residence of Nicholas Copernicus. The result of our stormy meetings was six issues of "3i" magazine, published in Olsztyn in 2007.[4]

One great reward for my work was a letter that I received from one of my students. Here are some excerpts: "... I was your student in 2003–2008. … I have used the period of the several months that have elapsed since my graduation till the moment came in which I decided to sit down at this so soulless machine to write you a letter, to deliberate about the five years of my studies at the Institute of Political Science at UWM. One of the highlights of what I discovered in my memory was meeting you, an event that influenced me as a still emerging human being. Inspiring conversations within the framework of the "3i" group, interesting classes and your approach to students permeated my mind and my soul. If one of the most important tasks of universities is to create a man in a way that makes him become a better and more valuable human being, then, in my opinion, you are a great exponent of this noble idea. I want to thank you for all of your work. I regret that I did not do it in person. On the other hand, perhaps the time which has passed has let me look at it all from a distance and allowed me to get rid of my pride and arrogance. Maybe because of this my thanks are sincere and coming from the depths of the soul. … To conclude my letter, I would like to send you warm greetings. Also, a particular reflection has come to my mind. The statement that I was your student is off the mark – I still am your student."

## Eco-Phenomenology of "Learning-To" – Case Study: The Priorities of an Ethical Entrepreneur

Part-time students who are entrepreneurs or managers are asked to submit written reflections on their everyday work. The submission of the reflections is voluntary. I enclose below fragments of a paper written by Monica Biełous, a political science student at the University of Warmia and Mazury:

> The family farm is an unusual kind of productive enterprise. … We inherited our farm from my husband's parents, who even in their retirement were still for few years helping and advising us during our work on the farm. We have over two hundred hectares of arable land and grassland. We grow wheat, canola and corn. We use some of these crops to make feed for dairy cows which we breed …
>
> It is important for my husband to have good understanding, and relations in general, with our employees. … The small number of staff members has given us a good opportunity to get acquainted with their family situations, as well as their aspirations and expectations related to work on our farm. On the farm, employees are actual co-workers, because all tasks which are entrusted to them are also performed by their employers – from sweeping the yard where machines are kept to using these machines while working in crop fields. Our employees have thus an opportunity to learn and improve their skills across a very wide range: from animal husbandry and cultivation of the soil, to handling increasingly complex machines and devices – and not only agricultural ones.
>
> Considering all the problems that occur while working on the farm and trying to motivate employees, I decided out of curiosity to apply the principles of the '3i' formula in our

"own backyard." I must admit that initially I was quite skeptical about this method. ... In a nutshell, the main idea of this method is that whenever there is a problem, we are to realize all the possible solutions, and search for inspiration in the world around us for at length until one conceives by using creative thinking one's own vision of the best organizational model of the business that one leads.

... For we are friends with many other owners of farms, and I have an opportunity to talk with them about how they deal with similar problems. Usually, the only solution used by them is that of terminating 'bad' employees and hiring new ones, often equally problematic ones. For me, however, it was not the right solution. My husband accidentally became a source of inspiration as he, being upset because of some misunderstanding at work, said that every employee should be put in the boss position once in order to be able to fully feel the burden of responsibility for an enterprise that provides for both the owner and the employees. Then I came up with the idea of using an empty livestock building, which we had purchased recently, as a place for rearing young dairy cattle. We decided with my husband to inform our employees that this project will be possible only if one of them takes full responsibility for taking care of the animals and the building, since my husband and I had already too many responsibilities to take up a new one .... This idea met with approval from the staff members, and one of them, a young, fledgling husband and father, willingly agreed to undertake the new task. It quickly became clear that such a commitment may cause great psychological stress. Nevertheless, soon it turned out that the satisfaction coming from his own, clearly defined workplace, together with the greater than previous importance of the performed tasks became for him not a source of discouragement, but of proper motivation. This significantly increased his willingness to work and, consequently, his work effectiveness....

When under constant pressure connected to external factors, it is not easy for one to always maintain appropriate relations between employers and employees during everyday work. However, my experiences in being an employer, as well as the experiences of other farm owners known to me, have showed that the most important thing is to find a way to reach an agreement with your employees. ... Being a 'boss', even of the smallest number of people, means great responsibility for employees' existence and the future of their families. The desire to increase productivity in business should go hand in hand with the desire to increase the workers' standard of living, constant improvement of their skills, and to make them take satisfaction and even pride in being our employee. In our local, rural environment, the reputation of the farmer is of crucial import. If the opinion of a particular farmer is flattering, then the job satisfaction received while being his or her employee is a sufficient motivation to work efficiently. One works on one's own reputation as farmer for many years, and often 'inherits' this reputation from parents, or even grandparents.

...The measure of success in dealing with employees is the fact of seeing the same satisfaction as shows on our own faces on their faces when they hold wheat grain in their hands during harvesting, even despite being tired from working in summer heat for many hours.

## Conclusion

Eco-phenomenology is a reflection on the human attitude toward the natural environment and the human being himself, within the framework of which one analyzes contemporary research methods. The phenomenology of Anna-Teresa Tymieniecka, which concentrated on "creative experience," has significant meaning for these analyses. According to Tymieniecka, phenomenological research should be focused on will, imagination, and creativity – cognitive powers that exceed the limits of pure rationality. Stefan Banach's Café Method of Research, characterized by "initial spontaneity," is an example of a research method that fulfills these criteria, being at the same time a form of protest against the routinization and bureaucratization of science. Banach's approach to research work is a source of inspiration and currently has been revived in such methods of research as Open Space Technology, the World Café, and the "3i" formula. Eco-phenomenology of scientific activity also places the process of education in a new perspective, proposing research and teaching methods that allow for the empowerment of students by making them active participants in the didactic process.

## Notes

1. The damages and injuries in the complete ecosystem of the Earth in the long run endanger the sustenance of human life, on all levels of Abraham Maslow's hierarchy of needs. A phenomenological approach shows that the instrumental approach to needs fulfillment provides only a narrow, limited segment of the entire, possible richness of the experience of fulfillment and satisfaction. If a person has an instrumental attitude toward needs fulfillment, the process of satisfaction is sustained only by those factors that are immediately responsible for the current process of fulfillment. Accordingly, for a person who attributes self-value to the sources of his/her satisfaction and so respects the being in itself of nature and of another person, satisfaction is derived from a wider segment of nature or the entire complexity of the other person. That person who in her/his relationship to nature and to another person takes into account the being in itself (the self-value) of the other person, evolves in him/herself also a special sensitivity to the complexity of the being of nature or the being of the other person: "By virtue of this attitude the person in question acquires a more intimate relationship to nature and to the other. In this more intimate relationship, many more factors participate in the process of her/his needs fulfillment than is the case when a merely instrumental attitude is taken. Here is a much richer experience of the other and of nature than one could have had via an instrumental treatment of the sources of satisfaction. … The satisfaction that a person has through an … eco-centric relationship to nature and to the other involves a much richer, 'multichannel' communication with the other person and with nature. An eco-centric … attitude unfolds complete horizons of experience. [Whereas] the instrumental attitude … cuts off complete layers from the experience of our natural embedment and of the entire personality of the other, therefore it cannot provide a principle or maxim for the guidance of one's life" (Marosan 2013, 148–150).
2. See Harrison Owen, 2015. Opening Space for Emerging Order. at: http://www.openspaceworld.com/brief_history.htm

3. A potential fifth principle is "Wherever it is, is the right place," which makes meeting participants aware and conscious of the fact that space is opening everywhere and all the time.
4. See: http://leksykonkultury.ceik.eu/index.php/3i; http://polskaprasa.republika.pl/tytuly/3i.html

## Works Cited

Abram, D. 1997. *The Spell of the Sensuous. Perception and Language in a More-than-Human-World*. New York: Vintage.

Barral, M.R. 1991. Creativity and the Critique of Reason. In Husserl's Legacy in Phenomenological Philosophies. *New Approaches to Reason, Language, Hermeneutics, the Human Condition*, Book III, Analecta Husserliana XXXVI, ed. Anna-Teresa Tymieniecka, 177–188. Springer Netherlands: Kluwer Academic Publishers.

Bergson, H. 1946. *The Creative Mind: An Introduction to Metaphysics*. New York: Philosophical Library.

Bombała, B. 2008. In Search of a New Model of Education. In *Education in Human Creative Existential Planning*, Analecta Husserliana XCV, ed. Anna-Teresa Tymieniecka, 383–393. Dordrecht: Springer.

———. 2010. Phenomenology of Management. *Leadership* [in Polish] Warsaw: Difin.

———. 2012 Phenomenology of Management – Didactic Aspects. *Management and Business Administration. Central Europe* (MBA.CE) 3: 51–60.

———. 2013. Phenomenology of the Management as the Eco-Empathic Leadership. In *Phenomenology and the Human Positioning in the Cosmos: The Life-world, Nature, Earth:* Book Two, Analecta Husserliana CXIV, ed. Anna-Teresa Tymieniecka, 203–218. Dordrecht: Springer.

———. 2014a. Anna-Teresa Tymieniecka's and Max Scheler's Phenomenology as the Ontopoietic Genesis of a Manager's Life. In *Phenomenology of Space and Time: The Forces of the Cosmos and the Ontopoietic Genesis of Life,* Book I, Analecta Husserliana CXVI, ed. Anna-Teresa Tymieniecka, 155–165. Dordrecht: Springer.

———. 2014b. From Harmel to Ouimet – Ethical Entrepreneurship in Practice. *Annales. Ethics in Economic Life* 17: 55–73.

———. 2014c. Phenomenology As the Epistemological and Methodological Basis of Management Sciences. *International Journal of Contemporary Management* 1: 150–172.

Brown, J. 2002. *The World Café: A Resource Guide for Hosting Conversations that Matter*. Mill Valley: Whole Systems Associates.

Brown, Charles, and Ted Toadvine, eds. 2003. *Eco-Phenomenology: Back to the Earth Itself*. Albany: SUNY Press.

Ciesielski, K. 2007. On Stefan Banach and Some of His Results. *Banach Journal of Mathematical Analysis* 1: 1–10.

http://leksykonkultury.ceik.eu/index.php/3i.

http://polskaprasa.republika.pl/tytuly/3i.html.

Marosan, Bence Peter. 2013. The Primal Child of Nature – Towards a Systematic Theory of Eco-Phenomenology. In *Phenomenology and the Human Positioning in the Cosmos: The Life-world, Nature, Earth:* Book Two, Analecta Husserliana CXIV, ed. Anna-Teresa Tymieniecka, 119–153. Dordrecht: Springer.

Moustakas, Clark. 1990. *Heuristic Research, Design, Methodology, and Applications*. Newbury Park: Sage.

———. 2001. *Phenomenological research methods* [Polish transl.] Białystok: Trans Humana.

Newman, John Henry. 1992. *An Essay in Aid of a Grammar of Assent*. Notre Dame: University of Notre Dame Press.

Owen, Harrison, ed. 1985. *Tales From Open Space*. Potomac: Abbott.

———, 2015. *Opening Space for Emerging Order.* at: http://www.openspaceworld.com/brief_history.htm

Ridley, Dale Scott, Paul A. Schutz, Robert S. Glanz, and Claire E. Weinstein. 1992. Self-regulated learning: The interactive influence of metacognitive awareness and goal-setting. *Journal of Experimental Education* 60: 293–306.

Rogers, Carl. 1983. *Freedom to Learn for the 80s*. Columbus: Charles E. Merrill.

Scheler, Max. 1986. *The Nature of Sympathy* (Polish transl.) Warsaw: PWN.

———. 1987. *Letters of Philosophical Anthropology and the Theory of Knowledge* (Polish transl.) Warsaw: PWN.

Selvi, Kiymet. 2014. Educational Paradigm Shift Towards Phenomenological Pedagogy. In *Phenomenology of Space and Time: The Forces of the Cosmos and the Ontopoietic Genesis of Life,* Book I, Analecta Husserliana CXVI, ed. Anna-Teresa Tymieniecka, 245–258. Dordrecht: Springer.

Tymieniecka, Anna-Teresa. 1988. *Logos and Life*. Dordrecht/Boston: Kluwer Academic.

———. 2011. *The Fullness of the Logos in the Key of Life.* Book I: *The Case of God in the New Enlightenment*, Analecta Husserliana C. (Polish trans.). Poznań: Wydawnictwo Poznańskie.

Ulam, Stanislaw. 1969. Wspomnienia z Kawiarni Szkockiej. *Wiadomości Matematyczne* 12: 49–58.

Verducci, Daniela. 2008. Education and the Ontopoietic Conception of Life. A Neoanthropological Perspective. In *Education in Human Creative Existential Planning*, Analecta Husserliana XCV, ed. Anna-Teresa Tymieniecka, 23–37. Dordrecht: Springer.

# Eco-Phenomenology: The Japanese Original Perspective in the Thought of Nishida Kitaro

**Valentina Carella**

**Abstract** Eco-phenomenology developed from the effort of a number of continentally-oriented philosophers exploring the thought of decisive authors in the phenomenological tradition, such as Husserl, Heidegger, and Merleau-Ponty, with the purpose of offering a different insight into environmental issues than those predominant in Anglo-American philosophy. This initiative has proceeded not only from Western scholars but has had a resonance also in the distant philosophical tradition of Japan. The present contribution seeks to deepen the thought of a central figure for Japanese phenomenology: Nishida Kitaro. Nishida, indeed, was the first to introduce Husserlian thought in Japan and to seriously dialogue with the German phenomenologist from a Zen-oriented point of view. The purpose of this effort is to throw light on the theoretical origins of modern Japanese (eco)phenomenological thought, in order to grasp also its differences with Western eco-phenomenology, as led by Nishida's Buddhist reading of Husserl's works.

**Keywords** Nishida Kitaro · Japanese eco-phenomenology · Intentionality · Consciousness · Ishikiserareta ishiki · Ishikisuru ishiki · Zen Buddhism · Basho · Historical world · Absolute nothingness · Environment

## Introductory Remarks

The general meaning of eco-phenomenology is almost self-evident. By eco-phenomenology we usually mean that contemporary philosophical tendency to refer to classical phenomenological works – Husserl, Merleau-Ponty, and Heidegger – to advocate for a different way to approach environmental issues, such as environmental ethics (Brown- Toadvine 2003, 73). But how much is this tendency diffused? Can it be considered only a fascination of European

V. Carella (✉)
Università degli Studi di Macerata, Macerata, Italy
e-mail: valentina.carella.studisuperiori@unimc.it

 
W. S. Smith et al. (eds.), *Eco-Phenomenology: Life, Human Life, Post-Human Life in the Harmony of the Cosmos*, Analecta Husserliana CXXI,
https://doi.org/10.1007/978-3-319-77516-6_24

continental philosophy? The answer is negative. Eco-phenomenology is, from a geographical point of view, a widespread phenomenon, crossing different borders and cultural traditions.

A very distinct eco-phenomenological approach is the Japanese one. When we engage with Japanese eco-phenomenologists, indeed, we are confronting something different from Italian, Russian, or even Australian eco-phenomenological contributions. As Anna-Teresa Tymieniecka and Michel Masson showed, "Japanese scholars are not experts in Japan of a philosophical method which originated within another tradition. This means that Japanese phenomenologists have not confined themselves to the role of mere interpreters of an external philosophical tradition: on the contrary, phenomenological method has been appropriated by Japanese scholars in the light of their specific cultural tradition, with a central reference to Zen Buddhism" (Tymieniecka and Masson 1979, viii). So here lays the distinctive feature of Japanese eco-phenomenology: it is a philosophical tradition in which Husserl's thought is reread within the traditional Japanese intellectual legacy.

Thus, in order to properly understand Japanese eco-phenomenological positions and to grasp their distinctive features, it is important to clarify the terms of the first dialogue between Japanese philosophy and Husserl's project. This is precisely the aim of the present article: to throw light on the way phenomenology was introduced in Japan, understanding some eventual variance that may descend from the Buddhist legacy, with the purpose of identifying some eventual differences from the occidental eco-phenomenological approach. More specifically, this article will focus on Nishida Kitaro's philosophical perspective. The reasons for this choice are shown in the following section. Section three will focus on Nishida's theoretical dialogue with Husserl, while in the fourth section we will try to understand what consequences descend from this theoretical dialogue in the comprehension of the human-environment relation and, consequently, in contemporary Japanese eco-phenomenology.

## The Phenomenological Relevance of a Non-Phenomenologist

Nishida Kitaro (1870–1945) is generally considered the first Japanese who built up a philosophical system in the proper sense. Before 1874, in which year Nishi Amane coined the word *tetsugaku*, it was impossible in Japan to find even an expression to translate the Western word *philosophy* (Piovesana 1968, 18–19). However, as Ōhashi pointed out, after 1874 even those who taught at the "tetsugaku department" were not really philosophical thinkers. Their works were just an eclectic mixture of Buddhist and Western philosophy, having as a central reference Hegelian idealism. The first who understood that philosophy could dialogue with Buddhism while remaining something different from it was Nishida Kitaro (Bertossa 2015, part 1). This is the reason why people used to regard Nishida as the first original philosopher in Japan: he was the first who clearly distinguished philosophy from Buddhism. Nevertheless, Nishida's philosophical position can still be defined, to quote Abe

Masao's definition, as a "Zen-oriented philosophy," as it was an attempt to investigate, from a philosophical point of view, the existence of an ultimate unitary absolute dimension, not transcending reality but encompassing all its contradictions (Abe 1990, xiii). This tendency to absolute unity is still the main common characteristic of the Kyoto School, as the followers of Nishida's thought are called.

Thus, we see that Nishida is not directly linked with contemporary Japanese phenomenology and eco-phenomenology; he cannot be considered the father of the Japanese phenomenological approach – or more precisely, not so in the same sense that we ascribe to Husserl the general paternity of phenomenology. Nishida, indeed, did not formulate a Japanese phenomenological method matching the Husserlian one with the Buddhist intellectual legacy. Neither he can be considered a phenomenologist, nor did he define himself in this way (Tymieniecka and Masson 1979, ix).

Nevertheless, it stands to reason that it is impossible to understand some characteristics of Japanese phenomenological debate while disregarding Nishida's thought. Nishida, indeed, was the first in Japan who understood the relevance of Husserlian phenomenology for contemporary philosophy and for this reason it was he who first introduced Husserl's works into Japan (Nitta et al. 1979, 8). Thus, even though Nishida is not directly linked with Japanese phenomenology, he certainly had an indirect influence on the development of the Japanese phenomenological trend. This is not only because he was the one who made the Japanese philosophical debate pay attention to Husserl's works. Moreover, this was because he was the first in Japan who seriously confronted the philosophical method rolled out by Husserl.

Therefore, in Nishida's works we can find the first philosophical attempt to reread phenomenology from a Buddhist and Zen-oriented point of view. The result was that, as noted by Tymieniecka and Masson, this dialogue let emerge issues that would become crucial in later debates in Japanese phenomenology (Tymieniecka and Masson 1979, ix). In other words, since Nishida was the first who confronted himself with Western phenomenology, he became the lens through which subsequent Japanese phenomenologists approached the main phenomenological issues. Here we see the reason why it is important to deepen understanding of Nishida's thought in order to grasp the main distinctive features of Japanese phenomenology. As a matter of fact, it is from apprehending Nishida's account of phenomenology that we understand some of the main thematic tendencies of Japanese phenomenological debate and, above all, the theoretical reasons why some lines of thought have prevailed over others.

## Nishida on Husserl Phenomenology: A Theoretical Dialogue on Consciousness

### *Appreciation of the Unity of Intentional Consciousness*

In the previous section we have seen that Nishida was the figure who elaborated the theoretical framework within which subsequent Japanese phenomenologists have read Husserl's works. The analysis of the main lines of this framework is the specific object of the present section. In other words, we are now going to focus on Nishida Kitaro's position on phenomenology, retracing the more relevant points of his dialogue with the Husserlian outlook.

Methodologically, we firstly need to say that Nishida, in all likelihood, did not know all the stages of Husserlian thought. His references, indeed, were always directed to the early major works of Husserl, specifically to *Logical Investigations* and the first volume of *Ideen*. As a result of this, we can argue that Nishida mainly focused on what Merleau-Ponty called the first period of Husserlian thought wherein Husserl's main concerns were with the themes of essence and intentional consciousness (Husserl 1950, 87). This became the pivot of the dialogue between Nishida and Husserl.

Nishida's position on phenomenological intentionality is a double-faced position. On one hand, as Jacynthe Tremblay has pointed out, Nishida appreciated Husserl's definition of intentionality as a good source of unity for the manifold and sometimes contradictory life of consciousness (Tremblay 2000, 59). In his first famous work, *Zen no kenkyū* [善の研究] (translated and published in English as *An Inquiry into the Good*), we read the following passage: "In order to clarify why the reflective activity emerges, we need to start from the evidence that, as I previously said, consciousness is originally a unique system whose natural state is a spontaneous developing and fulfilling. However during this process some contradictions and collisions emerge. … But also those states which contradict and collide with each other, from another point of view may immediately be considered the starting point of another and even bigger systematic development; … In this case we deal with a still uncompleted state of a grand unification … [as] behind contradictions and collisions stands the possibility of unification" [my translation] (Nishida 1965a, I 24).

Here Nishida does not directly echo Husserl's words on intentionality. However, Husserl's close readers will readily recall here the definition that Husserl gave of intentionality in the first volume of *Ideen*: "intentionality is what characterizes consciousness and what allows considering the stream of *Erlebnisse* in the unity of consciousness" [my translation] (Husserl 1950, 3, 283). Here we see that Nishida's view on consciousness is close to Husserl's: both think that consciousness is that operative unity which allows the subject to have different states of mind.

Moreover, if one dislikes indirect association, fearing the risk of mere juxtaposition, we can find in Nishida's following works direct appreciations of Husserlian intentionality as a good conceptual scheme that allows us to identify a subjective unity standing behind all the multiform and maybe contradictory stages of the stream of consciousness, while at the same time making them possible. As a matter of fact, in his important second work *Jikaku ni okeru chokkan to hansei* [自覚に於ける直と反省] [Intuition and Reflection in Self-consciousness], Nishida wrote: "Husserl interestingly distinguishes various worlds according to 'intentionality of consciousness': when we are in a mathematical attitude, the world of mathematics is there for us; when we adopt the natural attitude, the natural scientific world is there for us; and all these worlds are embraced by a Cartesian cogito" (Nishida 1965b, II 71).

Here we read, directly from Nishida's words, appreciation for phenomenological intentionality as it brings all possible different conscious attitudes into a unity. Thus, we can validly affirm that Nishida called on phenomenological intentionality because, since it unfolds a unitary vision of consciousness, it satisfied in a certain sense Nishida's Zen-oriented need to find the unity that stands behind the multiform and contradictory manifestation of reality, making them possible.

## A Still-Dualistic Consciousness

We have noted that Nishida shows almost complete appreciation of the concept of intentional consciousness. Now it is why this is "almost complete" that we have to focus on. Obviously this means that Nishida did not embrace *in toto* Husserl's vision of consciousness, that he had some skepticism about it. This less than complete satisfaction emerges in a letter Nishida sent in 1922 to his student Kiba Ryōhon, who had gone to Freiburg with the purpose of directly following Husserl's lectures. In this letter the father of the Kyoto School wrote: "How is phenomenology? Is it too difficult? In constrast to Rickert's School, which focuses only on logical structure, phenomenology refers also to some element of experience. … However, I wonder, if it continues to develop itself in this static manner, what would become of it. As Goethe made fun of Mendelssohn, would phenomenology end up killing the butterfly in order to capture its beauty?" (Yusa 2002, 179).

Through an evocative metaphor, Nishida tried to express his doubts concerning exactly the starting point of phenomenology: that is, the stiffening in the analysis of the structure of intentional consciousness. Nishida denominated this stiff view of consciousness with the Japanese idiom *ishikiserareta ishiki* [意識せられた 意識]. In his work titled *Ishiki no mondai o mushi* [意識の問題を無視] [The problem of consciousness] he explicitly linked this idiom with phenomenological intentionality; as he wrote, "Husserlian phenomenology clarifies the structure of *ishikiserareta ishiki*" [my translation](Nishida 1965f, VII 217). As Mayuko Uehara pointed out, *ishikiserareta ishiki,* together with other Japanese neologisms, presents some difficulties in translation, as it has no exact equivalent in Western languages (Berthon 200). However, even graphically, it is also possible for non-Japanese speakers to notice that the expression *ishikiserareta ishiki* is composed of nominal elements. The first, *ishiki* (意識), means precisely "consciousness," while *serareta* (せられた) is a nominal suffix often used to suggest the passivity of the term it is combined with. Thus, *ishiki-serareta ishiki* is a consciousness (*ishiki*) made passive (*serareta*) by consciousness itself (*ishiki*).

In order to clarify his terminology, Nishida borrowed Husserl's distinction between *noesis* and *noema*: intentionality – said the Japanese philosopher – is consciousness in the accusative position, consciousness intended as a *noema* (Nishida 1965d, V 435). This borrowing makes it easier to understand that by *ishikiserareta ishiki* Nishida means a kind of consciousness thematized by consciousness: consciousness not in itself but in the way it is thought by consciousness itself – a consciousness that is the object of consciousness. In this sense it is a *noema*. In other words, *ishikiserareta ishiki* is consciousness as it appears to and as it is understood by consciousness itself.

So *ishikiserareta ishiki* is an objectified unity of consciousness and from this evidence descends Nishida's critique of Husserl's approach: as phenomenology starts from the reduction on pure subjectivity, and as it is the subject intended as pure intentional consciousness, and as the latter is an objective kind of consciousness, phenomenology has an abstract starting point that prevents its catching the deepest original unity of reality. Thus, intentionality is not the deepest unity that Nishida was looking for. On the contrary, it still presupposes another unity in the background. As a matter of fact, every *objectified* item needs an *objectifying* pole: every *noema* – as Husserl already noticed – refers to a *noesis*.

As a consequence of this, Nishida affirmed the exigency to reconnect intentionality to the pole from which it was generated in order to reach a deeper unitary consciousness. Nishida, indeed, defined this process of seeking the deepest unity as a "transcendence in the noetic direction" (Nishida 1965d, V 84). This led to Nishida's concept of *ishikisuru ishiki* (意識する 意識). Comparing *ishikisuru ishiki* with *ishikiserareta ishiki,* it becomes evident that the difference between them, from a linguistic point of view, is contained in the suffix of the first occurrence of *ishiki*. Thus, in order to understand the proper meaning of *ishikisuru ishiki* and its relationship with *ishikiserareta ishiki*, we need to focus on the suffix *–suru* (する). In her analysis of the expression, Uehara remarked that, in contrast with the use of –*serareta*, *–suru* is used to transform a nominative element into an active verb (Berthon and Gossot 1998, 201). As a result, *ishikisuru ishiki* is consciousness apprehended in its activity, or better, is consciousness taken *as* activity. It is precisely this activity that allows for unity. Indeed, in *Intuition and Reflection in Self-consciousness,* Nishida recognized that every activity includes in itself the opposition between subject and object, as activity is a unique event with two sides, which are both parts of it: a subject who acts and an object which is acted upon (Nishida 1965b, II 71). As a consequence, *ishikisuru ishiki* – or consciousness as activity – is the unity that can hold together the consciousness reflecting on itself and the result of this reflection, objectified consciousness: *ishikiserareta ishiki.*

Here we touch the core of the dialogue that Nishida had with Husserl's phenomenological approach. The latter, with its concept of intentionality, cannot provide an adequate understanding of consciousness itself, as it provides a stiffening definition of it. Nevertheless, upholding a transcendental *Ego* is still a good way of unifying the multifold and sometimes contradictory states of consciousness. As Tremblay pointed out, Husserlian phenomenology can do nothing more than radically deepen the representative consciousness in the sphere of intelligible *noesis*; in effect, phenomenology analyzes the intellectual subjectivity and not the subject in itself, in its total being: subjectivity taken as activity (Tremblay 2000, 82). Proceeding from the evidence just uncovered, we can convey that Nishida considers phenomenology to be a good but provisional approach, as it maintains the dichotomy between the reflected objectified pole and the reflective pole. Thus, phenomenology needs to be overcome by a new philosophical perspective wherein all the dichotomies are encompassed in an original unity. That is to say, in the philosophical quest for the deepest unity, phenomenology can be considered simply as a phase, as a stopover from which we need to start searching again.

## Being in the World As Being the World

Phenomenology in Japan was read within the theoretical framework briefly presented above. Here it is important to clarify that this is not the Japanese phenomenological scaffold but rather the Japanese phenomenological background. As stated in the first section, Nishida was not a phenomenologist nor can be considered the father of Japanese phenomenology. However, because he introduced Husserl in Japan, he pinpointed those elements which may be critically considered from a Japanese Buddhist point of view. Succeeding Japanese phenomenologists could not but take into account Nishida's remarks on Husserl's general method. In this way it becomes clear, for instance, why Japanese phenomenology tends to radically stress the relevance of action over reflection (Tymieniecka and Masson 1979, ix; Dilworth 1979, 249–250).

Furthermore, it can be demonstrated that Nishida set up not only the theoretical scaffold of Japanese phenomenology but also the practical one, which is a specific concern for the hermeneutic of subject-environment dynamics. We have seen that phenomenology is considered by Nishida to be a provisional perspective which exposes a certain unity of the subject that cannot be considered to be the ultimate and deepest unitary fundament for which Nishida was looking. That deep unity of reality is what Nishida called Absolute Nothingness (or, taken in its positive sense, *ishikisuru ishiki*) (Nishida 1965f, VII 221). As ultimate unity it encompasses all the oppositional poles of reality, not excluding the subject-object opposition. As a consequence, both the subject and the external object have to be considered as different manifestations of the same actuality: in effect, Absolute Nothingness.

The first thing we can notice is that moving away from the formal intentional perspective of phenomenology does not imply only the primacy of action over reflection but, moreover and more radically, the denial of that relational unity which was a crucial point in Husserl, especially for the understanding of the subject-world relationship. As pointed out by Derrida in his famous introduction to Husserl's work *The Origin of Geometry*, indeed, from a phenomenological point of view, the absolute is neither objective nor subjective: the absolute is the pure relationship between the subject and the object, in which relation the subject and the object generate and hold each other (Derrida 1978, 203). As a result of this, phenomenology affirms the impossibility of reducing either the subject's constitution of the world to the objective structure of reality (as realism held) or that reality to the subject's constitutive disposition (the idealist position). Consequently, a phenomenological approach to environmental issues starts with acknowledgement that neither can the world be reduced to the constitutive acts of the subject – as Husserl noted in his *Ding und Raum*, the world is a transcendence for the subject – nor can subjective consciousness be shrunk to a mere receptor and recorder of an alleged objective structure of the world (Husserl 1973, 16, 17–19).

Now, Nishida's major work *Basho* 場所 [Place] opens with the following statements: "Saying that the object transcends intentional acts, which for us are inside consciousness itself, is unthinkable and not only for saying that our conscious content indicates an object, for it is even impossible to affirm that the object transcends our intentional acts.... In order to establish a subject-object relationship, there must be something else which contains both of them in itself" [my translation] (Nishida 1965c, IV 210). Here Nishida is quite explicit in his confrontation with Husserl's position: the relational unity between the subject and the transcendent object cannot be sustained from within consciousness (a first argument: *pars destruens*). Rather, it requires a third unitary dimension which stands behind them (a second argument: *pars construens*).

Thus, as with intentional consciousness, Nishida does not totally reject object transcendence, but rather brings it back to a deeper unitary vision. However, bringing the subject-object relation into a whole that contains both the terms of this relationship has a great consequence, which is contained in this brief remark extracted from *Zettai mujunteki jiko dōistu* [絶対矛盾的自己同一] [*The absolute contradictory self-identity*]: "the historical world is the place where the subject and the environment reciprocally oppose and determine themselves" [my translation] (Nishida 1965g, IX 152). The significance of this statement for the concrete relationship humans have with the world in which they live is not self-evident. We may grasp it by recalling again Husserl's position. It is well known that Husserl posited as the base of his new philosophical approach the act of *epoché*, describing it, in his *Ideas,* in the following way: "The whole prediscovered world posited in the natural attitude, actually found in experience and taken perfectly 'theory free' as it is actually experienced, as it clearly shows itself in the concatenations of experience, in now without validity for us;... it shall be bracketed" (Husserl 1950, 3, 62).

Thus, phenomenology starts by putting into brackets all possible considerations about the external world, about the environment: Husserlian phenomenology proceeds *as if* the world was not there for us. Precisely through this methodological expedient of *epoché,* did Husserl find the possibility of catching the intentional essence of the pure Ego (Husserl 1950, 3, 33–35). Thus, all evidence declares that in Husserlian phenomenology, even though we deal with a subject that needs reference to the external world to reach a proper personal constitution, we find a subjectivity that can be absolutely affirmed in itself simply by reference to its conscious activity (Husserl 1950, 4, 61). This latter, indeed, is, from the Husserlian point of view, the only thing that no one can doubt nor put into brackets: our most original and ineludible experience.

As previously told, Nishida declared exactly the contrary. As a matter of fact, he defined the subject as the *subjective pole* of a deeper reality. He states the impossibility of the subject's determining his essence by simple reference to itself, to its consciousness. Nishida is quite explicit on this point in his work *Watashi to nanji* [私と汝] [*I and Thou*], in which he writes: "We cannot but fall back into solipsism if we start from the consciousness of the individual self in a strict sense. Nevertheless the individual does not start from itself. In order for the individual to be born a ground is needed, where it can be born. In other words, an environment is needed" [my translation] (Nishida 1965e, VI 345). Thus, the subject cannot affirm itself only by referring to its conscious activity. For the subject to emerge, we need foremost reference to the environment. This, indeed, is precisely the material dimension that stands beyond the subject, in front of it and distinguished from it. Those characteristics of the environment are exactly what makes the environment so relevant in the determination of the subject, as "the I – affirmed Nishida in *Basho* – can be conceived only contra a non-I" [my translation] (Nishida 1965c, IV, 208). An environment is needed for the subject to reach its own being, as the environment is precisely the countering non-I in which the subject can understand itself as an I.

Therefore, the subject attains its own essence only by oppositional action before the environment. As a matter of fact, one can understand one's own status of being the subjective pole of reality only because there is an aspect which can be regarded as an objective pole. That is, an aspect opposed to it: a non-subjective aspect. The statement that the subject and the environment reciprocally oppose and determine themselves in the historical world means precisely this: that the subject cannot emerge without the opposition that is within the environment; the subject determines itself only with the reference to the environment.

Here we can grasp the practical relevance of Nishida's theoretical dialogue with phenomenology. Once intentionality and its associated relational unity with the external world have been rejected, consequently, as Tremblay pointed out in her *L'être-soi et l'être-ensemble,* there is no possibility of maintaining the subject's phenomenological transcendence vis-à-vis external world (Tremblay 2007, 63). In Nishida the subject cannot be said to stand without reference to the external world and such a position is in evident contrast with Husserl's closing statement of *Ideen II*, wherein he explicitly recognized that even without the environment there would still be a subject, impoverished in its spiritual life but still a subject, for it always subsists in intentional consciousness (Husserl 1952, 4, 297). Therefore, the theoretical differences between Nishida and Husserl are of a great relevance also for any environmental discourse, because they lead Nishida to deny that the subject can be understood apart from its environment. This means that the latter is recognized to have a greater role in human existence: since the environment determines the subject in its definition, the subject itself now assumes a topological nature instead of the ontological nature postulated in Husserlian phenomenology.

## Conclusions

This article was conceived with the purpose of throwing light on the philosophical outlines of the first encounter of Japanese thought with Husserlian phenomenology. For this reason it focuses on the philosophy of Nishida Kitaro, for he designed the route subsequently taken by Japanese phenomenology proper. This dialogue was retraced, first of all, in its theoretical outcomes, and then we tried to grasp the practical relevance of these for what concerns the subject-environment relationship. From a theoretical point of view, Nishida, proceeding from a "Zen-oriented" requirement of finding an ultimate unity that encompass all the contradictions of reality, rejects the transcendental subject as the ground of philosophy. The pure Ego is led back to the unity in which it is contained, and which is a subjective manifestation of Absolute Nothingness in its various expressions.

As a result of this, the subject needs the objective pole to attain its real self: here a transcendental subject who constitutes the external world with its essential intentional attitude has vanished (Husserl 1952, 4, 301), and a subject that understands itself as a subjective pole only because there is an external dimension which can be seen as an objective pole stands forth. While in Husserl we have a subject who constitutes the *Umwelt*, in Nishida it is that very environment which constitutes the subject, set in opposition to it within the unity of the historical world. Thus, can we sum up the practical outcomes of Nishida's dialogue with Husserl, noticing that, to the quite *anthropocentric* vision Husserl had of the human-world relationship, Nishida's Japanese thought is replying with a *biocentric* interpretation of that relationship.

This, of course, does not mean that Japanese eco-phenomenology today still displays a biocentric approach in the same way as does Nishida's philosophy. What is true, however, is that, since Nishida designed the framework of Japanese phenomenological debate, its starting point diverges quite radically from that of Western phenomenology. And so while the latter, after Husserl's death, had to face insinuations of transcendental idealism and thus, within the environmental debate, of anthropocentrism (Brown and Toadvine 2003, 73–74), Japanese phenomenology and eco-phenomenology had to explain how it could be possible to understand the relationship of the subject with the environment not in a biocentric way, once the relational unity between the subject and the environment had been broken by Nishida's thought.

## Works Cited

Abe, Masao. 1990. Introduction. Nishida Kitaro, *An Inquiry into the Good*, xv–xvii. New Haven: Yale University Press. Print.

Berthon, Jean-Pierre, and Anne Gossot, eds. 1998. *Japon Pluriel 3. Actes du troisième colloque de la société française des études japonaises*, 199–208. Paris: Éditions Philippe Picquier. Print.

Bertossa, Franco. 2015. Interview to Ōhashi Ryōsuke. *Asia.* Web. 28 January 2015. http://www.asia.it/adon.pl?act=doc&doc=442.

Brown, Charles, and Ted Toadvine, eds. 2003. *Eco-Phenomenology: Back to the Earth Itself.* Albany: SUNY Press Series in Environmental Ethics and Philosophy. Print.

Derrida, Jacques. 1978. *Edmund Husserl's Origin of Geometry: An Introduction.* New York: N. Hays. Print.

Dilworth, David A. 1979. The Concrete World of Action in Nishida's Later Thought. In *Japanese Phenomenology*, Analecta Husserliana 8, ed. Anna-Teresa Tymieniecka and Yoshihiro Nitta, 249–270. Print.

Husserl, Edmund. 1950. *Ideen zu einer reinen Phänomenlogie und phänomenlogischen Philosophie. Erstes Buch: Allgemeine Einführung in die reine Phänomenologie* [Ideas: General introduction to Pure Phenomenology and to a Phenomenological Philosophy. First book], Husserliana 3. The Hague: Martinus Nijhoff Publishers. Print.

———. 1952. *Ideen zur einer reinen Phänomenologie und phänomenologischen Philosophie. Zweites Buch: Phänomenologische Untersuchungen zur Konstitution* [Ideas Pertaining to a Pure Phenomenology and to a Phenomenological Philosophy. Phenomenological Investigations of Constitution.], Husserliana 4. The Hague: Martinus Nijhoff. Print.

———. 1973 *Ding und Raum. Vorlesungen* 1907 [Thing and Space], Husserliana 16. The Hague: Martinus Nijhoff. Print.

Nishida, Kitaro. 1965a. 西田幾多郎太全集 [*Nishida Kitarō zenshū*][*Nishida Kitaro Complete Works*]. Vol I: 3–200. 善の研究 [*Zen no kenkyū*] [*An inquiry into the good*]. Tōkyō: Iwanami Shoten. Print.

———. 1965b. 西田幾多郎太全集 [*Nishida Kitarō zenshū*][*Nishida Kitaro Complete Works*]. Vol II: 1–350.自覚に於ける直と反省[*Jikaku ni okeru chokkan to hansei*] [Intuition and Reflection in Self-Consciouness]. Tōkyō: Iwanami Shoten. Print.

———. 1965c. 西田幾多郎太全集 [*Nishida Kitarō zenshū*][*Nishida Kitaro Complete Works*]. Vol IV: 208–289. 場所 [*Basho*] [*Place*]. Tōkyō: Iwanami Shoten, 1965. Print.

———. 1965d. 西田幾多郎太全集 [*Nishida Kitarō zenshū*][*Nishida Kitaro complete works*]. Vol V: 123–185. 叡智的世界 [*Eichiteki sekai*] [*The Intelligibile World*]. Tōkyō: Iwanami Shoten. Print.

———. 1965e. 西田幾多郎太全集 [*Nishida Kitarō zenshū*][*Nishida Kitaro Complete Works*]. Vol VI: 341–427. 私と汝 [*Watashi to nanji*] [*I and Thou*]. Tōkyō: Iwanami Shoten. Print.

———.1965f. 西田幾多郎太全集 [*Nishida Kitarō zenshū*][*Nishida Kitaro Complete Works*]. Vol VII: 215–224. 意識の問題を無視 [*Ishiki no mondai o mushi*] [*The Problem of Consciousness*]. Tōkyō: Iwanami Shoten. Print.

———. 1965g. 西田幾多郎太全集 [*Nishida Kitarō zenshū*][*Nishida Kitaro Complete Works*]. Vol IX: 147–222. 絶対矛盾的自己同一 [*Zettai mujunteki jiko dōistu*] [*The Absolute Contradictory Self-Identity*]. Tōkyō: Iwanami Shoten. Print.

Nitta, Yoshihiro, Hirotaka Tatematsu, and Eiichi Shimomisse. 1979. Phenomenology and Philosophy in Japan. In *Japanese Phenomenology*, Analecta Husserliana 8, ed. Anna-Teresa Tymieniecka and Yoshihiro Nitta, 3–17. Print

Piovesana, Gino K. 1968. *Filosofia Giapponese Contemporanea*. Bologna: Patron. Print.

Tremblay, Jacynthe. 2000. *Nishida Kitaro. Le Jeu de L'individuel et de L'universel*. Paris: CNRS éditions. Print.

———. 2007. L'être-soi et l'être-ensemble. *L'auto-éveil comme méthode philosophique chez Nishida*. Paris : l'Harmattan. Print.

Tymieniecka, Anna-Teresa, and Michel Masson. 1979. Phenomenology in Japan: A Presentation. In *Japanese Phenomenology*, Analecta Husserliana 8, ed. Anna-Teresa Tymieniecka and Yoshihiro Nitta, vii–vix. Print.

Yusa, Michiko. 2002. *Zen and Philosophy. An Intellectual Biography of Nishida Kitaro*. Honolulu: University of Hawaii Press. Print.

# "Negative Seeing": Robert Smithson, Earth Art, and the Eco-Phenomenology of "Mirror Displacements"

**Ming-Qian Ma**

**Abstract** Mounted randomly in various geo-ecological loci in Yucatan, Mexico, Robert Smithson's earth art of the "Mirror Displacements" stages an eco-phenomenology characterized by "a wilderness of unassimilated seeing." His leitmotif of an "anti-vision" or "negative seeing," which paradoxically enables the world to appear counter-intuitively through the mirror displacements, presents itself as an artistic rendition of Jean-Luc Marion's phenomenon of givenness. Theorized as the third phenomenological reduction contra that of Husserl and Heidegger, Marion's phenomenology of givenness postulates a phenomenon saturated with intuition, which appears absolutely and unconditionally, beyond the limits set by the horizon and the transcendental I. In both Smithson's eco-phenomenology of the "Mirror Displacements" and Marion's phenomenology of givenness, the appearance of this saturated phenomenon is, as Marion epitomizes it, "invisible according to quantity, unbearable according to quality, absolute according to relation, irregardable according to modality."

**Keywords** Jean-Luc Marion · Immanuel Kant · Edmund Husserl · Robert Smithson · Phenomenology · Phenomenological reduction · Saturated phenomenon · Givenness · Eco-phenomenology · Earth art

## Introduction

Regarded as an "iconoclastic" artist who "has now come to symbolize the expansive, antiformalist movements that emerged in the mid-1960s and early 1970s" in America, Robert Smithson (1938–1973) occupies a radically unique position in the contemporary art arena (Flam 1996, xiii). Anti-anthropomorphic in sentiment and post-humanist in approach, his art projects, also referred to respectively as

M.-Q. Ma (✉)
Department of English, The State University of New York at Buffalo, Buffalo, NY, USA
e-mail: mma3@buffalo.edu



W. S. Smith et al. (eds.), *Eco-Phenomenology: Life, Human Life, Post-Human Life in the Harmony of the Cosmos*, Analecta Husserliana CXXI,
https://doi.org/10.1007/978-3-319-77516-6_25

"'earthworks' and 'earth art'," challenge the received theoretical paradigms and conceptual categories of phenomenology by foregrounding the phenomenality of the world itself, so much so that, as Jack Flam puts it, "the question of mind and 'nature' appears to be obliterated" (Flam 1996, xvii, xxiv). Smithson himself is eloquently explicit about an earth-oriented and eco-centered phenomenology that undergirds and permeates his art. In an essay titled, rather tellingly, "A Sedimentation of the Mind: Earth Projects (1968)," he thus describes the earth and mind comparatively:

> The earth's surface and the fragments of the mind have a way of disintegrating into discrete regions of art. Various agents, both fictional and real, somehow trade places with each other—one cannot avoid muddy thinking when it comes to earth projects, or what I will call "abstract geology." One's mind and the earth are in a constant state of erosion, mental rivers wear away abstract banks, brain waves undermine cliffs of thought, ideas decompose into stones of unknowing, and conceptual crystallizations break apart into deposits of gritty reason. Vast moving faculties occur in this geological miasma, and they move in the most physical way. This movement seems motionless, yet it crushes the landscape of logic under glacial reveries. This slow flowage makes one conscious of the turbidity of thinking. Slump, debris slides, avalanches all take place within the cracking limits of the brain. (Smithson 1996, 100)

Smithson's language therein makes a revealing statement; it delineates, by way of a hybridity of generic vocabularies and a plethora of mixed metaphors, a terrene image of two parallel worlds in the process of merging into one: a vast geo-ecological landscape on the one hand and a human mind on the other, with the latter sinking into and engulfed by the former. In this sense, Smithson presents his earth art, at once literally and metaphorically, as a phenomenological critique of the human mind in terms of perception, highlighting, in particular, the very polemics of limit or boundary that construct vision as the constitutive mechanism of phenomenological reduction and containment. It articulates, through a privileging of the enormity of the phenomenal world and its indifferently all-encompassing movements, a different phenomenology, a phenomenology that is "beyond the bounds of beholding" (Flam 1996 xxiv). Neither transcendental nor existential, Smithson's phenomenology is, as is so embodied by his earth art, material or ecological, the phenomenality of which lies in what the artist calls "undifferentiation" (Smithson 1996, 102). Of this, he writes:

> At the low levels of consciousness the artist experiences undifferentiated or unbounded methods of procedure that break with the focused limits of rational technique. Here tools are undifferentiated from the material they operate on, or they seem to sink back into their primordial condition. ... This entropy of technique leaves one with an empty limit, or no limit at all. ... The rational critic of art cannot risk this abandonment into "oceanic" undifferentiation, he can only deal with the limits that come after this plunge into such a world of non-containment. (Smithson 1996, 102)

Limits or boundaries are, in other words, schemata of constituting consciousness and dividing techniques of sufficient reason, and they are deployed to reduce the phenomenal world to rationally containable categories. By contrast, Smithson's earth art exhibits, as the artist himself states therein, an eco-phenomenology of the undifferentiated, which shows itself through "dedifferentiation," or de-demarcation, of phenomena: of mind from matter, of medium from the mediated, and of the rational from the irrational, among others (Smithson 1996, 103). With consciousness descending into, rather than transcending, the world in its primordial state, dedifferentiation ushers in "the oceanic," an uncontained and uncontainable "limitlessness" in which "all boundaries and distinctions [lose] their meaning" and "all notions of gestalt unity" collapse, therefore resulting in "one's inability to see" (Smithson 1996, 103, 110, 130).[1] Situated beyond the bounds of beholding, Smithson's earth art thus stages an eco-phenomenology which, as is demonstrated provocatively in his project of "Mirror Displacements," phenomenalizes an "anti-vision" (Smithson 1996, 119–133, 130).

## A Setting for Negative Seeing

Documented in "Incidents of Mirror-Travel in the Yucatan (1969),"[2] Smithson's earth project of "Mirror Displacements" consists of nine groups of mirrors, with 12 mirrors in each group (Smithson 1996, 119–133).[3] These groups of mirrors are distributed in various geo-ecological loci in Yucatan, Mexico: a field of ashes on a charred site of red soil between Uman and Muna (Smithson 1996, "The First Displacement," 120), red clay mixed with white limestone in a suburb of Uxmal (Smithson 1996, "The Second Displacement," 121), the side of a heap of crushed limestone covered with large clusters of butterflies near Bolonchen de Rejon (Smithson 1996, "The Third Displacement," 122), dry seaweed and eroded rocks on the beach of the Gulf of Mexico, south of Campeche (Smithson 1996, "The Fourth Displacement," 123), the lush jungle at Palenque (Smithson 1996, "The Fifth Displacement," 124), a high sandbank of the river Rio Usumacinta ("The Sixth Displacement," 127), a tentacled tree near Yaxchilan (Smithson 1996, "The Seventh Displacement," 128), the slope of the eroding Island of Blue Waters (Smithson 1996, "The Eighth Displacement," 129), and mangrove branches and roots near Sabancuy (Smithson 1996, "The Ninth Displacement," 131).

Smithson's description of "The First Mirror Displacement," which is the shortest of all, could be read as sketching out the fundamental features that characterize, most generally, all the "Mirror Displacements," outlining some of the basic properties of this earth project in terms of an eco-phenomenology of "anti-vision:"

> Somewhere between Uman and Muna is a charred site ... On this field of ashes ... twelve mirrors were cantilevered into low mounds of red soil. Each mirror was twelve inches square, and supported from above and below by the scorched earth alone. The distribution of the squares followed the irregular contours on the ground, and they were placed in a random parallel direction. Bits of earth spilled onto the surface, thus sabotaging the perfect reflections of the sky. Dirt hung in the sultry sky. Bits of blazing cloud mixed with the ashy mass. The displacement was *in* the ground, not *on* it. Burnt tree stumps spread around the mirrors and vanished into the arid jungles. (Smithson 1996, 120–121)[4]

Thus deployed, Smithson's "Mirror Displacement" operates on two levels simultaneously, as is signaled by the two words themselves in the name of the project.

*Metaphorically* (almost to the point of being a cliché), on the one hand, a mirror represents perception, a human construct intended for an ego-centered verisimilitude ("perfect reflections"). It is an anthropomorphic device in which one sees, by design, nothing but one's own exact duplicate. Moreover, the perfect reflection of one's intended image in the mirror occurs within and is rendered possible by the perimeter of the mirror ("twelve inches square"), which, as the mirror's delimiting framework, constitutes an all-enclosing horizon. A determining condition, according to which "phenomena obtain their meaning," this horizon is, as its "first determination" decrees, "*what allows phenomena to manifest themselves to us*," that is, to subjectivity (Geniusas 2012, 1). In other words, the mirror metaphorizes its own boundary as "a structure of determination that pre-delineates the purview within which each and every phenomenon appears," as "*what consciousness co-intends in such a way that what is co-intended determines the sense of appearing objectivities*" (Geniusas 2012, 7).

*Literally,* on the other hand, the mirror is indeed displaced in more ways than one. For one thing, it is physically removed, position-wise, from its established seat above the ground ("on it") to a spot in it ("into low mounds of red soil," "in the ground"), as Smithson makes emphatically clear, hence dethroning it from its transcendent vantage point. For another thing, the mirror, instead of being identified with subjectivity and sustained by consciousness, now finds itself, status-wise, entirely and unequivocally earthy, having become part of an amorphous expansiveness of crudity ("supported from above and below by the scorched earth alone," "followed the irregular contours on the ground," "placed in random parallel direction," "burnt tree stumps spread around the mirrors"). Moreover, with its reflecting surface covered by "bits of earth, thus sabotaging the perfect reflections of the sky," the mirror loses, function-wise, its capacity as a system of representation authorized by subjectivity, as a structure of determination authenticated in the name of self-reflecting consciousness.

Emerging out of such a "Mirror Displacement" is, then, a strange eco-environment, an alien phenomenal world of "Dirt hung in the sultry sky. Bits of blazing cloud mixed with the ashy mass." It unfolds, indeed, "an enchanted region," as Smithson himself so describes it, "where down is up," and where phenomena appear in manners against common sense and beyond apprehension (Smithson 1996, 119). Unintended and counter-intuitive, the appearances of phenomena in Smithson's "Mirror Displacement" thus articulate an eco-phenomenology of what Smithson also calls "negative seeing," one that defies the logic of vision, resists conceptual mapping, and flouts any attempt at constitution (Smithson 1996, 130).

## A Shared Vision

It can be said that Smithson's eco-phenomenology of "Mirror Displacements," as such, is pro*visioned* with philosophical underpinning by Jean-Luc Marion's "Phenomenology of Givenness"[5] as unfolded in his works *Réduction et donation*

(1989), *Étant donné* (1997), and *Du surcroît* (2001); herein, all quotations from these works will be from their English translations: *Reduction and Givenness* (1998), cited as *RG*; *Being and Givenness* (2002), cited as *BG*; and *In Excess* (2002), cited as *IE*.

Contra Husserl's transcendental reduction and Heidegger's existential reduction, Marion presents his "phenomenology of givenness" as the third phenomenological reduction, one that postulates the "originarily unconditional" givenness of phenomena (Marion 1998, *RG* 205).[6] Proclaimed as "'first philosophy' according to phenomenology," it posits that phenomena appear unconditionally, giving themselves, showing themselves, and starting from themselves alone (Marion 2002a, *IE* 23, 25; 2002b *BG* 70).

Both unconditional in origin and "unconditioned" owing to "its certitude and its automatic universality" (Marion 2002a, *IE* 23), givenness is, as Marion asserts, what constitutes "The phenomenality proper to the phenomenon" (Marion 2002a, *IE* 23, b, *BG* 19). It is a givenness whose ontological features Marion details thusly:

> Now, this datum gives itself to me, because it imposes itself on me, calls me, and determines me—in short, because I am not the author of it. The datum merits its name by its being a *fait accompli*, such that it happens to me, and in which it is distinguished from all foreseen, synthesized, and constituted objects, since it happens to me as an event. This unforeseen happening marks it as given and attests in it to givenness. Givenness does not indicate so much here the origin of the given as its phenomenological status. Better, most often, givenness characterizes the given as without cause, origin, and identifiable antecedent, far from assigning them to it. And it is sufficient that the given—the given phenomenon—gives itself starting from itself alone (and not from a foreseeing and constituting subject) in order that the fold of givenness is witnessed. The objection turns in this way to the confirmation of my thesis: givenness does not submit the given to a transcendent condition, but rather frees it from that condition. (Marion 2002a, *IE* 24–25)

It follows, then, that "*the phenomenon gives itself*," but only in its own "process of arising into appearing" (Marion 2002b, *BG* 68). In a rhetoric that resonates with Smithson's de-metaphorization of the mirror, Marion expounds further the unmediated appearance of the phenomenon:

> The phenomenon can appear as such, and not as the appearance of something else more essential to it than itself, in short it can appear without the lack implied by an in-itself or the withdrawal implied by a noumenon—and this is indeed the primary goal of phenomenology—only if it pierces through the mirror of representation. Appearing must thus remove itself from (if not always contradict) the imperial rule of the a priori conditions of knowledge by requiring that what appears force its entry onto the scene of the world, advancing in person without a stuntman, double, or any other representative standing in for it. (Marion 2002b, *BG* 69)

The unconditional givenness of the phenomenon as described here entails that, "In the strict phenomenological sense, *the phenomenon is no longer ... visible*" (Marion 2002b, *BG* 69).[7] Marion asserts this, and such a phenomenality of invisibility proper to the phenomenon has to be understood against the "three of its characteristics" of Husserl's "'principle of all principles'," a principle which states that "every originarily giving intuition [*Anschauung*] is a source of right for knowledge, that everything that offers itself originarily to us in 'intuition' [*Intuition*] is to be taken simply as it gives itself, but also only within the boundaries in which it gives itself there" (Marion 2002b, *BG* 184; Husserl 1998, 25).[8]

Whence arises, first and foremost, the issue of perception or intuition. Having freed itself from the rules of the *a priori* conditions of knowledge, which "is to say [from] intuition and the concept [that] determine in advance the possibility of appearing for every phenomenon," the phenomenon gives itself and shows itself forcefully as a *fait accompli* in the face of a subject (Marion 2002b, *BG* 181). In this way, the phenomenon becomes invisible because it would no longer "tie its fate to intuition," as Marion's "inverse hypothesis" states, and consequently appears as unintended (Marion 2002b, *BG* 197, 187). It "breaks through the frame, is abandoned to the world of which it now makes a part," Marion writes, and, as such, is unperceivable even to the "sufficient intuition" postulated by Husserl (Marion 2002b, *BG* 69, 184). For it is the "first characteristic" of Husserl's "principle of all principles," according to Marion, that intuition is never neutral, and—emphasizing that this intuition occurs "only within the boundaries in which it [every originarily giving intuition] gives itself there,"—that it is still a conditional intuition, however originarily sufficient Husserl claimed it to be (Marion 2002b, *BG* 184). More specifically, Marion contends, it "remains framed, inasmuch as it is intuition, by two conditions of possibility… assigned to every phenomenon;" and these two formal conditions of possibility, or the two other characteristics of Husserl's "principle of all principles," are "the horizon and the I" (Marion 2002b, *BG* 185, 179).

Secondly, it follows that, with its severance from intuition and its confines, the unconditional appearance of the phenomenon is invisible also in that it does not "admit limitation, de facto and de jure, by a horizon," hence freeing itself from "the priori limit of a horizon of phenomenality" (Marion 2002b, *BG* 187). For the horizon, being the "second characteristic" of the principle of all principles, functions to render phenomena manifest, granting them visibility. A structure of co-intending with consciousness, it frames intuition by providing the latter with a scaffold required by the "logic of penury" that "intuition obeys," Marion points out, a scaffold within the limits of which intuition "must first be inscribed de jure" in order that it give itself "within certain de facto 'boundaries'" (Marion 2002b, *BG* 185). Using "seeing an object, transcendent by definition" and the concomitant incomplete apprehension of that object as an illustration, Marion claims that the horizon is thus "the organization of all the successive lived experience around one single object," an organization in which "the known (the immanent lived experience already recorded) remain not only in memory, but be co-deposited within the same horizon as what still remains unknown (the lived experience yet to come), for the sake of a single intended transcendent object," and in which the known and the unknown "could simply be united… in one and the same objective intention" (Marion 2002b, *BG* 185, 186). Such being the case, Marion argues that "within the horizon, the unknown refers in advance to the known because it welcomes it and fixes it" (Marion 2002b, *BG* 186). Hence, a tautology under the mask of a paradox: inasmuch as the horizon is concerned, the logic of intuitive penury guarantees an intentional fullness, characterized by a "determinable indeterminateness" (Husserl 1998, 94).[9] Marion continues,

> The horizon in advance takes possession of the unknown, the un-experienced, and the not gazed upon, by supposing them to be always already compatible, compressible, and homogenous with the already experienced, already gazed upon, and already interiorized by intuition. The intention always anticipates what it has not yet seen, the result being that the unseen has, from the start, the rank of a pre-seen, a merely belated visible, without fundamentally irreducible novelty, in short a pre-visible. The horizon therefore does not so much surround the visible with an *aura* of the nonvisible as it assigns in advance this nonvisible to this or that focal point (object) inscribed in the already seen. (Marion 2002b, *BG* 186)

Functioning as the "[designation] of the object of all givenness" and the "assimilation of givenness to intuition" for "its intentional aim at an object," the horizon is thus a determining structure of a proactive vision of positivity (Marion 2002b, *BG* 187). In this sense, the "horizon of appearing always already seen, or at least visible" circumscribes a closure, or more accurately put, a foreclosure (Marion 2002b, *BG* 187). By leaving nothing unseen or invisible, Marion argues emphatically that the horizon constitutes a seeming "openness [that] would be equivalent to a visual prison, a panopticon broadened to the dimensions of the world, a panorama without exterior, forbidding all genuinely new arising" (Marion 2002b, *BG* 187).

Thirdly, the unconditional appearance of the phenomenon is invisible because it "precedes every other instance (including and, above all, the I)," an "I" who, as "the third characteristic of the 'principle of all principles'," is ultimately one with horizon. Marion makes it clear, the I is presupposed to be "as transcendental and as horizon" (Marion 2002b, *BG* 188, 187, 188). In other words, the phenomenon appears unconditionally, still "to us," granted, but "on its own basis to an I" without being constituted into an alienated phenomenon "by and on the basis of the I" (Marion 2002b, *BG* 187). Thus, Marion continues, the unconditional givenness of the phenomenon "obligatorily confiscates the function and the role of the *self*, and therefore can only concede to the ego a *me* of second rank, by derivation," thereby demoting the "I" to a "passive receptivity" (Marion 2002a, *IE* 45, 48): "In other words, the *ego*, deprived of transcendentalizing dignity, must be admitted as it is received, as an *adonné*: the one who is itself received from what it receives, the one to whom what gives itself from a first *self*—any phenomenon—gives a second *me*, the one of reception and of response" (Marion 2002a, *IE* 45). Moreover, Marion states, "without reserve or limits, the I must renounce every claim to the synthesis of objects or the judgment of phenomenality. In the realm of givenness, it no longer decides the phenomenon, but receives it; or else, from "master and possessor" of the phenomenon, it becomes its receiver" (Marion 2002b, *BG* 188).

Vis-à-vis these three characteristics of Husserl's "principle of all principles," Marion's phenomenology of givenness assumes, in turn, a different "hypothesis" (Marion 2002b, *BG* 189). An "inverse" one, as Marion has described it earlier, it is a hypothesis that Marion claims "would permit us to go to the limit in determining phenomenality and experiencing afresh what possibility means or can give" and would enable imagining "certain phenomena [that] could appear only by playing at the limits of phenomenality" (Marion 2002b, *BG* 189). It posits, imaginatively as well as rhetorically, an unconditional givenness of the phenomenon "without the limit (the principle of a horizon) or condition (the transcendental I)," a givenness "that is finally absolutely unconditioned (without the limits of a horizon) and absolutely irreducible (to a constituting I)," and whence the phenomena, which "would invert limit (by exceeding the horizon, instead of being inscribed within it) and condition (by reconducting the I to itself, instead of being reduced to it)" (Marion 2002b, *BG* 189).

Marion's "inverse hypothesis," stated as such, leads to a further one. Since it is the "logic of penury" inscribed in intuition that determines the limited possibility of the phenomenon, and since it is "the two finitudes of the horizon and the I [coming] together in the finitude of intuition itself" that delimit the appearances of the phenomenon, the very limit or finitude of intuition has, then, to be inverted (Marion 2002b, *BG* 197). For the "unconditioned and irreducible phenomena (if there are any) would become thinkable and possible," Marion argues, "only if a finally nonfinite intuition could secure their givenness" (Marion 2002b, *BG* 197). To this end, Marion posits an intuition that, contrary to the "poverty" that characterizes intuition, which is to say the penury that marks the "ideal adequation of intuition to intention," would "give *more, indeed immeasurably more*, than the intention would ever have aimed at or foreseen" (Marion 2002b, *BG* 199, 197). The result is an inversion from "the phenomenon supposedly poor in intuition" to "a phenomenon saturated with intuition," to what Marion calls "a saturated phenomenon" (Marion 2002b, *BG* 197).

Furthermore, crediting Kant with "a foretaste of ... saturated phenomenon," which "Kant formulates ... in a rare term: the aesthetic idea," Marion points out that, as with Kant's doctrine of "representation of an object according to a principle," the saturated phenomenon "too can never become knowledge, but for a contrary reason" (Marion 2002b, *BG* 197, 198). For, as Kant puts it unambiguously, the surplus of intuition is one "for which a concept can never be found adequate."[10] Marion, thus, explains:

> It is no longer a question of the nonadequation of (lacking) intuition leaving a (given) concept empty. *It is inversely a question of a deficiency of the (lacking) concept, which leaves the (superabundantly given) intuition blind.* As a result, it is the concept that is deficient, no longer intuition... *The excess of intuition over every concept... prevents the aesthetic idea from making an object visible.* It is important to insist on this: the failure to produce the object does not result from a shortage of givenness (as for the idea of reason), but well and truly from an excess of intuition, therefore from an excess of givenness. (Marion 2002b, *BG* 198)[11]

Elaborating further on Kant's position, Marion argues that the "excessive givenness" (Marion 2002b, *BG* 198) or the "aesthetic idea," that defines an "'inexponible representation',"[12] "can be understood as follows":

> Because it give "much," the aesthetic idea gives intuitively more than any concept can expose. To expose here equates disposing of (or organizing) the intuitive given according to rules. The impossibility of the concept arranging this disposition comes from the fact that the intuitive superabundance no longer succeeds in exposing itself in a priori rules, whatever they may be, but rather subsumes them. Intuition is no longer exposed in the concept; *it saturates it and renders it overexposed—invisible,* unreadable not by lack, but indeed by an excess of light... a surplus of intuition, therefore of givenness, over and above intention, the concept, and the intended... a saturated phenomenon will no doubt no longer constitute

> an object (at least in the Kantian sense), for it is not self-evident that objectivity has enough authority to impose its norm on the phenomenon. (Marion 2002b, *BG* 198, 199)[13]

That said, Marion, by way of an inverse reading of Kant that would exceed the Kantian categories, sketches out the properties of the saturated phenomenon as "*invisable* according to quantity, unbearable according to quality, absolutely according to relation, irregardable according to modality. The three first characteristics put into question the ordinary sense of horizon...; the last, the transcendental sense of the I..." (Marion 2002b, *BG* 199).

## Smithson's "The Mirror Displacements" as a "Saturated Phenomenon"

Foregrounding its eco-phenomenology of "negative seeing," Smithson's earth art of "Mirror Displacements" begins, not surprisingly, with a meditation and speculation on a horizon. "Driving away from Merida down Highway 261," Smithson writes, "one becomes aware of the indifferent horizon ... devouring everything that looks like something," a "closedness," that is, which imposes "restrictions on all forward movement," and in which everything is "imprisoned" (Smithson 1996, 119). The phenomenological question he then raises is "How could one advance *on* the horizon, if it was already present *under* the wheels?" (Smithson 1996, 119).[14] So formulated hypothetically, Smithson's question is Marionian in that the two prepositions of "on" and "under" therein have already presupposed a spatial and positional displacement: they signal, both denotatively and connotatively, the removal of the horizon as the "*unsurpassable* limit" from afar, always remaining "*relative in regard to* [*one's*] *current situatedness,*" to the immediacy of the unconditional appearance of the phenomenon, by erasing the elusive distance that constitutes the transcendentality of a horizon, to a spot of "tumult of 'de-differentiation'" where, as Smithson himself puts it in his documentation of "The Seventh Mirror Displacement," the horizon is "submerged and suffocated in an asphyxiation of vanishing points" (Geniusas 2012, 2; Smithson 1996, 110, 128). This spatial and positional displacement of horizon, which inverts the limit by exceeding the horizon, then turns the horizon into "something else other than a horizon," as Smithson claims; it unfolds, in other words, an oceanic "openness" of and to phenomenality (Smithson 1996, 119).

With the horizon thus displaced, Smithson's "Mirror Displacements" engage an "anti-vision" or "negative seeing" at the limit of phenomenality, where saturated phenomena appear unconditionally. In this regard, "The Eighth Mirror Displacement" presents itself as an illuminating case in point. Mounted precariously on the sandy slope of the "Island of Blue Waters," a slope that is "dropping, draining, eroding, trickling, spilling away" (Smithson 1996, 129), this "Mirror Displacement" is documented extensively as follows:

> Small bits of sediment dropped away from the sand flats into the river. Small bits of perception dropped away from the edges of eyesight ... Sight turned away from its own looking. Particles of matter slowly crumbled down the slope that held the mirrors. Tinges, stains, tints, and tones crumbled into the eyes. The eyes became two wastebaskets filled with diverse colors, variegations, ashy hues, blotches and sunburned chromatics. To reconstruct what the eyes see in words, in an "ideal language" is a vain exploit. ... Sight consisted of knotted reflections bouncing off and on the mirrors and the eyes. Every clear view slipped into its own abstract slump. All viewpoints choked and died on the tepidity of the tropical air. The eyes, being infected by all kinds of nameless tropisms, couldn't see straight. Vision sagged, caved in, and broke apart. Trying to look at the mirrors took the shape of a game of pool under water. All the clear ideas of what had been done melted into perceptual puddles, causing the brain to gurgle thoughts. Walking conditioned sight, and sight conditioned walking, till it seemed only the feet could see. Squinting helped somewhat, yet that didn't keep views from tumbling over each other. The oblique angles of the mirrors disclosed an altitude so remote that bits of "place" were cast into a white sky. How could that section of visibility be put together again? Perhaps the eyes should have been screwed up into a sharper focus. But no, the focus was at times cock-eyed, at times myopic, overexposed, or cracked. ... The eyes crawled over grains, chips, and other jungle obstructions. From the blind side reflections studded the shore—into an anti-vision. (Smithson 1996, 129-130)

Smithson's documentation of "The Eighth Mirror Displacement" here details what the artist refers to as "a wilderness of unassimilated seeing" (Smithson 1996, 129). Here, as elsewhere in Smithson's "Mirror Displacements," the inability of the "seeing" to be assimilated into an ideal adequation to intention, hence an intuitive "wilderness" exceeding the limit of the concept, results from the fact that the unconditional appearance of the saturated phenomenon is, as Marion has stated earlier, "invisable according to quantity." Defined as that which "*cannot be aimed at* [*ne peut viser*]," "meant, or intended" this invisibility of the saturated phenomenon, or this "impossibility" of the saturated phenomenon to be aimed at or intended, Marion explains, "stems from its essentially unforeseeable character [*son caractère essentiellement imprévisible*]" (Marion 2002b, 199, 363).[15] Take, for instance, the phenomenon of how "tinges, stains, tints, and tones crumbled into the eyes." Designating a specific relation between the phenomenon and the eye, the verbal phrase "crumbled into" delineates here the appearance of the saturated phenomenon from two perspectives. First, with its denotative meaning of "to break down into small crumbs" or "to fall asunder in small crumbs or particles,"[16] which "'de-structuralized' any literal logic" of the eye, it gestures toward a "wilderness," or what Smithson also calls the "indecisive zones" and "riddling zones," of the phenomenal world by connoting an overwhelming increase in "all kinds of nameless tropisms," an irresistible addition to the already limitless quantity of the phenomenon that is unassimilated and unassimilable by intention and representation (Smithson 1996, 128–9).[17] It thus inverts the limit or the logic of penury of intuition, whose foreseeability is predicated upon the principles of homogeneity and finitude that mobilize the "successive synthesis" (Marion 2002b, *BG* 200).[18] Marion writes,

> According to Kant, quantity (extensive magnitude) is declined by composition of the whole in terms of its parts. This "successive synthesis" allows for the representation of the whole to be constituted according to the representation of the sum of its parts. In effect, the mag-

> nitude of a *quantum* implies nothing more than the summation of the *quanta* that make it up. From this homogeneity another property follows: a quantified phenomenon is "(fore-) seen in advance [*schon... angeschaut*] as an aggregate (sum of the parts given in advance) [*vorher gegebener*]."[19] This sort of phenomenon would always be foreseeable, literally seen before being seen in person or seen by procuration, on the basis of another besides itself—more precisely, on the basis of the supposedly finite number of its parts and supposedly finite magnitude of each among them. (Marion 2002b, *BG* 200)

Secondly, with "Tinges, stains, tints, and tones" that "crumbled into the eye," thereby inverting its logic of poverty, intuition ceases to be limited by its concept, whatever that may be, and becomes both saturated and saturating. It thus finds its own freedom to "give more," as Marion has argued earlier, "indeed immeasurably more, than the intention would ever have aimed at or foreseen," rendering itself "overexposed," as Smithson himself puts it in his exposition, as if qua Marion, and hence invisible. In this way, "its excess can neither be divided nor adequately put together again by virtue of a finite magnitude homogeneous with finite parts," Marion argues, "since the saturating intuition surpasses limitlessly the sum of the parts by continually adding to them" (Marion 2002b, *BG* 200). Smithson concurs; and when referring to the excessive "views ... tumbling over each other," for instance, he asks rhetorically in the same vein, "How could that section of visibility be put together again?" As such, the saturated phenomenon of "Tinges, stains, tints, and tones crumbled into the eyes" no longer constitutes any object, and lends itself instead only to "an instantaneous synthesis," as Marion will make clear, one "whose representation precedes and surpasses that of the eventual components, instead of resulting from it according to foresight" (Marion 2002b, *BG* 200). Resonating with Smithson's phenomenological notion of objects as "the excrement of thought and language," as "phantoms of the mind, as false as angels" (Smithson 1996, 122), Marion then continues to detail this "instantaneous synthesis" and its concomitant invisibility in contrast to the "successive synthesis" and its self-evident visibility of the object:

> The [instantaneous] synthesis takes place without complete knowledge of the object, therefore without *our* synthesis. It is thus freed from the objectness that we would impose on it so that it might impose on us its own synthesis, accomplished before we could reconstitute it (a passive synthesis, therefore). Its coming forward precedes our apprehension, rather than resulting from it... it comes before our gaze at it, it comes early, before us. We do not foresee it; it foresees us. (Marion 2002b, *BG* 201)

Smithson's anti-vision, eco-phenomenalized as it is in "a wilderness of unassimilated seeing" in "The Eighth Mirror Displacement," likewise intimates that the unconditional appearance of the saturated phenomenon is "unbearable according to quality." Understood as "intensive magnitude," quality is what "allows intuition to fix a degree of reality for the object by limiting it," Marion points out (Marion 2002b, *BG* 202). However, contrary to the foresight built in the "successive synthesis of the homogeneous," anticipation of intensive magnitude operates "in a perception of the heterogeneous, in which each degree is demarcated by a dissolution of continuity with the preceding," Marion notes, "therefore by an absolutely singular novelty" (Marion 2002b, *BG* 203). Otherwise put, quality is none other than the more refined measure inscribed in and deployed by the logic of penury of intuition in ideal adequation to intentional aim at an object. But, Kant refines the measure further. Owing to his privilege of the "poor phenomenon," Marion observes, Kant nevertheless "approaches intensity only by strangely privileging phenomena of the weakest intensity, precisely where intensity is lacking, to the paradoxical point of basing it on the very absence of intensity, negation" (Marion 2002b, *BG* 203). Defining intensity "starting from its degree zero," such a Kantian paradigm, one of "a poor phenomenon, indeed one empty of intuition, definitely blocks, in metaphysics at least, every advance toward the liberated phenomenality of givenness" (Marion 2002b, *BG* 203). By contrast, the saturated phenomenon is unbearable; Marion explains:

> For the intuition saturating a phenomenon attains an intensive magnitude without measure, or common measure, such that starting with a certain degree, the intensity of the real intuition passes beyond all the conceptual anticipations of perception. Before this excess, not only can perception no longer anticipate what it will receive from intuition; it also can no longer bear its most elevated degrees. For intuition, supposedly "blind" in the realm of poor or common phenomena, turns out, in a radical phenomenology, to be blinding. The gaze cannot any longer sustain a light that bedazzles and burns. The intensive magnitude of intuition, when it goes so far as to give a saturated phenomenon, cannot be borne by the gaze, just as this gaze could not foresee its extensive magnitude. ... For not bearing is not simply equivalent to not seeing.... It concerns a visible that our gaze cannot sustain. (Marion 2002b, *BG* 203)

Foregrounding its anti-vision, Smithson's "The Eighth Mirror Displacement" dramatizes, perhaps most explicitly, such an intensive magnitude of the saturated phenomenon by unfolding "a wilderness of unassimilated seeing" in which, as the artist himself puts it, all "measure is dropped and incomputable" (Smithson 1996, 124). Facing the onslaught of the intensive magnitude from "ashy hues, blotches and sunburned chromatics" as well as "grains, chips, and other jungle obstructions," among many others, "the eyes" can no longer "see straight," Smithson acknowledges, as "sight consisted of knotted reflections bouncing off and on the mirrors and the eyes." The intensity of the saturated phenomenon thus passes, as Marion observed earlier, beyond all the conceptual anticipations of perception to such an elevated degree that "All the clear ideas of what had been done melted into perceptual puddles." The result is that "vision sagged, caved in, and broke apart." To his own tentatively-proposed solution that "perhaps the eyes should have been screwed up into a sharper focus," the artist responds with an immediate and resounding, "but no." For, however the eyes may have attempted to focus, they turn out to be either "cocked-eyed," "myopic," "overexposed," or "cracked;" the eyes are, in other words, blinded by the blinding intensity of the saturated phenomenon that they cannot see, much less sustain. Described figuratively by Smithson as "two wastebaskets," the eyes in the "Mirror Displacement" receive a similar but less expressive account from Marion when he says,

> Thus, the eye experiences only its powerlessness to see anything, excerpt the bursting that submerges it—almost metallic and vibrating—which blinds it. Thus appears the excess of intensive magnitude in the pure and simple impossibility of even maintaining it within the horizon of the visible. (Marion 2002b, *BG* 205)

That the unconditional appearance of the saturated phenomena is invisable according to quantity and unbearable according to quality leads, inevitably, to the third property of the phenomenon as such: it is "*absolute* according to relation," as Marion states, "which means it evades any analogy of experience" (Marion 2002b, *BG* 206). Smithson's earth art of "The Mirror Displacements" strikes one as radically unique precisely because, and in light of Marion's phenomenology of givenness, the unconditional appearance of this saturated eco-phenomenon is, simply, unprecedented. What is that, for instance, which "the eyes see" but can not be reconstructed "in words, in an ideal language"? If, following the general outline of Kant's definition of the principle of such analogies, experience is understood as "possible only through the representation of a necessary connection of perceptions," a connection that "will have to produce itself ... through concepts " that would connect "a priori" the "existence of objects" and "their relation in time,"[20] and if, as Marion has epitomized it, it is this connection as such that "permits three relations: inherence of accident in substance, causality between cause and effect, commonality among several substances," Smithson's "Mirror Displacement" is one that foregrounds the opposite: disconnectedness; "the distances between the twelve mirrors are shadowed disconnections," Smithson writes, "The mirror surfaces being disconnected from each other" (Marion 2002b, *BG* 206, 123–124, 128). It follows that the saturated phenomenon that appears in "The Mirror Displacement" is, then, disconnected from or un-analogous to any experience whatsoever, having exceeded the limit of concepts that would otherwise perform such connection.

Moreover, the saturated eco-phenomenon in Smithson's "The Eighth Mirror Displacement" defies the "three presuppositions" that Kant employs to establish the above-mentioned three relations permitted by connection (Marion 2002b, *BG* 206–207). First, mounted randomly on the irregular contours of the eroding sandy slope of an "unknowable zero island," the "Mirror Displacement" stages an "anti-vision" or "negative seeing" precisely because the saturated phenomenon does not appear "by respecting the unity of experience," as Marion explains, "that is to say, by taking place in a network as tightly bound as possible by lines of inherence, causality, and commonality that assigned to it, in the hollow as it were, a site" (Smithson 1996, 129; Marion 2002b, *BG* 207). The artist's question of "Where is the island?" implies an uncharted locale for the "Mirror Displacement" that is a far cry from "a site predetermined by a system of coordinates, itself governed by the principle of the unity of experience" (Smithson 1996, 129; Marion 2002b, *BG* 207). In this sense, Smithson's "anti-vision" declares, as does Marion's rhetorical question, "the possibility that a phenomenon might impose itself on perception without assigning it either a substance in which it resides like an accident or a cause from which it results as an effect, or even less an interactive *commercium* where it is relativized" (Marion 2002b, *BG* 207). Secondly, in Smithson's "The Mirror Displacement," "the mirror itself is not subject to duration," as the artist understands it, and hence is "timeless" (Smithson 1996, 122). With such "Timelessness," "The Mirror Displacement" breaks free from Kant's privilege of analogy and his presupposition that "all empirical time-determinations must [*müssen*] stand under rules of universal time-determination. The analogies of experience ... must [*müssen*] be rules of this description" (Smithson 1996, 121; Kant 1965, 210).[21] An analogy, which is instrumental in the construction of unity, and "by recourse to" which "this unity should be always be accomplished," as Marion analyzes it, functions, on the one hand, as the constitutive "procedure that allows [one] to secure temporal and conceptual necessity, therefore the unity of experience" and, on the other hand, as the sole executor of "the regulation of experience by necessity, therefore to assure its unity" (Marion 2002b, *BG* 207, 208). Since the mirrors are themselves timeless, the saturated phenomenon thus mirrored, such as "bits of 'place' … cast into a white sky," severs itself from any analogy of experience by stripping itself of the temporal horizon as the condition for an analogical, intra-temporal connectivity. Thirdly, Smithson's use of plurals in "The Eighth Mirror Displacement," such as "knotted reflections," "All viewpoints choked and died," or "views … tumbling over each other," corresponds to a similar usage in "horizons were submerged and suffocated in an asphyxiation of vanishing points." Situated within the "wilderness of unassimilated seeing," Smithson's usage of the plural here invokes Marion's question of how "phenomena exceed their horizon" and his proposed answer in a "hermeneutic of an infinite plurality of horizons," against which the "essentially and absolutely saturated phenomenon" appears absolutely (Marion 2002b, *BG* 209, 211). Marion writes:

> If the hermeneutic of an infinite plurality of horizons is by chance not enough to decline an essentially and absolutely saturated phenomenon, it could be that each perspective, already saturated in a single horizon (bedazzlement), is blurred once again by spilling over the others—in short, that the hermeneutic adds the bedazzlements in each horizon, instead of combining them. Then, not only no single horizon, but no combination of horizons, could successfully tolerate the absoluteness of the phenomenon, precisely because it gives itself as absolute, that is to say, free from all analogy with common-law phenomena and from all predetermination by a network of relations, with neither precedent nor antecedent in the already seen or foreseeable. In short, there would appear a phenomenon saturated to the point that the world (in all senses of the word) could not accept it. Having come among his own, his own do not recognize it; having come into phenomenality, the absolutely saturated phenomenon could find no space there for its display. (Marion 2002b, *BG* 211)[22]

That being the case, it could be argued, though, that the absolutely saturated phenomenon does find its display space in Smithson's "wilderness of unassimilated seeing" which exists indeed only on the outer-fringe of a radical imagination and, in which, "by giving itself absolutely," the saturated eco-phenomenon "also gives itself as absolute," as Marion contends, "free from any analogy with already seen, objectified, comprehended experience. It is free because it does not depend on any horizon. In every case, it does not depend on this condition of possibility par excellence—a horizon, whatever it might be" (Marion 2002b, *BG* 211–212).

Situated within "a wilderness of unassimilated seeing," Smithson's "anti-vision" or "negative seeing" also articulates, perhaps most literarily, that the absolute and unconditional appearance of the saturated eco-phenomenon is *irregardable* [beyond any seeing] according to modality. On this issue, Smithson's punning on "eye/I" kills two birds with one stone. For one thing, the literal absence of the first person singular "I" in the "Mirror Displacement," whose existence can thus be evoked only through punning, invalidates, by default, the "categories of modality" as the "operators of the fundamental epistemological relation to the I" who is the "power of knowing" and with whom objects must agree "absolutely if they are to be known" (Marion 2002b, *BG* 212). More specifically, insofar as phenomena are concerned, Marion explains:

> This agreement determines their possibility (therefore also their actuality and their necessity) to be and to be known as phenomena solely by the measure of their suitability to the I, for whom and by whom the experience takes place. ... The phenomenon is possible strictly to the extent that it agrees with the formal conditions of experience, therefore with the power of knowing that fixes them, therefore finally with the transcendental I itself. ... Far from showing itself, it is staged only in a scene set by and for an other besides it, actor without action, submitted to a spectator and transcendental director. (Marion 2002b, *BG* 212–213)

In this sense, the absence of the "I," or the erasure of an "objectifying intentionality," which otherwise constitutes the phenomenon into an object through the categories of modality, suggests, then, a "disagreement between an at least potential phenomenon and the subjective condition for its experience" (Marion 2002b, *BG* 213). This disagreement results from "the type of phenomenon that is exceptional by excess," so much so that this saturated phenomenon, being both "nonobjective" and "nonobjectifiable," "annuls all effort at constitution" by the "I," and, hence, the displacement of the "I" and its categories of modality (Marion 2002b, *BG* 213).

For another, the "eye," which features exclusively in "The Eighth Mirror Displacement," and which puns on the "I," is depicted as being purely passive. Either the saturated phenomenon "crumbled *into* the eyes," for instance, or "The eyes became two wastebaskets *filled with* diverse colors, variegations"; either "all viewpoints [were] *choked*," or "the eyes, *being infected by* all kinds of nameless tropisms."[23] From this perspective, the "eye" in Smithson's "Mirror Displacement" is also displaced, having been removed from the position of the "author" or "master and possessor" of the phenomenon, much as Marion lays out, to become the "receiver" or the "witness" (Marion 2002a, *IE* 113). Blinded by the overwhelmingly intense lights of the saturated phenomenon, this receiver is no longer able to gaze at the saturated phenomenon; "gazing, *regarder*," which means to "keep the visible thus seen under the control of the seer, exerting this control by guarding the visible in visibility," is the same as "transforming it [the phenomenon] into an object visible according to an always poor or common phenomenality—visible within the limits of a concept," thereby keeping the object "in an objected state for the I" (Marion 2002b, *BG* 214).

Thus deployed, Smithson's "Mirror Displacement" displays, artistically, the saturated phenomenon. It articulates, from within its "wilderness of unassimilated seeing," an imaginative hypothesis most succinctly verbalized by Marion when he says, "determining the saturated phenomenon as irregardable amounts to imagining the possibility that it imposes itself on sight with such an excess of intuition that it can no longer be reduced to the conditions of experience (objecthood), therefore to the I that sets them" (Marion 2002b, *BG* 215).

## Conclusion

"Only appearances are fertile," Smithson writes in his concluding remarks on his "Mirror Displacements" earth art; "they are gateways to the primordial" (Smithson 1996,132). With its denotative meaning of "pertaining to, or existing at (from) the beginning, first in time, earliest, original… fundamental, radical, elementary,"[24] the term "primordial" designates a state of phenomenality prior to the imposition of reduction, be it transcendental or existential, a state of undifferentiated phenomenality of which the only reduction, if there be any, would be the reduction of the phenomenon to its own absolute and unconditional givenness. In this light, "only appearances are fertile" precisely because they are the appearances of the saturated phenomenon, rich in intuition, giving itself, showing itself, and starting from itself alone, absolutely and unconditionally; hence, appearances are the portals to a world where, as Marion observes, "givenness would organize phenomenality universally without exception," where, that is, "all horizons are shattered" (Marion 2002b, *BG* 179; Horner and Berrand 2002, ix). In Smithson's eco-phenomenology thus demonstrated in his earth art of the "Mirror Displacements," his leitmotif of an "anti-vision" or "negative seeing" presents a post-humanist position that resonates intimately with Marion's phenomenology of givenness. For Smithson, "Art works out the inexplicable"; so does the phenomenology of givenness for Marion, for "it sustains itself" paradoxically, "not on differentiation, but dedifferentiation, not on creation but de-creation, not on nature but de-naturalization" (Smithson 1996, 132).

## Notes

1. In Smithson's text, the term "limitlessness" is put in quotation marks, and it occurs in a passing reference to Anton Ehrenzweig in Smithson's brief discussion of Tony Smith's artwork.
2. As Smithson makes clear, "the mirror displacements were dismantled right after they were photographed" (Smithson 1996, 132–133). Therefore, this study of Smithson's earth art of the "Mirror Displacements" is based on the narrative documentation of this particular earth project.
3. Smithson states clearly that each "Mirror Displacement," such as the First, the Second, the Third, the Fourth, the Fifth, and the Seventh, has 12 mirrors in it, but he does not specify how many mirrors there are in the Sixth, the Eighth, and the Ninth. While it is safe to assume, given the consistency in the number of mirrors used in the majority of the displacements, that there are 12 mirrors in each of the nine displacements, the actual number of mirrors involved is irrelevant as far as concerns the phenomenological issue covered in this paper.
4. Original emphasis. Unless otherwise noted, all italicization for emphasis in this paper is original.

5. The phrase "Phenomenology of Givenness" is taken from the subtitle of Jean-Luc Marion's book *Being Given: Toward a Phenomenology of Givenness*, which will henceforth be cited in the text as *BG*. Marion's phenomenological trilogy also include *Reduction and Givenness: Investigations of Husserl, Heidegger, and Phenomenology*, henceforth cited in the text as *RG*, and *In Excess: Studies of Saturated Phenomena*, henceforth cited in the text as *IE*.
6. For Marion's outlining of the differences between the three phenomenological reductions, see "Conclusion: The Figures of Givenness" in (Marion 1998, *RG* 203–205), especially pp. 204–205.
7. My emphasis.
8. Quoted in (Marion 2002b, *BG* 184).
9. Quoted in (Marion 2002b, *BG* 186).
10. Kant. *The Critique of the Power of Judgment* (Kant 2000, 218). Quoted by Marion (Marion 2002b, *BG* 198).
11. My emphasis.
12. Kant. *The Critique of the Power of Judgment*, (Kant 2000, 218, 219). Quoted by Marion (Marion 2002b, *BG* 198).
13. My emphasis.
14. My emphasis.
15. For the etymological definition of "invisible," derived from the word *viser* by the translator, see Note 41 on page 363 in *BG*.
16. *The Oxford English Dictionary*, 1989. Vol. IV, 81, 81.
17. Quoted by Smithson (Smithson 1996, 128).
18. Quoted by Marion (Marion 2002b, *BG* 200).
19. Kant, *Critique of Pure Reason* (Kant 1965, 199); quoted by Marion (Marion 2002b, *BG* 200).
20. Kant, *Critique of Pure Reason* (Kant 1965, 208); quoted by Marion (Marion 2002b, *BG* 206).
21. Kant, *Critique of Pure Reason* (Kant 1965, 210); quoted by Marion (Marion 2002b, *BG* 207–208).
22. Marion's "hermeneutic of infinite plurality of horizons" is what he considers to be the "third case" of phenomenological situations as regards how the phenomenon exceeds its horizon; and this "third case," in Marion's account, "redoubles the first two cases by lumping them together" (Marion 2002b, *BG* 211). For the details of the first two cases, see (Marion 2002b, *BG* 209–211).
23. My emphasis.
24. *The Oxford English Dictionary*, 1989. Vol. XXII, 489.

## Work Cited

Flam, Jack. 1996. Introduction: Reading Robert Smithson. In *Robert Smithson: The Collected Writings*, xiii–xxxv. Berkeley: University of California Press. Print.

Geniusas, Saulius. 2012. *The Origins of the Horizon in Husserl's Phenomenology*. Dordrecht: Springer. Print.

Horner, Robyn, and Vincent Berrand. 2002. Introduction. In *In Excess: Studies of Saturated Phenomena*, iv–xx. New York: Fordham University Press. Print.

Husserl, Edmund. 1998. *Ideen I. Ideas Pertaining to a Pure Phenomenology and to a Phenomenological Philosophy. First Book: General Introduction to a Pure Phenomenology*. Dordrecht: Kluwer Academic Publishers. Print.

Kant, Immanuel. 1965. *Critique of Pure Reason*. New York: St. Martin's Press. Print.

———. 2000. *The Critique of the Power of Judgment*. Cambridge: Cambridge University Press. Print.

Marion, Jean-Luc. 1998. *Reduction and Givenness: Investigations of Husserl, Heidegger, and Phenomenology*. Evanston: Northwestern University Press. Print. [Cited as *RG* in the text of this paper.]

———. 2002a. *In Excess: Studies of Saturated Phenomena*. New York: Fordham University Press. Print. [Cited as *IE* in the text of this paper.]

———. 2002b. *Being Given: Toward a Phenomenology of Givenness*. Stanford: Stanford University Press. Print. [Cited as *BG* in the text of this paper.]

Smithson, Robert. 1996. In *Robert Smithson: The Collected Writings*, ed. Jack Flam. Berkeley: University of California Press. Print.

*The Oxford English Dictionary*. 1989. 2nd ed. Oxford: Clarendon Press. Print.

# The Transcendental Philosophy of Krishnachandra: An Indian Approach to Human Life

**Koushik Joardar**

**Abstract** The present article humbly proposes that, inspired by Kant, one of the greatest modern Indian philosophers, Krishnachandra Bhattacharyya, was doing a sort of phenomenology in the name of "Transcendental Psychology" without knowing of the existence of Husserl and his works. The task of a philosopher or the reflecting consciousness, says Krishnachandra, is to practice a kind of regress towards transcendental subjectivity in order to realize the subject as freedom. At the final stage of this, the subject-object distinction vanishes altogether and thereby the Absolute is achieved. Krishnachandra, being influenced by Kantian-Hegelian philosophy and being committed to the Indian philosophical tradition, has advocated a special kind of phenomenology that is both descriptive and prescriptive. The goal of his transcendental philosophy is *mokṣa* (liberation).

**Keywords** Phenomenology · Transcendental psychology · Subjectivity · Freedom · Absolute · *mokṣa*

## Krishnachandra's Indian Setting

"All of Phenomenology is not Husserl, even though he is more or less its center," writes Paul Ricoeur (Ricoeur 2007, 3). Ricoeur then names philosophers of the phenomenological tradition with some of whose thought Husserl connects: Kant, Descartes and even Hume (but not at all the Hegelian philosophy denominated phenomenology). And, of course, there are the existentialists who come later and were influenced by Husserl.

Krishnachandra Bhattacharyya (1875–1949), popularly known as KCB in India, is not known to be a philosopher belonging to this tradition. Indian Philosophy is a different kind of philosophical tradition and, therefore, it is not to be expected that a philosopher of this tradition can contribute something original and significant to

K. Joardar (✉)
Department of Philosophy, University of North Bengal, Darjeeling, West Bengal, India



W. S. Smith et al. (eds.), *Eco-Phenomenology: Life, Human Life, Post-Human Life in the Harmony of the Cosmos*, Analecta Husserliana CXXI,
https://doi.org/10.1007/978-3-319-77516-6_26

the Philosophy of the West (just as it is not expected by Indian traditionalists that a thinker from the West can really contribute something to their Philosophy). Moreover, Krishnachandra is not very much known in the West, one reason for that being that his works are not published internationally. Nevertheless, I take this opportunity to present to you the thought of Krishnachandra, wherein two different traditions meet.

Traditionally, Indian Philosophy is not marked by individual thinkers but by schools of thought. There are six orthodox schools of thought established by six original thinkers who expressed their thought in the form of verse; all the other thinkers have been interpreters of the concepts of a particular school and critics of other schools. The same is true also with the unorthodox schools such as Jainism and Buddhism. Thus, in traditional India, the ideas of different schools have existed simultaneously over the years conflicting with each other. However, the situation changed in nineteenth and twentieth century colonial India. There are thinkers of this period who have both a traditional background and exposure to western thought, and are studied as individual thinkers. Krishnachandra is one of them.

## Krishnachandra's Kantian Inspiration

In *The Subject as Freedom* (SF), published in 1930, Krishnachandra Bhattacharyya calls for a new philosophical study, "Spiritual or Transcendental Psychology," to deal with subjectivity (Bhattacharyya 1983, 390). Like Husserl, KCB borrows the idea of the"transcendental" from Kant and uses it in a somewhat Kantian sense. At the very outset of his First Critique, Kant explains his sense: "I apply the term *transcendental* to all knowledge which is not so much occupied with objects as with the mode of our cognition of these objects, so far as this mode of cognition is possible *a priori*. A system of such conceptions would be called *Transcendental Philosophy*" (Kant 2003, 15). Despite all differences with Kant, Husserl's own transcendental theory is a continuation of the Kantian tradition. Krishnachandra, who made Kant popular in India, designates his philosophy as "transcendental" without explicitly explaining why he does so. However, KCB states his understanding of the word in its Kantian context clearly in his "Studies in Kant" (SK). "The 'transcendental' may be taken provisionally to mean what we are certain about as not objective. By 'object' is meant a content that is other than the consciousness of it. The transcendental then would be a content that is not distinct from the consciousness of it" (Bhattacharyya 1983, 663). Thus, the transcendental, first of all, belongs to the realm of subjectivity. Then, to repeat, it is not just consciousness, but a content of consciousness that is not distinct from the consciousness of it. This content may be called structure or modes of subjectivity. Krishnachandra's philosophy is transcendental because it deals purely with subjectivity and its modes.

## Psychologism?

But the word "psychology" that KCB employs to designate his philosophy is detested by phenomenologists. In fact, Husserl himself took a psychologistic approach in his *Philosophy of Arithmetic* and argued that numbers and the laws of mathematics are derived from the mental acts of experiencing. This psychologism, an attempt to reduce Logic, Mathematics, and Philosophy in general to Psychology, was vehemently criticized by Frege. Absorbing this criticism, Husserl rejected psychologism altogether and was convinced that experience can only give us contingent truth, whereas logical and philosophical truths are necessary. He even accused Kant of psychologism for presupposing "faculties" and "categories."

KCB also distanced himself from Psychology as a Natural Science. Psychology deals not so much with the subjective function as with the object as it is known. If psychological truth is abstraction from objects, Krishnachandra's Transcendental Psychology is about freedom from such abstraction. Empirical psychology takes an objective attitude, in which the object appears to exist beyond its relatedness to the subject, whereas the subjective attitude taken by transcendental psychology rejects this *beyondness* as meaningless. In fact, KCB denies the possibility of a metaphysics of the self because the attitude of metaphysics is objective and the self cannot be known objectively. "The attitude of metaphysics like that of the sciences including psychology is objective" (Bhattacharyya 1983, 387), he noted. Thus, the psychology of which KCB speaks is a special kind of psychology, one free from the objective attitude of its naturalistic counterpart. As the subject cannot be known from the objective attitude, properly speaking, there can be no metaphysics of the subject.

## The Ultimate Import of the Subject and Object Distinction

One of the major tasks of Krishnachadra's philosophy is to demonstrate the difference between the meanings of 'subject' and 'object', to clarify the notion of subjectivity as opposed to objectivity. To know the distinction between subject and object is to know the distinction between the meanings of 'subject' and 'object'. An object is always meant by a word that can be used by both the speaker and hearer to mean the self-same entity. When I utter 'table', the speaker may understand the same table that I am talking about. Thus, it is not only that words have meanings, but also that words are used by the speakers to mean objects. Objects are what are meant by the subjects in their using words. The word *this* may be taken as a symbol of what is meant. In contrast, the subjective, expressed through the word *I*, cannot be said to mean anything like an objective *this*. The speaker and the hearer can mean the same entity by using *this*, but *I* is never understood in that way. Nor is the word *I* meaningless, for it is understood by all. "As used, the term has a uniquely singular reference; but as understood, it is general in the sense the term *unique* is general" (Bhattacharyya 1983, 382). Whoever uses the word *I* refers to himself by it. *I* does not mean anything, but the subjective has an awareness of herself as the speaker of *I*. This is another peculiarity of *I*, that the act of speaking *I* is a part of understanding it, which is not the case with *this*. Thus, the subject is never an object.

The object is meant by the word *this* and subject is intended (but not meant) by the word *I*. However, sometimes the subject may be spoken of as an object as in "I am this" or "This is I." KCB rejects both these statements as unintelligible, holding that *this* can only refer to my body. As I am not my body, 'This is I' is a false statement but "I am this" is believed because my individuality is prior to my body. Whenever I relate myself to my body, I intend to say that I am the speaker of *I* individualized in this body. But I never identify myself with my body.

Both subject and object are known. The object is known as a meant fact, but subject is known not in that sense but 'is known in itself'. KCB criticizes the Kantian theory that the self is unknown but can only be thought. Rather, he holds the opposite view that "the subject is known though neither thought (meant) nor intuited" (Bhattacharyya 1983, 393). Clearly, for KCB, knowledge is something more than a synthetic *a priori*. The subject knows herself as she understands herself as the speaker of *I*, and this understanding is a "direct believing in something that is not meant but revealed as revealing itself" (Bhattacharyya 1983, 393). Thus, the subject is self-evidencing like Husserlian apodictic evidence, the non-existence of which is inconceivable. The question of the reality of self does not arise because that reality is self-revealed without being meant. In contrast, the reality of the meant – that is, the object – can always be doubted. Nothing meant is self-evidencing. Krishnachandra denies that the self can be thought of because only thinking about an object is possible.

## Some Cross-Examining

However, may we not question KCB's position that the self cannot be thought of, because are not we presently thinking about the self? Here one may recall Sartre's critique of the Cartesian cogito in the introductory chapter of *Being and Nothingness*. The consciousness which says *cogito ergo sum* is not actually the consciousness that doubts; rather, it is the consciousness that takes the primary consciousness of doubt as its object. Whenever a consciousness "thinks about itself," it is actually a reflective consciousness that takes primary consciousness as its object. "The reflecting consciousness posits the consciousness reflected-on as its object" (Sartre 1966, 12). This primary consciousness, in Sartre's terminology, may be called pre-reflective consciousness. It is not that only reflective consciousness is self-conscious, but that by nature, all consciousness is self-conscious. KCB's subject that cannot think of itself is pre-reflective consciousness, which is never meant. Nevertheless, we may say that his transcendental psychology is a reflective consciousness.

It is clear from KCB's contending that to speak of an unmeant object is meaningless. Now, is the relation between meant and what (or who) means, that is, between the object and the subject, necessary? An answer in affirmative will place one closer to the phenomenological tradition. But before answering, we need to understand Krishnachandra's concept of spiritual progress. After criticizing Kant with respect to the unknowability of the self, KCB suddenly brings in the concept of spiritual progress and states, "Spiritual progress means the realization of the subject as free" (Bhattacharyya 1983, 94). That the subject is essentially free is felt: "the subject is only known in itself and felt to be free or dissociated from the object" (Bhattacharyya 1983, 385). In Krishnachandra's usage, "freedom" and "dissociation" are almost synonymous, and the distinction between distinction and dissociation is of fundamental importance with him, the way in which two objects are distinct or an object is known as distinct from the subject, yet the subject cannot be said to be distinct but dissociated from the object. He again uses the words 'alienation' or 'dissimilation' for 'dissociation'. Freedom, then, is the presupposition behind the act of dissociation. However, KCB uses 'freedom' and 'dissociation' in the same sense; to feel free and to feel dissociated from an object is the same thing for him.

## Spiritual Progress in Stages

Now let us return to the concept of spiritual progress. Progress here means realization, as KCB points out. There may be some who experience freedom already. For those who need some specific activity for the realization that the subject is essentially free, he provides a method. The method has an inward direction or, as he puts it, it is a method of cognitive inwardising. This inwardising direction indeed sounds Cartesian or Husserlian, but KCB does not elaborate this method much except for stating that, "A method implies a series of consecutive steps for the realization of an end. The steps in this case should correspond to a gradation of subjective functions or modes of freedom from the object" (Bhattacharyya 1983, 394). The whole treatment of freedom is then description or explanation of these gradations, different modes of freedom that we have to realize step by step. The three broad stages in which freedom is (to be) realized are: bodily subjectivity, psychic subjectivity, and spiritual subjectivity.

The first stage is bodily subjectivity, in which the subject identifies himself with the body. This involves not only the perceived body but also a feeling of body or a felt-body, such that the bodily subject dissociates himself from the objects 'outside' the body. It is through the awareness of the body that the subject first becomes aware of its subjectivity and its 'distinction' from the worldly objects.

In the next stage, that of psychic subjectivity, the subject dissociates herself even from the felt body and identifies with psychic life, which actually comprises image and thought.

A negation of this stage leads to the third and the highest stage of spiritual subjectivity. In this stage, the freedom (to be) realized is freedom from thought; it too has three sub-stages, those of feeling, introspection, and beyond introspection. In the psychic stage, thought and content appear to be distinct. But in the awareness of feeling, the 'unthought content of feeling' is not distinct from feeling itself. Feeling,

KCB states, is consciousness of detachment from meaning, an awareness of content as unmeant. Introspection, as distinct from the psychic phase, is awareness of the subject as being expressible by the word *I*. It is I-consciousness itself. In introspection, the subject understands himself not through the meaning of *I*, but through the word itself. Krishnachandra prefers to say that introspection is self-revealing rather than self-knowing. But still, this is not true subjectivity. True subjectivity is beyond I-consciousness. It is not even expressible by any word. "The non-being of distinction is finally understood here and hence too the conception of the absolute self" (Bhattacharyya 1983, 449). What KCB may intend to say is that even in introspective awareness of I, there is an awareness of distinction. The awareness of I presupposes an awareness of other, and vice-versa. In the final stage of beyond introspection, the subject-object distinction vanishes altogether and the Absolute is achieved. This sounds very much Hegelian, indeed.

In every grade of subjectivity, KCB states that freedom is felt with respect to objects (or meant content) presented to it. Even in introspection, the subject can still be an object to introspection, and so KCB thinks that this is not "freedom itself." True freedom is achieved when the subject becomes absolutely free from any sense of objectivity, meant content, and duality. It is difficult, however, to understand how without any sense of duality in what sense the "absolute" still remains subject. KCB never makes it clear whether with the loss of objectivity or duality, subjectivity is also lost.

## Krishnachandra's Transcendental Psychology and Husserlian and Existentialist Phenomenology Compared

A comparison of the transcendental psychology of Krishnachandra and the phenomenology of Husserl and the Existentialists will yield such interesting observations as the following:

1. KCB takes a first person-account viewpoint and rejects the possibility of any unmeant object. His methodology, like that of Husserl, is introspection. The goal of his philosophy is to show that subject is essentially free. However, what he does is not only descriptive metaphysics but also a prescriptive vision. The aim of transcendental psychology is not only to show that the subject is free, but also to show the ways in which the subject may become free. It urges an effort to free oneself from meanings in order to realize one's true nature as freedom.
2. Freedom is achieved through negation. Negation means dissociation. In the first stage of the process of realization, the subject as a felt-body dissociates itself from objects outside; in the second, the subject dissociates from the felt-body and identifies itself with thought. In the final stage, thought is also negated and the subject frees itself from all kinds of meaning.
3. The negation of which KCB speaks is not Hegelian because it does not give rise to synthesis. It is not Cartesian because it is not a methodological doubt pursued

in order to bring back the whole world as certain. Neither is it Husserlian *epoché*, the suspension of judgments regarding whatever is other. The negation of Krishnachandra is best described as subject's dissociation from objects or objectivity in the process of its realization as the unmeant. It is a kind of denial – denial of the existence of objects, yes, but a denial of subject's association with objects.

4. Freedom is nothing but dissociation or freedom that expresses itself in the subject's dissociation from objects. Absolute freedom is achieved when the subject completely frees itself from object, content, or meaning. KCB seems to maintain that objects limit freedom. But he does not see that objects at the same time make freedom possible because, whether it is freedom to or it is freedom from, freedom is meaningless without any reference to objects. The final negation annihilates freedom.
5. The phenomenological concept of nothingness necessarily leads to the existentialist concept of freedom. Because consciousness is not a thing, it always has to engage something for its existence. Freedom is to choose between options, and "man is condemned to be free." The significance of this is that humans cannot choose not to choose. But the goal of Krishnachandra's transcendental psychology is the subject's freeing itself completely from objects. The freedom of which KCB speaks is not just from this object or that; it is in fact freedom from choosing between options. One may say that Krishnachadra's subject achieves its true self when it is free from even freedom.

## Estimation

In the last analysis, it can be said that Krishnachandra Bhattacharyya surely exhibits a Western style of philosophizing. His affinity for Kant is unconcealed. Unlike the traditional Indian thinkers of the past, he never speaks for or from the standpoint of any particular Indian school of thought. He is not a materialist like the Cārvākas for he does not think that mind is essentially matter. He is not a realist of the Nyāya-Vaiśeṣika school that believes that there really exists a mind-independent world, and the world as such can be known and be talked about. He stands far away from the standpoint of the Mīmāṃsā School because he is not concerned with the rituals and sacrifices prescribed in the Veda- Saṃhitās. It is sometimes said that there is a considerable influence of Vedānta on Krishnachadra's thought. He indeed wrote three significant works on Vedanta, specifically *Studies in Vedāntism*, *Saṅkara's Doctrine of Māyā,* and *The Advaita and its Spiritual Significance*. The subject's becoming absolutely free seems to resemble Saṅkara's unqualified Brahma. However, I do not think that KCB has rejected the world as being ultimately unreal, as Saṅkara does. The subject only has to negate the world in the sense of dissociation because it is unnecessary for and a hindrance to the realization of its true nature. Rather, as it seems to me, he is closer to the Sāṃkhya-Yoga school of thought (on which he also wrote extensively), which admits of Puruṣa (consciousness) and Prakṛti (material world) as two separate principles, and the goal of puruṣa should be to dissociate itself from Prakṛti. However, the world is not asserted by KCB so strongly as Sāṃkhya asserts it; he does not even think that it should be a philosopher's right concern to do other than explore the subject's consciousness. Moreover, in the works in which he gives his own philosophy, Krishnachandra hardly uses any Indian philosophical terminology.

But, despite all the above considerations, we cannot overlook that Krishnachandra's thought is also deeply rooted in tradition. One common characteristic of all the Indian schools except the Cārvakā school is attendance to the concept of *mokṣa* or liberation. For the Indian Schools, the world is full of sorrow and the same soul experiences birth many times in different lives to suffer worldly pains. However, they also believe that it is possible to liberate oneself from repeated birth and be rid of life's pains forever. The whole transcendental psychology of Krishnachandra Bhattacharyya seems to presuppose this concept of *mokṣa*. That is why, for him, absolute freedom for the subject is ultimately a cutting off from not only the world, but from any sense derived from the world. The goal of Krishnachadra's philosophy is freedom as such, the possibility of which is never envisaged by the existentialist phenomenologists of the West.

## Works Cited

Bhattacharyya, Krishnachandra. 1983. *Studies in Philosophy*. Delhi: Motilal Banarsidas. Print.

Kant, Immanuel. 2003. *Critique of Pure Reason*. Trans. J.M.D. Meiklejohn, New York: Dover Publications. Print.

Ricoeur, Paul. 2007. *Husserl: An Analysis of His Phenomenology*. Evanston: Northern University Press. Print.

Sartre, Jean Paul. 1966. *Being and Nothingness*. New York: Washington Square Press. Print.

# Henryk Skolimowski's Eco-Philosophy as a Project of Living Philosophy

**Anna Małecka and Katarzyna Stark**

**Abstract** The paper presents Henryk Skolimowski's concept of eco-philosophy understood as a 'living philosophy'—i.e., a universal philosophy focusing on life and its creative evolution. Skolimowski, considering himself an heir to Teilhard de Chardin, describes the world as a unity of multifarious mutually interdependent manifestations. The world constantly creates itself, generating new aspects of life, consciousness and sensitivity. Evolution constitutes a way to self-cognition of the primordial light, and humanity participates in the evolution of this fundamental light in the process of its actualization. In their existential efforts human beings achieve the level of spirituality and divinity within their evolutionary development. For the contemporary Polish thinker, the universe possesses divine attributes, and it is in life itself that spirituality originates.

Skolimowski strongly emphasizes the relation between human beings and nature. Introducing the concept of "ecological man," he claims that we have to treat the world as a sanctuary. Hence, the ethical postulate of his philosophy: we should approach life, including all people, all creation, and the entire cosmos with reverence. He also formulates an imperative of ecological responsibility for everybody and everything, as in the cosmic dimension everything is united.

**Keywords** Henryk Skolimowski · Eco-philosophy · Living philosophy · Sanctity of life · Ecological man · Ecological ethics

## Introduction

Although Henryk Skolimowski is not an eco-phenomenologist, his ideas accord much with eco-phenomenology: the ideas of the creativity of life, the unity of the cosmos, the role of humans in eco-development, a new spirituality, and the sanctity

A. Małecka (✉) · K. Stark
AGH University of Science and Technology, Krakow, Poland
e-mail: amm@agh.edu.pl; stark@agh.edu.pl

 
W. S. Smith et al. (eds.), *Eco-Phenomenology: Life, Human Life, Post-Human Life in the Harmony of the Cosmos*, Analecta Husserliana CXXI,
https://doi.org/10.1007/978-3-319-77516-6_27

of life. The author himself calls his philosophy "a living philosophy," for it is projected as part of life itself, as one of the outcomes of life evolution. Considering the original impact of his views on the development of ecological thought, Skolimowski is, and quite pertinently, regarded as a creator of Polish mainstream eco-philosophy (Pasek and Dyczewska 2012, 71).

The trend of eco-philosophy here analysed primarily grows out of criticism of the positivist cult of science and reason, which dominated human consciousness in the nineteenth and twentieth centuries, as well as from questioning the high status uncritically granted technology in various fields of life in contemporary civilization. Addressing of pan-technicism and a machine ethic accepting injustice and exploitation, the Polish philosopher sees risks to the human condition in technology, which has become a barbaric god to man.

According to the author of the *Living Philosophy*, it is an erroneous cosmology accepted within Western culture that is responsible for the present predicament of man and his relationship with the world. To quote Janusz Górecki commenting on Skolimowski's concept, this cosmology "consists in perceiving the world in the categories of the machine. This cosmology is characteristic for our Western contemporaneity, and its process of development has continued for ages. The basic assumption of Henryk Skolimowski's ecological philosophy is an interpretation of the world as a sanctuary, not as a deterministic machine" (Gorecki 2007, 209–210).

Within so-called deep ecology, the mechanistic outlook is strongly challenged and a holistic approach towards the world as a system is postulated. The holistic paradigm as applied in the description of the world and cosmos overcomes the scheme of René Descartes' dualistic philosophy, thus defending the close relationship of man with other forms of life. All existents—human beings, animals, plants, and elements—constitute the mutually interdependent manifestations of life itself. Thus the pro-ecological standpoint excludes the Cartesian psycho-physical dualism. In the view of the theory of evolution, this dualistic controversy becomes pointless. Analyzing Skolimowski's thought, Ignacy S. Fiut says that it is more pertinent to speak in terms of co-evolution in the development of nature and the human spirit (Fiut 2003, 68).

The author of *A Vision for the New Millennium* suggests supplementing the scientific approach with mystical experience and poetic description. In his opinion, science's presentation of the cosmic image of the world cannot provide us with a final certainty, and for this reason the choice of a particular cosmology has to be combined with a holistic and reverent attitude. "Eco-philosophy is a rational restatement of the unitary view of the world in which the cosmos and the human race belong to the same structure" (Skolimowski 1992, 2).

The author details the following as the main characteristics of his eco-philosophy:

1. An orientation towards life, as opposed to contemporary linguistically-oriented philosophy. In his view, all philosophy should be a "living philosophy," in effect incorporating affirmation of the sacredness of life.

2. Eco-philosophy means involvement for the sake of human values, nature, and life, in contrast to academic philosophy's concern with objectivity, abstraction, and bare facts. "Life is a phenomenon of commitment. In avoiding commitment, we are avoiding life" (Skolimowski 1992, 43).
3. The concept of a new spirituality as a higher level of nature's evolutionary development. Being inscribed in human nature, the spiritual and religious needs of man are not extinguished even in the contemporary era of the positivist cult of science. However, in the so-called new spirituality, the existence of a personal God is not usually accepted. As Beata Szymańska states, "a new spirituality means assuming some form of transcendence apart from religious institutions, and referring to complexes of elements adopted from various religions and philosophies, rather than to specific systems of belief" (Szymańska 2001, 95). In Skolimowski's broad understanding of the term, spirituality constitutes a crowning point of the advancing evolutionary process.

The roots of Skolimowski's eco-philosophy may be found—as the Polish thinker repeatedly emphasises—first of all in Teilhard de Chardin's cosmology, representing a Christian evolutionism in which an Omega Point constitutes the ultimate goal of evolution. Skolimowski also refers to Teilhard de Chardin's notion of the noosphere, which in the path of its historic refining leads to the supreme consciousness, that is, the actualization of the Omega Point.

For the author of the *Technology and Human Destiny*, cosmology thus understood is interpreted as metaphysics—no longer does it need any justification, but rather it justifies all other sciences. In the view of his new eco-cosmology, the universe possesses divine attributes, and our spirituality emerges out of the universe as an intrinsic aspect of it. What is more, as higher outcomes of evolution, we play a crucial role in the process of its further development. Eco-cosmology assumes that "the universe is home for the human race, and we are its stewards, custodians, and guardians" (Skolimowski 1992, 14). According to the author, his eco-cosmology relies upon seven pillars:

1. The anthropic principle. The universe is home for men; being the guardians of the universe, we are responsible for our own fate and for everything around us.
2. A precision of the anthropic principle stating that evolution constitutes a process of creative becoming. "Conceiving of evolution as a creative force, which ceaselessly articulates life in ever new forms of consciousness, is not only congruent with the anthropic principle, but also a necessary extension of it" (Skolimowski 1992, 18).
3. The participatory mind. The mind constantly generates knowledge out of the cosmic amorphous mass of initial data. This means that we co-create together with the cosmos.
4. The implicate order. A universe based on an implicate order is holistic. In the universe all elements are interdependent and mutually define each other.
5. A theology of hope. Hope is an affirmative expression of faith in the ultimate meaning of the cosmos.

6. Reverence for life. Empathy for, and authentic recognition of, the fraternity of all beings.
7. Eco-ethics, which postulates all values related to the worship of life, alongside responsibility for all living forms (Skolimowski 1992, 16–27).

## Evolution Towards Divinity

To quote the philosopher's own words: "The new cosmology (I shall argue) is not God-centred (as in the Bible); is not human-centred (as in a traditional humanism); is not matter-centred (as in the dying scientific-technological world-view), but is evolution-centred" (Skolimowski 1992, 98). Being evolution-centred, it is focused on light. Echoing the Bible, Skolimowski will say that in the beginning, there was light. Light is a symbol of life itself, of the primordial source of existence of all things. Evolution, interpreted as if in Hegelian terms, constitutes a self-cognition of this light. Man plays an essential role in the process of light's self-actualization.

Primordial light's evolution manifests itself in the following phases: photosynthesis; logosynthesis, wherein life is introduced into the sphere of reflective thinking in ethics, religion, art, and philosophy; and theosynthesis, in which light is transformed into sacred being. In other words, cosmic evolution is completed in four cycles, transcending the state of the primary light in the subsequent forms of matter, life, life creativity, and finally life divinity.

Through man, in his self-knowledge, light becomes aware of the evolutionary process of its self-actualization and aspires to the realization of the "potency to further intensification of divinity" (Skolimowski 2007, 24). Man has to realize that the actualization and intensification of divinity within himself constitute the crucial imperative of his life.

Evolution aims at increasingly complex and hierarchical structures, culminating first in the form of biological organisms and, in the end, in the form of human beings. Man is a protector of evolution and, at the same time, only a link in its development. We should feel at home in the universe because we constitute its crowning glory. We are a miracle in the evolutionary chain, because we are able to achieve the level of spirituality and holiness.

It is Skolimowski's assumption that evolution has its objective. And yet, the philosopher says that it is unpredictable, agreeing with Henri Bergson's picture of creative evolution and its indeterministic character. New forms of existence for light emerge in the evolutionary process so that, through human self-awareness and spirituality, light can attain self-knowledge and project "the direction in which the further track shall lead" (Skolimowski 2007, 20).

Divinity is not a gift received from an external God. Neither is God transcendent in relation to man. It is evolution itself which possesses the hallmarks of divinity, as divinity emerges out of the inside of life, constitutes the very essence of life and its ultimate meaning. God is "a radiation of constantly transcending life" (Skolimowski 2007, 174).

The establishment of institutional religions is a culminating point of the evolutionary stage of the cosmos' sacralization (theosynthesis). Man creates images and symbols of God, as well as "complex theologies," which often provoke interreligious disputes. Skolimowski characterizes organized religions as "light filters" which—though necessary at a given stage of humanity's development—have monopolized the primordial light for their followers, requiring obedience and in many cases also intolerance towards other religions. Skolimowski juxtaposes the external God (regarded as an object of worship) and the internal God discovered within oneself on the way to self-perfectioning and known without any mediation or exterior authority.

However, Skolimowski grants a separate ontic status to Gods in the process of evolution. He writes, "Gods are not subjective beings, or mere fictions of human imagination. They are real beings, representing the eons of evolution and transformation of light. The higher we climb the ladder of the sacred, the greater the potential for entities created by light" (Skolimowski 2007, 33). As Fiut emphasizes:

> The philosopher combines eco-theological questions with a reconstruction of the traditional understanding of transcendence. It is within this realm that he tries to find such solutions and revaluations of the dominant axiological models that would univocally justify *love* as a form of the religious ties between man and life on the Earth and in the Cosmos. He is even ready to claim that *in its ultimate consequence* religion *is a form of ecology,* conceived however as a fragment of a future *cosmic ecology*, alongside the present forms of culture (Fiut 2003, 197).

The concept of divinity is ingrained in the natural structure of the human psyche. The Polish philosopher postulates: "What I am proposing, on the other hand, is the natural conception of divinity, or the noetic conception of divinity, as the mind is the creator of all orders, including the divine or spiritual orders. Sacredness is an attribute of the mind, not an attribute of the cosmos. Only when we approach the universe with a reverential attitude and behold it with a mind that is sacred, do we find the universe sacred" (Skolimowski 1992, 234). This quotation summarizes the author's concept of the participatory mind, which co-creates together with the cosmos. That is why Konrad Waloszczyk interprets Skolimowski's ideas as "a kind of pantheising religion which is not rooted in revelation but in a cosmic mysticism, finding divinity in the cosmos, and the cosmos within the human soul" (Waloszczyk 2007, 9).

Human beings constitute fragments of divinity emerging in the evolutionary process, and, at the same time, "bringing Him [God] to being, so to speak, by actualizing the sensitivity-sacredness-divinity latent in us. God is at the end of the road. We are its awkward, dim, unpolished fragments for the time being" (Skolimowski 1992, 110). In the higher phase of theosynthesis, the cosmos itself summons the so-called "coryphaeuses of the sacred sphere"—monks, saints, pilgrims, yogins who achieve enlightenment and divine unity with the universe.

## The Ecological Person and his Ethics

Our eco-philosopher calls his ethics cosmic ethics or the ecological ethic. In this outlook the world is assumed to be a Sanctuary, and so is each of us. That is exactly why all life should be approached with reverence—including all men, all creation, and the entire cosmos.

Values come neither from God, nor are they purely subjective and conventional—their source lies in the very nature of an evolutionary process, in our becoming as a species, as moral beings, as carriers of various forms of sensitivity. Emphasizing the link between man and nature, Skolimowski introduces the concept of the ecological person defined as "a bundle of sensitivities which are in the process of continuous refinement" (Skolimowski 1992, 120). Loyalty, altruism, honor, love, and compassion are forms of our crystalised sensitivity.

The author of *Living Philosophy* argues that man, as a product of evolution at its present stage, is developing spiritually in every sphere of life: intellectually, ethically, and aesthetically. The attitude of reverence for life is styled eco-humanism or—as proposed by Fiut—the "new humanism" or "real humanism" (Fiut 2003, 153). For man, the Earth is a temple in which the ethical principle of moderation and eco-justice is to be followed. The principle of eco-justice is the principle of "justice for all and everything, which includes also the imperative of ecological liability for all and everything, mutually linked in the cosmic dimension" (Fiut 2003, 41).

However, Skolimowski's final words in this field seem to pass beyond the limits of ethics itself. As he writes: "When man really matures and becomes perfect, no ethics will be necessary. Because ethics are only to help imperfect man on his way to perfection" (Skolimowski 2007, 88). Thus the Polish philosopher introduces the concept of the spiritual superhuman, a man perfect in his divinity, who after revaluating values will be able to live without any moral imperatives, beyond all moral principles.

## Transcendence—Creativity

The entire evolutionary process is embraced by acts of transcendence. Skolimowski considers transcendence to be the "driving force of the development of the whole cosmos" (Skolimowski 2007, 75). He claims that a permanent aspiring to "go beyond," to transcend the limits, to build ever more complex structures, is ingrained in the nature of the cosmos. The continuous acts of transcendence occur in preconscious as well as conscious life. Thus, human life results from transcendence and, also, we ourselves give meaning to life in acts of transcendence, says the author of the *Living Philosophy*:

> Transcendence without God *a priori* given is possible, for transcendence stands for an ever-increasing perfection of our capacities and attainments. Indeed only this concept of transcendence is justified within a truly evolutionary perspective: transcendence that is void of

> an original God. If we assume God at the beginning, then transcendence stands for a process of curious retardation—of going back—and not for the process of going beyond, and beyond, and beyond until we reach pure spirituality" (Skolimowski 1992, 110).

The Polish thinker underlines the close relationship between creativity and transcendence: "Creativity is the energy of the cosmos, the energy that organizes transcendence and is its co-partner. Transcendence urges forward, and creativity organizes this urge in forms and structures" (Skolimowski 2007, 77). Consequently, the essence of transcendence is best expressed in all kinds of creativity. Owing to transcendence, we create a changing picture of the universe, exceeding the limits of biological existence and aiming towards the development of spirituality. It can be said that the co-occurrence of creativity and transcendence assumes the shape of the Aristotelian combination of form and matter, both on a cosmic scale and in the conscious acts of man.

## Future Prospects—The Third Millennium

At the threshold of the third millennium, our consciousness should be radically changed, Skolimowski postulates. We should break from materialism and consumerism, which challenged the significance of love and spiritually enslaved the people of the twentieth century. We are responsible "for the shape of the world, for its future, for its destiny" (Skolimowski 2007, 161). If man is to enter "the age of light, wisdom, and love," he must be spiritually reborn. Skolimowski declares that this will be only possible if humans accept and implement in their lives the assumptions of ecology constituting a new revelation, a great integration, a new holism, and a new unity of the world, knowledge, and meaning. The imperative of ecology indicates that we have to "perceive ourselves and all mankind in a completely different perspective—in the perspective of light, harmony, and great cooperation with the forces of the spirit and holiness. Spiritual rebirth is a condition *sine qua non* for the existence of the third millennium as an era of peace, love, and human coexistence of all with all" (Skolimowski 2007, 162).

In the new millennium, beginning in the twenty-first century which, in Skolimowski's words, will be an ecological century or will not exist at all, "we must have courage to look directly into the future's eyes, because the future is already emerging. We have to think in terms of a year 2500 perspective. With this in mind, we need to see each other as shepherds of the Earth, shepherds of all the achievements and evolution of mankind, responsible for the spirituality of future generations" (Skolimowski 1999, 236). In this vision of the third millennium, a new ecological spirituality becomes a target for the entire biosphere, for the subsequent stage of its development. The good shall be glorified, and the postulate of sanctity actualized. Humanity shall return to the roots of the teaching of the great religions, the essential goodness of everything, since "Buddhism, Hinduism, and Christianity, are religions of light, promoting the good and growing out from the source of the good" (Skolimowski 2007, 185). The eco-philosopher's religious aspirations belong to the vast scope of the New Spirituality movement. In a spirituality for the third millennium, he proposes to replace all existing monotheistic religions which—in his view—have completed their historic mission, with one universal religion. Reflecting upon Western culture, Skolimowski believes that "one of the possibilities is the combination of the main monotheistic religions—in a Religion of One God; so that the names: Judaism, Christianity, and Islam vanish" (Skolimowski 2007, 123). Therefore, in the vision of God understood as the light embracing the whole cosmos, his answer reads: "The God of the cosmos and of all people shall not be exclusive and vindictive. The God of the cosmos must be all-encompassing and merciful" (Skolimowski 2007, 122).

## Conclusion

Skolimowski looks optimistically upon the world, being convinced that, despite the risks posed by contemporary civilization, the age of the twenty-first century will be eco-friendly. He believes that in its development, "living philosophy" will achieve a universal character. He expects that it will not remain merely a project, but will become an immanent part of reality itself, fully revealing its spiritual aspects in the unbroken chain of evolution.

It is worthwhile to quote Skolimowski's own words concerning the future of philosophy: "The greatest reward, nay, even triumph of eco-philosophy will be (and just this probably applies to any new philosophy) its disappearing from the language as "eco-philosophy," and will be its melting with life to such an extent that no one will be aware of the fact that it is eco-philosophy" (Skolimowski 2001, 63).

## Works Cited

Fiut, Ignacy S. 2003. *Ekofilozofia. Geneza i problemy.* Kraków: Stowarzyszenie Twórcze Artystyczno-Literackie w Krakowie.

Górecki, Janusz. 2007. O książkach Henryka Skolimowskiego. Afterword in Henryk Skolimowski, *Geniusz Światła a Świętość Życia.* Seria Wydawnicza: Biblioteka Filozoficzna, 208–214. Warszawa: Oficyna Wydawnicza Vega.

Pasek, Zbigniew, and Agnieszka Dyczewska. 2012. Man and Nature. A New Project on New Spirituality. *Problems of Sustainable Development* 7 (2): 67–76.

Skolimowski, Henryk. 1992. *Living Philosophy: Eco-Philosophy as a Tree of Life.* London/New York/Ringwood/Toronto/Auckland: Penguin/Arkana.

———. 1999. *Wizje Nowego Millenium.* Kraków: Wydawnictwo EJB.

———. 2001. Powstanie i rozwój filozofii ekologicznej widziane z perspektywy osobistej. In *Wokół ekofilozofii. Księga jubileuszowa ofiarowana Profesorowi Henrykowi Skolimowskiemu dla uczczenia siedemdziesięciolecia urodzin,* ed. Andrzej Papuziński, Zbigniew Hull. Bydgoszcz: Wydawnictwo Akademii Bydgoskiej.

———. 2007. *Geniusz Światła a Świętość Życia*. Seria Wydawnicza: Biblioteka Filozoficzna. Warszawa: Oficyna Wydawnicza Vega.

Szymańska, Beata. 2001. Duchowość pogranicza. In *Człowiek wobec świata na przełomie wieków. Nowe i dawne wzorce duchowości,* ed. Marta Kudelska. Kraków: Wydawnictwo Collegium Columbinum.

Waloszczyk, Konrad. 2007. Słowo wstępne. Foreword in Henryk Skolomowski, *Geniusz Światła a Świętość Życia*. Seria Wydawnicza: Biblioteka Filozoficzna, 9–10.Warszawa: Oficyna Wydawnicza Vega.

# Part VII
# Eco-Anthropology: Sentience, Desire, Language, Creativity

# Small Talk with a Grape Vine: Presence and the Sensuous Depth of Being

**Lena Hopsch**

**Abstract** This paper's point of departure is the thinking of American philosopher David Abram pointing to how our body places us in in a common, intersubjective field of experience and enters into relations with other presences. Without sensuous experience there would be nothing to know, he states in his *The Spell of the Sensuous, Perception and Language in a More-Than-Human World* – establishing the fact that we do not perceive the world as a detached (body) subject from outside the world, but from within.

This idea was earlier developed by Maurice Merleau-Ponty, who depicts our carnal echo to the world. This very radical idea elucidated within the phenomenological tradition gives voice to a world experienced from our situation within it.

Here the emphasis is on how this thinking can lead us to call for a silent conversation with things by a bodily "I" in the world. We may ask ourselves: Is it first when we bend our back and point our nose to the soil in the act of cultivating the land that a silent listening can begin? Will this teach us a more resilient way of living?

**Keywords** David Abram · Merleau-Ponty · Carnal echo · Sensuous experience · Cultivating · Resilient living

## Introduction

Today there is a growing interest in a more sustainable way of life. Pollution, greenhouse gases, and stress deprive nature and humans of energy and resilience. Many of us ask ourselves if this is the way we want to live. Philosophy,

---

Photographer Michael Hopsch's work is represented at Gothenburg Art Museum, Sweden, and at several Municipalities and City Councils in Sweden.

L. Hopsch (✉)
Representative of the Faculty of Chalmers, Department of Architecture, Chalmers University of Technology, Gothenburg, Sweden
e-mail: hopsch@chalmers.se

 
W. S. Smith et al. (eds.), *Eco-Phenomenology: Life, Human Life, Post-Human Life in the Harmony of the Cosmos*, Analecta Husserliana CXXI,
https://doi.org/10.1007/978-3-319-77516-6_28

especially phenomenology, points to new ways of thinking and acting. In the shadow of First World War, Phenomenology grew out of a critique of how we are making use of technological achievements in such a way that we make ourselves homeless. Such usage is an act detaching us from lived, bodily experience. Through it we pretend we are independent of everything that surrounds us, even from nature itself. Instead, a phenomenological thinking may lead us to call for a silent conversation with things via a bodily "I" in the world. We may we ask ourselves: Is it first when we bend our back and point our nose to the soil in the act of cultivating the land that a silent listening can begin? Will this teach us a more resilient way of living?

The paper's point of departure is philosopher David Abram's describing how our body places us in a common, intersubjective field of experience and enters into relations with other presences. He shows in his *The Spell of the Sensuous, Perception and Language in a More-Than-Human World* that without sensuous experience there would be nothing to know, establishing the fact that we do not perceive the world as a detached (body) subject outside the world, but rather from within the world. This apprehension was similarly developed by French philosopher Maurice Merleau-Ponty, who depicts our carnal echo to the world. This very radical idea, elucidated within phenomenology, gives voice to a world experienced from our situation within it.

In the following essay we will look more closely at experience from our situation within the world and at how essence connects us to existence. Husserlian intentionality aims at a direct phenomenological experience, the act of experiencing as such being the means to reach this. First, we must return to the zero point of perception, our bodily presence in the world. Secondly, how the actions of the hand can connect mind, body and world will be discussed. Third, we will consider how our way of acting in the world is dependent on a silent conversation with the world that surrounds us – a world we need to both understand and take care of.

## The Bodily "I" in the World

The body places us in the "common, intersubjective field of experience" and, by our body, we "enter relations with other presences" (Abram 1997, 45). Without sensuous experience there would be nothing to know, as American philosopher David Abram points out. He directs us to the fact that we do not perceive the world as a detached (body) subject outside the world, but from within, in what Merleau-Ponty depicts as our carnal echo to the world. This is a very radical idea that "continued phenomenologically – that is, as we actually experience and *live* it – … the body is the very means of entering into relations with all things" (Abram 1997, 47). Abram, following Merleau-Ponty, wants to return and re-build a "truly authentic phenomenology, a philosophy which would strive, not to explain the world as if from outside, but to give voice to the world from our experienced situation within it" (Abram 1997, 47). Merleau-Ponty calls us to touch, sense, and taste the world – thus reaching the sensuous depth of Being. This thinking leads us to a call for a silent conversation.

## Silent Conversation

Bidding us to for a moment return to the Ancient idea of a rational intellect that severs us from the natural world and positions humanity to be the exploiter of other living beings, Abram asks himself if it is this idea that have brought us further and further away from a relational experience of life and from lived experience. We have to return to the act of experiencing the intentional object as such. In order to reach this, we need to put brackets around what we know in order to know anew: the operation that Husserl called the *epoché,* a suspension of judgment.

*Perception* is an active and constituent act wherein a conversation can take place. By this 'open activity' we orient ourselves in the world. This openness, receptivity, and creativity are what we mean when we talk of having a percept, engaging in perception. The intentional act is a part of perception. Perception is an event. With, for example, 'small talk' with other beings, we become rooted in the world in a Heideggerian sense (Ruin 2005, 38). Nature invites us to participate. Growing plants, the care of the farmer, is just such an 'open active' situation, one where contact with the living plants turns into a conversation with the situation (Ruin 2005, 52). This lived experience emerges over time. The time spent and the open field for exploring and understanding the needs of the plants in every situation turns into an act of silent conversation. In order to take care and be willing to preserve our surroundings, on which we are dependent, we need to understand our environs as an understanding mediated by the senses.

According to Merleau-Ponty, perception is reciprocity: an active and constituent act. "The sensible gives back to me what I lent to it" (Merleau-Ponty 2002, 249). Perception is to be seen as a wordless communication, an attunement within the duration of time. If we return to the act of cultivating the land, the percept of the vine can be seen as a similitude for the relation between man and world.

Cultivating can be seen as a relational act: taking care, giving fruit, drinking, becoming one with. In wine can be seen not only an intoxicant, but a tie to the soil minerals, the moisture, the particular wind of that year and the amount of sun, all of which are bottled. No wonder that bread and wine are seen as the flesh and blood of the divinity in a spiritual exchange between man and God. Between the grape vine and man-made culture there is a resonance – explored by the wine. Artistic imagination and philosophy can coincide when developing concepts. On October 3, 2009, the artist Spencer Tunick performed an installation in collaboration with Greenpeace, drawing attention to the effect that climate change is having on French wine production. Over 700 volunteers turned up at a vineyard near Macon, France to pose in their Adam and Eve costumes among the grape vines in a vineyard. The images of that are surrealistic but point directly to our dependency on the soil.

A phenomenological, direct experience of the world, as found in growing plants, can engage us in a resilient way of living. It can animate the world of the sensible, drawing attention to the event of perception – participation – in one others' existences: Grape vine and Man. David Abram writes that the act of perception is always unfinished and open-ended, as is the philosophy of Merleau-Ponty.

This thinking will be further investigated here through some notes made during walks at Weinberg in southern Germany appearing in the following section. We will look more closely at the experience of growing wine. The grape vine is regarded as a symbol for the relation between man and world – culture and nature. The product of wine, being grown and created by man and thus returned to man, is a product and an artifact that in itself unites essence and existence.

## Small Talk with a Grape Vine

*Walking notes, Southern Germany:*

*How connected the grape vine-covered, steep mountain is to the village down at its foot, to the buildings and the people inhabiting them. Growing and harvesting, in all this the performance of hand and habit.*

*To dress: to put on work clothes, to put your arms in the shirt sleeves, legs in the trousers, wrap your belt around your waist, put your feet, seconds ago touching the cold floor, into the rubber boots, a hat as protection against the sun, and lastly put your hands in the work gloves where each finger finds its place. The body that is enveloped in a (new) working skin.*

*A quick walk up the hill.*

*The garden shed where tools are being stored. Stonewalls, brick-roof, the small green wooden door where you have to bow to get in. A wooden handle to open the door by pushing it inwards is worn by all hands that have touched it over the years.*

*Uphill; sticks, wire, and plants in rows.*

*Taking care of the plant: trimming it in order to make it strong and able to carry the load of the grapes. Much time and effort is spent on this. The time spent with the plant makes one relate to it and the habit of trimming it accomplishes a unity of experience. First the tool in my hand, as an extension of action; how the trimming tool fits into the palm of the hand, how the fingers grip the handle and how it allows the trimming of the plant. A relation experienced by skin, muscles, and skeleton via sensory-motor experience and hapticity. Sensing.*

*The olfactory realm activated by the smell of the plant by trimming, the moist grapes touching your skin, the smell of herbs when cutting a bunch.*

*The smell of soil, of warmth in the air colliding with the cold of an early frost. The sensation of the wind. The temperature and the humidity of the landscape.*

*The light of the day, and the inner image of how to trim relating to an analysis of the situation of every performed cut, sensitivity to how every cut affects the vine, its form, and future performance. To be able to drink the wine during the winter when all nature is at rest and the only sun seen is the one bottled and stored in the wine itself. The idea of care (Sorge) for the plant and for one's own life.*

*When grasping the ceramic mug with its cold glaze, filled with sour, amber-colored liquid that, when drunk, invigorates the senses. Taste unites the gesture of the hand, the form of the cup, and the smell of the liquid contained within, experience being united (again) by that performance. It also links the drinker to the past actions of his hand, the feet on the ground, the relation to the grape vine as an existence in-the-world.*

*The act of drinking the wine relates our body to embodied actions and experience, a lived experience rooted in the-world-being of both the subject and the object where they coincide in the consummation of the wine. The wine being a part of one's own body and relating back to the first encounter with the grape vine on the Weinberg, situated in the agricultural, domesticated landscape.*

*It is a chiastic opening in time where man touches the plant and the plant touches back. It is also an intertwining of world-body and 'flesh'. Man is being thrown into the world and starts to take care of (transform) the environment, and to form it in relation to an inner vision, an interpretation of a bond between essence and existence bridged by time.*

*Both man and the plant are timely and temporal. They both play a part in a cyclic transition from grape to wine and body. They both have to face being timely beings and face their own endings. By this they share a destiny that relates them in the 'flesh'.*

*Imagination correlates to essence, unmediated apprehension to consciousness, extended attention that reaches out into the surrounding landscape and embodies itself as an atmosphere brought into the body of man taking care of the plant.*

*Communis, communion, communication.*

*Everything in the experience is drawn together by the performance of the hand. The hand becomes the nucleus of experience and performance among past, present, and future. The ability of the hand performs the act of trimming, changes consciousness, and triggers imagination. Conceptualize a whole chain of actions relating man and world, a give and take, a relation of communication and communion.*

*But, it all starts with the small talk, an experience in every-day life – a lived experience. Actions transform into habitual actions – a unity of experience and a quality of experience owing to the perception of all senses heightening the impact of the percept, transforming it into conscious awareness. This does not exclude the ability to share this experience; it can be shared with others, as the acts of drinking and eating unite people, and as the labor performed to make it possible to share wine and bread. "Shed for you, for you issued" - given a new meaning. I complete myself in things and others, in the encompassing earth"* (Abram 1997, 62).

*Relating it to memory, creating a space for feelings.*

*It all starts with the small talk, before it is translated into meaning, intentionally giving it significance.*

## Conclusion

In this essay I have followed a trait in philosopher David Abram's text that is inspired by the works of Merleau-Ponty. Abram points to how the way we speak of things reflects our understanding of our surrounding world and of nature, emphasizing language as the vehicle of thought. In my reading, I have put stress on sensory experience and how a sharpened, extended attention brings us in tune with the surrounding world. The relation between humankind and earth has been elucidated by some walking notes from hilly Weinberg in southern Germany.

Relating to nature through a deeper experience make us bond with it. Engaging in growing and consuming products of the land can help us experience the deep communality expressed by Merleau-Ponty when he points to the experience of "the acidity of lemon and the insidious stickiness of honey" (Merleau-Ponty 2008, 45–51). In this essay we have looked more closely at how essence connects us to existence; how Husserlian intentionality governed by the *epoché*, a direct phenomenological experience, is the means by which to obtain this connection; how sensuous perception governed by sensuous experience is at the very heart of knowledge. After having returned to the zero point of perception, our bodily presence in the world, we saw how the actions of the hand connected mind, body, and world. Taking care puts us in a state of active listening, showing how our way of acting in the world is dependent on a silent conversation with the world that surrounds us. This way of paying attention is needed for both understanding and taking care of our environment. Here we have seen this mediated by the experience of growing and tending a vine.

In his philosophy, Merleau-Ponty returns to descriptions of phenomena, as seen in *The World of Perception,* showing that simple everyday experiences are what connect us to our existence since, as he kens, there is no world outside the world. This world is the only one we can know anything about, he says. So, for a growing branch of eco-phenomenology a re-reading of his radical ideas is a fruitful acquaintance (Merleau-Ponty 2002, 398–399).

In a cry for a more sustainable way of life – a bodily "I" in the world experienced when we bend our back and point our nose to the soil – the act of cultivating the land teaches us to listen silently. This 'direct experience,' an active attention studied and articulated within phenomenology, can, teach us a more sustainable and resilient way of living together. This points to the usefulness of phenomenological thought and of a philosophical ecology, where, as Abram puts it, "'objective' rationality no longer detaches us from lived sensuous reality" (Abram 1997, 277). We can again start to listen.

To conclude, in his work Abram show us that we need to add another layer to the life-world model: besides our individual, social, and cultural contexts we need to consider the ecological realm in which we live.

**Photo, Michael Hopsch**

## Works Cited

Abram, David. 1997. *The Spell of the Sensuous, Perception and Language in a More-Than-Human World*. New York: Vintage Books. Print.

Merleau-Ponty, Maurice. 2002. *Phenomenology of Perception*. London/New York: Routledge. Print.

———. 2008. *The World of Perception*. London: Routledge. Print.

Ruin, Hans. 2005. *Kommentarer till Heideggers varat och tiden* [Comments on Heidegger's Being and Time]. Huddinge: Södertörns högskola, Södertörn Philosophical Studies. Print.

# Auditory Phenomena and Human Life: Phenomenological Experience

**Ineta Kivle**

**Abstract** The present study analyzes auditory phenomena from the point of view of hermeneutical phenomenology and shows their interconnectedness with the understanding of *man, hearing and listening* within the context of human life as the horizon of meaningful sonority and silence. The central questions to be answered in this study are these: What is experienced as sound and sonority? How does a human see himself in inclusion of his being from where he listens, understands, and speaks? The study explores the classical standpoints of Husserl's phenomenology and other philosophical apprehensions which confirm that auditory phenomena is not to be apprehended solely as an isolated horizon but as being permeated by the visible, the perceptible, and the comprehensible.

**Keywords** Husserl · Sound · Auditory phenomena · Sonority · Silence

## Auditory Phenomena Philosophically Considered

Philosophical interpretations of auditory phenomena[1] describe them using terminology[2] and explore them from standpoints that differ from the descriptions and methods used and standpoints assumed in other sciences and theorizing. Philosophers' interest in considering auditory phenomena as thinkable objects touches upon the differences and similarities between sound, music, voice, and speech, matters which are not to be found in the works of other theoreticians.

Auditory phenomena differ from each other.[3] Therefore, to give an adequate description of them, one needs to take into consideration both the way the phenomenon is given in experience and the way we make use of philosophical approaches. In any case, to clarify auditory phenomena philosophically means to relate them to the experience one is going through and to view them as thinkable objects.

I. Kivle (✉)
University of Latvia, Riga, Latvia
e-mail: ineta.kivle@lu.lv

 
W. S. Smith et al. (eds.), *Eco-Phenomenology: Life, Human Life, Post-Human Life in the Harmony of the Cosmos*, Analecta Husserliana CXXI,
https://doi.org/10.1007/978-3-319-77516-6_29

Phenomenological interest in sound is focused on intentional experience, on the way sound is intended in consciousness, while at the same time bearing in mind how meaningful phenomena relate to each other within the context of mental experience. Phenomenology substantiates sound as a subjectively grasped meaningful phenomenon, showing it as an intentional and temporal object, and this gives impulse for the interpretation of sonority within the context of both the life-world and the common intersubjective world. However, interpretations of voice and speech are more the findings of hermeneutics and existential philosophy. Phenomenological investigations, concerning the link between sound and music, predominantly incorporate art and music, whereas phenomenological hermeneutics opens up possibilities for interpretations of auditory phenomena in their historicity.

## Husserlian Phenomenology and Heideggerian Fundamental Ontology

The description of sound based on the phenomenology of Husserl reflects not only the structure of the given phenomenon – in other words, its objectivity and appearance – but also elucidates the structure of intentional and temporal consciousness and explains man's inclusion in the world of a meaningful temporal and spatial horizon. Husserl describes the world from the standpoint of his own authentic experience, being one of the first philosophers to approach this problem. He writes about the "surrounding world" (*Umwelt*) from the perspective of "I," and his description of sound (tone) follows from the underlying question: *How do I experience a sound in my subjectivity?*

In Heidegger's analysis of *Dasein* as *being-in-the-world,* the context of historicity comes to the fore, which is to move away from Husserl's pure "I," away from the transcendental *Ego. Dasein* as being in the world is included in the being from which one speaks, understands, and listens. Pondering *Dasein* elucidates the continuum of spatial temporality wherein one is one kind of being among others, but *Dasein* provokes distinguishing oneself from other beings by primary modes of existence, among them, care, mood, speech, and understanding. Exploring Heidegger's analysis of *Dasein*, speech appears therein as a mode of human being in the world from which one speaks and understands. Speech and voice are immanently connatural to human beings and show the sonority of language. Heidegger's questions concentrate on belonging to being – sound and sonority belong to being, speech is a mode of existence of *Dasein*, and sonority can be understood as a manifestation of truth in the audible *worldliness of world.* Heidegger's philosophical meditation on the *worldliness of world* enables understanding of the human being as being determined not by intentional experience but by openness to being and *listening in.*

In Husserl's philosophy, the world is viewed as a meaningful phenomenon that has various modifications: the practical world, the theoretical world, the ideal world, the world of the natural attitude, the surrounding world, and the life-world, to name a few. These constituted worlds are not separate phenomena of cognition, but meaningful complex phenomena that are always open to the future. In any case, the content of these constituted worlds is changeable. Husserl's *constitution of the world* and Heidegger's *worldliness of world* are given *a priori* for any human during his or her life. In other words, the *constitution of world* and *world worldliness* are immanent activities adhering to a human being as well as to the world within the context of Husserl's phenomenology and Heidegger's fundamental ontology. If in the center of Husserl's philosophy there is the "I" who constitutes the world, and there is an intermittent constitution of the world in subjectivity and intersubjectivity, then in Heidegger's fundamental ontology, *being-in-the-world* is viewed as existent in the network of the world and the relevant world is an event in *being-in-the-world.*

Conventionally, the hermeneutical and phenomenological approaches to auditory phenomena are influenced by two principles: first, Husserl's appeal "Back to things themselves!" opens up philosophical dimensions for understanding things as they are and allows for a reality to speak in its own form; second, Heidegger's principle "letting be" diversifies philosophical descriptions and interpretations of auditory phenomena, including everything that a human hears, thinks, and perceives as sonority.

## Environment Philosophically Considered

Regardless of the differences between Husserl's and Heidegger's views, common to both philosophers are empathy and tolerance in the explanation of philosophical questions. They acknowledge a special philosophical environment of listening to thoughts and things, understanding of one's inner self, and contemplation of the world. Such an approach gives an impulse for philosophical interpretations of the environment, which can be regarded as an intentional and meaningful realm of activity or as a permission to be. The philosophical environment can more or less be regarded as an exclusive sphere of thinking, writing, and talking, as a particular creative activity. Philosophically understood, the environment has resonance with a variously interpreted concept "territoriality," which is regarded as the place and space of a person's life, as the concept of a philosophical text, as the "territoriality of sonority." In this case, when a territory has been established and developed, it accommodates expressive qualities that can be compared to the taking of space and time or with marked horizons of hearing and seeing. These "territories" are differentiated from each other because they are insinuated through different meaningful activities – a concert has its own intentional territory of sonority, whereas a lecture provides not only for a meaningful intersubjective and audible place for the event, but also for a thinkable horizon and common intellectual experience. Such organized territories are characterized by their own rhythms, circumstances, and activities.

In *Ideas I* Husserl writes, "I always find myself as someone who is perceiving, objectivating in memory or in phantasy, thinking, feeling, desiring, etc.; and I find myself actively related in these activities for the most part to the actuality continually surrounding me" (Husserl 1982, 31).[4] Activities make their own environment and implement their own forms: the sonority of words and music make the event exactly this or that performance; painters express their experiences in painting; philosophers come forward with their own philosophical concepts. Creative activities combine silence and words as well as seeing, contemplation and thinking, substantiated so that a human constitutes related spheres of sonority, visibility, sensitivity, and understanding. Such dimensions as seeing and hearing, sonority and silence, speaking and listening, and understanding and reflection penetrate into any meaningful world of human life.

## The Husserlian Apprehension of Tones

In phenomenological experience sonority is contemplative and thinkable. We do not so much listen to a sound as grasp it by thinking and so give a description of auditory phenomena, showing their interrelations and distinctions from the point of view of meaningful perception – audibility. The phenomenological description of sound as an experienced phenomenon is based on the paradigm that rejects the subject–object division and confirms that we know the objective as it is given in subjectivity. When we listen to music, we experience its melody. However, if we think about music as a complex intersubjective phenomenon, we are in phenomenological attitude and practice philosophical thinking. This example shows that the meanings of the phenomenon are changeable: as an object the melody heard is the same, both when listening to it and when thinking about it, but each time the melody seems to be different because it is grasped by different intentions and different constituted meanings. Husserl's contemplation of tone, melody, or any other phenomena as thinkable objects elucidates a structure of formation of meaning. In Husserl's philosophy, alongside the description of tone as an object of time, the constitution of objectivity, and the appearance of a phenomenon, one finds a deep analysis of the structure of intentionality and internal temporal consciousness. Focusing on man's meaningful and fluxing consciousness, on the feasibility of perceiving sound, begins with the question "How?": How do I think when I think phenomenologically about the sound I heard? Consequently, a view on auditory phenomena is being developed, one looking for answers to two most essential questions: first, how does a sound present itself in man's hearing, listening to, and contemplation?; secondly, what do I know about consciousness at the moment of grasping sonority? A bilateral relation – sound and its sonority, on the one hand, and fluxing hearing and listening, on the other hand – illuminate the meaning of sonority and provoke another question: How do I grasp the meaningfulness of time in sonority? Husserl writes:

> Each tone has a temporal extension itself. When it begins to sound, I hear it as now; but while it continues to sound it has an ever new now, and the now that immediately precedes

> it changes into a past. Therefore at any given time I hear only the actually present phase of the tone, and the objectivity of the whole enduring tone is constituted in an act-continuum that is in part memory, in smallest punctual part perception, and in further part expectation. (Husserl 1991, Sec. 2.7)

The phenomenological method demonstrates that here is a horizon, one comprising only auditory phenomena, that is constituted in the way of phenomenological reduction by means of 'bracketing' everything that does not refer to sonority and hearing. According to this philosophical thinking, sound, and other auditory phenomena are beyond real sonority and free from any coincidences. The reduced phenomena display the phenomenological structure or, in other words, the most essential parameters of phenomena, namely, objectivity and appearance. Husserl writes, "A violin tone, in contrast, with its objective identity, is given by adumbration, has its changing modes of appearance" (Husserl 1982, Sec. 1.44.). In the same way, exploring the phenomenological method, Husserl suggests three components of internal time consciousness, such as primal impressions, retentions, and protentions. Considering sound as a temporal object, he shows the composition and structure of the flux of phenomenal time. Husserl's concentration more on the philosophical problem of time than on sound confirms the content of his phenomenological thinking. According to him, the description in detail of sound as a meaningful phenomenon lights up the fluxing structure of consciousness, human expectations, and openness to the future.

## *Lebenswelt* and Communication

Husserl's concept of the *Lebenswelt* gives feasibility to interpretation of auditory phenomena within the context of everyday life and substantiates that human life is understandable from what is heard and listened to, what is seen, what is touched. The various directions of our activities and intentions make our lives along with our environment and surroundings meaningful. The whole of our world becomes meaningful – a meaningful world. This world is defined by what one perceives and experiences with one's eyesight and hearing, whereas the life-world comprises cultural objects, things, and human experience, substantiating that a human lives in a combination of culture, things, and environment (*Kultur-Sachen – Umwelt*). The life-world is relative because it is given to us for a limited period of time. Different people live in different worlds.

In *The Crisis of European Sciences,* Husserl writes that among the objects of the life-world we also find human beings living with a common world meaning as within a particular horizon of intersubjectivity. Phenomenological explanation of social interrelations describes common experience and includes a sphere of communication that becomes apparent in the audible and visible world as speech, voice, emotions, art, and science. Speech and voice, being immanent in human existence, not only testify that communication is essential for our mutual interrelations but also impart understanding that inner speech and mute voice engage in a silent conversation with one's inner self.

Humans' ability to speak, listen to speech, and perceive silence differentiates them from other forms of life. Silence might be experienced as the absence of noise or the cessation of speech. Meaningful silence expresses itself as "deep silence," "true silence," the "silence of openness," and the like. Experienced silence, in its turn, is linked with sound and voice. The sonorities of spoken words and silence, being modulations of our ability to perceive auditory qualities in the ambient environment, do not counteract each other. The experience of silence as well as of sound extends to and characterizes a primordial faculty of a human being – to find oneself within one's inner self and listen to one's inner words or contemplate and listen to the environment around one. Sound and silence are intentionally linked – sounds flow, then they are interrupted by silence, thus providing for a specific rhythm of speech, talk, or any other performance of meaningful auditory phenomena. Likewise, we can listen to silence, overwhelmed by incoming sounds, because silence before speech manifests itself as an expectation of a voice. Meaningful silence adheres exclusively to the sounds that are determined by this particular moment of silence. In other words, sound touches on silence and, conversely, silence touches on sound. Absolute silence and absolute sound dwell in our imagination, but it is impossible to experience them, as a living body always meets with a meaningful silence and meaningful sonority. The constitution of silence forms intentions and sits within the phenomena of a meaningful environment where silence and sonority correlate with each other. Silence and sonority are mutually determined, based on the human experience of listening to speech, voice, music, or nature. Silence and sonority cannot be regarded as an act having absolute autonomy. This pairing leads to the relatively limited horizon of a common spatio-temporal situation that can be experienced only by those who share this particular situation and are incorporated into it. In this case, silence is an active human performance, one determined by rational, intuitive, and emotional intentions as well as by accumulated experience.

Auditory phenomena are not only heard, but are also grasped either visibly or kinesthetically. Speech includes linguistic muteness; sound is intentionally linked with silence. Auditory phenomena mark the horizons of audibility, where the unheard comes into being. The horizon of audibility widens and narrows, depending on what we are listening to. Auditory phenomena are not to be found solely against an auditory horizon; they are also permeated by the visible and the visually grasped. The world that is in such a way constituted includes both the sounds of music, nature, voice, and visually grasped things, ongoing activity, and other people.

The full description of auditory phenomena, therefore, presents human life as a flexible and variable horizon where hearing and seeing, and sonority and silence can be adequately constituted in subjectivity and intersubjectivity, while human life can be understood as an event in the world, elucidated in the continuum of spatial temporality. In connection with all this, Anna-Teresa Tymieniecka said that a human being should be considered as manifesting *"the human condition in the unity of everything that there is alive."* Human life as a flexible and variable horizon constantly moves, featuring the unity of the varied and inseparable directions of human life.

## Notes

1. In phenomenological discourse, sound, speech, and voice are called not audible but "auditory phenomena" whereas phenomenology reveals how sonority is constituted. The expression "audible phenomenon" is used more in the natural sciences and in the audio and information technology industries.
2. The phenomenological and hermeneutical approach to auditory phenomena are based on Husserl's programmatic statements of phenomenology, Heidegger's analyses of *Dasein,* and the methods of hermeneutical interpretation.
3. Sound, speech, voice, and music are organized auditory phenomena that differ from noise, crackling, or crashing. This study concentrates on sound and sonority. The phenomenology of music and voice requires a special philosophical approach and has been carried out by philosophers of sound and music Don Ihde, Thomas Clifton, Joseph Smith, Bruce Ellis Benson, etc. Music as the most complicated of organized auditory phenomena includes questions about art and historical tradition. However philosophical analyses of speech and voice is insinuated in the philosophy of language based on Wittgenstein's cognitions, Heidegger's ontology of language, Derrida's critical analyses of Husserl's views on language, etc.
4. All references to Husserl's writings follow this standard form – (1.31) indicates the 31st section of the first of his books in the Works Cited list.

## Works Cited

Husserl, Edmund. 1982. *Ideas Pertaining to a Pure Phenomenology and to a Phenomenological Philosophy: First Book: General Introduction to a Pure Phenomenology.* Husserliana, Collected Works 2. Trans. F. Kersten. The Hague: Martinus Nijhoff. Print.

———. 1991. *On the Phenomenology of the Consciousness of Internal Time.* Husserliana, Collected Works 4. Trans. J.B. Brough. Dordrecht: Kluwer Academic Publishers. Print.

# Animal Being Means Desiring: Subjectivity, Singularity, Diversity in Post-Human Life

**Roberto Marchesini**

**Abstract** This paper will demonstrate that desire is the most immediate expression of subjectivity, which transcends need and somehow gives it meaning—I live because I desire, I live by expressing desires that overwhelm me and shape my existence. All of an animal's endowments only serve to express desires. Desire colors the world: it fills the eyes of a child with wonder; it supports the chaotic games of a puppy; it gives meaning to the events of the world. If desire disappears, life falls into a vegetative atemporality. If desire decreases, life fades.

**Keywords** Subjectivity · Desire · Animality · Post-humanism · Animal-being

## Introduction

Animal being concerns people directly but neither as a counter-term from which to learn about the human condition by exclusion, nor as a regressive figure, a dark legacy that can come back from the phylogenetic abyss at any time and put humanity at risk. Rather, animal being is the very foundation of our subjectivity and our dwelling in the world through the vitality of desire. Animal being is non-equilibrium, a referential openness that rejects any ontological autarchy, and cannot be explained through a detailed and exhaustive investigation of the causalities operating in the here-and-now. Animal being is the continuous invention of the present, a step beyond the algorithm of causes. It transcends the phenomenon through epiphanic emergence and is a singular invention of existence.

The animal condition continuously synthesizes different 'times' and motives. Thus, it fulfills its being in the moment—its presence—by never being fully comprised in the immediacy of a function. Paradoxically, the animal can be said to be present—that is, to have a presence—because it cannot be fully explained by

R. Marchesini (✉)
Centre Study for Post-human Philosophy, Bologna, Italy
e-mail: estero@siua.it



W. S. Smith et al. (eds.), *Eco-Phenomenology: Life, Human Life, Post-Human Life in the Harmony of the Cosmos*, Analecta Husserliana CXXI,
https://doi.org/10.1007/978-3-319-77516-6_30

referring to the causal mechanisms operating at the exact moment in which it is considered. Animal being is, therefore, a continuous desire, the action of a tireless Penelope that relates proximate and remote causes, phylogenetic and ontogenetic motives, inherent motivations and contextual opportunities, in a diachronic and relational flux wherein animality manifests itself in this connective presence rather than in its function as such. The animal is subjective because it escapes the objectivity of the causes operating in its immediacy.

On the contrary, the animal deflects the situation according to coordinates of partiality that give the impression of a judgment that is subjective, because it is unique, and creative in its unpredictability: a judgment translatable as "attribution of a meaning-value" to the mere phenomenal givenness. Being present means having sovereignty over one's own condition and, in this sense, not being a slave of the present. However, this sovereignty absolutely needs endowments in order to act in the present. But then, it must not give to these endowments the autonomy and explanatory completeness to dictate the individual's coordinates of action. Therefore, one must not mechanize animal behavior—that is, claim that animal expression is dictated by automatic devices whose operation is per se necessary and sufficient to define animal behavior. The animal can be seen as a *Dasein* only, if any meaningful ownership is to be recognized and, therefore, the individual is considered to be the owner of its endowments. The individual uses its endowments; it is not passively guided by them.

However, in order to so see things, the model used to explain animal behavior from Descartes onwards must be profoundly changed. As is known, according to Descartes the isochronous and linear operation of our *res extensa,* the body, is not able to explain human behavior. However, our *res cogitans* counterpart, the transcendent principle that does not require a scientific explanation, compensates for this inability. It becomes obvious, therefore, that, from a materialistic point of view, the elimination of the *cogitans* requires an epistemological rethinking of our *res extensa*; otherwise one must resort to other dualisms called in to rescue an unpersuasive explanation. As soon as Darwin's principle of continuity is recognized (Darwin 1871), no meta-predicative paradigm of the explanatory mechanism can be maintained. In other words, our phylo-ontogenetic endowments, natural or learned, must no longer be considered to be automatic devices that control the individual (as a puppet) but, rather, as tools that the latter uses. Desire must be regarded as the master of subjectivity.

## The Animal Machine as Meta-predicative Condition

The Western tradition, aware of Descartes' idea that the animal is an 'automaton'—that is, a mechanical entity totally governed by triggers designed to produce functional determinisms—has given rise to a rich tradition of explanatory models, variants that are often antagonistic or conflicting with each other but which are actually subsidiaries of the same conceptual paradigm and perfectly consistent with its epistemological assumptions. Neither the European psychohydraulic model nor the North American chaining model, even within the debate in which they were opposed to each other during the first half of the twentieth century, question the mechanistic principle that deprives the animal of any expressive subjectivity. Animality is read as biomechanics; it does not really matter whether the trigger is instinctual or reactive, whether the automatism is informed by phylogenetics or ontogeny. Depriving the animal of its subjectivity means putting it at the mercy of chance and necessity and taking away from it any glimmer of a presence that is not a mere functional exercise.

Whenever one wants to focus on the meaning of an animal's conformity to standards, before considering its specific action, one must relate to this tradition and investigate its paradigmatic assumptions instead of its merely descriptive surface. This approach interprets its here-and-now with an implicit and articulate protagonism, using its own endowments rather than being acted through them. But what is its action field? In fact, to consider the animal condition in mechanistic terms does not mean evaluating its expression according to standards based on a multiplicity of ontopoietic and elicitative factors, which are not necessarily linear in their casual development. Foremostly, it means to deny any possibility that the animal controls its behavior. The animal condition, therefore, is not then seen as a full presence in the world: the animal is not free, it does not choose; it is not able to evaluate, to distance itself from the stimulating and instinctual magnetic drive; it does not decide but, simply, enjoys according to predetermined mechanics.

The evolutionary analysis based on adaptive specialization (whose buttress is ecological analysis, focused on the dynamics of sustainability of different biomes) forces one to face a plurality of forms that reveal precise ranges of expression—Baron Jakob von Uexküll's famous *umwelten* (von Uexküll 2010) or Niko Tinbergen's ethograms (Tinbergen 1953). However, in doing so, this analysis seems to distract us from the paradigmatic problem: what does animal being mean? I will analyze the choreograms of various species, the structuring processes of the different patterns, the Gestalt organization of the key signals, the multiple immersions of perception, the many faces of dietary, territorial, parental, and social behaviors, and so forth. I will delineate articulated catalogues of cultural occurrences, usage of tools, and expressive flexibility, reaching the threshold of consciousness without getting to the bottom of the basic assumption and, therefore, without questioning it. I repeat: *what does animal being mean?*

According to the tradition, every species is nothing but a form, a particular variation of this mechanical paradigm that is the theoretical foundation of the animal as *res extensa*—that is, a measurable entity that can, thus, be transformed into an ahistorical and necessitated algebraic formula. This leads to a very simple consequence: animals are machines—this is their common foundation—and each species is a particular model. Therefore, all the eco-adaptive descriptions become nothing but ways to describe the animal machine, which has as many forms as the functions required to dissipate energy in different environments and positions along the food chain. Descartes' principle remains in the background, but it firmly dictates the coordinates of the explanatory translation, regardless of the descriptive genealogy enunciating its model. The description of the machine becomes just another way to avoid the paradigmatic peculiarity of the problem or, rather, to avoid questioning the explanatory status quo.

This paradigmatic or meta-predicative implication allows one to shift one's attention to other details and thus avoid the real issue of what it means to be animal in the light of Darwin's continuist perspective: one limits oneself to the predicative analysis (Tinbergen's famous four ethological questions), thus shifting the discussion from animality as a condition to the species-specific adaptive and evolutionary repertoire; one focuses on the general features of the machine—whether chemical or thermodynamic, cybernetic or informatic—shifting explanation to energetic and functional configurations and away from the underlying explanatory paradigm. This trick or removal allows one to accept Darwin as topical product—that is, without having him interact with the systemic of the human cultural organism. Misled by the attention paid to descriptive features—the ethogram of the dog or cat—or by functional explanations, and this is so no matter which is right: whether Lorenz's psychoenergetics theory (Lorenz 1966) or Skinner's cybernetics (Skinner 1957), one forgets what Descartes left unsaid, which just may continually rule the basic ontological principle.

Martin Heidegger (Heidegger 1995) understood this aspect very well when he realized that the predicative explanation, still in vogue among the old humanists fascinated by the plasticity of the Vitruvian model, was not the crux of the matter because the difference between the human being and animal otherness was ontological and, thus, meta-predicative. The animal machine can perform actions, take part in some functions, be comprised in a period of time, and so on. But, if and while it remains a machine, it will never be really present, because a machine is in an isochronal state and does not own a *hic et nunc*.

As a result, the animal machine may have: fossil footprints of its past, but not memories; cascades of future events, but not projects; a stream of functions, but not existence; an end or shutdown of the process, but not death. Also, a machine does not have a real relationship with the world because, in reality, it only elaborates it. That is to say it translates inputs into output items; it enjoys or avoids the world, since proximity is not closeness and distance is never contemplation. The animal is functional dizziness, poor in world, but finding it useful for survival.

Thus, Heidegger clarifies Descartes' unsaid. This mechanical rendering of animality created an unbridgeable gap between humans and other species and, simultaneously, it definitively sanctioned humanity's operative freedom over the non-human universe. Since the seventeenth century, this paradigmatic development has been undoubtedly difficult and controversial. The point was to accentuate the dialectics of exclusion that was present only *in nuce* in the early humanists. Therefore, it is no surprise that in addition to the hypothesis developed after Descartes' death (which increasingly amounted to a reductionist approach to animal expressiveness), proposals to restore subjectivity to nature cropped up, albeit in an intermittent way. And yet, countless considerations privileged the reductionism to which Descartes offered a particularly effective paradigmatic crux—primarily, the posited autocratic and autopoietic operation of the human being as the world's sole protagonist, a principle that had already been active in the Western humanist metamorphosis for two centuries.

In addition, this mechanism was certainly in line with the spirit of its time, when scientific achievements came along with the geographical, social, and technological advances in a lavish unfolding of power and wonder. The machine, therefore, served as a great icon for the operational expansion of a humanity immersed in modernity. The anthroposphere spread through gears and hydraulic machines and so the non-human was swallowed inside these mechanisms as well. A further advantage was also given by a significant explanatory artifice: the call for an apparently explicit but, in fact, totally theoretical mechanics. On closer inspection, Descartes' mechanical hypothesis does not respect Popper's canons of falsifiability (Popper 1959) because, by not specifying what type of mechanism must regulate the animal function (that is, lacking a machine capable of reproducing the animal condition), it appeals to a dictatum that can only be accepted and not tested. However, this reductionist cryptotautology lends itself to a metamorphic dynamism in line with the technopoietic advance and every innovation, with its retinue of wonder and self-satisfaction. And so it finds itself explaining animality with a sort of suspension of critical judgment. Thus, it is able to conceal the artifice.

As I said, the form-image of each species—be it a dog or a cat—is not the only thing that distinguishes the descriptive principle of the machine. In fact, there is also the type of machine to which to refer, which is obviously in constant transformation by virtue of technological innovation. From Descartes onwards, therefore, it will be possible to see a series of "explanatory models" called upon to describe the functional mode of *automata-animals* without affecting the paradigm: their machine condition. The proliferation of explanatory models amounting to a catalogue of various automatisms and their triggers (predicative definition) has not questioned the basic paradigm called upon to explain the animal condition (meta-predicative definition). Despite discussion on the intelligence or consciousness of other species, this paradigmatic impasse is still in place today.

In other words, what changed in the many interpretative attempts proposed since the seventeenth century has not been the meta-predicative condition (being a machine) called upon to explain the animal condition but, rather, the types of automatism invoked to serve the explanatory function—that is, explaining how animality worked and to what kind of machine its working could be plausibly attributed. The resulting predicates of the animal changed, too. The animal, no longer made of wires and pulleys as Descartes claimed but, in rapid succession and by virtue of the feverish technological advance over the centuries, following the scientific revolution, it became a "chemical sensor" adjustable through tropisms, a "steam engine" for its impulsiveness interpretable according to thermodynamic principles, a "cybernetic mechanism" governed by complex recursions of retroaction, or a computer phylogenetically instructed by algorithms and heuristics.

## The Dialectics of Exclusion and the Anthropological Machine

In addition to this, other elements compose the framework of the animal being as it still appears to our eyes. Mechanistic reductionism was not the only instrument indicating how to interpret animality as a condition. It is possible to say that other traditions and assumptions have preceded it and accompanied it, resulting in a syncretistic view that made the animal an 'oppositional figure' with respect to the human condition. Therefore, to reflect on the animal condition means to enter into a convoluted and recursive labyrinth of preconceived attributions and interpretations. Animality has been promptly transformed into a dark dwelling on which to project one's fears or uncomfortable presentiments; a background against which to bring out the meta-predicative distinction of the human being; an amorphous category made up of apparent pluralities; a trampoline from which to jump into the hyperuranion; a pestilence-ridden territory that must be avoided through rituals of purification. To attribute a genealogy to this preliminary reading—that is, to determine a single source or a prevalent view—is very difficult.

It is possible to say that each outlook has contributed to the great building up of an ontological anthropocentrism pitched toward a disjunction between humans and animals. Overarching the mechanistic tradition, Western culture has stigmatized the animal condition and made it a counter-term of the human dimension. This gradual and complex process involved art, religion, and philosophy, providing the foundation of anthropopoiesis itself—that is, being able to perceive and recognize each other, to intercept one's own *telos* and strive, to direct ontopoiesis through anthropotechnic practices, to regulate codes of acceptance and rejection. Animality, thus, became in aspect a shore from which to depart but one, also, to keep in sight to see if the chosen route be correct, almost like a polestar showing which way *not* to go. As a premise or logical consequence (probably both, it is hard to say) an animal category—more simply, the category of "animals"—was created, one having predicates dissimilar to those attributable to humans.

In a sense, one can say that the path taken in our culture—albeit with moments of deviation and stagnation but also, conversely, with moments of sudden acceleration—was blazed by a distancing from a fictitious idea of animality, characterized by an arbitrary attribution of predicates that, though improper and categorically mistaken, are useful for the supposed oppositional operation. To understand this dictate, one should deeply analyze the dialectic of exclusion characterizing the humanistic project that aims to extract a metric-subsumptive image—that is, a universal of the human being—by rejecting any form of functional declination. The humanistic project basically purges the human being of any Epimethean features so as to give rise to two different genealogies.

Animal ancestry, the outcome of functional embodiment—that is, outcome of the performative rank—and Epimetheus's offspring, has nothing to do with human ancestry. In fact, the latter has a somatic, larval, and undeclined vitality and can, therefore, be applied as a unit of measure without a rank. Consequently, it is free and autopoietic, the offspring of Prometheus, inspiration, techne, and the performative outsourcing. Thus, it is able to subsume any performativity. The artifice of a double genealogy, which is anything but simple myth-making, influences the roots of Western philosophy, which, throughout its course, has repeated it like a chant, albeit in ever different poetic and semantic forms, over and over again. On closer inspection, this operation manifests itself as a tautology that is anything but cryptic: animals do not share anything with humans as they are, in fact, animals. Nevertheless, today it is difficult to abandon this false interpretation, so that the animal as a counter-term and otherness, rather than being a reality to discuss, becomes the very element of comparison and the background that allows the human being to emerge.

As I said above, this is a threefold error: first, the human being is separated, in the principal characters that makes him such, from the animal condition; second, the attempt is made to extract a categorical structure that involves denominators common to all heterospecifics in opposition to the human being; third, the predicates from which the human being wants to remove himself are then attributed, in an arbitrary way, to this category. Essentially, the human being is purged from animality. But actually, the predicative differences are interspecific. These predicates exist in *every* species (and, in a sense, even in every individual) and are, therefore, also valid for the human being. The latter is qualitatively different from the cat because of meta-predicative differences that concern the ontological dimension, and not the adaptive-ecological profile. Offering a diversity of predicates, the human condition is, then, elected as a deeper dimension, and, therefore, it is opposed to the very condition of animality instead of the individual non-human species.

Before focusing on the cognitive performativity or the levels of intentionality present in nonhuman species, therefore, one should discuss this topic and deeply analyze the paradigmatic model called on to explain animal expression. Adhering in a Ptolemaic fashion to the anthropocentric paradigm and to Descartes' artifice, any attempt to build a continuum between all species—which, on one hand, would bring out animal subjectivity and, on the other hand, would be respectful of the diversity of species—is vain or misleading. In fact, to speak of subjectivity means to accept Darwin's challenge to identify adaptive-ecological peculiarities, applying to the cognitive-behavioral endowments the same analytical categories used in anatomy or comparative physiology. However, one must be aware that, by so doing, we will sketch relative proximities (homologies) and convergences using adaptive semantics (analogies) but without making the traditional dichotomy between human and animal that is the basis of the humanistic project. Animality is a constellation that includes us rather than a counter-term in an anthropopoietic dichotomy.

I think that the question of animality should be addressed in two ways: first, through predicative analysis, starting from the fact that every animal singularity is a specific way to invent one or more worlds, on the basis of specificities that cannot be traced back either to an exclusive dichotomy making the animal a counter-term to the human or to the inclusive dichotomy that is anthropomorphism; secondly, through a meta-predicative analysis that reflects on what animal-being means (including, of course, human being), calling into question the mechanistic paradigm separating the physical *res extensa*, and that considers subjectivity not as a conscious or rational expression of being but as understanding in desire, as the intentionality of desire. By adopting both these different modes of analysis, we can nullify the humanistic operation that Agamben calls the "anthropological machine" so as to rediscover the plurality of the animal condition and, also, the common matrix of the animal-being that makes us part of the same constellation with the animals.

## A Return to Irrational Subjectivity

There is no doubt that animality concerns humans directly, especially after Charles Darwin's revolution (Darwin 1859). The English naturalist highlighted the continuity between living beings and, in deference to a populational logic of predicates and to a phylogenetic proximity of similarities and differences, he brought out a non-anthropocentric taxonomy based on genealogical levels of affinity rather than on essences disjointed by definition. Evolutionism—and the subsequent synthesis that links genetics, embryology, paleontology and computer science—breaks the ontological isolation of the human being and, as a consequence, any attempt to maintain the old humanist distinctions is destined to falter, precariously holding on to external ontopoietic matrices such as language, *techne*, and culture (Marchesini 2014b). Even if pushed to the depths of the human being by a tradition that refuses to give up Pico's dicta, animality forces its way in and gets to the core of the creation of identity and expressiveness, so that distancing is no longer possible; the human being drifts in the animal condition.

However, perhaps in response to the tsunami of Darwin's theory, during the twentieth century some models relaunched, albeit in different ways, Descartes' postulates so as to explain animal expression by mechanizing the animal's endowments. The instinct (or innate automatism) of the tradition of classical ethology, as well as the conditioning (or learned automatism) of behaviorism, are striking examples that still stand out in the explanation of animal expression. As one can see, there is no room for subjectivity in this explanatory model because instinct is conditioning. That is, the endowments are enough to explain what the animal does. A subjective presence cannot be given here, because there is no sovereignty over those endowments. The animal cannot have ownership of them because the model explaining the endowments is mechanistic and imperative for the subject. If endowment alone is responsible for the structure of what is expressed, subjectivity does not actually exist and is pure appearance. Therefore, there is no need for consciousness understood in terms of Brentano's principle of intentionality.

To admit an animal subjectivity, obviously, it is necessary to adopt a non-exhaustive account of the part of endowment in the explanation of behavioral expression. A subject owns its endowment and uses it within certain limits, which entails a plurality of usage and the possibility of change. If that endowment itself is drafted to explain exhaustively what the animal does, it inevitably is no longer an instrument of the subject—an endowment indeed—but becomes an automatic mechanism that controls the individual and turns the animal into a puppet. Therefore, the actual complexity of an expressed function does not free the animal from the explanatory and mechanistic paradigm inaugurated by Descartes. Rather, we must do away with any presumption of subjectivity to read animal behavior as the direct manifestation of an automatism. As long as one remains bound to this claim, there can be no expression (regardless of its complexity) that cannot be translated mechanistically, thereby nullifying the very essence of animal being, which is based on presence and not simply on an expressive function.

When we say that animals are not objects, we create an inescapable epistemological problem that cannot be bypassed using the *deus ex machina* of consciousness or the tautological principle of a capacity to grasp the world as such. The discussion of animal existentiality must not be based on a transcendent question of principle but, rather, on a profound rethinking of the animal condition. And, the problem is not—at least, not in the first instance—that of which model to use to describe and explain the expression of subjectivity but, rather, one of the premises that make subjectivity and, even before that, the *concept* of subjectivity possible. Only a being whose behavior is not predetermined by its structures can be subjective. In order to speak of subjectivity, the individual must be the owner of his own instruments and any element, be it innate or acquired throughout his existence, must be a tool and not an automatism. Being subjective means being able to use one's own endowments just like one uses a city map to guide one's itinerary.

But, then, if one wants neither to fall into the infinite regression of the *homunculus* nor to use other forms of pseudo-explanation that, on closer inspection, are nothing but subtle tautological subterfuges or forms of *petitio principii*, one must go back to reflecting on subjectivity. Also, one must admit that the predicative expressions of animal being show us a way to curtail subjectivity, an expressive "how" of the latter, even while not solving the problem completely. Every animal—and I obviously include the human being but, also, the different forms of the taxonomic identity that refers to the momentariness of life rather than individuality—expresses itself by using, in a free, creative, and co-optative way, its own endocrine, cognitive, metabolic, motoric, and sensory endowments. Now, it is obvious that if these endowments change, then the subjective expression will change as well. However, the power of subjectivity lies in using those endowments freely and creatively, not fully adhering to its performative dictates. Subjectivity is the desire that exposes the subject to the world and leads him to immerse himself in a constant condition of problematicity.

In fact, to desire means to have problems. The life of every living being is in a continuous state of danger. By 'state of danger' I mean the need and opportunity to reach a target that is not available and, also, the avoidance of any immediate risk or critical situation. The word 'danger,' thus, implies a problematic condition—that is, being inserted in the here-and-now as presented in a problem. The state of problematicity is, therefore, the very condition of an animal being, whose behavior—that is, its expressions acting in and upon the world—inevitably connote desire, a libidinal vitality. Dwelling on the expressive modalities of this permanent state of languor is certainly useful for understanding who faces us but not for fully understanding their state. This seems more like an emergent or systemic understanding, one which also transcends a single endowment rather than the whole set of them. Maybe it is precisely for this reason that animal being is etymologically characterized as that which transcends its phenomenal being.

That said, subjectivity always requires an epiphanic intersection with the world (Marchesini 2014a), if one considers Prigogine's simple principle of the singularity of reality (Prigogine 1993) in open opposition to the deterministic view, an old legacy of Laplace's demon. To desire, therefore, means to build a singular dialogue with reality, creating a negotiating plan with it—a range of possibilities in the virtual field—and not simply to adapt to predetermined conditions. Therefore, behavior should not be interpreted as the consummation of a drive or reaction to a stimulus but, rather, as the expression of one's lingering in desire, reaching ever-new levels of reality. If the world presents itself to the subject in a singular way, it is clear that, while similarities may be attributed to, every danger brings about new connotations so that, if subjects want to go along with their desires, they must necessarily be creative, visionary, and co-optative with respect their endowments.

Let us go back to the old paradoxes that Descartes elegantly "dribbled" into his dualism, transforming the living being in rotting meat. Consider what makes us subjective, vital *because* desiring, emerging because non-subsumable, unfaithful to ourselves and always ready to break and free ourselves from constitutive constraints. Paradoxically, this does not belong to us. We do not choose our desires, the emotions that run through our dreams, the languor that stirs our imagination, the feelings that mock our rationality, and the reasons that link us and orient us towards the world. All we can do is give them some fictional representation. However, we know very well that it is not the little ball that stirs the predatory instinct of our cat. The same is true for the many excuses that we give to explain the irrepressible urge that makes us unravel life like a bundle only to come to the end of it and then regret the ignorant languor that left us amazed as child.

There is more subjectivity in dreams and hallucinations than in the logical reasoning that we use to place desires in the drawer of thoughts and representations. Thus, to understand subjectivity we must return to the animating desire and stop pretending to have control over this life. We do not own it; it owns us.

## Works Cited

Darwin, Charles. 1871. *The Descent of Man, and Selection in Relation to Sex*. London: Murray. Print.

———. 1859. *On the Origins of Species by Means of Natural Selection, or the Preservation of Favoured Races in the Struggle for Life*. London: Murray. Print.

Heidegger, Martin. 1995. *The Fundamental Concepts of Metaphysics: World, Finitude, Solitude*. Bloomington: Indiana University Press. Print.

Lorenz, Konrad. 1966. *On Aggression*. London: Methuen. Print.

Marchesini, Roberto. 2014a. *Epifania animale. L'oltreuomo come rivelazione*. Milan: Memesis. Print.

———. 2014b. *Fondamenti di zooantropologia. La zooantropologia teorica*. Bologna: Apeiron. Print.

Popper, Karl. 1959. *The Logic of Scientific Discovery*. London: Hutchinson & Co. Print.

Prigogine, Ilya. 1993. *Le leggi del caos*. Rome: Laterza. Print.

Skinner, Burrhus Frederic. 1957. *Verbal Behavior*. New York: Appleton-Century-Crofts. Print.

Tinbergen, Nikolaas. 1953. *Social Behavior in Animals*. London: Methuen. Print.

von Uexküll, Jakob. 2010. *A Foray into the Worlds of Animals and Humans: With a Theory of Meaning*. Minneapolis: University of Minnesota Press. Print.

# The Language That (In)Habits Us

**Antonio Domínguez Rey**

**Abstract** Heidegger's concept of language as a house of being (*Wohnung*) and the construction of a house that is privacy open to the world (Lévinas) refer us back to the origin of consciousness. A person's realization that he or she is "of" something, opens the self to a native place that displays still unknown possibilities. The current concern about ecology also refers to the vital relationship involving the Logos of life at the point of the space-time intersection of human movement – that of the body – in one's living space. This point is germination, a germ, and establishes a mode of existential presence in the world: a qualified voice. These relationships occur prior to the fact of our knowing them. They precede us. They are the a priori of consciousness, but are registered in the Logos. To them belongs language, which allows us to enter them, analyze them, determine them, and discover the native, original locus of consciousness. Language announces the open state of matter (*apeiron*), the uncertainty of which is determined by building precisely the intimate home of the primary relationship of life, one's habitat. The current concern for Ecology renews the question of the fate of Man, who is none other than the creator of sense for existence. When we stop creating, we break the latent rhythm of the cosmos and the corresponding radiation of existence. Man amazes himself.

**Keywords** Ecology · Language · Consciousness · Self · Space-time · Relationship · Poetics · Emergence

## Nest and Horizon

The roots of the word *ecology* send us not only to the ancient Greek world, deriving from *oikos* (house) and *logos* (reason), but also put us at the origin of humanity. We return to the cave, the refuge and shelter of man in nature. Humanity's residing in

A. Domínguez Rey (✉)
Departamento de Lengua Española y Lingüística General, Facultad de Filologia, Universidad Nacional de Educación a Distancia, Madrid, Spain
e-mail: adominguez@flog.uned.es

 
W. S. Smith et al. (eds.), *Eco-Phenomenology: Life, Human Life, Post-Human Life in the Harmony of the Cosmos*, Analecta Husserliana CXXI,
https://doi.org/10.1007/978-3-319-77516-6_31

the world marks out a portion of it as a settlement. Under this consideration, humanity has not evolved much. Human evolution still involves settling and dependency on a position or locus – place – in the world. Therefore, considering ecology as a phenomenon of life and knowledge still refers to an existential and native need to have a place where we can be inscribed. What has changed is the perspective. Originally, man took refuge in a niche in the world and progressively adapted himself to it to fend off the dangers of nature and sustain and spread human life. Nomadic or sedentary, explorer, hunter, or grower of seeds, we investigated the vital environment and adapting it to our needs: rest, food, privacy, and grouping together. Nowadays, our place or existential nest has been expanded. The exploration of the environment includes not only the earth beneath our feet but also the Universe that is its environ. The whole cosmos is a nest.

Nevertheless, concern for the settlement's own position and for the extension of the known reality remains. Humanity is still concerned with our permanence and the fact of being settled down in a given ecological niche. We are no longer satisfied with Earth's horizon. We are worried about the relationship of the planet with other objects, beings, and dimensions that relate to and upon which we depend.

The natural refuge first searched out and then built has become problematic, for it no longer secures for us a position in any corner of existence. Concern has become a threat. The thrust made to dominate and exploit the environment is today a threat to the consistency, subsidy, and permanence of humanity. This is the opposite of what is intended. Humanity has already entertained thoughts of abandoning Earth and inhabiting other planets. However, the Earth does not move, says Merleau-Ponty. Whether we live on Mars or the Moon, there also there is land. Wherever man goes, there is land. One's very body is made up of minerals. It is mud. Relocation does not mutate us, not at least for the time being.

In the transition from nomadic to agricultural settlement, and from thence to the energetic exploration of the environment to discover the forces or potential of the visible and invisible, there was another remarkable clue. The notion of shelter, cave, or nest was extended to the body itself as a sign of the spirit that inhabits it: breathing, energy, work, fecundity, reproduction, and, above all, the ability to understand, to guess, to projecting beyond where one is. We go from here to there leaving a trace, a memory, a settlement, the reminiscence of objects inhabiting spaces. This becomes History.

As the cave and the body merge together in the experience of habit and living, with their living environment, humankind comes to depend on a sense of permanence and belonging to something. We take responsibility. The horizon expands. It becomes a project, and understanding in our environment proceeds from it. As item is related to item, there is engendered assembly, setting, screening, orientation, sense.

Considering the verbal roots of the word ecology makes manifest to us, therefore, the conditions and prefixes of inhabited place: the Space-Time (S-T) of Movement (Mv) and Mode position (M) of a person anywhere. Distance is a dimension in an objective field of realizations. When man is capable of building a hut that imitates a cave, a den, or a bird nest, he is ready to dominate the objects and the environment in which they are located. He knows how to manage, to relate the lengths, surface areas, and the volumes of bodies. He opens a halo of references and creates a habitat. The construction of a house is another breakthrough, because thereby humanity defeats the resistance and transforms the connection of materials: mud, containers, vegetation, straw, wood, skins, and bones. The house rises, elevating organic force, work, and vital elements: air, earth, water, fire; and other life powers such as reproduction and nutrition as well.

## Bachelard's Apprehensions

"Car la maison est notre coin du monde. Elle est … notre premier univers. Elle est vraiment un cosmos," writes Gaston Bachelard (Bachelard 1961, 32). The house is 'body and soul.' It summarizes the vertical orthogenesis of the horizon combining the resonance of the soil, air, blood, heartbeat, breathing rate, and the sky that covers it. Man surpasses himself by building a roof with dynamic openings. He opens volume and gets resonance, an *inside-out*, like life and body. This has created an ecosystem of multiple relations, a habitat, "the topography of our intimate being," summarizes Bachelard (Bachelard 1961, 27).

Ecology refers to the intimate, *intus*, that which is inward, internal volume. If we were to look only at the earth and foundations, we would speak of archaeology. The nest, the cradle, the house-body are already the metonymy and synecdoche of interrelated elements. A continent contains. It is part of the content and refers to all which belongs to it. We have here a fragmentary whole, for it announces that which it never gives or it becomes everything: an even permanence.

Both the metonymy and synecdoche of the habitat make possible the metaphor or exoticism of a building. The elements are structured and open up to the outside, which is the space where the inside dwells, creating life time. The house is a metaphor: it opens out to all abroad, in which it bases itself from its very foundations (*metonymy*), and representing it in its parts: walls, roof, door, windows, home – focus "fire" – (*synecdoche*). Actually, the house never holds or closes the external. It lets one see, it penetrates the air. It is the abode of Rhetoric, but through hermeneusis: the settlement of a primordial sense of life. This is Poetics. The elements and tools assembled into blocks of references and meanings do not necessarily attain sense.

The transition from Rhetoric to Poetics presupposes an original emergence of sense. The same everyday elements are handled, but the idea that originates the opening of sense happens at a singular moment. Once given, the environment is transformed. It changes countenance. It acquires a new face. What happens immediately is driven by other parameters and redirects up to what is previously known, as if it affects their cause and foundation. It discovers components or features previously unknown. Poetics is founded today as the permanent that transcends the phases of time and concerns us today even more than it did at the origin of time, because we know more and know better. The Relationship (R) of Space-Time in Movement has here a specific Mode of living as human according to a unique rhythm: the life cycles of nature, the alive, vital principle of actions. Nature and man inscribed in it depend on that correlation in vital elements.

The natural rhythm obtained in this way fulfills the habitat of residing here-and-now as the permanence or topicality of the living body. This means the same as joy, natural goodness never quite satisfied. It persists in existential concern, now as a loss of the acquired, and return to the shelter as a response to animal fear over subsistence in the environment. Joy is a counterpart of this concern and reveals the basic impulse of positioning oneself in the world to be one of tenure and maintenance ("main-tenance," says Emmanuel Lévinas). Dwelling is to remain in the here-and-now rhythmically vital carrying out. There is a balance of trends and tensions because sense also reveals itself to be positioned. This is, in turn, the affection, the condition, the Self: attachment of the self in what is being (*ens*, *entis*) or its participle of presence. What is being. That is, *to be* is, in fact, alterative, as it alters at all times, going beyond itself. Correlated with the elements and the instruments that manage them, being depends on them even as, other possible realizations of being other than itself present themselves. Here is the human aspect of the technique. It discovers a new horizon of human accomplishment. In this sense, *techne* belongs to Poetics, but not as Rhetoric, because here the instrument repeats only the use already revealed. Rhetoric disuses it, even by excess of formalism, to convert the form into a kind of or center of an *autotelos*.

## Lévinas' Apprehensions

For Lévinas, the essence of a house is misleading and "digs interstices in the continuity of the Earth" (Lévinas 1971, 184). It is *anterior posterius* for it touches on something prior, but only once it has been constituted as every kind of work. It delimits a space. It collects and hosts an opening without getting rid of the primary and permanent matter, the neutral, shapeless *il y a* or "there is" of existence: that which is thick, hard, rude, humble, and gross, impassive (Lévinas 1978, 91, 92). Further, the sound of basic speech, its atomic germs, the elemental component of human sound, the friction of phonic articulation one day unknown but then a signal and a sign of something to hand and then gone away. This is not the thematic wisdom of flesh in contact with the "there is" of elements and the place they occupy and inseparable from them. It is, rather, a visceral correlation with the world, rather than an entity (*beingness*?); a native sense was never born altogether themed. That house is only a transit, as is every theme, interstitium of Space-Time in biological and cosmic rotation (Movement, Mutation, Becoming).

This character of the ecological phenomenon maintains, in a formal suspension (*epoché*), the being of objects and its imminence in this transit. Here is, I believe, the neutral attribute of Lévinas' understanding of *il y a* or simple being there without any determination, or place. Again, this would also be something *anterior posterius*, intuited once we see or feel any existential irradiation, as we inscribe and enroll ourselves beating into existence. Yet, one wonders if there be a gap between the neutral and any original position or if, instead, we are to consider neutral, unthinkable, indefinite, every thought and concrete determination. The neutral would then be a subfloor or imprecise nowhere so that, without it, any precision would be impossible. Calling it neutral would be a deduction *a fortiori*, as it is not in any single element or compound of which it was one of its characteristics. Such forcing would damage the origin of the entire human position. This neutrality makes an attempt against the sense of the ecology in Poetics. The effort of a human being is never neutral, as it requires contact with the elements, industrious ingenuity, exchange, and human interactivity.

Lévinas attributes a feminine character to this mode of alterative presence or dialogue with the environment. The element relates to touching it, assembling it, opening a bend, considering structural dimensions, or making volume. This is also the case with the voice. It says outside and beyond one's speaking, it's always another of himself. In the footsteps of the feminine "you", resonate "les epaisseurs secrètes de l'être" (Lévinas 1971, 167). A material thickness, prior to the dimensions that circumscribe any object – length, width, height, and volume – is, thus, constituted. The corporeality of the world precedes its objectivity, resonating and vibrating. The body itself is a germinative resonance of qualified vibrations, in proximity to the Other, a "pre-geometric eidos of space" as the dwelling of others (Lévinas 1974, 96, 152 22n). Therefore, the house is "la présence discrète du Féminin" (Lévinas 1971, 185).

Lévinas also compares the way of alterative proximity or dialogue with the environment to the feminine. Any objective realization will be separation (*écart*) or footprint (*trace*) of that secret resonance. The house, or formed work, contains a thicket of being that is still a secret, one which is separate from the parallel, angular surfaces, or volume of the frame. This priority is a desire, inscribed in the vital tension following from man's transcendence of Eros, which would be only a figure of our existential difference with the Uni-verse. The human body, then, shows itself to be the located residence of an anonymous being whose thickness resonates. For Lévinas, body predates Space-Time as *içi* (here) and, of course, the knowledge that makes it objective as a thing. This is a constant activity and folds, into one's flesh, the process of awareness. Another matter is that it differs, so says the French philosopher, from intention, because the position of the body matches, he addss, its action (Lévinas 1978, 138). If there are secret depths of the Self in the resonance of the feminine body, this tends to something else, we say. It distends. It breeds. Intention presumes tension, such as the concept of the face (*visage*): the upwelling of the other in the anonymous thickness of being.

The house has become a fold of awareness, or the Self's niche, altered by figures representing the unnameable, unthinkable Other, diverse as it is. Ecology refers, and proceeds as a phenomenon, to consciousness and its discursive folds: the folding of the I-You relationship, into He or It, of every conscious position.

We have thus discovered an *anterior posterius* position, only now known as such, it having been disclosed. This position induces an *a priori* depth whose scope is lost and absent without memory, immemorial, because no link of consciousness comes to contain it. The horizon of being presupposes, in every entity, something previous, only thematically telling of it and falsifying even as stating it. This horizon opens a space without a precise location for hermeneusis. Such openness or waiting is time, stress, and event already, *Ereignis* or "event," energy and matrix power, the root of something that is or can be happening. Being creates the expectation of something Other.

This is what the language relationship is, the uttering of the expectant and spontaneous speech of man. We speak to the Other oneself, altered (*alterum*). And in talking about, we reveal the hidden, secret, and nascent. The interrelation of elements in the construction and maintenance of the house and of life with food presupposes, along with work, language. This happens resoundingly, emitting sounds that, in a moment, are recognized as a spontaneous emission induced by our interactivity and the work being carried out, discovering the secret relations of objects among themselves. Also, there is a previous Relationship (R). Its setting is the possibility of a link showing its components.

## Heidegger's Apprehensions

Language is an elated expression of the body's natural sound, moving and acting between the elements of the living world: close or far away, vibration, alterative disposition, energy induction, resonance, and the like. This living dimension of the elements, Heidegger calls poetry: "Poetizing is properly [and originally] to let something inhabit" and is building (Heidegger 2000, 193, 206). *Poieo*: Doing something with the voice, bringing to presence the hidden or possible something still unknown.

Heidegger also presupposes an *a priori* of being, although he denies anything previous to itself setting in "nothing" all that is not as much as Word. The problem of being relates to language. The *a priori* concerns the preconscious but is revealable at any moment. We trust that from our talking there emerges something different from what has already been spoken. To talk reveals. This is the truth of language: each of its instances captures the being of a thing as a Word. The thing is precisely a thing when it meets a word (Heidegger 1985, 154). Somewhat of the thing affects the Word holding it up among things, even if this is only for an instant. Something of the Word relates to the thing, being its revelatory mediation. This inter-nature affection founds the objectivity of knowledge, which is not necessarily representative of being otherwise. This representation is one among other important effects of that affection, since it guides us in the direction of the foundation of knowledge and the unveiling of being.

For Lévinas, this would be a contradiction in terms, because holding up a thing subject to its characteristics somehow without being made a suspension (*epoché*), as it presupposes the action of the unveiling. The reification of language could be absolute. What is not said, or only could be said, does not exist. The limits of being are those of language and vice-versa, says Wittgenstein. But, man assumes a position before the revealed and said. This takes away the thingness of the thing, and it cuts out the objectivity of what is objective with calm and moderate resignation (*Gelassenheit*). Only in this way, can what is built build (*Bauen*) and dwell (*Wohnen*) as word (*Wort*). Heidegger conditions human freedom to worrying care (*Angst* and *Sorge*), which reaches forward, gathering, in proximity and neighborhood in an unveiling that is *aletheia*. Listening to the rumor of being requires ultimately attending to its emergence. Therefore, Heidegger (Heidegger 2000, 183) conceives a step backwards (*der Schritt zurück*), clearing what covers the being in every entity as a way of *epoché,* or critical suspension, letting the hidden be expressed. A hearing suspension, or Mode of being that listens to the original, removes the representations. The original call claims us, insofar as it announces what is coming or becoming. This is freedom: to let something be in its arriving in its own way without violating it.

This is poetry. For this reason, not only do thought and language merge in the poetic function but also the poet finds an "open space" (*das Offene*) in which people may make the truth-dwelling house (*eine Behausung*) where the gods come "as guests" (Heidegger 1981, 148). There is an epiphany of the hidden or mysterious in the Earth where man lives. Our emplacement here is, then, like the hospitality of which Lévinas speaks, but is one that is more existentially committed. Hence, we see the care for the unity of the elements that Heidegger displays in his square (*Geviert*) formed by the Earth, the sky, the relationship with the divine – God – and the mysterious and mortal nature of man. The Earth is what "serving holds;" the sky, "light and twilight of the day, darkness and clearness of the night, hospitable and inhospitable weather." And, "The gods are the messengers of the divine that give us signs." And the mortals "are humans," obviously (Heidegger 2000, 152).

Who cares, keeps, candles, hosts. Welcoming is the "law" that founds the "house" where "the mystery of the being" is kept, and man looks after "the inviolability of the possible" (Heidegger 2000, 97). For this reason, language is also "the house of being" (Heidegger 1976, 313). This is, in turn, to inhabit the Earth, saving it from deterioration and welcoming the sky – light, water, power, horizon – waiting for a coming god, which means accepting death without knowing whether death is another stage of the coming or a possible resurrection of the Universe. The Universe, that is, all that is oriented to housing the last unit of life and understanding is to be inhabited poetically (Heidegger 2000, 155, 193).

Accordingly, language, a word, already bears in itself, says Jeanne Delhomme, who echoes Lévinas, what is real and, vice-versa, what is not. The word has not been brought or originated by the real, by its reference and the ideal representation that it contains, but bears the real thing (Lévinas 1976, 74). Thus, it is now the turn of poetry (Heidegger) and now that of ethics (Lévinas). The nature of being needs originating compliance and liability: one has to act according to one's own principles.

All that which is opposed to this naturalness interrupts the being diverting, forgetting, or sacrificing it. The vital tension or human time precedes on a quantified scale. The adequation of the Self to the manifestation of things, of the mind to an object – objective representation – which is the classical foundation of truth, retains in this alternative philosophical consideration the manifestation of that secret thickness which feminine being announces with its movement step by step through the interior of the house. This metaphysical objectivity reduces science to a technique of demonstration, prevailing the instrumental tools imposed (*Ge-Stell*) upon the position and anxious constancy (*Selbstand*) of being. According to Heidegger, science, especially techno-science, invalidates the relations of things to the original thing itself. It displaces things and obliterates the order of the fourfold correlation. The elements that constitute the original thing are united in the being of the thing and its dwelling "in something that is always dwelling," that is, the permanence of the four elements (*das Geviert*) (Heidegger 2000, 175).

## Logos and Polemos

Both Heidegger and Lévinas trust the logos of human permanence or ecology in the background of spontaneous nature and life. They are suspicious, however, of the abstraction and objectivity that the logos produces. The logos is for Lévinas, just as it is for Heraclitus and Heidegger – the latter following Parmenides – struggle, the clash of elements. It produces only a warlike tension: the logos has itself a need for violence to preserve itself from word facility (Heidegger 1983, 183). This violence, however unusual the thought might be, is also, as in Heraclitus, controversy (*polemos*: war), provoking a person's overcoming self to get the best out of his or her nature and explain it. This engagement outlines and develops what is unheard, not even said or thought (Heidegger 1983, 66). Therefore, the Good overcomes being as the Glory of an existence beyond the entity, the closed, totalitarian entity. This criticism relates to logos' totality, summarized thus by Heidegger following Heraclitus: Logos is "*the constant harvesting, the intrinsic togetherness of what-is, that is, the entity*" (Heidegger 1983, 139).

## Mathematics and Metaphysics

The set of entities as totality of being determines its mathematization, an identification of method and truth, which is to say the way of understanding the facts is by the application of concepts such as "taking," "appropriating something" in order to get to arrive at truth. This would be scientific rigor, the exact adaptation of the cognitive process to the known object. Now, not every object responds to mathematical precision. Hence, the adaptation of mathematics does not guarantee the truth of what is known: "if we mean by the strictness of a science its form and manner as it gets and determines knowledge appropriate to the object, then accuracy in mathematical sense does not necessarily establish the exactness of a science" (Heidegger 1996, 44). Heidegger quotes, as an example, the assimilation of historical knowledge into the mathematical method in the nineteenth century. There are sciences with different bases. By not regarding the evidence of each one, the yet unknown richness of reality is conditioned to its objective representation. On the presupposition of experience we impose a mode of perceptual constancy in accord with the calculated ends of objectivity. The process and its foundation become technical, for what science and technology respond to is the goal of metaphysics that conditions a being to its reasoned representation: a "ground that gives an account of the foundation, giving it reasons," and that, finally, prompts asking for explanations (Heidegger 2006, 66). The arrival of being to a given presence is broken down by the *im-position* (*Ge-stell* "enframing") of an incursion, because it reveals in its charge the power of being, its possibility, and forces the entity to appear improper or unsuited to its original form. "Im-position means the way of revealing that prevails in the essence of modern technology and which is itself nothing technological" (Heidegger 2000, 21). We should recall that the root "stell" of the German word *Gestell* (shelf, frame) has the same root as the verb "Stellen" (to put, to place) and the noun "Stellung" (position). It is related to the root "stall" (stable) and the Greek "stellein" (put in order, furnish).

This process involves a paradigm of opposed reductions, such as Self to subject and nature to object, the two poles of the world and, at the same time, body to thing in such a way that the correlative of the world would be:

subject – body: object – thing: World – God.

Such a proportion correlates with the contrast of theory and praxis, corresponding to the ontological difference of being and entity. It is understood, at the same time, to result from representation's opposition to revealing:

theory – praxis: being – entity: representation – revealing

The repeatable constancy of the form of representation imposes a criterion and submits to it whatever is manifest. Knowledge becomes technical and its form hides other manifestations, which is to neglect other looks, faces, presences of being. The fixity and profitability of the image of the world contained in the figures of representation awaken a desire to control understanding of them.

## Calculation and Control

This is the background of today's knowledge society: objectify forms and match them to dominate through mental representation of them and so control their will, as Schopenhauer would say. These forms clog the pores of the Self. They plot their differences: "everything fell on the same plane, on a surface that, similar to that of a blind mirror, does not reflect and no longer throws back anything" (Heidegger 1983, 49). They are forms that cover up and conceal the abandonment of being produced by calculation (*die Berechnung*), speed (*die Schnelligkeit*), and the irruption of the massive (*der Aufbruch des Massenhaften*). These are signs, in the technical order, of the reduction of language to statistical information; the reduction of culture to industry, as well as of science's reduction to "Science Inc."; the reduction of the land to resource; and the reduction of humans to "human resources." On the metaphysical plane, there follows, in turn, the instrumental reduction of spirit to intelligence; the flight of the gods and the values that announce them; the destruction of the Earth; overcrowding and suspicion before free creation (Heidegger 1983, 48).

## Archeological Considerations

What about this critical panorama? The opposed reductions of subject to object, of nature and body to thing, are the rotation effect of being on the internal tension in the cosmos and today's biotics. This is a response to the a priori a posteriori revealed. Each moment is a constant, current presence: the antecedant of the instant of an essence and a feeling is expressed in the prefixes of "pre-sent" and "pre-sentation." A presence is a tensional point of what happens emerging in pure actuality. It is a tension of the act of placing and stating in the world and, therefore, being specific, of maintaining an ongoing Relationship to the Other of the Self, and, according to a process or "internal plan," involving a decision of the person so placed. The divisions made come from the limitation and the marking out of permanent dimensions as the subject is assuming a position in the environmental realm – Space-Time-Movement (Mutation) in the precise Mode of an actual Relationship [S-T(R) Mv-M] – for it happens in tone of the speech, extension, thickness, roughness, density, latitude, depth, curvature, volume, resonance waves, and so on. Before objectification is imposed, there is an "unobjective reality," the dimensions of which are those of being, or of any objective entity that unfolds itself in accordance with the opening character of its presence. In this way, the world resounds in every entity – *on*, *ontos* "that which is being" – and this resonance is, as Henri Maldiney says, "an existential" (*Notes* 30). It depends on the *oikos* or habitat of matter in each human "space-time" or "instant-place." The form half-opens the Space-Time of the Movement or Mutation, as per case. It implies its own genesis, the "pre-spatial fields" of constitution (Maldiney 2012, 20, 21; Domínguez Rey 2014, 242–243), in an anticipatory prolepsis of Space-Time (Maldiney 2001, 85).

Existential emergence shows itself in the *forming form* of art and in biogenetic dynamism. It is a coeval event form. It is a non-uniform formation but, in fact, a correlated one because the preobjective reality of a Space-Time-Movement in the Mode of an emerging performance does not match, even though coeval, orthogenesis, cosmogenesis, and ecogenesis. The biotic process is one that we cannot retroject or introject. Such an operation is only relevant to consciousness. In biotic dynamism, there is no turning back. The components would not yield the same result if we were to retroproject them. However, the dynamism of knowing is in fact retro-projective.

Cognizant retro-projection assumes even what could seem a "retroactive illusion," as Slavoj Žižek thinks it does, in fact, in the alleged 'return to origins,' considering that the Event is a Fall or "the loss of some primordial unity and harmony which never existed" (Žižek 2014, 50). But, the illusion of consciousness is an opening to the world. For atomic physics, the vacuum or the interval between the existence of potential particles and the footprint of their effective registering in some medium, which phenomenon involves their disappearance, functions in line with the so-called theory of "broken symmetry" (Žižek 2014, 53). There are some similarities here with the way poetry functions, and even language. The presence of words is a trace of a phenomenon, one that refers to a constant and current source and is never completely symmetrical. This sloppiness, always critical, makes sense possible, bringing together the poetic elements caught in the light of a complex saying. Its components connote a significant density involving a permanent critical act, more a gap or interval than a vacuum. The generation of particles is to be interpreted precisely as the dynamic of Space and Time in poetic Mode. This energy is the future registered in the singular performance of an instant, no matter how long it lasts.

## The Correlative Nexus

The genetic function allowing retro-projection is abstract. It includes its own training and that, therefore, formed the objective world. This is a special object. Only man reaches it, and this does not happen out of an abstractive genesis. The philosopher Ángel Amor Ruibal named this phenomenon "the principle of abstractive realization." He frames it first and apparently in the perspective of the theory of *moderate realism*, which attaches to understanding a precise (we would say today essentially intuitive) and a comparative condition. This "double intelligence operation" presupposes "in the singular … an aptitude to be conceived by the abstraction of the singularity" (Amor Ruibal 2005, 300). This is a well-known theory. Amor Ruibal sees here, instead, a genesis whose hiding could disfigure its forming dynamism and the permanent relationship given in the assumption of a mental conception's suitedness to things "according to the mode of understanding." Here is something objective and intelligible, because man discovers *a posteriori* that there is a previous natural and transcendental relationship that implies a specific value (Amor Ruibal 1995, 331). This brings to light a real and correlative nexus that also reveals some experiences and qualitative pre-judgement: a "nexus of transcendental and prelogical relativity" (Amor Ruibal 1995, 224; 1934, 44).

The act of *cognoscens* is vital and requires a suitable quality in the mode of effective knowledge appropriate to the object. This correspondence is genesis, breeding. Their realization depends on the ecological niche or dwelling of being such that it shows itself knowing. Here is a mode of intuitive presence that entails *ipso facto* a reflection of relationship or a horizon of entity integration. This is a sort of epigenesis or overlaying that comes out as knowing thinking, like the epigenesis that Kant attributes to pure reason (Kant 1998, 204). Without this presence there would be no knowledge. It maintains the resonance of the formation, and so it is not isolated from the process, nor isolated when consciousness recognizes herself in the act of knowing.

This knowing is never absolute, not even when it individualizes itself. Individuation already presupposes a duality of perspective. It keeps differing in the unique act of existence. It is the other of itself in every act of life. This otherness belongs to the establishing horizon of knowledge. Therefore, we can affirm an absolute character neither of life nor of consciousness, as it claims and hints, according to Hans Reiner Sepp (Sepp 2010, 218–219). The individual is what is not divided from an entity whose elements gather, each one being, thus, the other of itself, already unrepeatable. Here is an inner and intensive relationship in such a way that the individual is, as Gilbert Simondon says, an "*active centre*" or *the being of a relationship* (Simondon 2005, 63). The absolute is absolved, and what is required is only what it contributes to making one's place in the world and giving it sense. In the case of repetition, it responds to being different, something exclusively singular.

## Ecological Depth and Height

Only thus, does awareness achieve what *deep ecology* (Arne Naess, Aldo Leopold, Bill Devall) calls the intrinsic value of eco-phenomenology, the intrinsic value of life, from whence all manner of natural and human relationships come. But, even so, bodiliness (phenomenological *Leiblichkeit*) is confined to the common S-T (R) Mv of the human M of existence, the orientation, physical and mental limit of each of us. The inside, there revealed as one's own inner, would be an exclusive knowing and sense of existence if it were absolute, as Sepp says. But this inner never excludes, since it gathers and assembles and has its own experience as it feels the other of its process, as Sepp recognizes, but always in reference to the Self of subjectivity (Sepp 2010, 221). This other is also always new. The newness of an existence, the being of which is continually altered, is likewise also the disappearance of the place it leaves. Attention to this newness is also poetic and, so, highest principle of ecology.

This justifies any "precautionary principle" because, as Hans Jonas puts forth, our world is threatened by the consequences of technology, a realization that should not mean, however, an excess of inhibition or fear of danger. That could lead to paralysis and a recession of humanity, remarks Gérald Bronner (Bronner 2014). Precaution is part of the horizon of responsibility, the foreseeing of the fundamental conditions of life and the requirements of knowledge.

## Human Ethical and Behavioral Aptitudes and the Natural Aptitude of the World

Understanding of the world depends on this horizon, as likewise does the responsibility or cosmic response of man to his nature. Ecology requires human nature's preservation, which indicates an ethical aptitude, that is, the counterpart of human behavior in the objective presence of things. This is a universal ethic that includes the economic reasons for the elements and relations that have arisen as constituting themselves under the sense that comes. The logos of ecology also includes the norm or law (*nomos*) of the house (*oikos*) or human constitution, its economy. The problem arises when the process of making dissociates itself from actual objects and obtains objectivity without the original reference that brings sense. Then there arises a commercial metaphor that hides the real object and makes it disappear. This happens in the real financial economy and the exploitation of money becomes a mega-metaphor of nature. This process reduces life to production inspired by the irrational consumption of apparent goods that are, in truth, unnecessary. Such a reduction limits the range of nature and substance to what amounts to only a technical way of discovery or simulation of something possible, without apprehending any real possibility of existence. It even predetermines and alters the way of presence.

We, thus, follow a deviation from the ontological principle that barely satisfies its own operation, if only because it touches tangentially the distinguishing action of knowledge. Here is one of our most dangerous illusions and the most pressing challenge facing eco-phenomenology. One speaks, then, of a natural object the qualifying attribute of which is a quantifying algorithm or its previous mathematical reason. This is the case, for example, with Chomsky's pragmatism, which reduces the semantic content of concepts to a syntactic formality without letting us know if the mental connection means something inherent to reality or if it is simply a purely circumstantial phenomenon. Rather, the latter. This presupposes that the reality itself is syntactic (Domínguez Rey 2014, 260–265). We here only outline the ethical basis of such an ecology.

The aptitude of human nature is more readily behavioral than intelligible. When intelligence perceives a form, it has already been acted upon by its germ and *gramma,* or revealing impression, of something previous, antecedent, predisposing. It recognizes itself by acting in a context of correlations and special resonant links, without which it does nothing, because the split would be a phenomenon of its operating mode. There is an apparent separation of consciousness here, from knowing that something is still more of itself, is growth and development. This is not necessarily a quantitative plus, but something new of its still-hidden entity. This so-called distance already implies the presence of some other objective, whose relationship is an ontological a priori. The definition of entities thus includes existential behavior. The act of knowledge differs becoming actual. The difference of being is existing, its constituent entity.

The natural aptitude of the world, originally given, presupposes a singularity of knowledge which never totalizes. It discovers previously hidden factors and precise moments of neat contact. Every last one fits into the preconstitutional environment and is part of it, but each one reveals something new, previously unknown, or created from a specific time. What is thus being revealed generates a specific impression that establishes new existential relationships. This even precedes the image that Heidegger puts before the representation and allows a presentification, as it is the origin, the opening of beginning, the starting.

## Being Becoming Language

Language, and the trace that it leaves in thinking as creative activity, belongs to the singularity of objective-subjective existential reflection. There is formed, thus, a realm of existential and cosmic resonance, for it contains an internal relationship of rhythmic genesis. It retro-projects itself as active thought. The point of creative constitution is what Heidegger calls "the house of being" to indicate "that in thinking being becomes language" (Heidegger 1976, 313). This happens because the human sound becomes a symbol of the complex action there carried out. The phonic act exemplifies this complexity. It is prelude to a multiple correlation of factors which are explained, then, once analyzed. It is something *anterior posterius*. It is singular, too, because it is only recomposed in its elements from some degree of constitution and differently so in each concrete language. You can imitate it, but without leaving these consolidations or going from them. The phonic act has an atomic character. It indicates even the presence of absence and the absent resonance of a presence, two modes of depth whose interconnection originates language, but which interconnection can also be disengaged, which throws it indefinitely into an abyss. Its resonance reaches, at the same time, several points of the virtual field of established correlations, and its constitution integrates also the environment [S-T (R) Mv: M] or a contextual relationship as reference, visible in the tone. The human tone reflects the vibration and resonance of the inhabited environment in an accurate place and time according to the Movement Mode or qualified Mutation produced at any moment.

## Parallels in Language Formation and Biotic Emergence

The phonic process responds, as a *phonetic type*, to the existential correlation of the "genetic centre" according to Amor Ruibal, as the singular phonetism of each human group reveals. This is not unrelated to biotic genesis, because the phonic articulates by comparing, which presupposes intuitive precision. The phonemic performance discovers the scientific hinge of the psychological, logical, and ontological principles. It is analogical and digital. It analogizes as it, likewise, digitalizes. This is possible owing to the superposition of timely sound waves on the qualitative simultaneity of the symbolic field there opened, articulating them. They originate a volume of resonances and vibrations in the brain, the dynamism of which correlates with the cell system, homologously, Ángel López García says, of the grammatical correlation. Transcription, translation, and replicative transfer operations ("type 3 T + R") occur in cells and in language (López García 2002, 106). Sound waves transcribe the correlated impulses that translate the intention or expressive, conscious or unconscious, impulse in syntagmatic units. Such motions transfer in addition the symbolic complex there induced, forming new more complex units, as syllables, words, clusters of these in phrases, and so on. In their forming, these units replicate their own parts with different functions. The replication is done, then, reproductively as someone assimilates a unit, using these repetitive units with similar aims.

Thus, language inhabits us, or rather, (in)habits us. The house (in)habits. Not only is it habitable, but it (in)habits us. When we enter into a fifteenth-century temple, we inhabit, and the realm of that age inhabits us. We are part of the continent within the singularity of S-T (R) Mv in a different existential Mode. Thus, the Earth is where we live. There is a continuous transformation of the pre-constitutional field to a plane of an always current actuality. Knowing how to interpret this correlation of what is continuously permanent is the highest degree of ecological ethics. This interpretive function happens only in consciousness. We went from indeterminacy and uncertainty to the concreteness and term of an instant in multi-radiated correlations. We integrate what we have lived and perceived in a horizon of existential inscription. So, everything here depends on there or elsewhere, as everything now is an irradiated retro-(hence-forth)-pro-jection of permanent actuality.

## The Founding of Ecology

The actual underlies any relationship of elements and their atomic germs. The positions vary, the extensions vary, but the genetic variation and mode of emergence are a stable dynamism. So, we venture explanatory models and macro system integrators and microstructures, as Conway Lloyd Morgan and Charlie Dunbar Broad made clear in the early twentieth century, as did Paul Oppenheim and Hilary Putnam in the mid-twentieth century and Jaegwon Kim (Kim 2006) from the twentieth into the twenty-first centuries. The problem of the relationship of the part or parts to the whole, with the reduction of this to those or not, dealt with emergence, happening, the principle of completeness or indeterminacy, with or without reflexive downward causation, which only states to us a categorical ontological a priori relationship of ontopoietic integration. There is a singular moment in which the objective-subjective relationship with the environment discovers an autonomous elation induced both by constituents not entirely known – though human sense remains – and the ability to influence those constituents by changing the relationship's process. Such a singular phenomenon is a *poem*. It brings one to a sense's *prolation* that ignores the accurate and immediate moment but, sensing, confirms it once it consolidates its origin. This is a phenomenon that involves continuous interpretation by moving among a factor of uncertainty, its concreteness, and gathering together on the horizon that it induces. This is something similar to what happens in the noun phrase and verb phrase between a name and the proposition in which it is inscribed, whatever it is: a noun, adjective, verb, pronoun, or marker. There is always an internal predicative relationship (R) that interprets (i) the springing up, formation of a name (n) and propositional (P) constitution according to principles (p) there revealed: (R) i n/p (P). The middle slash here indicates the procedural difference of indeterminacy and its concreteness at a given time. With the propositional (P) integration, which obtains in every science, we extend the outline we have exposed in other publications (Domínguez Rey 2012, 170, 236).

The concrete determination of something yet indeterminate assumes a progressive background of correlations between attributes and conceptual features which, in turn, reflects an underlying origin. The perceptive value of sensations influences this determination, as they are also interrelated by that rhythm on this prepoetic level. Karl Rahner refers to this when he describes, in a Heideggerian way, the background evoked by a word. The reality circumscribed by a name differs from it and from other realities, but its elements join each other, confronted and allied, as they refer to a single origin that harmonizes both unity and difference: the word frames the singular and always claims, therefore, an order itself not includable, the always precedent, constant *a priori* in the depths and under the bottom (Rahner 1960, 443).

Language that (in)habits us thus reflects a prior ontological inherency consolidated through overlays and so recovers what science, linguistic or not, analyses, then determining possible causes, reasons, inductions, principles: in a word, theories. The cause then comes to be an interpretive application of an ontological correlative foundation and *a priori*. It arrives late to the process, we would say, and reveals constituent priority. The logos of a being thus manifested already presupposes a discursive, ontopoietic, quantum character of action, because it is an irrepeatable singularity, though one replicable once controlled by techno-science (Domínguez Rey 2014, 33–34).

Ecology is, thus, founded. Concern for the dwelling of being does not obey only the giant machinery of techno-science. Rather it prompts worries about the amorphous silence of the designated ontopoietic priority. Gigantism would be only a shadow effect in the revealing muteness of the Uni-verse. That shadow darkens the basic and original relationship that Anna-Teresa Tymieniecka spoke of in her Bergen interview as being the essence of her philosophy: "our relationship to the earth and to the cosmos" (Torjussen et al. 2008). This relationship is "a shaping force" that individualizes: *ontopoiesis*.

So, we have a "genetic centre" of an "internal plan" (Amor Ruibal), an "active centre" of individuation (Simondon), and a natural "shaping force" (Tymieniecka). Its origin is in the rhythm of the Uni-verse, the waves of which are inherent correlations. Attention to this emerging rhythm saves the world from destruction, since both life and the habitat of man depend on it.

## Works Cited


Amor Ruibal, Ángel. 1934. *Los problemas fundamentales de la filosofía y del dogma. Tomo noveno "ninth."* (3° Póstumo). *El Conocer humano: (Función de deducción)*. Santiago de Compostela: Tipografía del Seminario Conciliar de Santiago (de Compostela). Print. 10 vols. 1914–1936.

———. 1995. *Los problemas fundamentales de la filosofía y del dogma I*. 1914. Ed. facsim. Santiago de Compostela: XUNTA DE GALICIA. Print. 5 vols. 1995–1999.

———. 2005. *Los problemas fundamentales de la filología comparada: Su historia, su naturaleza y sus diversas relaciones científicas. Segunda Parte*. Madrid: Fernando Fé, 1904–1905. Santiago (de Compostela): Universidad Pontificia 1905. Print. Ed. Facsim. Santiago: Consello da Cultura Galega. Print. 2 vols. 2005.

Bachelard, Gaston. 1961. *La Poétique de l'espace*. 3rd ed. Paris: PUF. Print.

Bronner, Gérald. 2014. *La Planète des hommes. Réenchenter le risque*. Paris: PUF. Print.

Domínguez Rey, Antonio. 2012. *Mundo, Texto, Contexto: Intersticios. (Génesis Discursiva)*. Madrid: UNED. Print.

———. 2014. *El Gramma Poético: Germen Precientífico del Lenguaje*. Barcelona: Anthropos. Print.

Heidegger, Martin. 1976. *Wegmarken*, GA Bd. 9. Frankfurt am Main: Vittorio Klostermann. Print.

———. 1981. *Erläuterungen zu Hölderlins Dichtung*, GA Bd. 4, 148. Frankfurt am Main: Vittorio Klostermann. Print.

———. 1983. *Einführung in die Metaphysik*, GA Bd. 40. Frankfurt am Main: Vittorio Klostermann. (1976). Print.

———. 1985. *Unterwegs zur Sprache*, GA Bd. 12. Frankfurt am Main: Vittorio Klostermann. Print.

———. 1996. *Einleitung in die Philosophie*, GA Bd. 27. Frankfurt am Main: Vittorio Klostermann. Print.

———. 2000. *Vorträge und Aufsätze*, GA Bd. 7. Frankfurt am Main: Vittorio Klostermann. Print.

———. 2006. *Identität und Differenz*, GA Bd. 11. Frankfurt am Main: Vittorio Klostermann. Print.

Kant, Immanuel. 1998. *Kritik der reinen Vernunft*. Hamburg: Felix Meiner Verlag. Print.

Kim, Jaegwon. 2006. *Trois essais sur l'émergence*. Paris: Ithaque. Print.

Lévinas, Emmanuel. 1971. *Totalité et infini. Essai sur l'extériorité*. 2nd ed. Den Haag: Martinus Nijhoff. Print.

———. 1974. *Autrement qu'être ou au-delà de l'essence*. Den Haag: Martinus Nijhoff. Print.

———. 1976. *Noms propres*. Montpellier: Fata Morgana. Print.

———. 1978. *De l'existence à l'existant*. 2nd ed. Paris: J. Vrin, Paris. Print.

López García, Ángel. 2002. *Fundamentos genéticos del lenguaje*. Madrid: Cátedra. Print.

Maldiney, Henri. 2001. Existence: Crise et création. *Henri Maldiney entouré de ses amis. André du Bouchet. Roland Kuhn. Jacques Schotte. Existence. Crise et Création*. Jean Pierre Charcosset pr. La Versanne: Encre Marine: 73-113. Print.

———. 2012. Notes sur le rythme. *Henri Maldiney: Penser plus avant....* Actes du Colloque de Lyon (13 et 14 Novembre 2010). Jean-Pierre Charcosset comp. Précédés de trois textes d'Henri Maldiney: *Sur le vertige. Notes sur le rythme. Rencontre et ouverture du réel*. Chatou: Les Éditions de la Transparence: 17-22. Print.

Rahner, Karl. 1960. *Das Wort der Dichtung und der Christ*, Schriften zur Theologie IV. Einsiedeln: Benziger. Print.

Sepp, Hans Rainer. 2010. Grundfragen einer phänomenologischen Oikologie. *Acta Universitatis Carolinae, Interpretationes, Studia Philosophica Europeana* 1: 218–219. Print.

Simondon, Gilbert. 2005. *L'Individuation à la lumière des notions de forme et d'information*. Grenoble: Jerôme Millon. Print.

Torjussen, Lars Petter, Johannes Servan, and Simen Andersen Øyen. 2008. A*n Interview with Anna-Teresa Tymieniecka*. Web. 27 August 2008. http://www.phenomenology.org/Bergen-Interview.htm.

Žižek, Slavoj. 2014. *Event: Philosophy in Transit*. London: Penguin Books. Universidad Nacional de Educación a Distancia (UNED), Madrid (Spain). (Revision of translation by Marga Calvo Arribas).

# Phenomenological Elucidations Carried Out by Constructing a Phenomenological Language

**Erkut Sezgin**

**Abstract** This paper is about what is actually being touched, with its power of transforming awareness, by the deconstructive tracking and clearing, by the animation of auto-hetero affective movement, of the internal manifestation co-signifying the background, and the spontaneous manifestation of which is animated by Wittgenstein's "inventive" "inceptual" thought experiments by constructing a phenomenological language. Thereby, we are allowed to track the structuring of the intentional forms of expressions of our truth beliefs and certainties to the point of uncovering their hidden confusions as confusions resulting from a missing sense of awareness of the internal co-signifying connections of the verbal event. Recovering, and appropriating, the freedom of this new sense of awareness, goes hand in hand, by appropriate tracking and experimenting, with the recognition of the internal co-signifying connections of the "verbal event" as Heidegger calls it. The elucidation of the "manifest in the event with its internal connections" is carried out by Wittgenstein by constructing a phenomenological language applying the principle of recognition of what is essential and inessential in our language if it is to represent. This practice also clarifies what remains unclear when using terms such as "verbal event," "clearing," "opening of the site" and related terms coined by Heidegger to thematize the central problem.

**Keywords** Auto-hetero affective movement · İnternal vs. external connections · Transversal vs. longitudunal movements · Proto-phenomenon of expression · Verbal event · Deconstructive elucidation of the structure of intentional self experience · Clearing and opening of the site · Appropriative response · Co-signifying manifestation

E. Sezgin (✉)
Independent scholar, Ataköy, Istanbul, Turkey

 
W. S. Smith et al. (eds.), *Eco-Phenomenology: Life, Human Life, Post-Human Life in the Harmony of the Cosmos*, Analecta Husserliana CXXI,
https://doi.org/10.1007/978-3-319-77516-6_32

## Wittgenstein's Excavation of Language

The phenomenological elucidations of Wittgenstein are carried out by constructing a phenomenological language. These elucidations work by animating an auto-hetero affective movement that tracks by recollecting its own manifest ground as signifier and signified in interplay, and in weighing the co-signifying connections of sense differences. These shifts are experimented with and tracked, for recognition of what is essential and inessential for the representation of sense differences, through the external definitions of naming and by describing pictures and rules to be followed and used as means and ends, as tools and techniques uncovering historical intentionality. Thereby the self-experience and imagination of the actors are so conditioned by the rules and pictures of the historical language-game(s) in which they are trained and educated that they misunderstand and misconstrue the logic of naming, describing, and experiencing the truth and certainty of what is meant by the use of names in particular, and of words in general.

That is, in fact, a misunderstanding, a misconception, in particular a betrayal worked by epistemological theories with essentialist or metaphysical implications. These are not simple misunderstandings and misconceptions remaining limited to themselves but they involve a deep misunderstanding of the "self reality" of everything or, in Heidegger's terms, of the different "beings" of everything represented or pictured. The being of "I-self" as well as the self reality or the being of everything and anything meant to be experienced is misapprehended as if it were self-demonstrative proof of its own sense difference and identity, of its own truth and certainty.

That is a failure of self-understanding that spreads like an infection so that one also misjudges the self experience of different historical forms of human life, as characterized "mythological" as opposed to the "rational scientific," while the differences are rooted in the different ways of experiencing life with different systems of rules, of pictures and narratives that are intertwined and woven by means of the agreement of practices in *practice*. "Practice," in the latter sense, differs from the understanding of practices carried out intentionally as represented by the sense understood in its intended sense of "practice," agreement as to which is decided by the practitioners as if practices are subject to their intending and willing, by their agreeing to start a practice and to stop it when the end is reached, effectively by carrying out a practice as a means to an end. Thus, that "practice" is understood in an intended sense, as a practice learnt, ruled, and structured. The agreements and disagreements as to rules are acquired by learning and following the rules of the uses of pictures, the sense differences of which are learnt as being internally connected with their manifest co-signifying surroundings. The intentional habits of reasoning, structured by learning and following the rules of pictures, and defined by external definitions of sense differences as means and ends, presuppose for their own possibility the spontaneous practice of learning and following sense differences as a *learning that originates as a verbal event* in co-signifying the manifestation of internal connections in simultaneity. Here, the latter sense of "practice" needs to touch and respond appropriately in its manifest sense. That is to say, to respond without the intervention of intentional reactive habits of interpreting the sense of practice and action, so as to grasp the manifest medium of the co-signifying proto-phenomenon in which intentional self experience is configured to be shared and ruled according to the intentional memory, anticipations, and reactions of imagination, along with its intertwining and interweaving with the narratives and practices that are formed and ruled to be practiced as rites and rituals by which to unfold and change historically.

## Wittgenstein's Resort to Eastern Philosophy

Therefore, while "practice" in its ordinary intentional sense seems to be experienced as if it be subject to intentional willing, meaning, carrying it out in a way depending on the intentional choice of practice, practice in the latter sense requires the spontaneous animation of an auto-hetero affective movement tracking how an intentional sense of practicing with choices, willing, meaning with the use of pictures of language, are ruled and structured by being internally-connected in co-signifying connections with their manifest surroundings. Therefore, the latter sense of "practice" is connected with the manifest sense of a transversal movement that requires animating and recovering in order to track the structuring of practice into the uses of pictures that are ruled and defined to be followed with logical steps and segments diachronically, longitudinally, as means and ends, tools and techniques of intentional action with memory habits. This manifest sense of the movement of life, as it is not subject to being sensed and experienced by means of the representational use of language, is conveyed by negations of representational language as "action in inaction" or vice-versa by Eastern philosophies. The elucidations of the internal signifying connections of the Uses of pictures, as being essential to the representations of sense differences by external definitions that name and describe by constructing a phenomenological language, allows our sharing the very insight expressed by the Zen master Hui-neng about Use: he points out that no-thing, or the "I-self," has a self-identity and self-reality to be understood and experienced with a privileged sense of being, it being self-demonstrative, proof of its own sense difference and identity, and self-representative of its own sense in isolation from the internal co-signifying connections of sense differences in what is manifest. Thus, he points out:

> From the first not a thing is. The body is no-body without its Use, and the body is the Use. To be itself is to know itself. By using itself, its being is demonstrated, and this using is, in Hui-neng's terminology, 'seeing into one's own Nature'. Hands are no hands, have no existence, until they pick up flowers and offer them to the Buddha; so with legs, they are no legs, non entities, unless their Use is set to work, and they walk over the bridge, ford the stream, and climb the mountain. (Suzuki 1993, 42)

Similarly, Wittgenstein points out that "Every sign *by itself* seems dead. *What* gives it life – In use it is *alive*. Is life breathed into it there? – Or is the *use* its life?"

(Wittgenstein 1968, 432). Deprived of this sense, we all continue not only to misjudge and misunderstand the self experience of the actors of other historical language-games but, also, our own very self experience of our *selves,* as well as the self reality of what we mean, believe as true and certain as being self-demonstrative of its own sense difference and identity in space and as space. Thus, we are also led to believe scientific truth to be established by observing, experimenting, and modeling reality through scientific observation and experiment, what is presented and taught still in physics classes, as if the self essence or "structure of matter" were modeled by a picture of "atoms," "photons," and the like:

> We have been told by popular scientists that the floor on which we stand is not solid, as it appears to common sense, as it has been discovered that the wood consists of particles filling space so thinly, that it can almost be called empty. This is liable to perplex us, for in a way we know that the floor is solid, or that if it isn't solid, this may be due to the wood being rotten but not to its being composed of electrons. To say, on this latter ground, that the ground is not solid is to misuse language. For even if the particles were as big as the grains of sand, and as close together as these are in a sand heap, the floor would not be solid if it were composed of them in the sense in which a sand heap is composed of grains. Our perplexity was based on a misunderstanding; the picture of the thinly filled space had been wrongly *applied.* For this picture of the structure of matter was meant to explain the very phenomenon of solidity. (Wittgenstein 1969a, 45)

How do these elucidations of Wittgenstein, by constructing a phenomenological language, contribute to the project of Husserl's program of saving philosophy from the naïveté resulting from the "the rationality of a 'lazy reason' which evades the struggle to clarify the ultimate data (*die letzten Vorgegebenheiten*) and the goals and directions which they alone can rationally and truthfully prescribe"? (Husserl 1970, 16).

## The Phenomenological Wittgenstein

The way to self-understanding requires, therefore, insightful understanding or phenomenological awareness of manifest phenomena, that co-significations are essential to learning, acting, meaning, naming, represening, defining rules and pictures of language by defining external connections. This means that the *site* of manifest internal co-signifying connections, as essential to the naming connection, needs to be elucidated. That "site" was earlier formulated in the *Tractatus* as the essential nexus of language and world effected by means of the internal connections by which signs of language and objects of world are configured or concatenated together to mirror a form with a possible sense as a fact both of language and world. This mirroring was formulated as a result of an segmented form of the elementary form of the expression of an elementary proposition. That is to say, not as a form intended, meant, conceived, willed as subject to the intentional habits of action of the epistemological subject but, in accord with an earlier formulation in the *Tractatus*, by the form of the expression of an elementary proposition which manifests itself by co-segmenting, as being internally connected by the essence of language and world whereby signs are manifest by being strung or concatenated together to mirror a possible configuration to be seen, sensed, and to combine with other elementary form of expressions in truthful functional combinations falling in their sense somewhere between contradictions and tautologies. The possible senses of the elementary form of expressions are here supposed to be determined by "the possible configurations" determined by the essence of language and the world, postulated as "the configurations of 'objects'." Thus, neither "objects" nor "signs" make sense and can be meant, in isolation from the proto-form of the expression that manifests and segments, to configure a form with a possible sense as a possible fact of language and world that is configured in accord with the essence that connects language and world.

This was the earlier formulation of the naming connection in the *Tractatus*. The elucidation of the nexus of internal connections, as being essential to the naming connection by defining pictures by external ostensive definitions of naming and by describing rules and pictures to be followed and used as means and ends for intentional use, is at the center of Wittgenstein's later elucidations throughout. The principle of elucidation is formulated as being based on the *recognition* of what is essential and inessential in our language if it is to represent. In his *Philosophical Remarks,* he declares, "A *recognition* of what is essential and inessential in our language if it is to represent, a recognition of which parts of our language are wheels turning idly, amounts to the construction of a *phenomenological* language" (my emphases) (Wittgenstein 1975, 51). He continues, "I should like to say, if there were an external connection no connection could be described at all, since we only describe the external connection by means of the internal one. If this is lacking, we lose the footing we need for describing anything at all – just as we can't shift anything with our hands unless our feet are planted firmly" (Wittgenstein 1975, 66).

I have emphasized "recognition" in that elucidation comes by means of assembling and configuring reminders requiring an appropriate response in touch with the deep stream of the movement with the manifest surroundings, for the reminders, really to *work* as reminders with the significance of awakening touch, need to *collide*, rock, and possibly to slacken the hardened and tightened knots of our stiffened intentional habits of experiencing and reasoning about the surrounding world, the premises of which are expressed by our gestures of meaning, showing the sense differences represented and ruled by external definitions. They appear, therefore, to trigger the intentional habitual reactions of imagination associated with them as if they are experienced as self-demonstrative proof of their own truth and certainty in the meaning and showing act of the use of "this" and "that." Even G.E. Moore's famous gesture of showing his hands in his Proof simply betrays the failure of *appropriate* awareness of the manifest internal connections in which our use of limbs and words operate in coordination, to be ruled, to be experienced with a truth and certainty that manifests itself appropriately and spontaneously as being independent of our willing, meaning, proving reasoning habits intentionally! Thus, the failure of awareness not only betrays an intellectual failure but, more than that, it implies and betrays inappropriate reasoning, meaning, sensing, and imagining habits about everything represented to the point of confusing and identifying their self-identity or reality. They are open and apt to be further confused owing to further projections of naming and describing them in epistemological theories by picture constructions based on external definitions with essentialist implications that, in turn, alternate the reactions of essentialist imagination between the two poles of suppositions of essences or their absence! Between the suppositions of grounds and groundlessness! That is a confusion and puzzle resulting from the inappropriate reactions of an intentionality with a missing sense of awareness of the manifestation of the proto-phenomenon of expression that needs to be appropriated. This is specifically a missing appropriate response that tracks the structuring of intentionality starting from the co-signifying and con-figuring of the verbal event, starting from the spontaneous segmenting of the proto-phenomenon and tracking its modulations, which are ruled and structured into the forms of expressions of intentional self-conscious learning, meaning, sensing, and experiencing of one's surroundings. This is a tracking of what is diachronically meant to be followed and used intentionally back to the internal co-signifying manifestation of the sense differences signified synchronously, in simultaneity, transversally, and vice-versa in simultaneity.

Therefore, the *appropriate* elucidations must be carried out by means of appropriate responses in resonating and coordinating with the manifesting, segmenting, con-figuring of the form of expression in interplay with its co-signifying with the sense differences of its manifest surroundings, hence, to the point of tracking the conditioning and reacting of the intentional habits of meaning and showing, which are structured by actors being trained and educated with the rules and pictures of the language games defined and modeled by defining external definitions. The appropriate response is auto-hetero affective awareness and movement, which must be awakened by "inventive," "inceptual" thinking. These are Heidegger's qualifications, necessaries if thinking is to start philosophy anew for the "clearing" and opening of the "site" (as The Essence of Truth) to elucidate the manifest ground, which Heidegger attempts to elucidate in terms of the Greek experience of "*physis*," "*aletheia*," *ekstasis-poiesis*. The possibility of the opening of the Site of the Essence of truth depends on the unfolding of the verbal event. For the essence of truth, for Heidegger, depends on the very possibility of responding in appropriate resonance with the manifest surroundings in *ek-statik Ek-sistence,* the appropriate response that would characterize Dasein's original way of being in-the-world. Thus, the spontaneous unfolding of the verbal event, if it is tracked in its caesuras of rifting and ridging in echoing and interplaying with co-signifying surroundings, would allow our tracking and bridging the intentional self-conscious structuring of reasoning, experiencing, meaning, sensing as segmented longitudinally, or diachronically as the Event which, temporalized as space and fragmented by ordering and representing sense differences and by external definitions, is so effaced, forgotten, suppressed, as not to be sensed and responded to by our intentional habitual, mechanized self-conscious habits of speaking, reasoning, acting. Here is a mechanization and a conditioning, acquired by actors' being born into a historical language-game wherein they are trained, educated to operate and follow the rules of applying the pictures of a language as means and ends, without, however, any sensing of the impact and effect that those the rules and pictures, followed and held fast to, have upon their imagination and emotional life. As a result, the actors are bewitched in the way that strengthens and spreads the bewitchment and confusions so that they become reflected in the implicit or explicit suppositions and presuppositions of epistemological theories. Philosophy, however, requires being initiated before any theorizing by projecting pictures if it is to be philosophizing in the proper and appropriate sense. Hence, the urgency of elucidating the site of the manifest internal co-signifying connections that are essential to any representation through defining pictures based on external connections. Thus do Wittgenstein and Heidegger come to respond to the urgent matter of elucidating and clearing the site essential to learn, mean, speak with the rules and pictures of the historical language game(s).

"Philosophy is the will to return to the beginning of history and thus is the will to surpass itself" (Heidegger 2012, 31). This project of Heidegger's for starting philosophy with an authentic opening of the site of the essence of Truth, is better achieved by means of such thought experiments as are found in Wittgenstein's writings, which are inventive and inceptual, since they work by animating the spontaneous auto-hetero affective movement required for the opening and clearing of a site. The assembling and configuring of his reminders, by means of thought experiments, is carried out by allowing us also to track and experiment along with Wittgenstein's reminders the structuring of intentionality and imagination and the memory reactions and habits of actors following their training and education in applying and operating within rules and pictures defined by external definitions of naming and describing as well as by the narratives and rules of historical-cultural language-games. Here, the reminders always operate through movements of auto-affective sensibility animated to track the manifest significations transversally, that is to say, in the co-signifying interplay of simultaneity, *syn*chronously as well as longitudinally, *dia*chronically, while recollecting and assembling the internal signifying connections by colliding and contrasting Wittgenstein's reminders with our deep "forgetfulness."

## Restoring Missing Awareness Inceptually

This is not an ordinary "forgetfulness of something once 'known,'" later to be remembered again as a "representation" to be used again as a representational tool of intentional habits! No. What is recollected and assembled is tracked as internally-connected to restore a missing sense of awareness, which is also deconstructive and transformative of our self-conscious structured reasoning habits. But, the intentional self-conscious habitual sensibility of the actors may have been so separated and alienated as not to be able to sense, experiment, track, and recognize appropriately the reminders assembled and recollected as such, other than to react to them as nonsense, as self-contradictory (as exemplified by Russell's later reaction to Wittgenstein's writing), given their own logic and reasoning habits, the common sense of which is resting on the premises maintained as self-demonstrative proof of their own truth and certainty. These also happen to be the premises maintained and shared implicitly or explicitly by the arguments and reasoning of epistemological theories with essentialist implications.

Thus, the reminders assembled not only restore a missing sense of awareness of the manifest internal connections presupposed as the possibility of naming and describing rules and pictures by external definitions as well as operating with them but, also, expose all the confusions extending to the point of misunderstanding and misrepresenting the reality of everything, including what is pictured and modeled by scientific practices! That misunderstanding is shared by the scientist and the epistemologist, who respond with the same reasoning and analyzing habits of argumentation by applying and following picture projections, which then create pseudo-puzzles and paradoxes that follow from the confusions that ensue from the misapplications and mis-projections of pictures, as exemplified by popular understanding of the physical reality modeled and experimented upon by the physical sciences. The hidden nonsensical suppositions and confusions of these sciences are exposed by Wittgenstein's elucidating the site, the con-texture of internal signifying connections, as essential in representing sense differences as are external definitions.

Here, noting and recognizing the pseudo-paradoxes goes hand in hand with the recognizing, experimenting, tracking, and recollecting of the manifest internal co-signifying sense differences to be awakened at a touch through the collision of the unprecedented authentically extra-ordinary and the structure of the ordinary. Hence, they go hand in hand with the simultaneous touch of sense *transcendent and immanent,* namely, with the *ek-statis poiesis* touch of sense in a simultaneous rocking and resonating worked by *transversal* movements animated as the auto-hetero affective sense of movements by means of constructing phenomenological languages inventively, *inceptually*. They always start from scratch, from the verbal event animated by auto-hetero affective movement, tracking its modulations and ruling in interplay with co-signifying fissures, caesuras, turns, and twists with sense differences signified. Hence, the "site of the Essence of Truth," how truth and certainty are manifested and configured with its manifest surroundings, is opened, tracked, and always kept opened, like a scar fissured with co-signifying consequences. This is from his *On Certainty*:

> It is as if I were to see a painting (say a painted stage-set) and recognize what it represents from a long way off at once and without a slightest doubt. But now I step nearer: and then I see a lot of patches of different colors, which are all highly ambiguous and do not provide any certainty whatever.
> It is as if "I know" did not tolerate a metaphysical emphasis. (Wittgenstein 1969b, 481, 482)

Here is a stream of spontaneous movement always started from a scratch by inventive inceptual thought experimental movements, from numerous different starts, to clear and open the *site* presupposed in learning, meaning, defining the sense differences by external definitions of naming and describing pictures as representations. These are, thus, tracked by the means of auto-hetero affective movement, animated as a transversal movement in responsive touch with a synchronous manifestation co-signifying the phenomenon along with the structuring of intentional self-conscious forms of expression, ordered and ruled by learning and following the uses of pictures defined as "causes" and "effects," as pictures to be followed and operated with consequences as means and ends. Thus, intentional memory actions and reactions, anticipations, which are ruled and structured by learning and sharing the rules and pictures defined by the external definitions of naming and describing one's surroundings and to be followed and operated both longitudinally and diachronically, are elucidated at the site of the manifest internal co-signifying connections opened by the transversal and longitudinal auto-hetero affective movements animated by inception.

Hence, by configuring the surroundings through a phenomenological language, by inventively starting from scratch, Wittgenstein's thought experiments open the *site* of manifest *proto-phenomenon,* of *the verbal event* as Heidegger thematizes, in a co-signifying web of interrelations so that we are allowed to track and experiment with the possibilities of the configurations of the sense differences synchronously resonating with the manifest simultaneity of transversal movements, as well as track diachronically the movements of intentional memory action structured by being trained and educated to share and follow the rules and pictures defined by external definitions of naming and describing rules and pictures.

This thought experiment comes from the *Blue and Brown Books*:

> If we had a sensation of toothache plus certain tactual and kinaesthetic sensations usually characteristic of touching the painful tooth and neighbouring parts of our face, and if these sensations were accompanied by seeing my hand touch, and move about on, the edge of my table, we should feel doubtful whether to call this experience an experience of toothache in the table or not. If on the other hand, the tactual and kinaesthetic sensations described were correlated to the visual experience of seeing my hand touch a tooth and other parts of the face of another person, there is no doubt that I would call this experience "toothache in another person's tooth." (Wittgenstein 1969a, 53)

This thought experiment is so remarkable and significant for its elucidating the sense of the "internal connections between a sensation and its manifest surroundings" by configuring the manifest surroundings in the medium, of which, also, the manifest proto-phenomenon is streamed to modulate in co-signifying connections of sensing and meaning recognizing the sense differences; thus, to track and recollect awareness of the manifest unfolding of the surroundings, the signifying configurations of which also configure the possibility of sensing and meaning of the intentional form of the expression as a proto-phenomenon of expression.

These phenomenological thought experiments, prompting, experimenting, and tracking the modulations of the manifest through auto-hetero affective movement in co-signifying connection with its manifest surroundings, are also essential to understanding how rules of language, intentionality, and self-conscious memory and imagination, anticipation, and association of means and ends operate by being being internally connected and being intertwined with the rules and pictures defined by the external definitions and narratives of historical language-games. In other words, the phenomenological site, opened as such, opens a manifest field without a "subject" and therefore without an "object" too, the "object" being a picture that exemplifies the standpoint and world horizon of intentional self-consciousness as it is structured by actors who have been trained and educated by the pictures of language set to be followed and used as the means and ends of intentional memory in action.

This state of affairs is not elucidated appropriately by any description of the meaning and naming connection defined by associating a "sensation" and a "behavior," or "behavior as the expression of a sensation," which are external definitions, inappropriate pictures with their associated, imagined senses as representations made from the intentional standpoint. That is to say, such are not the appropriate descriptions that are made in the resonating response following on an appropriate sense of touching the manifestation of life. Oddly enough, the appropriate sense of response is recovered only by configuring the manifest signifying surroundings of the proto-phenomenon of expression in unaccustomed and unexpected ways of configuring the surroundings as standing against and in collision with the intentional customized habits of meaning that focus on sensation. Thus is the actual manifest proto-phenomenon of a verbal event recovered, to be tracked with its spontaneous sensing and meaning as these are internally connected with one's configured surroundings, the appropriate awareness of which is impossible through intentional self-conscious habits of meaning, showing, experiencing one's surroundings.

## Appropriate Response to Wittgenstein's Thought Experiments in Language

The elucidation of the rules of language, although extended into the manifest root proto-phenomenon by Wittgenstein's thought experiments, seems not to have been properly and *appropriately* responded to so far. Instead, the configurations of his thought-experiment reminders have been deprived of their appropriate phenomenological insight into the roots of rules and the proto-phenomenon of the "verbal event" as thematized by Heidegger, while "socio-cultural," "instrumentalist," and "conventionalist" understandings of the "rules" and of "language games" have been found more convenient to understand and convey in academic writing, to serve the purposes of teaching culture, an industry in the service of educating and informing public culture.

The term "transcendental," in Husserl's usage, can be taken as clarifying the manifest phenomenon of expression presupposed to be essential—not in the sense of the "essentialism" that has usually been ascribed to Husserl and criticized appropriately, but "essential" in the sense that Wittgenstein brings out in his constructing a phenomenological language, as the possibility of meaning, saying, speaking, reasoning by using the pictures defined as means and ends by external protocols for naming and describing. The phenomenological aspect of Wittgenstein's reminders, assembled and configured to track the internal co-signifying connections of sense differences signified and the manifest sense touched through the numerous thought experiments he initiated, seems to me to be the most *overlooked* or, better put, the most un-responded aspect of his *writing*. That is to say, as such *writing* as it is, the texture of which is characterized by its own creative, "inceptual" "spontaneous," "transversal" movements of rifting and ridging, while recollecting and assembling his reminders in configurations of co-signification of the internal connections of sense differences as different possible ways of meaning, showing, experiencing the manifest surroundings. Thereby such differences are made to collide with our structured intentional habits of meaning, reasoning, showing the sense differences that are expressed by our gestures of meaning, of showing (like Moore's self-demonstrative gesture of showing his hands!), as they are experienced, meant, shown as self-demonstrative proof of their own truth and certainty. This texture of his writing (*dichten*), textured by thickening and tightening the internal signifying connections of the sense differences that must be responded to and followed appropriately in *touch* with a like awakening of auto-hetero affective movement, seems to me to have thus far failed to find its *appropriate* response on both sides of Europe. This failure is owing to a lack of appropriate response in the analytical Anglo-Saxon ways of reasoning, arguing, theorizing, and habits of writing, on the one side, and in the phenomenological movement of the continent, on the other side. Though the phenomenological movement in philosophy on the continent is focused on the deep question of the "opening of the site" whereby the intentional self-experience of truth and certainty are expressed to be shared and ruled historically, it seems to have fallen short of responding to Wittgenstein's ways of writing the spontaneous, inventive, inceptual movements that are especially appropriate for the opening of the site, even more appropriate than Heidegger's approach, which frequently emphasizes and the importance of "appropriate" response to the "verbal event" in its program. The same can be said about the French auto-affective tradition, which is written more in dialogue with Husserl and Heidegger than with Wittgenstein. Jacques Derrida and others refer to the auto-hetero affective "trace" of the proto-phenomenon of expression, of the "verbal event," as an "archi-writing (*archi-écriture*), which must be appropriately sensed and responded to as originating the sense differences of characters as signified, in-*scribed* and ex-*scribed*. But then, it happens that the writers, who seem to respond by coining these appropriate terms for the unfolding of the verbal event rightly, instead of focusing so as to be appropriately tuned and ready to resonate, so as to touch with appropriate movements of writing such as found in that of Wittgenstein, Husserl, or Heidegger, who are dedicated to opening and clearing the transversal and longitudinal sense of touch, which is meant to prompting the awakening of deconstructive awareness as well as its appropriation. They have been distracted and scattered, losing focus on the site that requires opening, as well as on keeping it open by an appropriate writing with appropriate sense and response.

That dedicated response is actually carried out in Wittgenstein's writing, and by Merleau-Ponty, notably in his work in progress *The Visible and the Invisible*. The latter great text was responded to poorly by Jacques Derrida, who fell short of doing justice to this "rich text," although he qualifies it as such in his *On Touching—Jean-Luc Nancy*. Such failures, it seems to me, are owing to losing focus of the site, which requires responsive and inventive animating movements of writing, which requires sustained animated tracking of its own modulations of sensing, which means keeping in touch with one's co-signifying configurations with manifest surroundings. As pointed out by Maurice Blanchot, "There is always a risk that reading, instead of animating the multiplicity of transversal routes, reconstitutes a new totality from them" (Blanchot 1992, 51).

This gives the reason for my writing here of Wittgenstein with reference to Husserl's and Heidegger's projects dedicated to the "clearing," to the "opening of the site of the Essence of Truth," whereby historical intentionality with the rules and pictures of historical language-game(s) is meant to be clarified in its structuring and behaving. How this site is opened by Wittgenstein's thought experiments, in the way in which it is opened for himself and for his readers by "inventive," "inceptual" thought experiments that allow a tracking in touch with the auto-hetero affective movement in spontaneous response to the manifest phenomenon has always been the focus of my philosophical writing. This is such writing as responds to the cue extended by the movements of writing, manifested as a transversal wave of the verbal event which is simultaneously affected and modulated as a longitudinal wave with co-signified consequences that result in diachronic learning, and as following the sense differences in order to mean and represent them by external definitions.

That may be the gift of a Selfless sense of awareness to be appropriated by responsive possible readers in order to keep the way open to the gift of the clearing and keeping the site opened. This requires an appropriative response of the awakening of auto-hetero affective sensibility in the tracking of its unfolding and modulating its ruling and structuring as an intentional self-conscious experience shared with others, while keeping in touch with co-signifying configurations with its manifest surroundings, maintaining a sense of awareness of the movement appropriated to respond both transversally and longitudinally as well as synchronously and diachronically, which may then be extended as a gift to be responded to by the generations coming.

## Works Cited

Blanchot, Maurice. 1992. *The Step Not Beyond*. Trans. Lycette Nelson. Binghamton: SUNY Press.

Heidegger, Martin. 2012. *Contributions to Philosophy (of the Event)*. Trans. Richard Rojcewics and Daniela Wallega-Neu. Bloomington: Indiana University.

Husserl, Edmund. 1970. *Crisis of European Sciences and Transcendental Phenomenology*. Evanston: Northwestern University Press.

Suzuki, D.T. 1993. *The Zen Doctrine of No Mind*. Ed. Christmas Humphrey. Weiser Books.

Wittgenstein, Ludwig. 1968. *Philosophical Investigations*. Oxford: Blackwell.

———. 1969a. *Blue and Brown Books*. Oxford: Blackwell.

———. 1969b. *On Certainty*. Oxford: Blackwell.

———. 1975. *Philosophical Remarks*. Oxford: Blackwell.

# Phenomenology After Conceptual Art

**Andrew Chesher**

**Abstract** The reception of phenomenology in art criticism reached its apex in the mid-1960s in its application to Minimalism in the United States. The focus was on the embodied, direct perceptual experience of Minimalist sculpture, but in light of Conceptual art's 'dematerialised' practices which developed as the decade progressed, the interest in phenomenology waned. This paper looks at the history of this reception and presents Merleau-Ponty's late ontological work as a corrective to an inadequate understanding of phenomenology in critical discourses on art at the time. It argues that the late Merleau-Ponty offers tools for an effective critique of early conceptualism's idealism, as well as a basis on which the 'dematerialised' and dispersed ontology of the art work shared by both Conceptual and more recent Post-Conceptual practices can be investigated.

**Keywords** Phenomenology · Language · Perception · Sense · Intertwining · Aesthetics · Art theory · Art history · Conceptual art · Minimalism · Maurice Merleau-Ponty · Renaud Barbaras · Rosalind Krauss

## Phenomenology After Conceptual Art

In his article "A Sedimentation of the Mind: Earth Projects" the artist Robert Smithson described a visit to the slate quarries of Bangor-Pen Angyl in Pennsylvania. Contemplating a wall of rock reflected in a murky lake, he wrote of how "all boundaries and distinctions lost their meaning in this ocean of slate and collapsed all notions of gestalt unity" (Smithson 1996, 100). Smithson's article was published in September 1968 in *Artforum*, an art journal that between the mid-1960s and mid-1970s carried a series of historically important statements by Robert Morris, Sol LeWitt, Mel Bochner, and Joseph Kosuth, among others. We are on the cusp of

---

A. Chesher (✉)
Chelsea College of Arts, University of the Arts London, London, UK
e-mail: a.chesher@chelsea.arts.ac.uk

 
W. S. Smith et al. (eds.), *Eco-Phenomenology: Life, Human Life, Post-Human Life in the Harmony of the Cosmos*, Analecta Husserliana CXXI,
https://doi.org/10.1007/978-3-319-77516-6_33

Conceptual art's heyday in New York, the city from which Smithson ventured out on field trips to sites in the industrial and suburban margins. Among his companions on this particular outing are the sculptor Nancy Holt (his wife), and Dan Graham, whose photo-text work "Homes for America" (published in *Arts Magazine* two years prior to this trip) was to become a canonical early example of conceptualism. On other trips the company included the chief practitioners of Minimal art: Carl Andre, Donald Judd and Robert Morris. The latter two were not only active makers, they were also artists who wrote and whose writings have continued to have more or less as much influence as their works. However, there is a difference between Judd and Morris on the one hand and Smithson and Graham on the other. Whilst the former would not have seen their writing as art, the latter explicitly did.

Minimalism and Conceptual art were so closely related in this milieu as to resemble siblings, if not twins in some respects. There was no significant generational divide between their early proponents. Nonetheless, there was an historical and philosophical caesura between them, and this can be located symptomatically in their attitudes to "notions of gestalt unity" and the phenomenological stance implied in Smithson's phrase. Phenomenology's critical stock was high during the first half of the decade, but fell as the focus shifted from the situated perception of the object in Minimalism to the dematerialisation of the object with Conceptual art. What this paper looks at are phenomenology's vacillating theoretical and critical fortunes in this context. It does this with the aim of speculating on the possibility of bringing phenomenological ideas back to bear on subsequent Post-conceptual art.

In an interview Robert Smithson gave the year after publishing "Sedimentation," the artist spoke of considering "the facile unitary or gestalt ideas part of the expressive fallacy" (Lippard 1977, 89). It would be reasonable to suppose that, at least in part, he was responding to the ideas espoused by his friend Robert Morris who held that the human body determines the perception of scale and should be the measure for sculpture, a viewpoint which itself was based on the phenomenological idea that the world gave itself to be grasped as a gestalt (Morris 1995, 11). Smithson is diametrically opposed to Morris here, as he writes: "You just have to deal with the fundamentals of matter and mind, completely devoid of any anthropomorphic interests" (Lippard 1977, 89). By and large, it seems to be Morris's version of phenomenology that Smithson is rejecting here.

However, some things about Smithson should make us think twice about the prevalent view that there is an incompatibility between phenomenology and conceptualism. Primarily, Smithson did not reduce art to information or concept, nor to the discourse or text of post-modern criticism; he resists such reduction because "the interaction between matter and mind" is what his work was about (Lippard 1977, 89). It is true, nonetheless, that this interaction certainly is not synonymous with that which phenomenology classically entertained; it is not a question of intentionality and intentional objects. Regarding his site visits and the 'nonsite' gallery works he subsequently produced abstractly mapping them, Smithson spoke of being embarked on "a return to the origins of material, sort of a dematerialization of refined matter" (Lippard 1977, 87). He was not seeking a concept, but a material reality, one which is both mirrored by and mirrors the mind that is part of it.

The more militant Conceptual artists, as Peter Osborne has recently pointed out, sought to eliminate the aesthetic dimension of art (Osborne 2013, 47–51). Smithson was not among them. In fact, he rejected an art dissolved into ideas for the reason that "it only deals with the mind and it has to deal with the material too." (Lippard 1977, 89). If Post-conceptual art, which Osborne argues is how contemporary art is best categorised, has given up on Conceptual art's idealism, then the return of an aesthetic dimension was nonetheless an achievement of conceptualism, in as much as, in failing to rid itself of it, Conceptual art "demonstrated the ineliminability of the aesthetic as a *necessary*, though *radically insufficient*, component of the artwork" (Osborne 2013, 49). It is necessary in Osborne's view as "all art requires *some* form of materialization; that is to say, aesthetic – felt, spatio-temporal – presentation." (Osborne 2013, 48). Yet, it is insufficient because Post-conceptual art cannot be defined on the basis of its aesthetic form – that is to say, its medium. To do justice to Osborne's substantial theorising of Post-conceptual art would require a different essay from the current one. Suffice it to say that he would not countenance a return to phenomenology. Nonetheless, his affirmation of the necessity of a "felt, spatio-temporal" dimension is indicative of what would make phenomenology's return into critical discourse desirable, just as his recognition that this aesthetic presentation need not be either singular or specific indicates a challenge for a phenomenologically-oriented approach.

The question remains as to what form of phenomenology would be desirable to see return. The second aspect in Smithson that should give us reason for reconsidering a phenomenological perspective also helps on this count, as the mirroring of mind and matter on which the artist speculated echoes the late Merleau-Ponty's talk of the intertwining of sensing and the sensed. Smithson envisaged the interaction of mind and matter as a metaphysical maelstrom, the one folding into the other in an infinite and abyssal regress. But, this is not a purely intellectual notion; Smithson is describing an experiential infinity, an infinity of experience. When Merleau-Ponty writes that the "veritable infinity" is not the negation of being but "infinity of the *Lebenswelt* and not infinity of idealisation" (Merleau-Ponty 1968, 169), the proximity between him and Smithson starts to become apparent. The lifeworld for Merleau-Ponty is precisely that ground in which subject and object, self and other, sensed and sensing are intertwined with one another such that neither can, alone, found the possibility of the other. Without wanting at all to conflate Smithson's artistic stance with Merleau-Ponty's philosophical position, I wonder whether there is not perhaps greater affinity between the disorientation of entropy the former evoked and the pre-objective world that the latter sought to expose than would at first appear when judged both by Smithson's own attitude to phenomenology and by phenomenological criticism's historical eclipse at the moment his work emerges. The reason for this eclipse, and the possibility that it is based on an inadequate reception of phenomenology, is a large part of what the current paper will occupy itself with. It does this, as I say, because my purpose is to explore the possibility of engaging phenomenology again within art theory. To do so, starting from where phenomenology sank behind art's critical horizon seems as good a point as any.

One aspect of Conceptual and Post-conceptual art that presents a challenge to phenomenology is that it is not, for the most part, identical with any singular aesthetic object. The work of artists like Smithson or a British contemporary such as Richard Long, and of recent artists for whom that generation has been a decisive influence like Francis Alÿs, is often dispersed across various media. Ultimately, Conceptual and Post-conceptual practice is not unified within a single object, but might appear as a photograph at one point, a text the next, and as film or happening further on. The walking that plays a large role in the practices of Long and Alÿs is symptomatic. It is integral to their work, but cannot be presented as an object for direct contemplation, or even as a situation that can be participated in apart from at the moment in which it happens. Long's practice, for example, has had as its consistent centre of gravity since the late 1960s the treks he performs through landscapes in various parts of the world. This walking may be represented by a photograph or a map, a series of words or a sculpture made of materials collected along the walk's course, but none of these tangible products are in their own right 'the work'. The latter exists as the links between each of these discrete instances, visible and legible indices of the walk, which, though physical while it happens, is, as Long notes, "afterwards invisible" (Long 2007, 26). Texts, photos, arrangements of stones and mud become the work to the degree they operate as traces of this absent referent.

Walking or travel has had a similar status in Alÿs's work. For his piece *The Loop* (1997), the artist travelled from Mexico, where he lives, to San Diego for the exhibition that had commissioned the work. He did so, however, by circumnavigating the globe, crossing Australia and Alaska on the way, to get to the U.S. and nonetheless avoid crossing the border between Mexico and the States. The piece was recorded in the form of a simple map of the route Alÿs took hung on the wall, and in the format of a postcard, appropriately enough, handed out to visitors and bearing an image of an ocean, the curvature of the globe just perceptible on its horizon but no land visible, and below it a brief text describing the work. Often Alÿs describes the form his work aspires to exist in as rumour. He has spoken of endeavouring to create "a very schematic structure" in his work, "so that the project can travel as a rumour or story even while the event or performance is happening." Viewers should be able to "imagine it without having to witness it 'live', or having access to visual documentation" (Alÿs et al. 2004, 81). Like Long, then, who has written that "the freedom to use precisely all degrees of visibility and permanence" is important to his practice, Alÿs wants to detach the work from a specific moment of perceptual encounter, for which reason it is "composed of episodes, metaphors and parables" (Ferguson 2007, 11). Shot through with invisibility and absence, dispersed across space and time, not to mention across the diversity of media they may employ from the photograph to 'rumour', these practices clearly stretch the resources of a phenomenology that focuses on an embodied, direct perception of the object, which, as we will see, was the most prevalent art critical understanding of phenomenology contemporary with the rise of Conceptual art. So, we need to ask whether there is another phenomenology that might be more appropriate.

I will suggest that it is within the late Merleau-Ponty that an alternative may be sought. Jean-François Lyotard, who was more obviously and openly influenced by Merleau-Ponty than most other well-known French philosophers of his generation, wrote that "Merleau-Ponty would not have been a great commentator on Cézanne if 'Cézanne's doubt' hadn't been his own." (Lyotard 1991, 187). Merleau-Ponty's essay "Eye and Mind," (Merleau-Ponty 2007c) published a couple of decades after his first substantial foray into aesthetics in "Cézanne's Doubt" (Merleau-Ponty 2007a) moved on to discussing an artist, Paul Klee, whose art is more clearly detached from the nineteenth-century Realism that Cézanne's was still working through. Nonetheless, the analyses in this later essay are still far from being such that they could be readily extended to either Conceptual or Post-conceptual art. What was it, then, that Lyotard was indicating Merleau-Ponty shared with Cézanne? Perhaps the ability and propensity to decompose the perceived into the event of perception. The latter, perception, as Lyotard noted in an earlier work, "Merleau-Ponty strenuously placed … under the authority of the body" (Lyotard 2011, 55). Indeed "it is the body and it alone," Merleau-Ponty writes in *The Visible and the Invisible*, "that can bring us to the things" (Merleau-Ponty 1968, 136). "The thickness of the body," more precisely, is "the sole means I have to go unto the heart of the things" (Merleau-Ponty 1968, 135). So, Lyotard was right, in as much as the body as "an *exemplar sensible*" (Merleau-Ponty 1968, 135) is of central importance to Merleau-Ponty's account of perception to the end, just as perception itself is to his ontology. This thread of Merleau-Ponty's thought, though an integral part with fundamental importance to his way of doing phenomenology, is nonetheless not the most promising place to pick up at for current art theory. This is because embodied perception was central to, and perhaps overemphasised in, the reception of phenomenology within Minimalist criticism in the mid-1960s, after which 'phenomenological experience' falls into neglect as the emphasis shifts from perception and the body to language and idea as the decade unfolded.

But, this is to read Merleau-Ponty's late work in light of his earlier *Phenomenology of Perception*, rather than the earlier in light of the later work. When we do the latter, we notice that in *The Visible and the Invisible* Merleau-Ponty sees the body as *but* an *exemplar*; it is only a "variant" of the flesh, "that carnal being, as a being of depths," even if it is "a very remarkable one" (Merleau-Ponty 1968, 136). What changes in the late work, is that sentience and the sensed cease to repel one another as if they were two theoretical magnets and become, instead, both referred to being, of whose 'thickness' or 'depth' they are two variants. This new constellation is clearest in "The Intertwining" chapter of *The Visible and the Invisible*, wherein the author writes that "belongingness to one same 'consciousness'" is not "the primordial definition of sensibility," which we should rather understand as "a carnal adherence of the sentient to the sensed and the sensed to the sentient" (Merleau-Ponty 1968, 142). As the end of "The Intertwining" chapter shows, Merleau-Ponty envisaged a revised integration of language and cultural forms with perception within this new ontology. He did not follow up on this suggestion, his work on the book being broken off before he could; however, he had been laying the groundwork for some time, starting in the essay "Indirect Language and the Voices of Silence" some 10 years before. It is this development, picked up and elaborated upon over recent decades (for example, in Renaud Barbaras's phenomenology of life and of desire and Ted Toadvine's eco-phenomenology), that with its potential for integrating the discussion of culture, language and history with that of perception, indicates, I suggest, the sort of phenomenology required.

## Conceptual Art After Phenomenology

The Conceptual artist Joseph Kosuth published a now famous article in 1969 entitled "Art after Philosophy." The philosophy in question was largely Logical Positivism. Kosuth's article was an attempt to define what was 'conceptual' about Conceptual art, and both to extend the idea of art as idea to all art *per se*, as well as to designate those practices that transformed an essential in-itself conceptualism of art generally into a strong, for-itself, Conceptual practice. The somewhat strident views expounded in Kosuth's article were not fully subscribed to by any of the other key figures at the time, and, although the argumentation is certainly not without interest, the essay has received its fair share of theoretical critiques over the years, including those to be found in Thierry de Duve's *Kant After Duchamp* and Rosalind Krauss's *A Voyage on the North Sea* (Duve 1996; Krauss 2000). Strident or not, Kosuth's thesis was, on one point, not particularly partisan. It reflects the general view held by Conceptual artists and their apologists that phenomenology was largely irrelevant to art – irrelevant, at least, to the art they made or espoused. Kosuth is explicit on this point. Right at the outset, in his second paragraph, he writes that "'continental' philosophy need not seriously be considered here," specifying in a footnote "existentialism and phenomenology" as the continental philosophies he is setting aside (Kosuth 1999, 159). The one philosopher mentioned by name in this footnote is Maurice Merleau-Ponty.

For Kosuth's milieu, Merleau-Ponty's name was almost synonymous with phenomenology. His *Phenomenology of Perception* was first published in English translation in 1962 and it is to this book, either directly or indirectly, that most allusions to 'phenomenology' by artists and critics in the United States during the remainder of the decade can be referred. The key delineations of Merleau-Ponty's book are very clearly the basis for Robert Morris's argument in his essay "Notes on Sculpture, Part II," first published in 1966. Firstly, Morris argues that the object of sculpture is relative to the perception of the viewer, who is to be conceived as a situated and embodied subject. Secondly, the sculpture is considered to be a function of the total situation it is encountered within, which includes lighting and positioning. Morris's "Notes on Sculpture" essays are, beside another sculptor's essay, Donald Judd's "Specific Objects," arguably the most important manifestoes of Minimalist art; they proselytise for an art based on the body, its movement and situation, just as *Phenomenology of Perception* argues that perception, transcendence towards the world, is to be traced back to the embodied subject engaged in its tasks. The discovery of Merleau-Ponty by contemporary critics Annette Michelson and Michael Fried preceded Morris's, though Morris seems to have been the only exponent of

Minimalist sculpture to have actually read the Frenchman's work (Meyer 1998, 178); and it was Morris's recourse to phenomenology that was soon followed up by sympathetic critics, prominent among whom was Rosalind Krauss.

In her article "Allusion and Illusion in Donald Judd" (Krauss 1966), published in the same year and same journal as Morris's "Notes on Sculpture, Part II," Krauss was already using phenomenology as an analytical tool. The culmination of her efforts to bring Merleau-Ponty together with Minimalism came a decade later, however, in her first major work, *Passages in Modern Sculpture*, which, as Hal Foster put it, "gives us a minimalist history of modernist sculpture" (Foster 1996, 42) stretching 50 or so years from Auguste Rodin's "Gates of Hell" to Morris himself, alongside Robert Smithson, Richard Serra and Michael Heizer in its final chapter. Again, it is *Phenomenology of Perception* among Merleau-Ponty's writings that is most commonly referenced by Krauss during this stage of her career. Her reading of that work is largely congruent with the use Morris makes of it, but with one significant difference. On the one hand her interpretation seems to be a classical reading of Merleau-Ponty's *Phenomenology*, in as much as she talks of phenomenology having re-characterised "perception as a function of intentionality, as the simultaneous cause and result of the viewer's *prise sur le monde*" (Krauss 1981, 262). On the other hand, Krauss reads the sculpture to which she applies phenomenology as effecting a radical de-centering of the subject. If Krauss sees this de-centering as being consistent with Merleau-Ponty's *Phenomenology*, as indeed she seems to, then the version of the book she gives us is unconventional. A subjectivity that is defined as an intentional grasping of the world is more centered than not, even if it is defined as finding itself in its activity in the world, as it is in Merleau-Ponty's book. Indeed, while one can find support for a notion of de-centered subjectivity in *Phenomenology*, the perceptual subject presented in that book as a whole is a unified one. Let us be clear, however: I am not that Krauss misunderstands the book, but that she seems to be projecting its analyses and themes into another interpretative context.

Krauss's discussion of Michael Heizer's *Double Negative* (1969) is symptomatic. Heizer's earthwork, consisting of two ramps cut into opposite sides of a ravine in Nevada, is vast: almost half a kilometre in length from one side of the valley to the other. It offers no visual figure for the viewer on the ground and, so, as Krauss says, "the only means of experiencing the work is to be in it." Here, we are not far at all from Morris's phenomenology: the work is constituted for him, too, in the interaction between the viewer and the object. "We can only stand in one slotted space and look across to the other," Krauss writes, and continues: "Indeed it is only by looking at the other that we can form a picture of the space in which we stand" (Krauss 1981, 280). This last sentence is key to Krauss's phenomenological interpretation of sculpture and, indeed, to her interpretation of phenomenology. In terms of the reading of sculptural practice, the work is seen as coming into being in the viewer's encounter with the object, but crucially, at the expense of the autonomous identity of both viewer and object. In terms of phenomenology's interpretation, as the viewers discover themselves through the other (that is to say, through the object and the others' viewpoints of it), intersubjectivity comes to displace subjectivity. Krauss goes as far as to describe *Double Negative* as a metaphor for "the self as it is known through its appearance to the other." This is an inversion of the classic Husserlian understanding of the other apprehend via "analogical appresentation"; that is, the other known through analogy with the self. This account of the other is already critiqued by Merleau-Ponty in his *Phenomenology of Perception*, in the chapter entitled "Others and the Human World," wherein he says that "reasoning by analogy presupposes what it is meant to explain" (Merleau-Ponty 2012, 368). The argument implied here is that analogy, though it might be able to loan the other the appearance of being a self, cannot establish it as *other*. The identity of the self cannot function as an analogy for the other's alterity. So, here it is the alterity of the other that Merleau-Ponty is saying is apparently explained and yet actually presupposed. Eventually, this critique of Husserl will move from the margins to the centre of Merleau-Ponty's philosophy. The priority of intersubjectivity that will emerge in the later work, however, had not yet been made explicit in his *Phenomenology*, where the subject's interaction with their environment is, for the most part, presented as constituting it as a world of phenomena for a subject that is largely unified in its embodied existence.

In the late 1970s Krauss wrote a string of essays on photography and Surrealism, in which psychoanalysis and semiotics came to displace phenomenology. At this time, she was just starting to hone a notion of the unconscious that would eventually be brought back to bear critically on certain phenomenological themes in her work from the early 1990s on. This is at least part of the context that Krauss anticipates in her slightly eccentric reading of *Phenomenology of Perception*. In her 1983 essay "Richard Serra, a Translation," however, she returned to the phenomenological themes and exposition developed in her earlier book. Central to this essay is another work that is sited, like Heizer's *Double Negative*, in the landscape. Richard Serra's *Shift* (1970–1972) consists of around a quarter of a kilometre of concrete walls that zig-zag down an undulating field in Canada. The top of each wall is level, but eventually disappears within the swell of the terrain, across the crests of which the walls form a series of parallel tangents. The position of the walls was determined by the course Serra and a companion walking down the field took while trying to keep one another in sight, so that the resulting work is plotted from the way in which one part of the terrain is revealed as the viewer moves across it while another part is concealed. The viewer's movement in relation to the structure is, thereby, made the subject of the work, although it is only 'present' as a background against which the relation between the land, the walls and the viewer's body is perceived. *Shift* is then a work that, again, can only be experienced when the viewer is moving through and around it, i.e. while they are physically encountering it: the work was conceived, Krauss succinctly puts it, "as a network of perspectives that … would constantly define one's vision of the object in terms of one's relation to it" (Krauss 1985, 267). When Krauss then describes Serra's sculpture as having the "chiasmic trajectory" between seer and seen as its subject (Krauss 1985, 270), she in effect makes explicit the radicality of her reading of Merleau-Ponty's *Phenomenology*. Although nowhere in Krauss's writings of this period does she, to my knowledge, mention Merleau-Ponty's unfinished last book *The Visible and the Invisible*, her use of the term 'chiasmic' makes it clear that she is reading the *Phenomenology* in light of Merleau-Ponty's later work, because the 'chiasm' is a concept that Merleau-Ponty had not yet settled on in the earlier book. This later ontology, in which sensing and the sensed are mutually implicated within one another, would seem, therefore, to be the other part of the context into which Krauss recasts the embodied subject of Merleau-Ponty's earlier work. The art works to which she applied such ideas, however, were ones in which an embodied viewer and their direct, active perception was foremost: the work as object is perceived as literally present, just as the viewer experiences their own presence in relation to it. At least, this would be a relatively standard reading of Judd, Heizer and Serra in the vein of Morris's Minimalist theorising; but Krauss insinuates an absence into the heart of this presence, one that sets up and sustains it. Although Krauss abandoned phenomenology as she moved her focus to other forms of practice and other critical goals, we may ask whether the chiasmic relationship between absence and presence she started to outline within Heizer and Serra's work might not also be adapted to 'dematerialised' practices.

## The Rejection of Phenomenology

In part, Morris's interest in the moving body of the viewer in relation to sculpture came from the involvement he had with experimental dance at the time. Later, his then wife, the choreographer Yvonne Rainer, spoke of how Minimalism suppressed emotion (Meyer 2009, 152). This was undoubtedly partly to do with its own repressed roots in Greenbergian Formalism, whose search for a proper, if historical, essence to art it unwittingly reproduced in attempting to outbid it. Equally, though, phenomenology also tended to bracket feelings in an attempt to get to 'existentials', as Heidegger called them, the *a priori* categories of experience. Yes, in *Being and Time* (Heidegger 1962) Heidegger famously prioritises feeling (*Stimmung*, mood) ahead of understanding as the basis for our openness to Being. Nonetheless, only the single, specific feeling of anxiety is deemed adequate to the task of inaugurating an authentic openness for the subject. As Giorgio Agamben points out in his essay "Passion of Facticity," love is mentioned but once in Heidegger's *Magnum Opus*, and then only in a footnote (Agamben 1999, 185). Indeed, love, judging by Rainer's subsequent work to the Minimalist moment, is probably not far from what she had in mind when using this word 'emotion'; love as a complex emotional relation to another person, to an *other*. Alterity brings me to another exclusion of both Minimalism and phenomenology – at least the phenomenology of *Phenomenology of Perception* and Heidegger's *Being and Time* – namely, gender. Heidegger's conception of *Dasein* is, as Jacques Derrida showed, neutral as to sexual difference: gender simply "does not belong to the essential structure of Dasein" (Derrida 2008, 10). Given that Heidegger explicitly designates embodiment as secondary to *Dasein*, this is, perhaps, not surprising. More surprising, though, is the absence of any serious discussion of gender in Merleau-Ponty's book seeing as it is an account of the primacy of embodiment. However, I am digressing.

What led to Conceptual art's rejection of phenomenology was not so much questions of emotion, gender or sexual difference, at least not in the first instance, but rather more the fact that phenomenological criticism seemed to miss, and not only miss but also actively cover over, the ideological structure of the art world. The move to other intellectual pastures was motivated by an awareness of what is essentially the same "gap between individual and phenomenological experience and structural intelligibility" that, as Fredric Jameson notes, had caused realism to lose its critical potential as an aesthetic mode at the end of the nineteenth century. Jameson continues: "to put it more simply, if in the newly decentered situation of the imperialist network, you live something strongly and concretely, it is unintelligible, since its ultimate determinants lie outside your own field of experience" (Jameson 2007, 241). With Conceptual art this suspicion steadily develops: beyond the phenomenologically experienced work of art there are determining structures and conditions that remain invisible. In fact, works of art themselves are suspected of concealing the framework which subtends them: the work of art is an "illusion," the French artist Daniel Buren wrote in 1970, which "cancels out its viewpoint (the Museum/Gallery) … making the latter pass for a vague neutral frame" (Buren 1973, 45). Here it is the self-effacing structure of ideology, perhaps most influentially summed up in Roland Barthes's early critique of myth delivered in the late 1950s, which is seen to have escaped the phenomenologist's grasp, being as it is beyond direct perceptual experience.

"This is what the dominant ideology wants," Buren wrote, "that what is contained [the work of art] should provide, very subtly, a screen for the container [the museum or gallery]" (Buren 1973, 38). An obvious response to this situation, then, is to get rid of the work of art as a discrete and containable, essentially collectable and curatable object, as many artists aspired to do at the time. The critic Lucy Lippard was one of the first to recognise this tendency, for which she supplied the enduring if contested label "dematerialized art" in an article first published in 1968 (Lippard and Chandler 1971, 259). Among the many exhibitions mounted during the 6 years referred to in the title of Lippard's subsequent anthology of Conceptual art spanning the years 1966–1972, was one organised in New York by the dealer Seth Siegelaub in 1969. *January 5–31, 1969*, as the show is known, was one of the first significant exhibitions focused on the nature of Conceptual art (Osborne 2002, 29). It is also one whose form well exemplifies Lippard's term, as Siegelaub conceived of the catalogue - composed of a collection of the artists' statements and photos relating to the work - as "primary," whereas "the physical exhibition," he said, "was auxiliary to it" (Lippard 1977, 125). On the face of it, once art has been 'dematerialised', whether it be through use of language in the form of instructions, proposals and descriptions of works, or through other forms of information (another buzzword of the time), phenomenology was no longer relevant to the critical appraisal of these practices as they displaced the situated, embodied viewer Morris had in mind.

Siegelaub included four artists in his 1969 exhibition. One of the four was Kosuth. Another was Lawrence Weiner, eight of whose works were included in the catalogue. While the catalogue itself was displayed in one of the exhibition's two rooms, a selection of the textual works it contained were exhibited in material form in the other. Included in both rooms – that is to say, included in both the catalogue in its textual form as well as in material form in the 'auxiliary' exhibition – was Weiner's *A 36" x 36" removal to the lathing or support wall of plaster or wallboard from a wall* (1968). The title here describes precisely what was physically manifest in the second room. What is important, however, is that Weiner saw the two – textual statement and material realisation – as equivalent. The catalogue also contained his "Statement of Intent" in which he wrote, talking of such textual propositions, that "the piece need not be built," thereby implying that the proposal was sufficient itself for his 'piece', the work, to be said to exist (Alberro 1999, xxii). So it is possible to say, as Weiner himself maintained, that, whether it is the textual statement one reads or a physical manifestation, whether the statement is read in a catalogue or on a gallery wall, and regardless of the physical structure of the wall the work is materialised upon, it is the *same* work (Lippard 1977, 129).

## From One Phenomenology to Another

One way to account for this would be to follow Kosuth and say that Weiner's work is an idea. "Works of art are analytical propositions," Kosuth wrote, "A work of art is a tautology in that it is a presentation of the artist's intentions, that is, he is saying that a particular work of art is art, which means, a definition of art." (Kosuth 1999, 165). For Kosuth, therefore, it did not matter whether the work took the form of a text or object, because ultimately the idea is to be received independently of its material form, its 'morphology' as he referred to it. This conception would clearly lead us, as it did Kosuth, away from a situation in which phenomenology could find much resonance or purchase. It is a conception of the work that, being founded on the logical operation of self-definition with its aspiration to transparency and clarity, would seem to lead inevitably to an attempt to deny or bracket out the ambiguity of phenomenal existence. This was unsuccessful, as the failure of Conceptual art to eliminate the aesthetic alluded to at the beginning this paper indicates. As a theoretical position, it is vulnerable to the same critique Merleau-Ponty applies in *The Visible and the Invisible* to the dichotomies set up by Husserl and Sartre between consciousness and being. In this book Merleau-Ponty continued in intensified form his project, begun in the *Phenomenology,* to correct the idealist notion he finds in Husserl of a consciousness that "is defined by its presence to itself, its immanence" (Barbaras 2004, 56). If the divide between it and being were absolute, in the way such a conception of consciousness would imply, then the in-itself of being could not become the for-itself of consciousness. That is to say, conceived of as mutually exclusive opposites, the connection between being and consciousness evident in intentionality becomes inexplicable. Merleau-Ponty suggests, therefore, that they must rather be conceived as intertwined, each carrying the implication and potential of the other within it. Even if, in his subsequent career, Kosuth showed a more nuanced approach, in his early essay it is a similar immanence that is envisaged for art to that which Husserl envisaged for consciousness: the subsumption of a spatio-temporal form under an immaterial idea, and the reduction of the differences between such manifestations to the identity of the idea, is a similar idealism. Osborne pinpoints the problem with characteristic lucidity, although, as I have already indicated, his is not a phenomenological approach: Conceptual artists like Kosuth misconstrued art as having an "ideational ontological purity," and, in as much as they based their own practice on this conception, they were labouring under a "self-misunderstanding" (Osborne 2013, 109).

Richard Long visited Siegelaub's *January Show* and was impressed by Weiner's work. Nonetheless, as Smithson had done, he later distanced himself from the idea of "replacing the object with language" (Wallis 2009, 48). Indeed Weiner's "Statement of Intent" can be read this way. It can be understood as claiming that the work and the textual proposition are ultimately synonymous. Weiner writes in it that his pieces may be made – by himself or someone else – or remain textual statements, all of which are "equal and consistent with the intent of the artist." It does, indeed, sound as if he were saying that the statement is sufficient in itself, its realisation being merely 'auxiliary'. Weiner spoke in another context of using language "in an attempt to get across *only* the content, in the most concise package," and yet it is not at all certain that Weiner's work actually replaces "the object with language" (Lippard 1977, 130). Take, for example, Weiner's work *Many colored objects placed side by side to form a row of many colored objects* (1979), which has been realised repeatedly over the years in vinyl, neon or painted lettering inside galleries and in exterior spaces. Firstly, and most obviously, Weiner's text, whether printed on a page or stencilled on a wall, is an object, a row of letters, and it refers to itself as such. It alludes then to the fact that "the transcendence of the sign toward [its] sense," as Barbaras writes, "never abolishes the sign's materiality, never reaches a transparent meaning" (Barbaras 2004, 53). Weiner's words, as all written words, not only refer beyond themselves and derive their ability to do so from belonging to a linguistic system, but are also material inscriptions with phenomenal thickness. Secondly, Weiner's text also, as again all language arguably does, bears a reference to a phenomenal world, even if that world is no more than potential. The very fact that the letters forming the words of the statement *could* also themselves count as "colored objects placed side by side," makes the irreducible referential potentiality of language, as well as that referentiality's ambiguity, apparent.

The use of language, text as a medium, has nonetheless been seen as central to conceptualism's attempt to distil art's "ideational ontological purity." However, despite the impression of transparency it can at times produce, language is not synonymous with ideas, which is borne out by our brief reading of Weiner's *Many colored objects*. Indeed, in "Indirect Language and the Voices of Silence" Merleau-Ponty spoke of "an opacity of language." What he meant by this phrase was that "nowhere does [language] stop and leave a place for pure sense" (Merleau-Ponty 2007b, 244). There is no direct presentation of sense in language, pure or otherwise, because the sense of signs is always absent where they are present. It is absent from the sign inasmuch as when we are looking for the sense of signs we must go elsewhere, whether it be to their syntax or context, or the background system, or usages out of which they arise as meaningful signs. It follows that if by 'idea' such a 'pure sense' as Merleau-Ponty spoke of is meant, then text cannot be its repository. Indeed, much and possibly the preponderance of Conceptual art that takes textual form plays with the ambiguity of language as well as its materiality, the Belgian conceptualist Marcel Broodthaers being surely the ironic master in this respect. Weiner's use of language too seems to revel in the medium's suggestive vagueness and its opacity. So, rather than loading the identity of the work into the linguistically expressed idea, his 'statement of intent' could also simply, and equally well, be read as putting all degrees and variety of realisation on an equal footing. When Long subsequently says that "knowledge of my actions, in whatever form, is the art" he is, consciously or not, echoing this alternate reading of Weiner (Long 2007, 26).

At this point, it is worth recalling that, for Merleau-Ponty an object, or rather a phenomenon, is not at all a definite thing but, rather, is inherently ambiguous. In this respect at least, language and phenomena are not ultimately distinct. Both are characterised by a comparable incompletion. From Merleau-Ponty's point of view then it is possible to say that Weiner is not wrong when, in conversation with Siegelaub and the other artists who participated in the *January Show*, he refers to the infinity of language (Lippard 1977, 132). A sign, a word, is incomplete in itself and cannot, therefore, "be conceived of as a positive entity" (Barbaras 2004, 52), because it is a compound of presence and absence. The meaning of each present word is established against a background of language that is absent; it is this background that constitutes the word as meaningful, though it be as intangible as the word is demonstrably and materially present. "The sign," as Merleau-Ponty puts it, "makes sense only insofar as it is profiled against other signs," and its sense exists only in the form of a horizon toward which the sign tends (Merleau-Ponty 2007b, 244). Although the sense of the sign does not emanate from within it – although, that is to say, its sense is not immanent to it – a sign's sense is nonetheless intrinsic to its identity as sign. As well as alluding to something very close to this in his reference to language's infinity where "there is no edge," Weiner also contrasts it, however, to the material discreteness of objects, his examples being paintings and picture frames (Lippard 1977, 131). Undoubtedly, there is a distinction to be made between words and things, between the word 'painting' and a particular painting by Cézanne, say. In terms of Merleau-Ponty's phenomenological perspective, however, it is not a distinction between something infinite and immaterial on the one hand and a finite and material positivity on the other, as Weiner seems to imply. The compound of presence and absence found in the sign is also to be found in the phenomenon.

In the *The Visible and the Invisible* perception is no longer presented, if it ever was in Merleau-Ponty's earlier work, as the intending of a unified noema. Rather, each apparently discrete perception is seen as being actually "a certain node in the woof of the simultaneous and the successive" (Merleau-Ponty 1968, 132). Merleau-Ponty takes as an example a direct perception of redness, which turns out to be constituted of a network of other potential reds and red things. The directness of the perception, therefore, becomes at the least complicated by its implication with potential perceptions: it is not "a quale, a pellicle of being without thickness" but, rather, "emerges from a less precise, more general redness" (Merleau-Ponty 1968, 131). Each discrete perception Merleau-Ponty thinks of as having this halo of sense, a virtual linkage with other potential perceptions, hence, "this red is what it is only by connecting up from its place with other reds about it, with which it forms a constellation" (Merleau-Ponty 1968, 132). Without this absent horizon, the present perception would not be what it is. It would indeed not *be* at all: each perception is 'animated' by a sense (it is *red*, a *tree*, and so on); it has an aspect, and without this it would not even be perception, let alone this particular perception. So, in Merleau-Ponty's later thought language loses its claim to ideal transparency and phenomena theirs to unalloyed presence. At this point phenomenology as it was understood in the context of Minimalism has been surpassed.

Let us get back to Weiner. It was necessary to point out that his statements *are* objects despite their textual form and, also, to add that equally as language they are *not* objects, if by that is meant literal, material presences. Now we can go a further step and say that material objects themselves in their meaningful appearance are also more than merely present objects. Even as objects given to vision, their meaningful appearance is reliant on a sense that is not present in them or finite. (In the case of art objects, this absent horizon of sense is more salient than in everyday perception as the background they come into being against is necessarily historical.) On this ontological level, the level of sense, Weiner's statements would indeed be the 'same' as their materially realised counterparts. While his text works derive much of their effect from their referential ambiguity, and despite this being undoubtedly distinct (in ways I will not elaborate on here) from the perceptual ambiguity one experiences when encountering sculptures such as those of Serra or Heizer, both text and object produce an experience neither clear nor obscure, neither fully present and immediate nor entirely general and abstract. Perceptual and linguistic horizons are different, but the compound of presence and absence remains on an ontological level. One thing that perhaps could be said in this context about their difference is that Weiner's works are not synonymous with their medium in the way that Minimalist objects seem to be. This notwithstanding, is it not possible to think about Weiner's work, or that of other Conceptual and Post-conceptual artists, in this revised phenomenological framework? If so, it will because it is significantly distinct from that of classical phenomenology. It will no longer be focused on the constitution of the world in the intentional acts of consciousness, even if the job of constitution migrates to the body's intentional grasp of the world, as it does in the earlier Merleau-Ponty taken up by the likes of Morris and Krauss.

One of phenomenology's most significant discoveries, it is often said, was that made by Husserl of the transcendental nature of intentionality. That is to say, the *noema* (or sense) is not of the same order as the thing that appears in it (the sensed). The bracketing of the latter allowed the former to be shown as the form of its appearance. This is the result of the Husserlian reduction: to split the intending from the intended. However, Merleau-Ponty's late philosophy, Barbaras tells us, "proceeds entirely from the decision to reconceive intentionality as an originary 'reality', to recognise the irreducible and in some way unrendable character of the intentional fabric" (Barbaras 2004, 170). Rather than sense, therefore, being conceived as being entirely on the side of consciousness, giving form and identity to the inert, intended thing, in the late Merleau-Ponty it is thought of as being woven with the sensible (Barbaras 2004, 163). The nature of this intertwining of sense and the sensible, however, is that sense is presented in the sensible as "a certain absence": "the sense of sense lies in never being present in person," Barbaras writes (Barbaras 2004, 163; 52). As we have seen, sense is thus a horizon that must remain elusive and absent so as to be itself, just as the phenomenon must continue to unfold towards its absent horizons so that it can be what it is. In sum then, sense as horizon, as Merleau-Ponty thinks it, is what both enables the phenomenon to appear and keeps its appearing from completing itself, whereas in Husserl, on the contrary, the noema is conceived as a "unity of sense" (Barbaras 2004, 51; 171). In the picture supplied by late Merleau-Ponty the world's sense is not forged in the crucible of the intending subject, nor in the body's intentional grasp; if intentionality is able to give sense, it is because it has already received it.

In *Discourse, Figure*, which contains his most extensive reckoning with Merleau-Ponty's thought, Lyotard spoke at one point of walking "the same path as Merleau-Ponty, but in the opposite direction" (Lyotard 2011, 54). That is to say, rather than attempting to show that language can be related to the thickness of the phenomenal world beyond it, Lyotard meant to show in his book how phenomenal distance ("the beyond-Logos") already inhabits language undermining its transparency. It should be clear by now that, even if not explicit in Merleau-Ponty's late work, such an approach does not seem to be entirely at odds with the ontology it puts forward and the analyses it contains. Lyotard also clearly had sympathy for phenomenology's aim of going "beneath the realist view of the constituted or the given" and in particular Merleau-Ponty's attempt to do so, even if he believed that "sooner or later one will have to give up phenomenologizing if one wants to reach … this something that is not constitutable." In place of phenomenology, it was what he termed "deconstruction" that was needed (Lyotard 2011, 54–5). Perhaps what makes it possible to say that Merleau-Ponty is still doing phenomenology in *The Visible and the Invisible* and related texts is not only that he is *not* performing a deconstruction of conceptual oppositions as Lyotard does, but that he seeks to reappraise and refashion such phenomenological concepts as intentionality. In his late essay on Husserl, "The Philosopher and his Shadow," he speculates on how the ontological approach he is now taking to "unveil the pretheoretical layer," which he likens to an "archaeology," might transform phenomenology's core concepts: "Does it make no changes at all in our conception of noesis, noema, and intentionality – in our ontology? After we have made this descent, are we still entitled to seek in an analytics of acts what upholds our own and the world's life without appeal?" (Merleau-Ponty 1964, 165). This is, indeed, a big change from Husserl and his own *Phenomenology*, where intentionality is by and large explored as a subjective act.

As we have seen, in his late work Merleau-Ponty has moved on from thinking of intentionality as being explicable finally as the subject's grasp of the world. It is certain that he no longer thought of the world as primarily constituted in our intentional acts, as these themselves had a background, the soil of the lifeworld, that was "not constitutable" in them. Ultimately, going back to follow up the ramifications of these two simply stated conceptual developments, largely missed in Merleau-Ponty's reception within Minimalist discourse and in the denigration of phenomenology by proponents of Conceptual art, is what is required to re-establish phenomenology's relevance for Post-conceptual art. Merleau-Ponty spoke of the need to renounce "the bifurcation of the 'consciousness of' and the object," which the Husserlian concept of intentionality implies (Merleau-Ponty 1968, 141). What he refers to as the 'consciousness of' here is intentionality conceived of as immanent to the subject and founded on an ideal "unity of sense" independent of any actual object. So might we not, if we were to take up again and elaborate upon Merleau-Ponty's alternative version of intentionality and its rejection of an immanent, unified sense, be walking in the same direction as Smithson and his rejection of "facile" ideas of "gestalt unity" in the ontology of the work of art, too?

## Works Cited


Agamben, Giorgio. 1999. The Passion for Facticity. In *Potentialities*, 185–204. Stanford: Stanford University Press. Print.

Alberro, Alexander. 1999. Reconsidering conceptual art, 1966–1977. In *Conceptual Art: An Critical anthology*, ed. Alexander Alberro and Blake Stimpson, xvi–xxxvii. Cambridge: MIT Press. Print.

Alys, Francis, et al. 2004. *The Modern Procession*. New York: Public Art Fund. Print.

Babaras, Renaud. 2004. *The Being of the Phenomenon: Merleau-Ponty's Ontology*. Bloomington: Indiana University Press. Print.

Buren, Daniel. 1973. *5 Texts*. New York/London: John Weber Gallery. Print.

Derrida, Jacques. 2008. Heidegger's Hand (*Geschlecht* II). In *Psyche, Inventions of the Other*, vol. 2, 27–62. Stanford: Stanford University Press. Print.

de Duve, Thierry. 1996. *Kant After Duchamp*. Cambridge, MA: MIT Press. Print.

Ferguson, Russell. 2007. *Francis Alys: The Politics of Rehearsal*. Los Angeles: Armand Hammer Museum. Print.

Foster, Hal. 1996. The Crux of Minimalism. In *Return of the Real: The Avant-Garde at the End of the Century*, 35–68. Cambridge: MIT Press. Print.

Heidegger, Martin. 1962. *Being and Time*. Oxford: Blackwell. Print.

Jameson, Fredric. 2007. Rimbaud and The Spatial Text. In *The Modernist Papers*, 238–254. London: Verso. Print.

Kosuth, Joseph. 1999. Art After Philosophy. In *Conceptual Art: An Critical Anthology*, ed. Alexander Alberro and Blake Stimpson, 158–177. Cambridge, MA: MIT Press. Print.

Krauss, Rosalind E. 1966. Allusion and Illusion in Donald Judd. *Artforum* 4: 9 (May 1966): 24–26. Print.

———. 1981. *Passages in Modern Sculpture*. Cambridge, MA: MIT Press. Print.

———. 1985. Richard Serra: a Translation. In *The Originality of the Avant-Garde and Other Modernist Myths*, 261–274. Cambridge, MA: MIT Press. Print.

———. 2000. *A Voyage on the North Sea: Art in the Age of the Post-Medium Condition*. London: Thames and Hudson. Print.

Lippard, Lucy. 1977. *Six Years: The Dematerialisation of the Art Object from 1966 to 1972*. Berkeley: University of California Press. Print.

Lippard, Lucy, and John Chandler. 1971. Art and Dematerialisation. In Lucy Lippard, *Changing: Essays in Art Criticism*, 255–276. New York: E. P. Dutton & Co. Print.

Long, Richard. 2007. In *Selected Statements and Interviews*, ed. Ben Tufnell. London: Haunch of Venison. Print.

Lyotard, Jean-François. 1991. The Sublime and the Avant-Garde. In *The Inhuman: Reflections on Time*, ed. Jean-François Lyotard, 89–107. Stanford: Stanford University Press. Print.

———. 2011. *Discourse, Figure*. Minneapolis: University of Minnesota Press. Print.

Merleau-Ponty, Maurice. 1964. The Philosopher and His Shadow. In *Signs*, ed. Maurice Merleau-Ponty, 159–181. Evanston: Northwestern University Print.

———. 1968. *The Visible and the Invisible*. Evanston: Northwestern University Press. Print.

———. 2007a. Cézanne's Doubt. In *The Merleau-Ponty Reader*, ed. Ted Toadvine and Leonard Lawlor, 69–84. Evanston: Northwestern University Press. Print.

———. 2007b. Indirect Language and the Voices of Silence. In *The Merleau-Ponty Reader*, ed. Ted Toadvine and Leonard Lawlor, 241–282. Evanston: Northwestern University Press. Print.

———. 2007c. Eye and Mind. In *The Merleau-Ponty Reader*, ed. Ted Toadvine and Leonard Lawlor, 351–378. Evanston: Northwestern University Press. Print.

———. 2012. *Phenomenology of Perception*. London: Routledge. Print.

Meyer, James. 1998. Der Gebrauch von Merleau-Ponty. In *Minimalisms: Rezeptionsformen der 90er Jahre*, ed. Sabine Sanio, 178–189. Ostfildern: Reihe Cantz. Print.

———. 2009. The Minimal Unconscious. *October* 130, Fall: 141–176. Print.

Morris, Robert. 1995. Notes on Sculpture, Part 2. In *Minimal Art: A Critical Anthology*, ed. Gregory Battcock, 228–235. Berkeley: University of California Press. Print.

Osborne, Peter. 2002. Survey. In *Conceptual Art*, ed. P. Osborne, 13–51. London: Phaidon. Print.

———. 2013. *Anywhere or Not at All: Philosophy of Contemporary Art*. London: Verso. Print.

Smithson, Robert. 1996. In *Robert Smithson: The Collected Writings*, ed. Jack Flam. Berkeley: University of California Press. Print.

Wallis, Clarrie. 2009. Making Tracks. In *Richard Long: Heaven and Earth*, 33–61. London: Tate Publishing. Print.

# The Human Condition, Nature, Power and Creativity. Philosophical Anthropology and Eco-Phenomenology in the Context of Biopolitics

**Massimo Mezzanzanica**

**Abstract** The vision of an eco-phenomenology that Anna-Teresa Tymieniecka sketches in her Bergen interview and which is based on the concept of the ontopoiesis of life has some significant parallel outlooks of twentieth-century German philosophical anthropology. I will try to show here the value of the outlooks of Arnold Gehlen, Helmuth Plessner, and Adolf Portmann and of the idea of a philosophy of nature, biology, and the body which they imply, and that against the background of today's reflections on science, life, and politics. In a situation where politics touches more and more on the human body and life (as pointed out by Michel Foucault, Giorgio Agamben, and Roberto Esposito), and where the development of the life and computer sciences and of medical and information technologies makes possible radical transformations of the human being—which developments seem to override the traditional distinctions between man, animal, and machine—philosophical anthropology can still be for us an important heuristic instrument. Its conceiving of man as an open and undetermined being, one who is at the same time natural and artficial and who builds and conducts his life through the power of imagination, can interact with the perspective of eco-phenomenology in its considering the human being to be a relational or, in Tymieniecka's words, "as a human condition within the unity of everything there is alive."

**Keywords** Philosophical Anthropology · Foucault · Biopower · Gehlen · Handlung · Plessner · Homo absconditus · Tymieniecka · Ontopoiesis · Adolf Portmann · *Grenzcharakter* of living being

M. Mezzanzanica (✉)
Università degli Studi di Milano, Milan, Italy

 
W. S. Smith et al. (eds.), *Eco-Phenomenology: Life, Human Life, Post-Human Life in the Harmony of the Cosmos*, Analecta Husserliana CXXI,
https://doi.org/10.1007/978-3-319-77516-6_34

## Analysis of Biopower and a Critique of Anthropology

Although the concept of biopower is present in the work of other scholars, it was Foucault who for the first time used the concept as a global interpretative key to understanding different phenomena that are characteristic of modern society and politics. Foucault considers biopower to be a new form of power, one which arose and developed between the seventeenth and nineteenth centuries, a period that he refers to as the "classical age" (Foucault 1970). This power is no longer a negative, repressive and coercive one. On the contrary, it is a positive kind of power that is exercised over life to manage, secure, improve, and extend it. This new kind of power of life unfolds on two different levels: first, the "political anatomy of the body"; second, the "regulation of population." That is to say, it applies itself both to the individual body, treated like a machine with the aim of controlling it and making it more effective, and to the body of the population, that is, to "the body imbued with the mechanics of life and serving as the basis of the biological processes: propagation, births and mortality, the level of health, life expectancy and longevity, with all the conditions that cause these to vary" (Foucault 1990, 139). In this sense, biopower implies at the same time "the administration of bodies and the calculated management of life" (Foucault 1990, 140). But, according to Foucault, it is important to distinguish between discipline, which is based on the objectivizing approach of the human sciences, and biopower:

> Unlike discipline, which is addressed to bodies, the new nondisciplinary power is applied not to man-as-body but to the living man, to man-as-living-being; ultimately, if you like, to man-as-species. … So after a first seizure of power over the body in an individualizing mode, we have a second seizure of power that is not individualizing but, if you like, massifying, that is directed not at man-as-body but at man-as-species. After the anatomo-politics of the human body established in the course of the eighteenth century, we have, at the end of that century, the emergence of something that is no longer an anatomo-politics of the human body, but what I would call a "biopolitics" of the human race. (Foucault 2003, 242–43).

For Foucault, biopower is radically different from traditional sovereignty, which is based on a political contract. On the one hand, it is not exercised over subjects of rights but over living beings, and it tends to replace the norm with the law. The normalization of bodies is an essential issue thereof. On the other hand, its form is no longer the "right of death," but interest in fostering life. The right to impose death, which characterizes feudal forms of power, has now changed to a right to live. From power "to make die and to let live" it becomes a power "to make live and to let die" (Foucault 2003, 241). The right to kill—manifested in war and genocide—is now the dark side of a power in the service of maximizing life. Its motivation lays in the protection of the life of the population and it acquires an immunological connotation: "If genocide is indeed the dream of modern power, this is not because of the recent return to the ancient right to kill; it is because power is situated and exercised at the level of life, the species, the race, and the large-scale phenomena of the population" (Foucault 1990, 137).[1]

Foucault shows that the development of biopower has had deep historical, anthropological and epistemic consequences. In fact, in his opinion, it implies a deep change in the relationships between life and history. Life enters into the sphere of history, that is, it is being built up and shaped through the mechanisms of power and knowledge, and it is made submissive to political technologies that are, at one and the same time, forms of subjectification and objectification of the self. To characterize this interference between life and history, Foucault uses the concept of "bio-history." Here emerges, as Roberto Esposito observes, the peculiarity of Foucault's concept of life, its in-betweenness: "Life as such doesn't belong either to the order of nature or to that of history. It cannot be simply ontologized, nor completely historicized, but is inscribed in the moving margin of their intersection and their tension" (Esposito 2008, 31). Therefore, Foucault sees the meaning of biopolitics "in this dual position of life that placed it at the same time outside history, in its biological environment, and inside human historicity, penetrated by the latter's techniques of knowledge and power" (Foucault 1990, 143).

When the species, that is, the natural and biological dimension of man, becomes the real issue at stake in political strategies, mankind crosses the "biological threshold of modernity": "For millennia, man remained what he was for Aristotle: a living animal with the additional capacity for a political existence; modern man is an animal whose politics places his existence as a living being in question" (Foucault 1990, 143). As a consequence of this process, Western man becomes conscious of the political importance of the body, which always belongs to a social space: "Western man was gradually learning what it meant to be a living species in a living world, to have a body, conditions of existence, probabilities of life, an individual and collective welfare, conditions that could be modified" (Foucault 1990, 142). With this transformation are also connected the changes in the order of classical *episteme,* which engendered the problem of man in his specificity of living being, going from the natural history of the classical age to modern biology. While natural history considers, according to a classifying approach, every single living being by putting it in an order of representation, modern biology sees it as a contingent manifestation of a general law of life based on "quasi-transcendental concepts" such as "nature" and "human nature" (Foucault 1970, 363). But, Foucault considers the concept of human nature to be the product of a historically-built paradigm of the modern *episteme* and sees it as a form of naturalization that is connected with the development of the human sciences. "Man is an invention of recent date and one perhaps nearing its end," because the paradigma of human sciences will be not an eternal one (Foucault 1970, 386). In his archeology of the human sciences Foucault denies, on one hand, the scientific value of the concept of human nature, considering it only as a simple "epistemological indicator" that, in the history of knowledge, has served to "designate certain types of discourse in relation to or in opposition to theology or biology or history" (Elders 2011, 7). On the other hand, he points out its historical character with an aim to criticize its political implications. In his debate with Chomsky, Foucault states that concepts such as human nature and justice have constituted themselves in our civilization, in our type of knowledge, and in our philosophy. Therefore, they cannot be considered to be eternally valid.

The death of man, announced by Foucault as an implication of Nietzsche's proclamation of the death of God and of the upcoming birth of the *Übermensch,* can be therefore understood as the end of human nature; that is, it can be understood as the sunset of a historical image of man and as the outcome of the field of meaning that was structured by this image. In his "Introduction to Kant's Anthropology," in which we can find the first formulation of his critique of the reduction of the transcendental to the empirical that is typical of modern philosophy, Foucault writes: "The trajectory of the question *Was ist der Mensch?* in the field of philosophy reaches its end in the response which both challenges and disarms it: *der Übermensch*" (Foucault 2008, 124). The Foucauldian critique of anthropology is, therefore, conditioned by a Nietzschean vision of history as genealogy or as effective history (*Wirkungsgeschichte*). Such a conception, denying every constant and stable element in the life of man, asserts the historical character of every aspect of man, including those aspects pretended to be natural such as the body:

> We believe, in any event, that the body obeys the exclusive laws of physiology and that it escapes the influence of history, but this too is false. The body is molded by a great many distinct regimes; it is broken down by the rhythms of work, rest, and holidays; it is poisoned by food or values, through eating habits or moral laws; it construct resistances. "Effective" history differs from traditional history in being without constants. Nothing in man—not even his body—is sufficiently stable to serve as the basis for self-recognition or for understanding other men. (Foucault 1984, 87)

In contrast with post-Kantian philosophies and with the human sciences which, in his opinion, naturalize the transcendental, conceiving it as a real quality of man, Foucault aims to reactivate the project of critique, which he understands in the first instance as archeological analysis of discursive formations and, then, as genealogical analysis of the nexus of power and knowledge and, further, as reconstruction of the constitution of effects of truth that are, at the same time, effects of power. The critique of anthropology, therefore, represents the theoretical condition for both the analysis of the technical mechanisms of subjectification, which leads on to the social construction of man, and for the analysis of the biopolitical regime, wherein man becomes a variable of government interventions in the population.

## Bioeconomy, the Post-Human

Foucault emphasizes the relationship between bio-power and the development of the capitalistic economy, which could consolidate itself only through the insertion of bodies into the productive apparatus and thanks to the adaptation of population phenomena to economic processes—making necessary "the investment of the bodies, its valorisation, and the distributive management of its forces" (Foucault 1990, 141). To the present day, in the situation of post-Fordist capitalism, this power crosses over into a new form of accumulation which some scholars describe as "bioeconomy" or as "cognitive capitalism": a form of accumulation that more and more stimulates the vital faculties of the individual (in particular, communicative and relational capacities) into following an economic imperative and which tends to the transformation of existential activities into economically productive relationships (Fumagalli 2007). As noted by Thomas Stewart (Stewart 1998), the digitalisation of valorisation processes transforms knowledge into a productive factor, which creates a model of work and production that some scholars call "anthropogenic" (Fadini 2013, 18). While highlighting the nexus between life and economy, the Foucauldian analysis could not yet account for the phenomena, which emerged only in recent times, that concern the configuration of the body in the situation of the bio-economy. Foucault's analysis presupposes, in fact, the idea of an integral, unitary, delimited, and unfragmented body, which precedes the technical interventions to which it is subjected. Instead, the developments of the life and information sciences, of the neurosciences, and of the cognitive sciences, together with the application of the various technologies that invest life and the body, are producing today the image of a fragmented body, one which represents a field for new and more intensive bio-political interventions. In the biological field, research into molecular biology, genetics, and embryology creates the conditions for the emergence of life technologies such as genetic engineering, genetic therapy, and medically assisted procreation, which are radically challenging the idea of human nature by replacing the reproduction or generation of life with its production.

In medical science, the possibility that the organs of a man can continue to live in the body of another man seems to change the relationship between life and death, and this relationship has become more problematic after the development of techniques of resuscitation, which makes uncertain the natural border between life and death by making the latter the subject of a scientific definition or of a legal regulation.

In such phenomena seems to loom a new cartography of life that dissolves well-known boundaries and univocal separation lines through the opening of "spaces of exception" in which "bare life" is by now subjected to human technical control (Agamben 1998). The effects of cognitive and technical developments in the information sciences are equally disruptive. The development of artificial intelligence and the amplification of human intelligence through information technology have produced a cognitive and informational virtual space called the info-sphere or cyber-space. This virtual space, already prefigured in Theilard de Chardin's concept of the noosphere (Teilhard de Chardin 1959), produces, on the one hand, a radical transformation of human experience, and develops, on the other hand, a cognitive symbiosis between man and machine, one that lays the ground for the emergence of the "post-human" (Marchesini 2002). In this regard, different scholars look to the body from the point of view of its technical transformation and artificialization, by envisioning the cyborg, a "cybernetic organism," an ensemble of flesh and technology, a human being with prosthesis and electronic implants, as the expression of this new post-human condition. For example, Donna Haraway, analyzing what she calls the "informatic of domination," highlights the transformation of corporeality in this context by asserting the need to rethink the bodily experience in the realm of virtuality, where it seems to erase traditional oppositions such as those between mind and body, human and animal, organism and machine, public and private, man and woman, nature and culture, primitive and civilized (Haraway 1991, 163). In this situation the body is affected by a radical transformation; it is no longer a natural givenness, but it becomes a complex field in which different codes (from genetic to informational ones) cross. From another point of view, Giuseppe O. Longo considers "symbiont homo tecnologicus" to be the result of the symbiosis between man and technology (Longo 2003). Here technology is no longer an external prosthesis but invites a process of hybridization with the biological from which can arise unexpected evolutionary outcomes. Assuming the coming of this new evolutionary stage, Longo also stresses the risk of a disembodied post-human or an informational reductionism where, owing to information's absolute prevailing over material support, the biological body will become superfluous (Longo 2007).

## The Human Condition Between Nature and Artifice

Prospects such as those here described have stimulated renewed interest among philosophers in anthropological questions, questions about human nature and life, the relationship between nature and culture, and the role played by technique in human life and its limits. Polar opposites in this discussion are Jürgen Habermas—who, opposing man and technique, refers to the concept of human nature to denote the risks of a "liberal eugenics" and to assert, in a Kantian sense, the need to submit biotechnologies to the judgment of human reason (Habermas 2003)—and Peter Sloterdijk, who sees man as the outcome of anthropogenic mechanisms and considers today's biotechnological developments as the more recent result of anthropotechniques that produce a "domestication" of the human being through the creation of the "clearing" (*Lichtung*) in which man can develop (Sloterdijk 2001). From this point of view, humanism is criticized and deconstructed as an immunological dispositive which separates man from both technique and the animal. Considering (with Gehlen) technique to be an essential issue of human culture, Sloterdijk underlines the almost infinite plasticity of human beings and asserts the value of an ecological "homeo-technique," understood as a technology used for operating on materials that are of the same ontological quality as the operator, in contrast to a contra-natural and metaphysical "allo-technique" that treats materials as being of a separate ontological quality. But what emerges from Sloterdijk's discussion is the dark side of humanistic anthropo-technique, which seems to foster a situation, in many respects disturbing, in which humanity will be able to transform its genetic characteristics, realizing a transition from "birth fatalism" to "prenatal selection" (Sloterdijk 2009).

In different ways, both bio-power and techno-science seem to consider nature as a cultural product or to shift the boundaries between nature and culture and between the natural sciences and the cultural sciences, effecting a culturalization of the former and a naturalization of the latter (Bröckling and Schöning 2004, 13–14). The paradigm of twentieth-century German philosophical anthropology seems to offer theoretical tools to understand this situation. With the rise of beings such as stem cells, clones, cyborgs, robots, and computers, philosophical anthropology, which denies the possibility of fixing a predetermined essence of man, acquires a new relevance. What—and who—is a man? What distinguishes a human being from other living beings and from inanimate things? What connect them? Is it possible to speak of an essence of man or a human nature? What is the relationship between nature and the history of man? On these questions, philosophical anthropology seems to be able to converge with eco-phenomenology and with the theory of complexity in an attempt to construct a theory that considers at the same time both the natural and the cultural side of man, as well as to think, as Edgar Morin says drawing on Adolf Portmann, of the empirical and epistemological connection between man and animal, going beyond mutual disjunction and reductionism—a task which can be accomplished by complex thinking able to see the animality in the human being and to consider human nature to be simultaneously unique and dual, as a "uniduality" (Morin 1999, 5).

The need to consider the human being in the context of nature is a central issue in Tymieniecka's thought. Developing Husserl's phenomenology in the direction of a phenomenology of life, Tymieniecka aims to go beyond the field of pure transcendental constitutive phenomenology by showing that the "conscious" (*das Bewusste*) is rooted in lived experience of the corporeally conscious (*Leiblich-bewusste*) (Tymieniecka 1971, 2–3). In this transformation of Husserl's transcendentalism it becomes important to reconstruct the genesis of pure and abstract forms of thought starting from the generative level of experience, that is, in the "genetic progress of empiria" which "surges from and obviously stems from origins in nature that are physiological" (Tymieniecka 2011, 4).[2] Consequently, human subjectivity is placed in the environment and is no longer considered as a "demiurge" that is the origin of life and of the "life-world," but as cooperating in the genesis of life and nature, of the life-world and of the order of life (Tymieniecka 1986, 10–11). It is, therefore, an autonomous subject no more, but is part of the creative process of life, which Tymieniecka calls "ontopoiesis," emphasizing its ecological or eco-phenomenological meaning, consisting in the fact that "ontopoiesis reaches to the very germ of ecology: development and genesis." As a product of ontopoiesis, human being "is an ecological fruit" which "is formed by the earth and sucks the juices of the earth." This means that there is a relationship between the "self-individualization of life," which is the "basic instrument of autopoiesis," and the "laws of the cosmos and the earth" (Torjussen, et al. 2008).

There is a relationship between the "ecological" recognition of the value of every form of individual life in the cosmos and the importance that Tymieniecka ascribes to imagination. Life is not merely constitutive but, also, creative, and this appears through the emergence of "imaginatio creatrix of evolution." Thus, imagination is related to the central position of individuality in life:

> So, the central point of my earth and cosmos situating of life is accentuated by the emergence of *the imaginatio creatix of evolution* which allows our type of individuals to grow into their surroundings, in a very special way, getting above them. To my mind, individualization is the very special way in which life originates. Individualization is intimately related to the origin of life. Life originates precisely by progressive individualization. But I would not introduce types of measures of what is higher and lower. For me it is the unity of things which plays its role in each sphere. There would be no spiritual life without various phases of life preceding it and participating in it. Ours is a condition within the unity of everything alive, which depends on earthly and cosmic laws.

This leads to an evolutionary perspective that repudiates every form of anthropocentrism and, therefore, looks with suspicion at anthropology. Instead of speaking of man and of humanity, Tymieniecka prefers to speak of the "human condition," using an expression that we find also in Hannah Arendt and Helmuth Plessner. The expression the "human condition" means the "*very central point* that *human being can not be considered in itself as such, that there can be no anthropology that considers human being as such, in the middle of other things almost by chance*." Tymieniecka stresses that the "human condition" is situated "*within the unity of everything there is alive*." That means "the human being unfolds and generates in a mutual contributive relation to all the other living beings" (Torjussen, et al. 2008).

This critique of anthropology does not affect Gehlen's and Plessner's philosophical anthropology, which, though attempting to describe the characteristics of the human being, spurns anthropocentrism and considers man within the realm of living being, highlighting its constitutive openness and underlining the importance of imagination and creativity in man's life.

Gehlen develops all these themes through an attempt to build an "elementary" anthropology that considers the biological dimension as being constitutive of the human being, without falling for this reason into reductionism. Deriving man from the animal, according to a naturalistic or biologistic point of view, would prevent, in his opinion, our understanding at all the peculiarity of the relationship of man with nature, its being a global plan of nature, a unique project, never attempted otherwise.

Gehlen, in trying to overcome the contrast between body and soul, nature and spirit, the thing in itself and the phenomenon, and showing that sensorimotor functions are already soaked in spirituality, as thought, representation, and imagination are rooted in them, develops an *ante litteram* systemic approach based on the concept of a reciprocal connection between the parts of an organic totality. By this approach, different aspects of man (body, sensations, movement, language, thought, imagination) do not have to be considered to be isolated or in a hierarchical relation but, instead, have to be considered in their reciprocal interrelation or reciprocal retroaction as components of a system that is, in turn, set in the context of a social, cultural, and natural world.

In Gehlen's opinion, it is the concept of action (*Handlung*) that allows one to overcome the dualism between nature and spirit. This concept makes it possible to point out the open character of man. As a being who acts, man is, according to the expression of Nietzsche, "an unfinished creature," "a not-yet determined animal (*das noch nicht festgestellte Tier*). He is, therefore, a task for himself; he is the being that is constituted in such a way that he has no metaphysical or biological essence and he has to conduct his life (*Lebensführung*). Finding empirical evidence in the researches of Louis Bolk and Adolf Portmann, Gehlen bases this conception on the analysis of the morphological constitution of man, who, contrary to other animal species, is characterized by his retarded organic development, his lack of instincts, and the "hiatus" between his drives and their fulfillment. Rather than tagging man with deficiency (I recall that Gehlen distanced himself somewhat from the idea of man as a "deficient being" after criticisms were made of this concept by several authors, and in particularly by Konrad Lorenz), the openness and plasticity of man can be read in a positive way, as an effect of his lack of instinctual specialization and overabundance of drives.

Three aspects are connected with the indeterminateness of man: first, his openness to the world; second, the function played by imagination in his life; and third his relational character. As a being devoid of instinctual specialization, man does not live in a specific environment but is open to a world which, in turn, is not an "objective givenness" but must be interpreted. He is, therefore, an interpreting being who takes a position toward himself and others and who can define himself only in an indirect way, through confrontation with the non-human world. To identify himself, man has to identify with the other and, then, return to himself according to a logic that is explicated by Hegel's dialectic, G. H. Mead's concept of "taking the role of the other," and cybernetic description of nonlinear systems. As an undetermined being, man has to stabilize, discipline, and give himself a form through institutions and technique. He can do this thanks to imagination, which allows the right distance from the pressure of present. For this reason, man can be characterized not only as a rational animal, but also as an animal that has imagination. The relational character of man appears in the assimilation of the internal and external world through language, which forms, according to a thought of Novalis, an "external internal world," one in which there is a circular relationship between the person and the world (Gehlen 1988, 248). We can interpret the internal world from the external world, and vice versa, only because we experience both as being reciprocally related, and language is the hinge of this connection between impression and expression. By virtue of expression we can conceive of living beings as being animate. In fact, expression puts the individual in relationship to the world and society. If we can conceive of living beings as an internal/external world, this is by virtue of expression, which opens outs from the individual and alienates him from himself even in the internal relationship, thus allowing socialization. Expressivity is, therefore, an essential property of man.

An anthropology of non-anthropocentric cast is explicit in Plessner who, to point out the open and indefinite character of the human being, speaks of "homo absconditus" (Plessner 1983). From this angle of view, his position is not so different from Foucault's. If the latter believes that we can inquire into the human being outside the epistemological framework of anthropology, Plessner takes the humanistic viewpoint to the extreme by stating that it implies a self-relativization and, therefore, the end of the idea of a natural superiority of European culture in comparison with non-European cultures and so acceptance of competition from other "possibilities of being man" (Plessner 1981, 185–186).[3]

Constitutive of humanity is the co-belonging of self and other: "Mensch-sein ist das Andere seiner selbst Sein" (Plessner 1981, 225). This means that it is not obvious that Western man has to be considered the fundamental anthropological form, and that anthropology becomes a sort of "anti-anthropology" (Lindemann 2004, 27). This conception of humanity is grounded by the principle of inscrutability (*Unergründlichkeit*) or of open questions (*offene Fragen*), which Plessner derives from Wilhelm Dilthey and Georg Misch and which expresses the inexhaustible and, at the same time, accessible to understanding (*Verstehen*) nature of historical and spiritual life and, therefore, the openness of the question of man, which Plessner defines as an "open question" (Plessner 1981, 175–185).

According to the principle of inscrutability, man can get to know his essence only through experience of history, that is, by considering our various achievements over time. The essence of man identifies itself, in accordance with the perspective of *Macht und menschliche Natur* (Power and Human Nature), with his "power." Man, as power and, therefore, possibility and openness, is the "schöpferische Durchbruchstelle seiner geistigen Welt," the "produktive 'Stelle' des Hervorgangs einer Kultur" (Plessner 1981, 160, 149). The relationship of different cultures with the "creative life ground" (*schöpferischen Lebensgrund*) makes it possible to recognize the equal rights of all cultures and to avoid absolutising one unilateral image of man (Plessner 1981, 186).

This non-anthropocentric (and non-eurocentric) vision is reflected also in Plessner's seeing philosophical anthropology as a philosophy of nature or a biophilosophy that in inquiring into the being of man places him in the context of living forms. This philosophy considers the human being as a part of nature; and it sees nature as a component of man. In Plessner's opinion, it is not possible to have an anthropology without an analysis of life and the body. We find, then, in Plessner what is lacking in Foucault, that is, a philosophic analysis based on the transcendental, phenomenological, and dialectical categories of the biological dimension of the human being.

In *Die Stufen des Organischen und der Mensch* (The Levels of the Organic and Man), a work that includes his philosophy of nature, Plessner develops a logic of living forms that tries to identify "organic modals" and laws of the organization of life, considering man, beyond Cartesian dualism, as an organic unit, "psychophysically indifferent or neutral" (Plessner 1975, 121, 70). To overcome Cartesian dualism we do not have to proceed from man but from life. Plessner starts his analysis from the fact that "living things" are given in perception in a different way than are "nonliving things." He shows that it is the mode of relating to the environment (*Umwelt*) that defines the characteristics of living being and that allows one to specify the differences between different forms of life. His fundamental theory is that a living thing is a boundary-setting thing.

Living things distinguish themselves from nonliving ones through their boundary nature (*Grenzcharakter*). While, in inorganic bodies, the body and the border are totally extraneous, that is, the body ends where the surrounding medium begins, in organic bodies the border belongs to the body and, therefore, the body detaches itself from the medium. It does not begin where the medium ends, but it is independent. Beings of this last type are positional beings. The concept of positionality, which defines the mode of the relation with environment, allows one to reach a level that is phenomenologically originary and neutral compared to the ontological distinction between *res extensa* and *res cogitans*. Proceeding on this platform it is possible to reconstruct different levels of living being, from plant to animal to man.

Taking a phenomenological and not a classifying approach, Plessner does not speak of animals and vegetables, but of an open form that is immediately inserted in the surrounding world, and of a closed form that is inserted in a mediated way in the environment of life. He opposes the closed form, typical of animals, to the open form of vegetables. And if, in the ambit of closed positionality, the positionality of animals is centric (where the individual has no relation with its positional centre because it coincides with it, and is not aware of it), the positionality of the human being is excentric—for man lives at the same time as the centre and the periphery of his own positional field; he is able to access the center of his positionality and, therefore, he can take a detached view of himself. As an "I," he is able to assume some distance from the body and the surrounding world. But although he is able to assume distance from his organic and natural dimension, a human being is not able in any way to leave it. He is bound within his body and his psyche, yet he is also without location, beyond all relations to space and time, based on nothing. The excentric form of positionality expresses, on the one hand, the passive, material character that man has in common with animals, and preserves, on the other hand, the creative character of man in opposition to every naturalistic reductionism.

This condition of man manifests itself also in the fact that, according to "the duplicity of aspects" that characterizes the experience of himself and of the world, man, alone among living beings, has two types of relation with his body, defined by the expressions "being a body" and "having a body," where the first term identifies the living dimension of *Leib* and the second one the objective dimension of *Körper*. As an excentric being, man can realize himself only in the modes of "natural artificiality" and of "mediated immediacy." This means that, as an "open essence" (but not as a deficient animal as with Gehlen), humanity needs to complete itself through culture. Man must also discipline and regulate his life in an artificial way. Because of the difference between *Leib* and *Körper*, man needs also to express himself in an immediate way, but because of his excentric positionality, he can do that only in a mediated way. Man is also characterized by an essential doubleness: on the one hand, he must discipline and control himself, that is, he has to give himself a form; and, on the other hand, he is driven to break his previously achieved form through expression, going beyond every constraint that closes his openness.

The Swiss biologist Adolf Portmann has developed the Plessnerian view that defines the organic using the concept of a border. Studying the "appearance" (*Erscheinung*) of living forms in the light, Portmann states that the fundamental characteristics of living beings do not reduce themselves to those that an evolutionistic approach puts in the first place, in particular to those that serve the self-preservation of individuals and of species. More important is the phenomenon, which according to Portmann is present already in the life of plants, of each organism's showing itself to light and, thereby, relating itself to the surrounding environment. The peculiarity of living beings consists, therefore, in the phenomenon of self-presentation (*Selbstdarstellung*), which has a relational, expressive, and aesthetic connotation, and which refers to the importance of the surface of the organism, thus, to its border (Portmann 1960). According to Portmann, living beings are originarily foreordained to visibility and their outer surface is by nature the ambit that constitutes the condition of the sociality of the living being. So, if the Plessnerian analysis of excentric positionality shows that man is more than a being that can be controlled by the cultural dimension that Foucault calls "discours," and that his life has its roots in the natural dimension of body, Portmann's interpretation of self-presentation shows that life is more than genomes and that, already, at the levels of plants and animals it should not be interpreted in a naturalistic way because expression and relation represent essential characteristics of it (see Fischer 2004, 70–71).

**Acknowledgements** I would like to thank my friend Dr. Claudio Muti (Milan) for his help in the correction of my English text.

## Notes

1. Esposito explores the relationship between biopolitics and the paradigm of immunization (Esposito 2008, 45–77).
2. See Verducci, *La fenomenologia della vita di Anna-Teresa Tymieniecka*, (Verducci 2012, 81–82).
3. See Lindemann's "Reflexive Anthropologie…," (Lindemann 2004, 25–26).

## Works Cited

Agamben, Giorgio. 1998. *Homo sacer. Sovereign Power and Bare Life*. Trans. Daniel Heller-Roazen. Redwood City: Stanford University Press. Print. Translation of *Il potere sovrano e la nuda vita*. Torino: Einaudi, 1995. Print.

Bröckling, Ulrich, and Matthias Schöning. 2004. Disziplinen des Lebens? Einleitung. In *Disziplinen des Lebens. Zwischen Anthropologie, Literatur und Politik*, ed. Ulrich Bröckling, Benjamin Bühler, Marcus Hahn, Matthias Schöning, and Manfred Weinberg, 9–19. Tübingen: Gunter Narr. Print.

Elders, Fons, ed. 2011. *Human Nature: Justice vs Power. The Chomsky-Foucault Debate*. London: Souvenir Press. Print.

Esposito, Roberto. 2008. *Bios. Biopolitics and Philosophy*. Trans. Timothy Campbell. Minneapolis: The University of Minnesota Press. Print. Translation of *Bíos. Biopolitica e filosofia*. Torino: Einaudi, 2004. Print.

Fadini, Ubaldo. 2013. *Il futuro incerto. Soggetti e istituzioni nella metamorfosi del contemporaneo*. Verona: Ombre corte. Print.

Fischer, Joachim. 2004. Leben—das 'grenzrealisierende Ding'. Philosophische Anthropologie als Doppelkorrektiv von Genombiologie und Biomachtdiskurs. In *Disziplinen des Lebens. Zwischen Anthropologie, Literatur und Politik*, ed. Ulrich Bröckling, Benjamin Bühler, Marcus Hahn, Matthias Schöning, and Manfred Weinberg, 61–71. Tübingen: Gunter Narr. Print.

Foucault, Michel. 1970. *The Order of Things: An Archaeology of the Human Sciences*. Trans. Alan Sheridan. New York: Random House. Print. Translation of *Les mots et les choses: Une archéologie des sciences humaines*. Paris: Gallimard, 1966. Print.

———. 1984. Nietzsche, Genealogy, History. Trans. Donald N. Bouchard and Sherry Simon. In *The Foucault Reader*, ed. Paul Rabinow, 76–100. New York: Pantheon Books. Print. Translation of "Nietzsche, la généalogie, l'histoire," in S. Bachelard et al., eds. *Hommage à Jean Hyppolite*, 145–172. Paris: Presses Universitaires de France, 1971. Print.

———. 1990. *The History of Sexuality. Volume 1: An Introduction*. Trans. Robert Hurley. New York: Vintage Books. Print. Translation of *Histoire de la sexualité 1. La volonté de savoir*. Paris: Gallimard, 1976. Print.

———. 2003. *"Society Must Be Defended": Lectures at the Collège de France*, 1975–1976. Trans. David Macey. New York: Picador. Print. Translation of *"Il faut défendre la societé." Cours au Collège de France 1975–1976*. Paris: Gallimard-Seuil, 1997. Print.

———. 2008. *Introduction to Kant's Anthropology*. Trans. Roberto Nigro and Kate Briggs. Los Angeles: Semiotext(e). Print. Translation of *Introduction à l'Anthropologie. E. Kant, Anthropologie du point de vue pragmatique*, 11–79. Paris: Vrin, 2008. Print.

Fumagalli, Andrea. 2007. *Bioeconomia e capitalismo cognitivo. Verso un nuovo paradigma di accumulazione*. Roma: Carocci. Print.

Gehlen, Arnold. 1988. *Man: His Nature and Place in the World*. Trans. C. McMillan and K. Pillemer. New York: Columbia University Press. Print. Translation of *Der Mensch. Seine Natur und seine Stellung in der Welt*, ed. K.-S. Rehberg. Arnold-Gehlen-Gesamtausgabe, Textkritische Edition unter Einbeziehung des gesamten Textes der 1. Auflage von 1940, Bd. 3. Frankfurt a. M.: Klostermann, 1950/1993. Print.

Habermas, Jürgen. 2003. *The Future of Human Nature*. Trans. Helga Beister and William Rehg. Cambridge: Polity Press. Print. Trans of *Die Zukunft der menschlichen Natur. Auf dem Weg zu einer liberalen Eugenik?*. Frankfurt a. M.: Suhrkamp, 2001. Print.

Haraway, Donna. 1991. *Simians, Cyborgs and Women. The Reinvention of Nature*. New York, London: Routledge. Print.

Lindemann, Gesa. 2004. Reflexive Anthropologie und die Analyse des Grenzregimes. Zur Aktualität Helmuth Plessners. In *Disziplinen des Lebens. Zwischen Anthropologie, Literatur und Politik*, ed. Ulrich Bröckling, Benjamin Bühler, Marcus Hahn, Matthias Schöning, and Manfred Weinberg, 23–34. Tübingen: Gunter Narr. Print.

Longo, Giuseppe O. 2003. *Il simbionte. Prove di umanità futura*. Roma: Meltemi. Print.

———. 2007. Riduzionismo informazionale e postumano. *Annuario Kainos* 2: 43–55. Print.

Marchesini, Roberto. 2002. *Post-human. Verso nuovi modelli di esistenza*. Torino: Bollati Boringhieri. Print.

Morin, Edgar. 1999. De l'animalité et de l'humanité. Introduction. *Revue européenne de sciences sociales* 115: 5–7. Print.

Plessner, Helmuth. 1975. *Die Stufen des Organischen und der Mensch. Einleitung in die philosophische Anthropologie*. Berlin/New York: de Gruyter. Print.

———. 1981. *Macht und menschliche Natur. Ein Versuch zur Anthropologie der geschichtlichen Weltansicht*, 135–234. Gesammelte Schriften V. Frankfurt a. M.: Suhrkamp. Print.

———. 1983. *Homo absconditus*, Gesammelte Schriften VIII, 353–366. Frankfurt a. M.: Suhrkamp. Print.

Portmann, Adolf. 1960. *Neue Wege der Biologie*. München: Piper. Print.

Sloterdijk, Peter. 2001. "Domestikation des Seins: Die Verdeutlichung der Lichtung." In Peter Sloterdijk, *Nicht gerettet: Versuche nach Heidegger*, 142–234. Frankfurt a. M.: Suhrkamp. Print.

———. 2009. Rules for the Human Zoo: a Response to the *Letter on Humanism*. Trans. Mary Varney Rorty. In *Environment and Planning D: Society and Space* 27: 12–28. Translation of "Regeln für den Menschenpark: Ein Antwortschreiben zu Heideggers Brief über den

Humanismus." In Sloterdijk, Peter. *Nicht gerettet: Versuche nach Heidegger*, 302–337. Frankfurt a. M.: Suhrkamp, 2001. Print.

Stewart, Thomas A. 1998. *Intellectual Capital. The New Wealth of Organisations*. New York: Crown Business. Print.

Teilhard de Chardin, Pierre. 1959. *The Phenomenon of Man*. Trans. Bernard Wall. New York: Harper & Brothers. Print. Trans. of *Le phénomène humain*. Paris: Éditions du Seuil, 1955. Print.

Torjussen, Lars Petter, Johannes Servan, and Simen Andersen Øyen. 2008. An Interview with Anna-Teresa Tymieniecka. *Phenomenological Inquiry* 32: 25–34. Web: 18 April 2014. http://www.phenomenology.org/Bergen-Interview.htm.

Tymieniecka, Anna-Teresa. 1971. *Die Phänomenologische Selbstbesinnung*, Analecta Husserliana I, 1–10. Dordrecht: Reidel. Print.

———. 1986. Tractatus Brevis. First Principles of the Metaphysics of Life Charting the Human Condition. In *The Phenomenology of the Metaphysics of Life Charting the Human Condition II, The Meeting Point between Occidental and Oriental Philosophies*, Analecta Husserliana XXI, ed. Anna-Teresa Tymieniecka, 33–51. Dordrecht: Reidel. Print.

———. 2011. Transcendentalism Overturned. Life's Geo–cosmic Positioning of Beingness. In *Transcendentalism Overturned. From Absolute Power of Consciousness until the the Forces of Cosmic Architectonics*, Analecta Husserliana CVIII, ed. Anna-Teresa Tymieniecka, 3–10. Dordrecht: Springer. Print.

Verducci, Daniela. 2012. *La fenomenologia della vita di Anna-Teresa Tymieniecka. Prove di sistema*. Roma: Aracne. Print.

# Part VIII
# Ecology of the Human Mind and Human Relations

# The Concept of Life in Ludwig Binswanger's Phenomenological Psychopathology

**Anna Piazza**

**Abstract** Ludwig Binswanger's work consists of applying the phenomenological method in psychology, thereby distancing the discipline from the epistemological status of a natural science, which tended to grasp the human being by resolving him into something *objectifiable*. He makes use of Heidegger's concept(s) of *Dasein* and *In-der-Welt-sein*, which allows him to understand the alienated individual as one who found, in "alienation," a specific modality of being in the world, and he takes advantage of Husserl's transcendental analysis, seeing *intentional* consciousness as the essential nucleus of the psychopathological method. In this regard, the starting point for observation is the interior story of the psychophysical individual as totality of *Erleben*. From there, two points have to be stressed: on one side, the noematic ensemble (the essential structure of consciousness), and on the other side, the peculiarity that the concept of *life* obtains here, and always as intentionally-directed life.

Binswanger realizes as well an "inversion" of method, insofar as he notices how the *Erlebnis* and knowledge of it is the original phenomenon: on the basis of this knowledge a "science of life" can be constructed; it in turn cannot be used to explain the phenomenon itself, however. My paper interrogates Binswanger's critique of particular historical conceptions of the human being and life, for example, that of Aristotle, and that of Freud's "*homo natura*." I bring out his idea of the psychic life of man, as uniqueness and historical unity, one "determined by presence," by considering his analysis of Heraclitus' *Weltanschauung*, wherein is achieved an idea of man in his "cosmic aspect," also understood as "worldly transcendence."

**Keywords** Phenomenology · Psychopathology · Consciousness · Life · Love

A. Piazza (✉)
Department of Humanities Universidad Francisco de Vitoria, Madrid, Spain

Instituto Schuman de Estudios Europeos Universidad Francisco de Vitoria, Madrid, Spain
e-mail: anna.piazza@ufv.es



W. S. Smith et al. (eds.), *Eco-Phenomenology: Life, Human Life, Post-Human Life in the Harmony of the Cosmos*, Analecta Husserliana CXXI,
https://doi.org/10.1007/978-3-319-77516-6_35

## Introduction

Ludwig Binswanger's *Daseinsanalyse*, which took off in the first half of the last century, represented a qualitative leap in psychiatry, one meaning the overcoming of reductive biological categories thanks to the application of the phenomenological method in its practice. *Daseinsanalyse* showed, namely, how psychic illness could not be regarded as a "natural event" anymore, but should, instead, be recognised as one of the possibilities of the human condition—that is to say, characterised by the same fundamental structures, like corporeality, temporality, and being-in-the-world with others.

In 1913, Karl Jaspers introduced a new approach to psychopathology by considering mental illness beyond its possible inductive and causal explanation, that is, by considering the patient according to the totality of his biography. Jaspers was the first to realize that psychopathology ought to abandon the explicative ideal (*Erklären*) instituted by natural science, an idea that tends to lead every phenomenon back to a point of view or a rule. He thought that psychopathology should make itself clear about its implicit procedure, which was a reductive one. For Jaspers, being able to explain (*Erklären*) something did not at all mean being able to understand (*Verstehen*) it completely. Jaspers says in his *General Psychopathology,* "The exclusion of philosophy would nevertheless be disastrous for psychiatry: firstly, if we are not clearly conscious of our philosophy we shall mix it up with our scientific thinking quite unawares and bring about a scientific and philosophic confusion" (Jaspers 1972, 769).

As Binswanger later noted, it is necessary to go back up to the philosophical premises that are acting behind psychiatry and psychoanalysis, meaning specifically the conceptual model derived from Descartes, who divided man into soul (*res cogitans*) and body (*res extensa*), a dichotomy that Binswanger calls "the cancer of all psychology and psychiatry up to now."[1] This methodological division was particularly convenient for the goals of science, which can exercise its power precisely in the quantitative and measurable order of *rex extensa*, reducing, thereby, the psychic to an epiphenomenon of the physiological.

As Jaspers "invited" practitioners to a descriptive psychology that could consider the human person as a whole, he became one of the great forerunners of the "psychiatry of existence," which prepared the ground for the host of psychopathologists who, like Binswanger, would receive and mature the insights of Husserl and Heidegger (Callieri and Ales Bello 2002, 680).

If Jaspers specifically introduced into psychology the conception of a human existence which is "life in a world with everything alive" (Jaspers 1972, 759), meaning one not "readable" through the scientific paradigm, phenomenology for its part assumed the role of suggesting that our perception and description of the self is conditioned by our conception of nature (or reality), and that this conception, in turn, also cannot be justified by the scientific method on which it is based (Needleman 1967, xvi).

## Phenomenological Perspectives: Husserl and Heidegger

Binswanger explicitly took advantage of the fertile perspectives offered by Husserl and Heidegger. He pointed out how the phenomenological method allows the understanding of the patient's "world of meanings" and makes it possible to reconstruct the intentional unfolding of the singular existence that is a person, conferring a continuity of sense to its vicissitudes (Callieri and Ales Bello 2002, 680).

He affirms: "It was Edmund Husserl's great achievement to have shown, after Brentano, just what this 'phenomenological' method is, and to have indicated what an enormous vista it opened for research in the various sciences" (1967a, 207). For Binswanger, Husserl's doctrine of intentionality offered a way to understand the unitary relation between transcendental subjectivity and transcendental objectivity, claiming a shift from a 'theoretical' description of the psychic process in a subject to acknowledgement of the structures of "intentional consciousness," as "directedness toward something." Hence, Binswanger saw engaging intentional consciousness as the essential nucleus of psychopathological method and of the "eidetic science of consciousness," in effect, phenomenology, as being the essential foundation for all inquiry into illness.

Now, this had to be done by clarifying the concept of *intuitio* (Ricci Sindoni 2002, 658). Binswanger writes:

> Naturally in our conception the term intuition does not refer—and this has to be underscored—to a sense intuition, a visual one for example, does not concern the immediate contents of exterior or interior perception: in our modal acceptation it is counterpoised to the mediateness of non-intuitive thought, devoid of intuitions and indirect. The technical term used by Husserl, who intends thereby to distinguish it from sense intuition, is *categorical intuition*, or *vision of the essences* or *phenomenological intuition*. (Binswanger 1955)

Binswanger notices how this act goes infinitely beyond natural perception. It builds up *above* that, because it grasps forms of consciousness and not natural events as natural science does. Hence, the consideration of intentional acts firstly provides insight into essential structures and secondly a valuable methodological approach: the starting point of psychopathological observation was no longer the functional modality of the psychophysical individual, but rather the interior story as the fullness or totality of *Erleben*, in Jasper's words "life in a world with everything alive."

I should explain, briefly, how Binswanger makes use of an *objective* phenomenology to study the constitutive structures of the existence of the ill person, differentiating himself thereby from Jasper's *subjective* phenomenology. In this we can see just how he applies the phenomenological method to psychiatry.[2] Binswanger starts from Husserl's affirmation in his *Formal and Transcendental Logic* that, "The real world exists, only on the continually delineated presumption that the experience will go on continually in the same constitutional style" (Husserl 1969, 252). Phenomenological investigation leads to the fact of "the world's transcendence," a transcendence relative to the Ego, which means that "any straightforwardly constituted objectivity (for example: an Object belonging to Nature) points back, according to its essential sort (for example: physical things *in specie*), to a correlative essential form of manifold (actual and possible) intentionality … which is constitutive for that objectivity" (Husserl 1969, 246). This correlation allows an analysis of the structural, constitutive moments of this universe—for example, the temporal synthesis—and from here an analysis of the structural moments of the schizophrenic world too. As Binswanger finds, one can actually understand the "plot" of "transcendental filaments" in no way better than by considering them in their own deficiency, meaning in those "experiments of nature" called psychoses. In schizophrenia, as well as in melancholy and mania, the continuity or consequentiality of experience is compromised, and, with it, the practicality of the vital process. For instance, the "end of the world" feeling of schizophrenia is the most convincing object lesson of this transcendental presumption and is seen thereby to be full "loss of reality."

Furthermore, Binswanger readopts Husserl's designation of the temporal intentional moment of consciousness called, following Augustine, *protentio*, *retentio,* and *praesentatio*. These moments normally integrate and correlate themselves into a unity, allowing human existence to "historicize" and so keep its singularity. The maniac form of existence, instead, is not able to articulate itself within meaningful relationships with others, as well as to behave bearing in mind the past and the possibilities of future. It actualizes itself in a dissociated here and now, without thematic temporal or intersubjective perspectives, so the *praesentatio* sinks into a de-finalised and de-historicised solitude. In melancholy, in contrast, the temporal synthesis is differently defective: *retentio* and *protentio* are not weakened or nullified but *protentio* is always leaking into retentive moments and *retentio* always leaking into protentive ones. For instance, the expressions "If I did" or "If I did not" are *empty possibilities*: the past has no possibilities anymore, and what is free possibility retires into the past, meaning that constitutive protentive acts automatically become *empty intentions* (*Leerintentionen*). In this way, *protentio* fails to actually engage itself with the other moments and loses its theme, having nothing more to produce but an "empty future." Not being able to achieve a proper theme means that it is impossible to achieve a proper *praesentatio* as well.

What Binswanger means to highlight here is how illness is not a mere material, psychological fact, but is rather transcendental. The normal progression of experience actually involves a transcendental presumption and "prevision," while in the characteristic manner of maniac or melancholic experience the real world loses fullness or power. This "checkmating" of experience emerges not just at the level of temporal synthesis but even at the intersubjective level, *in effect* at the level of *appresentation*, meaning the constitution of a common world.

Even if Binswanger makes use of Husserl's 5th Cartesian Meditation to explain psychotic experience, where he analysed the constitution of the "otherness" of human beings, the main categories employed in his research are Heidegger's *Dasein* and *In-der-Welt-sein*. In his essay "Heidegger's Analytic of Existence and Its Meaning for Psychiatry," after commenting Husserl's great phenomenological discovery of intentional consciousness, Binswanger critiques him:

> Nevertheless, this consciousness was still suspended in the air, in the thin air of the transcendental ego. The … 'fundamental' accomplishment of Heidegger consisted not only in stating the problematic nature of the transcendental possibility of intentional acts. What he

> did, in addition, was to solve this problem by showing how the intentionality of consciousness is grounded in the temporality of human existence, in the *Dasein*. Intentionality in general is only possible on the basis of "transcendence" and is thus neither identical with it nor, conversely, does it make transcendence possible. Only by referring intentionality back to the Dasein as transcendence or being-in-the-world and only, therefore, with the inclusion of the transcendental ego in the actual Dasein, was the ... question posed as to the whatness of the beings that we ourselves are. (Binswanger, 1967a 207)

According to Binswanger, Heidegger's significance for psychiatry is not to be related only to the verification of the *ontological structure* of Dasein, as it also questions the actuality, possibility, and limits of the horizon of understanding or world-design of psychiatry in general. This concerns the effort of psychiatry to understand itself as science because, according to him, science is autonomous with regard to what, in its terms, can be *experienced*; but, to a certain extent it has to be referred to philosophy, meaning that the self-understanding of science, considered as the articulation of an actual store of ontological understanding, is possible only on the basis of a philosophical, ontological outlook.

Before deepening the question of science as it works as a particular "construct" of human existence in Freud's psychology, we can approach Heidegger's account in this regard. According to Binswanger, Heidegger presents the opportunity to go behind the conceptual horizons presumed by every scientific frame—that of nature and that of culture—approaching man's basic function of understanding Being as the "establishing of ground": a transcendental function. While Husserl's philosophy also made it possible for Binswanger to clarify the eidetic and original forms of consciousness, with Heidegger, he could come to understand the sense of being of the human as a whole, a question which precisely is not one answerable applying the scientific method alone. As Heidegger affirms, the being of man cannot be ascertained by a "summative enumeration" of body, mind, and soul. "What is needed is to return to (subjective) transcendence, to the Dasein as being-in-the-world, even while constant attention is being accorded its objective transcendence" (Binswanger 1967a, 211).

Hereby, we get closer to the topic of this presentation, because it is not that science does not consider life as a whole, but it views this whole primarily as a biological one, and its every further observation takes place at the level of factual objective "relations." Heidegger's analytic of existence, instead, though dealing with categories that could appear distant from a proper consideration of human nature, gives us an approach to Dasein, thanks to which the problem of nature can be reached. Firstly, Heidegger sees Dasein as situationally attuned existence *among beings* and, secondly, Binswanger notices that the consideration of Dasein as *thrownness*, directs, explicitly or not, our attention to Dasein in its bodiliness. These concepts offer the possibility of understanding man as both a creature of nature and a socially determined or historical being—"and this by means of one ontological insight, which thus obviates the separation of body, mind and soul" (Binswanger 1967a, 212).

Herein emerges a crucial question for the whole of psychopathology. The *thrownness* of Dasein reveals its *facticity*, its being set in existence without having freely decided upon it, meaning its being as a creature ultimately determined, effectively enclosed, possessed, and compelled by beings in general. Thus, the possibilities of being-in-the-world are *withdrawn*, which, on the one hand, can mean experiencing a "powerlessness" while, on the other hand, according to Binswanger, lends to Dasein its power: "for it is this that first brings before the Dasein the 'real', graspable possibilities of world-design" (Binswanger 1967a, 212).

In this sense, transcendence not only means a striding or swinging of Dasein toward the world, but it is also limitation, or better the possibility of projects and their realization in the space of its finitude. And, in the opinion of Binswanger, "only he who scorns these limits—with Kierkegaard's terms—is at odds with the fundamental conditions of existence, can become 'neurotic', whereas only he who 'knows' of the unfreedom of finite human existence and who obtains 'power' over his existence within this very powerlessness is unneurotic or 'free'" (Binswanger 1967a, 218). Hence, the sole task of psychiatry should lie in assisting man toward this "power," or possibility of his existence.

This clarifies further the profit of Heidegger's analytic of existence for psychiatry. Seeing the human being in the first place as an "open project," as life which articulates itself in possibilities, means a drifting away from the difference between being "of sound mind" and "of unsound mind" adopted by the psychiatry that still adhered to natural categories, and our being able to consider mental illness as "*particular declinations of the fundamental or modal structure of being-in-the-world as transcendence*" (Binswanger 1947, 23).

Thus, Dasein itself become the horizon or the context of the "medical" interrogation, in so far as Dasein itself produces progressively the conditions of the person's own world-experience. As Binswanger affirms, "This world-design did not manifest itself before the traumatic event occurred; it did only on the occasion of that event. Just as the a priori or transcendental forms of the human mind make experience only into what experience is, so the form of that world-design had first to produce the condition of the possibility for the ice-skating incident in order for it to be experienced as traumatic" (Binswanger 1958, 205).

To conclude, if Husserl helped Binswanger by focusing on the essential structures of psychopathological *Erlebnisse*, Heidegger did so by regarding the more general "how" of the patient, meaning *the global mode of his being*.

## From Freud's *homo natura* to *homo vita*

Binswanger's critique of Freud involves to some extent a more general critique of science as a system striving after a total explanation of phenomena. As we just saw, Jaspers began criticizing the scientific method in psychopathology by promoting realization of the difference between *Erklären* and *Verstehen*, one that particularly aimed at pointing out the method's functionalistic procedure: the tendency to be satisfied with theories because they "work." Freud himself notes this aspect at the beginning of his piece "Instincts and.

Their Vicissitudes": "The true beginning of scientific activity consists rather in describing phenomena and then in proceeding to group, classify and correlate them. Even at the stage of description, it is not possible to avoid applying certain abstract ideas derived from various sources and certainly not the fruit of new experiences only" (Freud 1915, 60). This affirmation shows how even the philosophical work accomplished by a master of the "school of suspicion," which should have indeed "unmasked" the sublimation and rationalization of instincts operated by moral instance and society, actually pursued a kind of "scientific ideal," meaning a rationalization itself. Hence, the distinction between *Erklären* and *Verstehen* can be equated with the distinction between phenomenology and philosophical system: each philosophical system, in fact, even Freud's, comes with its own pre-established criterion by which something is recognized as an entity, as something by reference to which all other entities encompassed can be "explained" (Needleman 1967, 33). Phenomenology instead attempts to be presuppositionless, being accountable for the givenness and the "how" of phenomena.

According to Binswanger, in the case of Freud, the methodological organizing principle of which he makes use to explain the totality of human dynamics is that of *homo natura*. This could be understood in the sense of the primitive natural man or the newly born infant, in any case, as a concept required by reflection. As Binswanger notes, indeed, what we have in it is not an actual man but an idea, a necessary requirement of scientific, biological reflection and reduction. "Both biological ideas treat man—with regard to his genuine historicity, his capacity for ethics, culture, religion, art—as *tabula rasa*" (Binswanger 1967b, 154). But, with Locke as well as with Freud, the idea of the mind as a *tabula rasa* comes up when scientific knowledge confronts certain limits. "Seen from the vantage point of the totality of human experience, the *tabula rasa* is thus a symbol of a particular negation, the expression of a dialectic boundary line" (Binswanger 1967b, 155). But, if this symbol is made into a reality and treated as the real beginning of human history, Binswanger observes, then a complete reversal of the historical connection of nature, history, and myth is worked. Natural science then inserts the product of its constructions—the idea of *homo natura*—at the beginning, converting natural development into history and, then, taking this nature and history as the basis upon which myth and religion are to be explained.

In the end, Freud succeeded in demonstrating a "mechanism at work," thereby creating the possibility of mechanically "repairing" it, with the psychoanalytic technique of unmasking and annulling repression and regression by means of the transference mechanism. But at the base of this procedure, again, there is a precise concept. The entire mechanism of the psychic apparatus of *homo natura* is set in motion by the *wish*, here equalized with instinct because, in this term, wish strives for pleasure as such. But this vision, says Binswanger, contradicts the experience of man in actuality, who strives for possession or *Erleben* of a particular thing that bring him pleasure.

At this point, Binswanger develops an "anthropological critique" of the theory of sublimation, according to which essential potentialities of human existences, like religion, ethics, art, are explained as reflections of childhood anxiety, a father complex, or external compulsion and introjection; more generally, higher forms of human experiences are explained by the pleasure principle. But in Binswanger words, "no criterion for judging a form as lower or higher can be mined out of instinct or the pleasure principle as such. For pleasure is just as much of an abstraction as force or power" (Binswanger 1967b, 175). The matter that Binswanger is concerned to underline is the relation of pleasure to human existence. His project at this point is to make the objective principle and mechanism of pleasure "anthropologically retroactive," that is, to retrieve into Dasein what phenomenologically is fundamental to it. In this way, he detects how Freud elevated the pleasure principle, *one* and *only* one particular mode of human existence or being in the world. Hence, here is reached the central point of Binswanger's critique of Freud: in his "natural psychology" what is being scientifically studied is, namely, never the whole man, but an object. But, "everywhere we find something that overflows and bursts the bounds of such psychology" (Binswanger 1967b, 169). What, indeed, works constantly and silently behind the constructions of this psychology, presupposed as self-evident and just as self-evidently being bracketed out, is *existence as ours*, the presence as presence. As Binswanger affirms, "when this self is objectified, isolated, and theorized into a ego, or into an Id, Ego and Super-ego, it is thereby driven out of its authentic sphere of being, namely existence, and ontologically and anthropologically suffocated" (Binswanger 1967b, 171).

Binswanger here recalls Heraclitus and his conceiving of the human being in his "cosmological aspect," which conception, instead of passing by the problem of the self, *seeks* the person's own self, offering an anthropology that finds its basis in cosmology and theology. He notes, firstly, that what generally constitutes the peculiarity of the Greeks is the fact of their "becoming conscious of the essential and universal rules of man," meaning a constant effort to reach a normative elaboration of man (Binswanger 2007 99). In particular, Binswanger shows with Jaeger how Heraclitus' man becomes a part of the cosmos and is as such subordinate to the rules of the whole, while, at the same time, carrying consciously that rule thanks to his own spirit, being, and so is able to participate in divine wisdom.

Therefore, what Heraclitus did, according to Binswanger, was seek presence at the point closest to philosophical awareness, meaning in one's *own presence*. Although this is actually the next thing to us, as Heraclitus perceived, we live separately from it in a multiform dispersion. The peculiarity of Heraclitus was establishing the purpose of existence in this quest, which, at the same time, means a participation in a transcendental world of wisdom and divine truth, what nevertheless has to fulfill in a historical process. The presence itself remains historical.

Binswanger quotes some fragments of Heraclitus, wherein he says that "all people have a claim to self-knowledge (literally, self-ascertainment) and sound thinking" even if "many people do not understand the sorts of thing they encounter!" (fr. 116, 17). Here, he underlines how Heraclitus introduced a precise concept to describe how man can cut his own path, that of *phronesis*: "Sound thinking (*sophronein*) [is] a very great virtue, and [practical] wisdom (*sophia*) [consists in our] saying what is true and acting in accordance with [the] real constitution [of things], [by] paying heed [to it]" (fr. 112). According to Binswanger, despite the equalizing here of *phronesis* and *sophia*, sound thinking is not to understand through "logical thought" but through something more complex having the sense of "right thought" or of man's "being present to himself." Man is in the "situation" of the *phronesis* just as he can recollect himself from the dispersion of existence, being able to act and decide for himself. This would be, for Binswanger, the meaning of Heidegger's ontological proposition: "Dasein exists. Furthermore, Dasein is an entity which in each case I myself am. Mineness belongs to any existent Dasein, and belongs to it as the condition which makes authenticity or inauthenticity possible."

Heraclitus' account was, therefore, showing how self-knowledge and the becoming himself of man coincides with the accomplishment of universal knowledge: "for those who are awake there is a single, common universe, whereas in sleep each person turns away into [his] own, private [universe]" (fr. 89). So, the meaning of the fragments 115 "Soul possesses a *logos* (measure, proportion) which increases itself," and 45, "One would never discover the limits of soul, should one traverse every road—so deep a measure does it possess," indicates "the norm of cosmological happening," which is *logos* itself.

Thus, first of all, Heraclitus, despite the modern differentiation between eternal and factual truth or between rationalism and empiricism, advocates one knowledge and one life wherein the subjective and the cosmological aspect share the same "fight," so that the individual vocation is to be fulfilled in the "common universe" and not in a private one. Secondly, aware of the dissonances of life, Heraclitus does not resolve them in a system. Other fragments of his say "everything flows" and "the path up and down are one and the same." The cosmological life is stoked thanks to the contraries that, through the fight, diverge and then transmute one into the another. This fire, this cosmological happening, is the objective and material expression of the norm and principle of any event, as well as an expression of *logos* and divine rule.

These observations allow us better to understand Binswanger's critique of Freud and the difference between science, or a philosophical system, and phenomenology, or love. Heraclitus' *logos,* as we saw, reveals and affirms in contrasts just as love does, remarks Binswanger. He quotes: "As are the discord of lovers, so are the discords of world. The conciliation is through the fight, and everything which is separate meets again." In this statement is also met again the "overcoming" of Husserl through Heidegger, for an analysis of the rigid structures of consciousness does not suffice for psychopathology anymore, analysis wherein the counselor's relationship with an "alive" patient is forgotten. If the psychiatrist is oriented to understanding human beings in their totality and in the *koinonia* of this totality with more ontological potentialities, then he reaches beyond the purely "theoretical" and factual knowledge of man and is directed toward transcendence itself as being-in-the-world and being-beyond-the-world. He is involved in a relation rooted equally in "care" and love.

## Notes

1. This cultural model for Binswanger has to be found even in Dilthey's differentiation between the humanities (*Geisteswissenschaften*) and natural science (*Naturwissenschaften*).
2. This analyses effectively the interior states of patients.

## Works Cited

Binswanger, Ludwig. 1947. *Über die daseinsanalytische Forschungsrichtung in Psychiatrie. Ausgewählte Vorträge und Aufsätze, I: Zur phänomenologischen Anthropologie*. Bern: Franke Verlag.

———. 1955. *Über Phänomenologie. Ausgewählte Vorträge und Aufsätze, II: Zur Problematik der psychiatrischen Forschung und Problem der Psychiatrie*. Bern: Francke.

———. 1958. The Existential Analysis School of Thought. In *Existence: A New Dimension in Psychiatry and Psychology*, ed. Rollo May, Ernest Angel, and Henri F. Ellenberger. New York: Simon & Schuster.

———. 1967a. Heidegger's Analytic of Existence and its Meaning for Psychiatry. *Being-in-the-world, Selected papers*. New York: Harper Torchbooks.

———. 1967b. Freud's Conception of Man in the Light of Anthropology. *Being-in-the-world, Selected papers*. New York: Harper Torchbooks.

———. 2007. La concezione eraclitea dell'uomo. *Per un'antropologia fenomenologica. Saggi e conferenze psichiatriche*. Trans. Enrico Filippini, ed. Ferruccio Giacanelli. Milano: Feltrinelli.

Callieri, Bruno, and Angela Ales Bello. 2002. The phenomenologico-existential approach to Psychopathology. In *Phenomenology World-Wide, Foundation—Expanding Dynamics—Life—Engagements*, ed. Anna-Teresa Tymieniecka. Dordrecht: Kluwer Academic Publisher.

Freud, Sigmund. 1915. Instincts and Their Vicissitudes. *Collected Papers*, Vol. IV. Trans. Cecil M. Baines. London: Hogarth Press.

Husserl, Edmund. 1969. *Formal and Transcendental Logic*. Trans. Dorion Cairns. The Hague: Martinus Nijhoff.

Jaspers, Karl. 1972. *General Psychopathology*. Manchester: University Press.

Needleman, Jacob. 1967. Preface. In *Being-in-the-world, Selected Papers*, ed. L. Binswanger. New York: Harper Torchbooks.

Ricci Sindoni, Paola. 2002. Ludwig Binswanger, the Inspiring Force. In *Phenomenology World-Wide: Foundation—Expanding Dynamics—Life-Engagements*, ed. Anna-Teresa Tymieniecka. Analecta Husserliana 80. Dordrecht: Kluwer Academic Publishers. 657–664.

# Meaningless Life: The Role of Clinical Phenomenology in Understanding the "Being in the World" of Psychiatric Patients

**Giulio Lo Bello**

**Abstract** The paper focuses on the use of phenomenological psychiatry in researching the life-world of psychiatric patients. Although the subjective experiences of patients can be fundamental for understanding the inner life crisis and revealing a meaning that could make these experiences understandable, they are often undervalued in clinical psychiatry for their not being considered useful for diagnosis.

The personal attitude of people affected by schizophrenia is often considered as simply a manifestation of psychosis, the epiphenomena of the mental illness. However, as we can see in the work of Minkowski and Binswanger, the behaviour of patients is the result of an inner struggle with the external world. Patients try to make coherent constructions compatible with their criteria of sense and meaning. These attempts often lead to spurning actions based on "common sense," to isolation from society, and to avoiding the influence of external reality.

Phenomenological method applied to psychiatry allows concentrating attention on subjectivity by investigating the role of values and emotions in the mental illnesses. The explanations made by patients could be seen as a peculiar way of defining oneself, one's own "being in the world." Inquiry into patients' narratives could open valuable views on the concepts surrounding the ideas of Being and life.

**Keywords** Phenomenology · Meaningless life · Philosophy · Psychiatry · Schizophrenia · Mental disturbance

## Introduction

The relationship between philosophy and psychiatry has undergone many changes through time, and the last decade has represented an extremely rich evolution of this connection. Philosophy has always been strongly connected with medicine, and the

G. Lo Bello (✉)
Faculty of History and Philosophy, University of Latvia, Riga, Latvia

 
W. S. Smith et al. (eds.), *Eco-Phenomenology: Life, Human Life, Post-Human Life in the Harmony of the Cosmos*, Analecta Husserliana CXXI,
https://doi.org/10.1007/978-3-319-77516-6_36

most visible results were obtained in ethical fields by transforming the relations between doctors and patients, by defending human and civil rights, and by forcing medicine to rethink its attitude towards the subjects of its care (Fulford et al. 2007).

Nowadays, psychiatry is on the verge of a crisis, the importance of which can be compared to the one that occurred during the 1970s when the antipsychiatry movement questioned the existence of madness and the value of psychiatry, claimed to be the accomplice of the abuses perpetrated in many medical institutions (Szasz 1974).

The current improvements realized in the neurosciences are opening again the debate over mind-brain dualism, and scientific discoveries stand begging for a theoretical framework that could give them valuable interpretation. The worldwide recognized DSM (Diagnostic and Statistical Manual of Mental Disorders insert note) has been criticized by several of its authors for the cramping consequences of its application, especially the reduction of the fundamental importance of doctor-patient relationship to a sterile check for symptoms (Frances 1994).

The complexity of this field of research is created by the ways in which psychiatric disorders affect a person's experiences and sense of self. Mental illness affects the patient's personal beliefs, feelings, and behaviours and social relations both inside and outside the family. It may interfere with the ability to create logical connections, with thoughts that are apparently not meaningful impairing our acting role as moral agents.

## The Role of Phenomenology in Psychiatry

Phenomenology and hermeneutics have had an important role in the development of alternative or complementary interpretations of mental illnesses, operating together with other fields like medical anthropology, sociology, and psychoanalysis. Those disciplines reshaped the practice of mental assistance and ethical codes, and in the case of Italy reshaped mental health legislation as well (Basaglia 1973). The involvement of experts from other fields of study became mandatory for the development of effective treatments.

The opinions of psychiatrists are still divided on the label of schizophrenia, ranging from denying even its existence, to considering it as a mere comprehensive category, to attempting to prove its biological origin and brain location. Nonetheless, an axiological dimension of schizophrenia is a component of human suffering that medical institutions have often disregarded.

The actual situation of impasse is the result of many factors; the most influential are the partial failure of biological reductionism and the criticism made by the biopsychosocial method. The first lacks in its not providing us with a physical origin and a location in the brain's cellular system for mental illnesses; and the second dispenses too generic formulation, one unable to integrate in a useful way the different perspectives coming from sociology and psychology (Frances 1994; Ghaemi 2011). These positions have been the leading ones for almost 30 years, but currently an increasing number of professionals demand a change of paradigm.

Similar needs for renewal in mental health care led Karl Jaspers (Jaspers 1997, orig. 1913) to search for new models as he reconsidered mental disturbance. Jaspers' interest in phenomenology originated in the field of psychiatry at the beginning of the twentieth century, when, as described in the previous paragraph, the influence of positivist ideas had led us to consider the brain to be like any other bodily organ. Therefore, the brain was thought to be susceptible to scientific investigation and to be without any psychological aspect.

Jaspers attempted to elaborate a different way of approaching mental illnesses and their expressions. The phenomenological method aimed to move forward from the scientific perspective while retaining a meticulous conception of investigation. The important heritage Jaspers left includes this change of focus in the subject of research: from outer signs to the inner life of patients, their subjective experiences. Introspective mental states become the aspect to be described without relying on theories built up remote from actual observation and so through participation in the other person's state of mind.

The analysis proposed by Jaspers is qualified as *subjective*; and psychiatrists perceive a limit to knowledge of the patient's inner state. The term subjective refers to the possibility of confronting the illness only through apprehending the personal experiences made manifest in a particular case, and a concomitant realization that this does not deliver a general understanding of schizophrenia and its symptoms.

The use of the term phenomenology has seen important differences regarding its meaning and application. Therefore, the description of its role in psychiatry should begin with a brief clarification of its main significances. In fact, contemporary psychiatry's employment of the term phenomenology reflects its original Greek meaning of "that which can be seen," namely, the visible appearances of a world we can investigate. This usage recalls the origins of the psychiatric discipline in which accurate clinical observation shaped its evolution and attitude as a medical science. The philosophy of the Enlightenment played an important role in the adoption of the assumption that the causes and the cures of the mental disturbances could be discovered through rigorous investigations. The positivistic foundation of psychiatry is greatly represented by Philippe Pinel, Chiarugi, and Tuke, whose clinical observations and experiences yielded the first classifications of mental disorders. Michel Foucault's further analysis brought to light contradictory aspects of a thus-far praised moment in the history of psychiatry (Foucault 1965). Even if the chains of Pinel's patients had been loosed, the asylums and the conditions of internment still made clear how medical interest in madness prevailed over the ends of care and the alleviation of suffering.

This paper investigates the concept of phenomenology in relation to psychiatry, as phenomenology is explicitly defined in the studies of Edmund Husserl (Husserl 1936) and Martin Heidegger (Heidegger 1962, orig. 1927). The foundation of knowledge in Husserl's phenomenology has to be reached with critical thinking through absolute evidence. Several elements of his thoughts are discussed in the course of the paper underlining their relevance to the reviewed theories. Owing to this theoretical requirement, nothing can be considered unquestionable. The scientific method's assuming that the world exists autonomously and is open to investigation only creates a naive idea of knowledge. The discoveries provided by the sciences were considered to be the mere accumulation of data. But where the reality is grounded on a subjective perception, and based on this assumption, can change through time, that invalidates the theories involving substantial forms able to remain uninfluenced for a long time.

Here consciousness is considered to be the real subject of knowledge, and in order to focus on this actuality, Husserl's phenomenology involves an *epoché*, the suspension of judgement about common ideas of the world, the bracketing of already-existing beliefs about the world. Individual consciousness is taken as the starting point together with studies of the structures and ways of its existence. Apprehending individual consciousness is key to moving forward in developing a science focusing on human beings, their role in the world, and their role for others, namely, describing in which way Being can actually exist.

Husserl's ideas were revived in Heidegger's works, in which we can acknowledge a different methodology which is more appropriate to the issues of psychiatry, thus advancing a different approach in both clinical practice and theoretical framework. Heidegger's contribution is immensely valuable because of the temporal and historical dimensions he has added to the ontological description of the human being. The concept of Dasein and its character of being a project in the world (with historical and bodily dimensions) brought out the deep compatibility between this philosophy and psychiatry. Heidegger's contribution to phenomenology is so valuable because of the historical dimension he added to ontological investigation. The human being is not only a simple presence in the world; the concept of Dasein transforms one into a being existing by projecting oneself in the world with a historical dimension and an active role vis-à-vis the surrounding elements. Dasein has been integrated into the complex of psychiatric analysis, revealing unexpected chances in formulating new interpretation of madness. The new point of view generated by this synergy shows the deep compatibility between the existential conception of the world and the psychiatric vision of how minds act in reality. The profile of Dasein was applied in order to discern possible meanings inside the erratic behaviours of patients.

Nonetheless, when a phenomenological approach was first adopted in the analysis of schizophrenia, particularly by psychiatrists like Binswanger (Binswanger 1953) and Minkowski (Minkowski 1970), several changes in practice and theory occurred and new interpretations arose. We must define the degree of influence of the phenomenological method in the search for understanding madness.

The attempt to overcome Jaspers' limits on how far comprehension can penetrate subjective experience was primarily made by Ludwig Binswanger. His theory postulated a common horizon of interpretation wherein mentally sane and insane people share their individual ways of being in the world, having different dynamics but still falling within the boundary of sense. The characteristic of this *objective* phenomenology is a firm belief in being able to investigate the structure of this altered intentionality and bring understanding into delusional worlds.

The further studies of Eugene Minkowski and Wolfgang Blankenburg (Blankenburg 1980) represent two significant examples of how phenomenology can be applied in schizophrenic interpretation, even though this particular approach clashed with with Jaspers' convictions. In particular, Minkowski argued that schizophrenia could be seen as a breach in the connection between the self and the world. The loss of vital impulse is the cause leading to the above-mentioned separation, and is not a deficit of higher cognitive abilities. The *epoché* emerges as a barrier that bars the conscious self from the common perception of reality. The application of phenomenology brought relevant meanings to the concept of schizophrenia as a disturbance of the self, understood as *Husserlian* intentional consciousness. Minkowski sets forth a new vision of madness characterized by several positive aspects in which holistic and dynamic views *humanize* the mentally disturbed. The separation between the brain and mind lacks a unifying explanation by which to integrate the observations made and data obtained. According to this objection, mental disturbances can be described as particular conditions of Being. A complete understanding of the human being passes through a hermeneutical moment that allows medicine and psychiatry to see the patient as a unity of body and mind in the light of an ontological anthropology. Moving from biological considerations to phenomenological ones entail attitudes we should not undervalue: the evaluation of distorted or impaired human actions as a peculiar form of being in the world can actually facilitate the reduction of social stigma.

## System of Diagnosis and the Phenomenological Method

Currently, the private life of patients has been used only to prove that paranoids has paranoia and a schizophrenics are unable to make logical nexus in their tales. These aspects do not count as valuable signs or symptoms obtained through physical or behavioural observation. The standard clinical diagnosis is not interested in what a patient is describing or how those descriptions are made. Chadwick considers that "one of the terrible problems of pre-emptively biomedical approaches in schizophrenia research is that patients feel that clinicians are not really listening to their experiences, while professionals have a conceptual schematic architecture within that biases them to see their patients' difficulties in organic, objective, materialistic but not heartfelt, subjective, experiential terms" (Chadwick 2007).

The term meaningless used in the title of this research paper refers to the peculiar characteristics of schizophrenia. In a common sense terms, they are described as a loss of reason, but more specifically as symptoms allowing us to recognize a disturbance primarily dealing with the problems of speech production and the perception of the reality. Those aspects of schizophrenia constitute the limit point for understanding the patients, but this assumption could mislead one to consider the delusional production to belong to the commonly shared sense of the world on which we unconsciously rely in everyday life. In this particular condition we can recall Heidegger's notion of ontological difference and consider the situation in which the essence of the structure is confused with its content. The narrations of patients often contain substantial alteration of space, time and physical structures; focusing only on verbal expressions would not provide meaningful results. Therefore, we should consider those expressions to be produced in a different context in which the common perception of the being in the world is altered. The consequences of total disinterest in the outer world and consideration of its existence belong to a different ontological level. Sass (Sass 1992) claims that "instead of issues existing within the world, the delusions in question primarily involve the overall ontological status of the world itself."

## Personal Values in Phenomenological Analysis

Personal values represent the guidelines of human action within society. Specifically, behaving according to shared concepts of the positive and the negative rules our relations with others. The aspect of this brought out in this research paper refers to the importance of ethical values in psychiatry from the patients' points of view, by studying the mutual influence and relations between moral values and mental disturbance. The main focus of the study follows from the experiences of schizophrenic patients. Research shows that schizophrenia affects the sphere of common sense as well as the way a subject perceives and responds to external stimulation. The degree of importance that philosophy has in the field of psychiatry was indicated by the actual *state of art* in medicine and by the potential that phenomenology could lead to the sharpening of alternative interpretations of mental disturbances. The ethical issues of psychiatric patients belong to the very same aspects of subjective life that are not considered in the biological paradigm. Owing to this situation, research in the ethical field is often disregarded.

Nevertheless, the first consequence of putting to the side the inner life of patients is a narrowing of the medical vision, which is then mainly concentrated on impersonal abstractions. This is useful for generating new explanations and theories or revising previous research, but it does not focus on lived life. As Mullen underlines, "the enquiry is about the presence or absence of the indicators of disorder not about an exploration of the patient's life and state of mind" (Mullen 2007).

Delusional behaviours within society are labeled as unreasonable and crazy, and still today such an image constitutes a negative stereotype. Life and relations with other selves represent a major aspect of our existence. A sense of presence is achieved and maintained through encountering the presence of the people around one. Individuality and self-definition are established and defined by being with others and being for others. A crisis in subjectivity ruins the ability to maintain the position we have established in the social fabric; such situations are often experienced by schizophrenic patients and, therefore, several areas of investigation are introduced with an aim to prove the importance of and need for better understanding of those domains in order to get a brighter picture of the meanings accompanying psychotic and delusional actions.

Values do not simply disappear, leaving a patient unable to understand what can be the outcome of an action. The transformation of the evaluating system suggests different scenarios. Interpretations differ as to what status should be accorded the *eccentric* behaviours assumed by the schizophrenic: whether to consider them as choices made to go against the common standards holding in the social environment or to maintain, as in Binswanger's point of view, that the *eccentricity* belongs to the ontological level of schizophrenics' selves, that it is their particular way of being in the world.

## Conclusion

This paper, through a historical and methodological path, has shown the rise and affirmation of a strong cooperation between medicine and philosophy. Any further research of this topic should also include a section on the value of hermeneutics when applied as a complementary discipline in the process of interpreting mental psychosis, and so reinforce apprehension of the importance of philosophical conceptualization. During the seminar at Zöllikon, Heidegger expressed the wish that philosophy break out of philosophical circles to become a useful tool for humanity, especially for those in need of help.

This important lesson about the common mission of philosophy and medicine, as sciences concerned with the human being and overall *for* the human being, stands as a reference point indicating the direction that research should take, provoking evaluation of the direction actually taken, and constantly reminding one to question the results obtained.

Our philosophizing cannot be concerned only with the development of a theoretical structure justifying psychiatric theories about the mind and madness. Philosophy is not simply about providing a framework within which to puzzle together all the items provided by applied science, but is also bringing new interpretative keys to both theories and practices. Diagnosticians and therapists receive a most valued assist in their cooperation with philosophers through the latter's alertness to the importance of the human side of the patients' experiences, of which they are always reminding, providing a necessary balancing corrective to the theoretical abstractions of psychiatry.

The research on mental illness presents us with a complex reality where different and sometimes even contrasting positions face each other. The insights of phenomenology are critical for overcoming this standing impasse, for rethinking psychiatric concepts so as to avoid a dogmatic involution, and for maintaining a balanced viewpoint among the different perspectives. At the same time, phenomenology is an instrument for making medical professionals aware of the real necessity of patient-centered treatment, by switching their focus from theories to the subjects of their treatments and by reinforcing protecting the civil and human rights of patients. These truly best practices prompt us to question the methodologies we apply and the results we obtain, thus leading to critical evaluation that makes psychiatry more conscious of its strengths and weaknesses.

## Works Cited

Basaglia, Franco. 1973. *Che cos'è la psichiatria?* Torino: Einaudi.

Binswanger, Ludwig. 1953. *Grundformen und Erkenntnis menschlichen Daseins* (The Foundations and Cognition of Human Existence). Zurich: Niehans.

Blankenburg, Wolfgang. 1980. Phenomenology and Psychopathology. *Journal of Phenomenological Psychology* 11 (2): 50–78.

Chadwick, Peter K. 2007. Peer-Professional First-Person Account: Schizophrenia from the Inside. Phenomenology and the Integration of Causes and Meanings. *Schizophrenia Bulletin* 33 (1): 166–173. London: Routledge.

Foucault, Michel. 1965. *Madness and Civilization*. London: Tavistock Ed.

Frances, Allen. 1994. Preface. In *Philosophical Perspectives on Psychiatric Diagnostic Classification*, ed. John Z. Sadler, Osborne P. Wiggins, and Michael A. Schwartz. Baltimore: Johns Hopkins University Press.

Fulford, K.W.M., T. Thornton, and G. Graham. 2007. *Oxford Textbook of Philosophy and Psychiatry*. Oxford: Oxford University Press.

Ghaemi, Nassir. 2011. The Biopsychosocial Model in Psychiatry: A critique. *Existenz* 6 (1.): (Spring 2011): 1–8.

Heidegger, Martin. 1962. *Being and Time* (1927), Trans. John Macquarrie and Edward Robinson. London: SCM Press.

Husserl, Edmund. 1936. *The Crisis of European Sciences and Transcendental Phenomenology*. Trans. D. Carr. Evanston: Northwestern University Press.

Jaspers, Karl. 1997. *General Psychopathology* (1913) vols. 1 & 2. Trans. J. Hoenig and Marian W. Hamilton. Baltimore: Johns Hopkins University Press.

Minkowski, Eugène. 1970. *Lived Time*. Evanston: Northwestern University Press.

Mullen, Paul E. 2007. A Modest Proposal for Another Phenomenological Approach to Psychopathology. *Schizophrenia Bulletin* 33 (1): 113–121. Oxford: Oxford University Press.

Sass, L.A. 1992. Heidegger, Schizophrenia and the Ontological Difference. *Philosophical Psychology* 5 (2): 109–132.

Szasz, Thomas. 1974. *The Myth of Mental Illness*. New York: Harper & Row.

# An Ecological Perspective on the Helping Relationship

**Antonio De Luca**

**Abstract** This essay reflects on the helping relationship starting from the contributions of phenomenology [Edmund Husserl, Edith Stein, Max Scheler, Anna-Teresa Tymieniecka] to the problem of intersubjectivity and from phenomenological psychiatry [Karl Jaspers, Eugène Minkowski, Ludwig Binswanger, Bruno Callieri] on the clinical and therapeutic aspects of a therapeutic relationship. Clinical data will be examined in the light of the contributions of religion as well to clinical interpretation.

**Keywords** Relationship · Suffering · Intersubjectivity · Therapeutic relationship · Ecology of relations

## Suffering and Encounter

The Italian poetess Alda Merini wrote, "The soul does not feel pain. The only pain that could match the soul is its exile, that of unintentional failure."[1] There are many forms of exile: it happens when you fail to meet each other and, above all, the Other. The Homeland for Maria Zambrano is not a Place, but a Time, the time of my story, the story that I can live when meeting with each other and the Other. According to Buber, it is from a relationship that we obtain the meaning of things, even in tragic situations, whereas for Romano Guardini one must be able to live an inner freedom in order to meet the Other, along with the means, the independence, and the freedom to withdraw. It is on the "between-us" that we base our understanding of the world, of ourselves, and of others.

The experience of suffering, even if it be a cipher of our love, can generate appropriate responses or can provoke mild or severe misunderstandings, even such as may also lead to hurting oneself, even going so far as to kill oneself or others.

A. De Luca (✉)
University of Calabria, Rende, Italy
e-mail: adelucapsy@gmail.com

 
W. S. Smith et al. (eds.), *Eco-Phenomenology: Life, Human Life, Post-Human Life in the Harmony of the Cosmos*, Analecta Husserliana CXXI,
https://doi.org/10.1007/978-3-319-77516-6_37

Not even a possible and imminent death, nor the loss of a loved one, nor serious violence or illness can *directly* cause the kind of overwhelming suffering that defines a serious pathological reaction. It is not an event that *directly* determines a particular type of lived-experience (*Erlebnis*). Husserl has explained this very well: isolation cannot be the basis on which to build or rebuild the meaning of things. Suffering arising in loneliness and the impact of learning of things can create a psychological disease. This is what happened to Oedipus when he discovered the truth just before being blinded: he saw and recognized the truth, but as he could not bear the extreme loneliness, he could not create conditions for restoring the meaning in life at that moment. His only escape was blindness.

If our lived experience of isolation can blind us, this is therefore an authentic, fundamentally spiritual, relationship – Edith Stein's point of view – one attuned to the flow of life itself – Tymieniecka's understanding – on which one can build the meaning of things and thus the appropriate manner for addressing suffering.

It is not the event but the failed encounter with the other that, in our lived experiences, grants us the capacity to feel anxiety, abandonment, disparity, and isolation from others and the Other. The relationship may become more important than the event itself, even though the event may be tragic. A person faced with a dramatic situation could kill another or kill oneself, or slowly die or, propitiously could deal with what has happened because a person has been authentically met. It is in the relationship with the other that different paths open. In particular, in a helping relationship the journey will be infused with hope. In any case, evil sets in we do not consider the other as person, and it also resides in the continuation of a manner of relating through resentment, anger, and frustration. This could lead to new hatred, discomfort in a relationship with the other, and so on.

We are a *poetic encounter*. What is the structure of the encounter? How could serious suffering be supported? According to Binswanger, the psychotherapeutic encounter happens because such meeting is given to humans as a possibility and it is not forced. This spiritual encounter between persons can help them endure in any lived experience and cope with any situation, even tragedy. The reliability, the testimony, the consistency, the humility, the availability towards others, the lived experience of freedom and the responsibility towards the others are the structural elements of a spiritual encounter. In every human being the need emerges to find themselves and others, to heal the fractures in the stories of each. This relationship could muddy things for them and obscure what they can do, because it can create serious misunderstandings over the direction to be taken[2] or the path that would allows one to proceed with hope, despite everything.

In a helping relationship what we have to understand there is no need to justify one's asking for help and understand as well that helping is not sharing opinions nor taking the place of the other who asks for help in their choices. The helping relationship arises in already complex situations of great suffering and becomes even more delicate, decisive, and important in situations that seem to be hell and are most inaccessible places of lived experience. If poetic and spiritual encounter will make possible tolerance of these situations, wider awareness, taking responsibility for life, both the lived experience of hope and planning is therefore necessary to create such encounters during this relationship, particularly in the helping relationship in extreme situations.

Insomnia: one should remain sleepless too, assuming Levinas' point of view, able to admire and caress the Face of the other and the Other. This can happen also when one appears injured, swollen, or even dirty. It is necessary to let each revive differently.

Helping one who attacks another is not easy. It is sometimes necessary to go to hell and look for profound humanity, which might still remain. Orpheus *singing praises to hell*, according to Rilke, implies warning of all the anguish to be met, all the abasement of what can be achieved in a human being. A person's freedom is a place where even God does not enter, the threshold at which God has stopped. Freedom itself highlights the problem and prompts how it might be used, according to Jaspers, in relation to the person's story and vision of the world.

The helping relationship arises therefore as a relationship that needs to become *ecologically* pure, spiritually clear, authentic, true and radical, keeping in harmony both the understanding of nature and being itself, whether or not there will come from it a new proposal even after the devastation of an incident or tragedy. After the destruction, only the ruins of existence[3] can overcome and survive the rubble of loss and anguish.

The experience of suffering becomes meaningless when you live in Exile, when a human being does not encounter the other or meet others physically, not encountering anyone authentically in the spiritual sphere but living in absence.[4] This can be the starting point for lived experiences of destruction for oneself and for others. This absence could become pure evil, in which, however, some may find themselves while believing that Exile is their existence.

Exile is better than nothing.

## Exile: The Story of Charles and the Absence of Spirituality in Others

If a person becomes an *object* and not a *subject*, it can become useless as a source of physical pleasure and he comes to live with himself and the other as a thing in a psychopathological scene that is difficult to deal with, to elaborate, and to overcome.

Charles is a 50-year-old man who looks older than he is. He has a full beard and weighs 125 kg. He comes from a foreign country, and when he was very young he was adopted by an Italian family who had no biological children. He does not remember anything about his life before the adoption, and nobody has ever told him anything. The way in which one usually has family ties through which one may recall the past is unknown to him. The interweaving of personal history with the family life to which one belongs weaves the story so familiar to everyone. Charles, however, did not belong to a group where consideration of the person could emerge. He has never lived in a family, as a loved one, part of a real family.

He remembered the pain and anger that slowly and gradually increased in him from an early age. The adoptive father abused him with a whip, which abuse remained firmly in Charles' mind, while his foster mother restrained him tightly so that he would not escape the lashes administered to his back. According to them he had to be "adjusted" and repaired, as if he were pieces of iron that had accidentally become crooked and not like an electrical appliance that one would repair using a delicate and precise strategy. Among the objects of abuse there also could be marginalized and "useless" ones.

In pre-adolescence Charles began working in a factory, and there he suffered physical and sexual violence as well from a group of workers. When he tried to talk with his adoptive father, he received no answer, no help or understanding. During his adolescence, he tried writing to girls hoping to have contact or a relationship with them, but it never led to anything developing. In small towns, in a provincial area, it often happens that accidentals or differences are greatly emphasized. He had become fat, too fat, and was teased for his weight as well as for being the "son of nobody," as he was dubbed by his classmates. With regard to his size, he began to understand how he could instill fear in others. When he beat up boys and saw the fear in their eyes, he realized that he could finally defend himself and gain respect. He could finally see in the eyes of others the terror that was his own perennial traveling companion in his exile from others. Therefore, as he became stronger, everyone began to be afraid of him. Thus arose the first complaints, which were resolved in a short time without any consequences until 1 day when he ran into a former classmate. It was late afternoon, and he began to quarrel. He killed her. In certain moments, he tells of the blood of that young girl, tragically killed by a fat young man, on an afternoon like many other afternoons.

How to reach a human being in Exile? What kind of relationship could be of help in such circumstances? What project can be proposed now? What spiritual encounter can one search for when things cannot be changed? How to help a person who is guilty of such a terrible, tragic, seemingly senseless, and absurd action? How we can understand without ever justifying or explain such a tragedy without decreasing his liability in respect to the murdered girl and her family who all suffer from this horror?

## The Homeland. Aurora and the Encounter with the Other and with Others: The Meaning of Suffering

Aurora was a young woman like many others: she was smiling, loved, and living with others. She studied medicine and wanted to become a doctor, but she never actually worked as a doctor although she dearly wished to. She suffered greatly for a long time from Marfan Syndrome, or Ehlers Danlos Syndrome, and died at the age of 24. She returned home shortly before her death to write a book, one which is not a script, an essay, or an autobiographical book.[5] It is conscious pure witness, one that not unlike her age, represents her early enthusiastic years. Aurora tells this part of her experience. She thought these might be the last moments of her young life but also transcribes her dreams, her plans, her troubles, even if she knew she was writing on the sand that shortly would be washed away by a wave.

She narrates still an open dialogue with the Other, God, truth woven concretely within a radical and authentic *telos*. She opens this dialogue in her most acute pain, even in the tragic discovery of her own limitations, in this case in her body, amid the silence of her own anguish. This experience allows Aurora to welcome illness and suffering in her last moments of life, so that she could talk about the small gestures of affection and consideration she received from others and her personal helper in particular. She lived through this as a gift. In the collision against the limits of her life, she recovers consciousness of and true love for oneself, others, life, God.

Fundamental questions of existence emerge.[6] She speaks of life and her Homeland where she meets herself and others, her history, finding herself, including the future and hopes. The illness becomes to her an hymn of joy: a stubborn, strenuous defense that can give her the same life and light that soothes every wound. In those writings she talks about our capacity to achieve love and to be able to go beyond every obstacle, including sickness and death. She reminds us that man is not only a body, prevailing in the pursuit of pleasure and the avoidance of pain, and it is important to underline that none can live only thanks to nourishment; nourishment alone is not sufficient.

This was clear to Aurora, what with her having serious health problems such as gastrointestinal and cardiac disease. The radical dimension and depth of our being is fulfilled in our spiritual existence – however brief life be, however painful it be to live captive to time and the limits of physical space – in a dialogue and authentic relationship with others. According to Aurora, human beings subsist in their *surplus,* which exceeds those limits,[7] any limits, and tries to find and reach transcendence, authenticity. Nonetheless, the personal spiritual sphere needs to be experienced and understood, and requires the spiritual sphere of another. One can live in the Homeland, in authentic encounter with each other and with the Other.

For Aurora the quality of life, even if it is sometimes so dreadful as it is in a serious illness especially involving disability and a precarious condition, does not coincide with the dignity of human being. Dignity belongs to a human being in every life situation, any condition of existence – even in a tragic event involving say amputation. It is hard to understand and make others understand this ultimate reality of human beings. Aurora spent her final years asking the same questions that Job asked, while accepting her physical illness, and her painful experience led to her dying with awareness, in the light of a radical relationship with the Other.

## The Poetic Encounter, the Helping Relationship and Ecology of Relationships

So there is an approach to death that does not suffocate life, and to a life already buried by a radical failure to meet others. Suffering, therefore, is not only the cipher of love. It makes a person responsible before himself and others. It allows one to accept existence itself radically, to the point that we can still live and love, despite everything, or sees us sink slowly into the abyss of isolation where no one seems to reach us, neither the other nor the Other.[8]

The real encounter with the other happens in the spiritual sphere, the same where prayers are uttered, that has been analyzed by Eugene Minkowski[9] as the fundamental dimension of life, one that requires sincerity, depth, humbleness of mind, recognition of Other, having to rise in radical and authentic Dialogue and so regain life's fertility and creation. Again and again, Art and Poetry exceed the materiality of things in the same way an authentic encounter goes beyond the data of the time and circumstances. In a helping relationship, the *setting* is mental and it comes to life in the authentic encounter. It is not certainly to be found in physical situations. You can meet the other in extreme conditions, if there emerges humility, openness, availability to the radical witness Tymieniecka spoke of in pondering these same essences of the spiritual life. I can physically build the best setting and yet not listen to the other; I could simply not understand lived experiences, feelings, or meet him or her. The meeting is a poetic encounter. To live the poetry of a moment can be sensed, welcomed in a human being, but it cannot be caught in a planned methodology, in a "morbid rationalism" in Minkowski's meaning, or in an ambitious, even if it is legitimate, planned, and anticipated project. Poetic encounter happens during the openness of a humble search. This shows how a relationship needs to develop radically and authentically for each one. The true word is witnessed.

Before a limit (in effect, of oneself, others, and physical existence) there appears the ultimate essence of human being. A person arrives fully conscious and therefore completely open to a fundamental and authentic dialogue with others, with the Other, and with life, as happened with Aurora. She faced the disease and death with a life relationship that became a new dawn, toward a light that never fades, even in face of dark existence.

Otherwise, a human being can arrive before limits as happened with Charles, at a serious misunderstanding of things and a radical change for oneself and for others. A complete closing in on oneself and unattainability could lead to murder, to the deletion of oneself and others. In this case, oneself and others are experienced as if they were objects, as bodies-objects, in a present without past experience – *hic et nunc* – without any project, being confined, marginalized, exiled from the materiality of things that sustains non-deterioration and continued existence and material reiteration. There is no transcendence when only matter is considered; the self is then imprisoned in the precariousness and fragility of a non-story, in a non-spiritual existence, like a rolling stone in a desert. In the materialism of the object/not subject there is eliminated the *surplus* of the human being, that which is deeply human and is part of our uniqueness, we being a unity of body-psyche-spirit, as Edith Stein's exposes. This eliminates the Homeland and its History, and so begins an Exile and therefore Evil.

In any relationship, even in that by which help is extended in extreme situations, one must seek spirituality in the *between-us* expressed by proximity, availability, understanding of the lived-experiences and emotions, your own and the other's, without any eliminating personal responsibilities. Slowly, life can rise and rise again, as it were coming out from beneath a rock that crushes lives and hopes. This can happen even in a desert, where there will be still a yearning for life.

Each person is unique and universal at the same time, and everybody has within themselves the possibility of living death prematurely (that is, either physical death or social death) or of living and regenerating life, contributing to the creation (*poiesis*) in the world of harmony, to be sought with cosmos and with others, as Tymieniecka reminds us.[10]

In a helping relationship, a human being is therefore still subjected to the a standing ethical judgment, as Guardini makes clear. Those who are involved (as a patient or as a helper) are called to respond to the same opinion: the helper cannot judge the way you understand a distressful situation, because one does not know how to react or how one would react in a situation of suffering, lived *in extremis*. This is the reason – according to St. Augustine – why we cannot judge. That is why Husserl's "suspension of judgment" becomes de facto not only an epistemological issue, but above all an ethical issue and therefore a clinical issue, mindful of the relapses that could occur in a relationship with oneself and others.

There is no suffering that can be addressed in spiritual encounter with the other. This must be accomplished not only according to the sacred consideration of the suffering one but also to aid places involved with the time that one offers for personal reflection, or those that offer help in searching for harmony and creative ability despite all.

There is therefore an indispensability to the human being: to be in relationship with another is essentially *being-for-others*, and that extends to our *being-to-love-and-to-be-loved*. We search for the other, and if the other does not meet one, then one falls into the materiality of things so that subjects become objects, even among the most useless objects. There is emptiness in the soul, in the words of their own story, and in interpersonal sharing. The absence of inner dialogue with the other breeds evil and shutdown. Here the reduction of a human being to an object is possible. We search the other in any way and yet there is inaccessibility and the loss for the other of his or her freedom amid our limitations.

We are a *poetic encounter*. Poetry – which for Novalis is the foundation of reality, of truth and thought – knows sometimes how to exceed and transcend pain, knows how to overcome the heaviness of living and the trivialization of everyday lived-experienced that can in some cases overwhelm the profound humanity of our being unique, of our being unitary, an individual person in relationships with others. We are limits that transcend their very existence. If there is no poetry in a tragedy, there is the possibility, however, that poetry can let us pull back from the brink of the absence of the other (into which we have fallen) when in Exile we no longer see a familiar face. To be within poetry is to live the deep inner dialogue with each other that not even death can suffocate. I continue to love despite the loss, or I can carry on for myself or a loved one. Living the word of silence and understanding each other is useful for grasping deep motivations without confusion. It is to live into *Us* without losing *You* or *Me*. We can push to live with openness to mutual recognition without diminishing the responsibility of each. This responsibility is to live in mutual consideration of ourselves as persons, the silence of a listening that is not that of a stone, while "powerful" as Georg Trakl sees, is not omnipotence. Hell is the absence of the other and the missed encounter with the other that generates the missed meeting with oneself.

According to Husserl, it should be emphasized that lived-experienced in consciousness has the features of evidence, truth, and reality. So, how can lived-experience take shape? How is it composed and recomposed? It takes place in these particular moments: a particular flow, a particular narrowness or breadth of the horizon, the meaning of the past and an opening and closing towards the present and the future, hope, emptiness, anxiety, excitement. Those emotive experiences are mediated by the intersubjectivity that inter-lays the foundation of this particular vision of the world.

A personal lived-experience is born in an interpersonal meeting, even if sometimes it is unintelligible to ourselves or to others. In the depth of ourselves there is the other.

A relationship cannot be maintained separate from the one kind of meeting that must be fulfilled and understood within the spiritual sphere. Probably there is no other type of inter-human encounter. Invisible and impalpable support sustains the visible world and the therapy, all forms of psychotherapy. It can be *caring* (in the spiritual sense) that enables and starts in certain situations a cure (even a physical cure). Or it is this *surplus* that gives meaning to the finite as transcendence, to the story, which itself is not beyond the material limits. The person exceeds in the spiritual sphere, in the human spiritual sphere. Reality, according to Martin Buber, includes participation so that the sense of things can be achieved through the living presence of the other, who comes alive within a mutual confirmation of reality's existence. Therefore, a relation with the other can create an extreme meeting of love, hope, but also endless suffering when we live with ourselves and each other as objects, as an objectification of the subject.

Charles, during his suffering, does not meet the other, and at a certain point he does not search for anyone, neither another or the Other, nor, moreover, the meaning of what has happened to him. He soon reacts with bitterness, resentment, and frustration, reaching the consequence of an unjustifiable murder. The helping relationship in this case must take into account the choices, the limits, the inner freedom of choosing a route and having responsibility for an action. This freedom and responsibility makes us human, as Guardini affirms. Charles could decide to review his life or to do nothing at all and find himself isolated and experiencing the world and relations *in his present world* as a not in common held and unshared world. A suggestion could generate a slow recovery, but it should be considered that a specific way of living and relating could be more decisive than one (radical) change, whether it be conscious or not. A cell in some cases can become a dwelling from which one cannot escape, exiled to one's own world without any other.

Aurora was put through severe and radical suffering by her illness, yet she recognized the overflow that nature has in a relationship, a relationship with others and with the Other. Although suffering, Aurora is able to love again, despite her condition. Affliction, which has radical impact, could make one withdraw from the world or could transform life through a deep opening to life itself.

In any case, suffering itself becomes a means to continue to love, realizing that overindulging oneself to help find the lost harmony, as much as possible, with oneself, others, and life as Tymieniecka proposed in her Husserlian understanding of philosophical existence.

## Notes

1. A. Merini (2000). *L'anima innamorata*. Milan: Frassinelli, p. 100.
2. Cf. A. De Luca (2003). *Frammenti di esistenza. Per una psicologia fenomenologica ed. esistenziale*. Foggia: Bastogi; A. Ales Bello, A. De Luca (eds.) (2005). *Le fonti fenomenologiche della psicologia*. Pisa: ETS; A. Dentone, A. De Luca (eds.) (2006). *Le fonti esistenziali della psicologia*. Pisa: ETS; A. De Luca (ed.) (2009). *Verso una psicologia fenomenologica ed. esistenziale*. Pisa: ETS; A. De Luca (2010). "Toward a Phenomenological and Existential Psychology," Book 3. Analecta Husserliana CV. Dordrecht/Heidelberg/London/New York: Springer; A. De Luca (2011). *Tra le rovine dell'esistenza. Sofferenza Psicoterapia Ripresa*. Rome: Edizioni Universitarie Romane; A. De Luca, A. M. Pezzella (eds.) (2014). *Con i tuoi occhi. Sull'intersoggettività*. Milan: Mimesis; De Luca A., ed. 2017. The understanding of the experience between psychopathology and psychotherapy, Agathos, An International Review of the Humanities and Social Sciences, 8, 2, 161–183.
3. Cf. M. Zambrano (1955). *El hombre y lo divino*. Madrid: Fundación Maria Zambrano; A. De Luca (2011), *Tra le rovine dell'esistenza. Sofferenza Psicoterapia Ripresa*. Rome: Edizioni Universitarie Romane.
4. See the studies of M. Buber and within the psychopathological ambit the studies of L. Binswanger, E. Minkowski, F. Basaglia, in particular: F. Basaglia, "Su alcuni aspetti della moderna psicoterapia: analisi fenomenologica dell'incontro," *Rivista di Freniatria* 78: II, 1954, 239–263), and of B. Callieri, in particular, B. Callieri (1993), *Percorsi di uno psichiatra*, Rome: Ed. Universitarie Romane. Cf. also A. De Luca (2011). *Tra le rovine dell'esistenza. Sofferenza Psicoterapia Ripresa*. Rome: Edizioni Universitarie Romane.
5. Morelli A. (2011), *Come un libro aperto. La mia semplice testimonianza*, Verona: Monastero del bene comune.
6. For Emil Cioran, in his *The Fall into Time*, you cannot yield to the temptation to think that our troubles are not useful for anything. Illness has the dual aspect of annihilation and revelation which opens to us ultimate realities sometimes invisible.
7. Max Scheler also speaks of *surplus*.
8. Important is Psalm 139, 8: "If I make my bed in the depths, you are there."
9. See E. Minkowski (1933), *Le temps vécu, Etudes phénoménologiques et psychopathologiques*. Paris: D'Artrey. Minkowski discovered prayer as "total internalization experienced," which reaches the roots of being, and at the same time as "total extrospection experienced," which arises when accepting and

embracing the entire universe. Minkowski speaks of a concrete "abstract" that reaches to the roots of being even as hope and expectations are somehow overtaken by prayer that fully embraces the flow of being.

10. See A.-T. Tymieniecka (1988a) *Logos and Life, Book I: Creative Experience and the Critique of Reason*. Dordrecht-Boston-London: Kluwer Academic Publishers, 1988. Analecta Husserliana XXIV; Tymieniecka, A.-T (1988b), *Logos and Life*, Book II: *The Three Movements of the Soul.*, Analecta Husserliana XXV, Dordrecht-Boston-London: Kluwer. See also D. Verducci (2012), *La fenomenologia della vita di Anna-Teresa Tymieniecka*, Rome: Aracne.

## Works Cited

Ales Bello, A., and A. De Luca, eds. 2005. *Le fonti fenomenologiche della psicologia*. ETS: Pisa.

Basaglia, F. 1954. Su alcuni aspetti della moderna psicoterapia: analisi fenomenologica dell'incontro. *Rivista di Freniatria* 78 (II): 239–263.

Callieri, B. 1993. *Percorsi di uno psichiatra*. Rome: Edizioni Universitarie Romane.

De Luca, A. 2003. *Frammenti di esistenza. Per una psicologia fenomenologica ed esistenziale*. Foggia: Bastogi.

———., ed. 2009. *Verso una psicologia fenomenologica ed esistenziale*. Pisa: ETS.

———., ed. 2010. Toward a phenomenological and existential psychology. In *Phenomenology and existentialism in the Twentieth Century*, Analecta Husserliana CV, 281–299, ed. Anna-Teresa Tymieniecka. Dordrecht/Heidelberg/London/New York: Springer. https://doi.org/10.1007/978-90-481-3785-5_18.

———., ed. 2011. *Tra le rovine dell'esistenza. Sofferenza Psicoterapia Ripresa*. Rome: Edizioni Universitarie Romane.

———., ed. 2017. The understanding of the experience between psychopathology and psychotherapy. *Agathos, An International Review of the Humanities and Social Sciences* 8 (2):161–183.

De Luca, A., and A.M. Pezzella, eds. 2014. *Con i tuoi occhi. Sull'intersoggettività*. Milan: Mimesis.

Dentone, A., and A. De Luca, eds. 2006. *Le fonti esistenziali della psicologia*. ETS: Pisa.

Merini, A. 2000. *L'anima innamorata*. Milan: Frassinelli.

Minkowski, E. 1933. *Le temps vécu. Etudes phénoménologiques et psychopathologiques*. Paris: D'Artrey.

Morelli, A. 2011. *Come un libro aperto. La mia semplice testimonianza*. Verona: Monastero Del Bene Comune.

Tymieniecka, Anna-Teresa. 1988a. *Logos and Life*. Book I: *Creative Experience and the Critique of Reason*, Analecta Husserliana XXIV. Dordrecht: Kluwer Academic Publishers.

———., 1988b *Logos and Life*. Book II: *The Three Movements of the Soul.*, Analecta Husserliana XXV. Dordrecht: Kluwer Academic Publishers.

Verducci, D. 2012. *La fenomenologia della vita di Anna-Teresa Tymieniecka*. Rome: Aracne.

Zambrano, M. 1955. *El hombre y lo divino*. Madrid: Fundación Maria Zambrano.

# Experience of the City: An Eco-Phenomenological Perspective

**Velga Vevere**

**Abstract** The field of ecological philosophy is a quite young field, encompassing Erazim Kohak's environmental ethics and Arne Naess' concepts of deep ecology and ecosophy, David Seamon's phenomenological ecology, Ted Toadvine's, Charles S. Brown's, David Wood's eco-phenomenological investigations, and, above all, Anna-Teresa Tymieniecka's phenomenology of life. At the same time, the theme of urban phenomenology taps deeply into Martin Heidegger's conception of authentic dwelling (disruptions of dwelling may be exposed as symptoms of ecological crisis, and authentic dwelling as a possible remedy for that). Jean-Paul Sartre's notion of the For-Itself (especially its spatializing/spatialized character), Walter Benjamin's explication of the phenomenon of flaneurism (city strolling), Michael de Certeau's analyses of everyday practices, and Juhani Pallasmaa's phenomenology of architecture open up interpretative pathways for future investigations in the field.

**Keywords** Tymieniecka · Ecological phenomenology · Phenomenological ecology · Deep ecology · Dwelling · Flaneur · Urban phenomenology

We live in the world that is permeated by a sense of multiple crises—political, economic, social, cultural, and ecological. Commonly, this makes us question our place in society and nature. At the same time, this rather grave situation fosters the development of ecologically-oriented thinking on how to stay alive and retain our essential humanity. Ecological phenomenology, in this sense, is a way of reflecting upon the situation in terms of the human condition and of deepening our understanding of the surrounding world, of ourselves and of ourselves within this world—the world past, present, and (it is to be hoped) future.

> Although concern for the environment was already expressed during the nineteenth century, urgent concern with our 'environment' is relatively new, owing to the grave threats that human use and abuse of technology is posing to living nature, living creatures, and our own

V. Vevere (✉)
University of Latvia, Riga, Latvia

 
W. S. Smith et al. (eds.), *Eco-Phenomenology: Life, Human Life, Post-Human Life in the Harmony of the Cosmos*, Analecta Husserliana CXXI,
https://doi.org/10.1007/978-3-319-77516-6_38

> vital life resources. The debates around these and many other connected issues are too well known to merit enumeration. It seems that all public and personal interests are involved: scientific, communal, economic, aesthetic. Earth science, environmental science, the various stripes of politics, etc., each have their say. A common meeting-point is sought in vain. And yet such a point is not out of sight. The very notion of the 'environment' at stake offers an essential point of departure for an adequate approach. (Tymieniecka 2005, xxxi)

So writes Anna Teresa Tymieniecka in her inaugural lecture for the studies gathered in Analecta Husserliana LXXXIV, *Phenomenology of Life. Meeting the Challenges of the Present-Day World.* The statement brilliantly captures the current condition of man—of the city or countryside resident, the inquirer or one who seeks diversion, the loner or the socialite—since the environment is our common dwelling place, inescapable and inevitable as such. The issues of the environment, of the body-world relationship, have been the focal point of philosophical and scholarly investigations for many years. Still, the field of ecological philosophy is a quite young field, encompassing Erazim Kohak's environmental ethics and Arne Naess' concepts of deep ecology and ecosophy; David Seamon's phenomenological ecology; Ted Toadvine's, Charles S. Brown's, and David Wood's eco-phenomenological investigations, and, above all, Anna-Teresa Tymieniecka's phenomenology of life. At the same time, the theme of urban phenomenology taps deeply into Martin Heidegger's conception of authentic dwelling (disruptions of dwelling may be exposed as symptoms of ecological crisis, the authentic dwelling as a possible remedy for that). Jean-Paul Sartre's notion of the For-Itself (especially its spatializing/spatialized character), Walter Benjamin's explication of the phenomenon of the *flaneur* (city stroller), Michael de Certeau's analyses of everyday practices, and Juhani Pallasmaa's phenomenology of architecture open up interpretative pathways for future investigations in the field.

The concept of dwelling, in its turn, ties together eco-phenomenology and city experience since, as we walk through city spaces, we are met with distinct sights, sounds, smells, tastes, and textures that help us to define a particular city, as well as enriching our experience of walking through such a space. These sensations should not be seen as separate, isolated feelings but, rather, the total effect of these sensations should be compassed in a phenomenological description.

> If man becomes frightened in the presence of nature, it is because he feels that he has been trapped in an immense, amorphous, gratuitous existence which completely freezes him by its gratuitousness. He no longer has his place anywhere; he is planted on the earth without a goal, without a raison d'être like heather or a clump of broom. On the other hand, in the middle of a town he feels reassured because he is surrounded by precise objects whose existence is determined by the part they play and which have a value or a price attached to them like a halo. They show him the reflection of the thing that he wants to be —a justified reality. (Sartre 1962, 106)

## Phenomenological Ecology and Ecological Phenomenology

Although, at first sight, it may seem a simple shift in vocabulary, the concepts of phenomenological ecology and ecological phenomenology, it seems, exhibit two different approaches grounded in different theoretical presuppositions, effectively, causality and intentionality. In other words, natural scientists and environmentally-concerned philosophers strive to apply phenomenologically-generated methods to analysis of natural occurrences, detecting the causal relations; and phenomenologists' concern lies in viewing things as they appear in their appearing in our experience. According to C. S. Brown and T. Toadvine, eco-phenomenology may bridge the gap between ecology and phenomenology. They write:

> The intersection of ecological thinking with phenomenology, the momentum that drives each toward the other, begets a new cross-disciplinary inquiry: eco-phenomenology. Eco-phenomenology is based on a double claim: first, that an adequate account of our ecological situation requires the methods and insights of phenomenology; and second, that phenomenology, led by its own momentum, becomes a philosophical ecology, that is, a study of the interrelationship between organism and world in its metaphysical and axiological dimensions. (Brown and Toadvine 2003, xii–xiii)

Still, before turning to basic concepts and currents that characterize today's eco-phenomenology, let us first take a pause to reflect on phenomenological ecology; more precisely, let us reflect on Arne Naess's ecosophy (deep ecology) and Erazim Kohak's ecological ethics. Phenomenological ecology is concerned with all lived relationships and interrelations, describing the ways that things, living forms, people, events, situations, and worlds come together to make environmental and human wholes (Seamon 1993, 16).

Speaking of a paradigm shift in ecological thinking, Arne Naess sharply distinguishes an anthropocentric approach oriented towards pragmatic values and the use of natural resources (even though responsible use) and deep ecology, which takes up a holistic stance, that is, recognizes the interconnectedness of everything that is alive. Naess offers an eight-point outline of deep ecology in his article "The Deep Ecology. 'Eight Points' Revisited" (Sessions 1995, 213–221). Therein he sets forth that the fundamental claim of deep ecology is that everything rests on diversity: "I do not attach inherent value to species or families (as classes or sets of beings with more than one individual or element) but to diversity, itself" (Sessions 1995, 217). Moreover, he recognizes that everything (be it alive or not) has its intrinsic value regardless the value assigned to it by humans. Deep ecology has its political undertones as well, since movement adepts (especially during the turbulent 1960s and 1970s) have joined rather diverse eco-political movements and, according to Naess, cooperation should rest on three basic requirements—peace, social justice, and ecological sustainability. On the basis of these premises, Naess develops the notion of the ecological self:

> What is the practical importance of this conception of a wide and deep ecological self? When we attempt to defend Nature in our rich industrial societies, the argument of our

> opponents is often that we are doing it to secure beauty, recreation, and other non-vital interests for ourselves. Our position is strengthened if, after honest reflection, we find that the destruction of Nature (and our place) threatens us in our innermost self. If so, we are more convincingly defending our vital interests, not merely something 'out there.' We are engaged in self-defense. (Sessions 1995, 232)

In other words, our innermost selves are woven into the fabric of all things around and, perhaps, this is an epitome of the ecological self. The scope of questions involving selves and nature form a special sphere of knowledge—ecosophy. "So an ecosophy becomes a philosophical world-view or system inspired by the conditions of life in the ecosphere" (Naess 1989, 38).

Erazim Kohak, a Czech philosopher and writer, presented another phenomenological take on ecology in his seminal work on the environmental ethics, *The Embers and the Stars*. The underlying concern of the whole book is modern man's total preoccupation with technology and resulting loss of relations with the natural environment. On one hand, this motif coincides with philosophy of the early American transcendentalists (Henry David Thoreau and Ralph Waldo Emerson) in their opposition to the advance of technology; on the other hand, Kohak distances himself from their romanticism, admitting that we "must approach nature anew, undertaking no less than a phenomenology of nature as the counterpart of our moral humanity" (Kohak 1984, 22). As a remedy for the spirit of reductionism, he proposes an 'ecological ethics' and a 'moral sense of nature,' made possible by "the radical opening of our life and thought to the world of others, human, animate, inanimate, in the integrity of its otherness and the meaningfulness of its being" (Kohak 1984, 213). Commenting on his earlier phenomenological endeavor, Kohak later notes that instead of the concept of the 'moral sense of nature' the concept of the 'moral significance of nature' would be of better use, since non-human nature needs to be noted, not just used, overlooked, or taken for granted. "I am convinced that we need to find ways of becoming aware of ourselves as one of many species, distinctive but not privileged. The continuity of humans with other animals and all of nature is to me a very basic experience—we are one species among many, not masters of all we survey or the source of all meaning and value. That I think crucial. We belong."(Cohen 1998, 36) In other words, experiential knowing of the world starts with admitting ourselves to be an integrative part of everything, and the best research tool for this is the phenomenology of lived experience. This entails at least two consequences regarding our relation to the environment. First, is recognition that the nonhuman world that is autonomous of humans and not for that devoid of value; therefore, moral considerations apply not only to human transactions with humans but also to human dealings with the rest of the world because that which is there is intrinsically valuable. Secondly, there is the realization that the world as our environment is not something that simply happens to us but is, rather, a way we constitute what-is as a meaningful, value-laden wholeness by our purposive presence (Cohen 1998, 211). Thus, phenomenology becomes the *radical* acceptance of responsibility. The most important apprehension here is that this is responsibility *before the fact,* not *after the fact.* This alone makes possible the introduction of the concept of ecological ethics rooted in the transcendental phenomenology. "What matters now is for philosophy to challenge humanity to accept the responsibility of its freedom. In their moments of greatness, both phenomenology and ecological philosophy do that" (Cohen 1998, 218).

The deep ecology of Arne Naess and Erazim Kohak, as well as Kohak's ecological ethics, represent a trend that can be broadly described as phenomenological ecology. As such, it "must be responsive to all lived relationships and interconnections, examining and describing the ways that things, living forms, people, events, situations and worlds come together to make environmental and human wholes" (Seamon 1993, 16). Thus, the starting point here is environmental or ecological thinking wherein phenomenology serves as a useful and productive tool for description of worldly experiences. In other words, phenomenological ecology could provide the way to conceptualize the intuitive insights of environmental researchers. Ecological phenomenology (eco-phenomenology), in turn, was born out of phenomenologists' regard for natural phenomena in their universal interconnectedness and conceptual opposition to naïve naturalism. In order to characterize the stance of eco-phenomenology, we have to recall the grounding propositions of deep ecology (eco-phenomenology largely rests on the critique made by deep ecology). These have been summarized by David Wood in the following way:

- There is a gap between the obvious impact made on nature by humans and hidden processes beneath the surface; in order to raise sensibility, mass ecological education is necessary;
- There is a gap between each individual impact and the resulting collective consequences;
- The insight that the deep interconnectedness by which everything depends on everything may mean disastrous natural consequences from disturbances, worst case scenarios;
- In some cases, this could call for drastic political measures up to the suspension of democratic institutions altogether. (Brown and Toadvine 2003, 231)

Critical reconsideration of these statements makes possible bridging the conceptual gap between obvious and hidden natural processes through the phenomenological reduction of our everyday experiences and meditation on the role of boundaries in the constitution the thinghood.

> The intersection of ecological thinking with phenomenology, the momentum that drives each toward the other, begets a new cross-disciplinary inquiry: eco-phenomenology. Eco-phenomenology is based on a double claim: first, that an adequate account of our ecological situation requires the methods and insights of phenomenology; and second that phenomenology, led by its own momentum, becomes a philosophical ecology, that is, a study of the interrelationship between organism and world in its metaphysical and axiological dimensions. (Brown and Toadvine 2003, xii–xiii)

In other words, the common ground for ecological phenomenology and phenomenological ecology is their focus on description of human experience as such, even as they retain their respective modes of questioning and theoretical explication.

Anna-Teresa Tymieniecka's project of phenomenology of life is the unique vantage point from which to enter the realm of eco-phenomenology. Her conception of the self-fueled, ontopoietic development of everything-that-is-alive is depicted in the four epic volumes of her *Logos and Life.* Book One of this series is devoted to the creative experience and the new critique of reason; Book Two is devoted to the three movements of soul (a radical examination of reason, the discovery of the finiteness of life, and the aspiration to pass beyond finitude); Book Three, *The Passions of the Soul and the Elements in the Ontopoiesis of Culture,* is inspired by the elemental rhythms of nature embodies in light and sea as metaphors. Book Four, *Impetus and Equipoise in the Life-Strategies of Reason*, pursues the strategies of logos in the ways it takes to establish its manifestation in life and its existential context (Tymieniecka 1988a, b, 1990, 2000). As the editor-in-chief of the Analecta Husserliana book series, Anna-Teresa Tymieniecka proposed ever-new topics and facilitated collective elaborations in the field of phenomenology of life, marking a turn towards eco-phenomenology. Let us mention just a few of these themes: human positioning in the Cosmos, the passions of the Earth, and phenomenology of space and time in human existence, among others. Tymieniecka puts forwards three propositions in this regard:

> Although each living being fashions through its way of life its very own, unique environment, yet three points have to be raised: 1) there is a specific way of coexistence of types and individuals within a life-network, for it is not restricted to casual encounter but varies in degrees of mutual involvement in innumerable primary nourishing, generative, and "symbiotic" processes. This amounts to what we could call in general a "symbiosis of life," which entails a sensibility of a sphere of rootedness among all living creatures; 2) all living beings share the indispensable conditions for life, conditions that the planet earth is fulfilling to this effect. ... We might live on the "surface" of earth, but we are utterly grounded in earth's situation; 3) it is not nature's organic vital conditions only that shape and sustain the environment of living beings. There is also the profound dimension of the specific sphere that the living human being encircles owing to the enactment and creative spread of the Human Condition as the ensuing human valuative, deliberative, calculating, planning, and deciding mind infused by the intellective, moral, and aesthetic sense-giving powers turns its attention to its world. (Tymieniecka 2005, xxxii–xxiii)

We may conclude that this programmatic statement marks the turn towards eco-phenomenology based on the apprehension of the self-evolving of life processes and the creative abilities of human reasoning.

## Urban Phenomenology

While the relation between eco-phenomenology and ecology is readily plausible, an elaboration involving city experience calls for still more explanation. Urban life, with its hectic rhythms, movements of people, and vast and enclosed architectural spaces, has always fascinated writers, poets, movie directors, and philosophers; let us mention just a few of them: Charles Baudelaire, Fritz Lang, Walter Benjamin, Jean-Paul Sartre, Martin Heidegger, Michael de Certeau, and Jean Baudrillard. Jean-Paul Sartre summarizes the feeling: "A city is a perpetual creation: its buildings, smells, sounds and traffic belong to the human kingdom. Everything in it is poetry in the strict sense of the term. It is in this sense that the electrically operated advertisements, neon lights and cars which about the year 1920 roused the wonder of young people were profoundly Baudelairean. The great city is a reflection of the gulf, which is human freedom" (Sartre 1962, 44). The city as a site of habitation, thus, becomes an object of phenomenological investigation. One such example is the phenomenology of architecture as developed by architects Juhani Pallasmaa, Christian Norberg-Schulz, Steven Holl, and Alberto Gomes-Perez. Their insights revolve around such themes as the haptic (sensuous experience) experience of architecture and *genius loci* (the spirit of a place). Pallasmaa, Holl, and Gomes-Perez explore the impact of architectural forms upon the human senses; phenomenology for them is an introspective philosophical approach to the basic phenomena of consciousness: "the relationship between the experiential qualities of architecture and the generative concepts is analogous to the tension between the empirical and rational. Here the logic of pre-existing concepts meets the contingency and particularity of experience" (Holl et al. 2006, 42). According to Pallasmaa, however, experiences of the built environment are perspectival, distinct, and incomplete; phenomenology seeks to incorporate these differing perspectives. Moreover, building and city provide the horizon for understanding and confronting the existential human condition. Buildings direct our consciousness back to the world and to the sense of ourselves within the world, therefore, their ultimate meaning lies beyond architecture (Pallasmaa 2005, 11). In contrast, Norberg-Schulz's phenomenology of architecture seeks to disclose the spirit of the place, where a place is regarded as qualitative, total phenomenon, irreducible to its properties. "Being qualitative totalities of a complex nature, places cannot be described by means of analytic, 'scientific' concepts" (Norberg-Schulz 1979, 7).

Phenomenology, on the contrary, allows approaching everyday phenomena in their complexity and ambient quality. To do that, we have to take several steps:

1. To make a distinction between natural and man-made phenomena (landscape and settlement);
2. To describe everything in horizontal-vertical (earth-sky) and outside-inside categories, introducing space as an existential conception;
3. To determine the character of things, that is, the way *how* things are. Norberg-Schulz here offers phenomenological description of spaces.

Here, the author draws heavily on Martin Heidegger's phenomenology of built artefacts and, more precisely, on his conception of dwelling. For Heidegger, there is close link between dwelling and thinking, between dwelling and man's situatedness within the world: "Spaces open up by the fact that they are let into the dwelling of man. To say that mortals *are* is to say that *in dwelling* they persist through spaces by virtue of their stay among things and locations" (Heidegger 1997, 101). The nature of human dwelling is fourfold. First, we, humans, dwell in the sense that we set the earth free in its own presencing. Second, we receive the sky as sky, namely, we let the day and night run their courses. Thirdly, we do not place our unrealistic hopes onto the shoulders of vicinities, lest we blame them for our failures. Fourth, we are a being capable of death as death, which is to say, of accepting our mortality and eternal responsibility for it (Heidegger 1997, 98). To dwell means to exist by one-self, to exist amidst others (under the objectivizing gaze of others), to exist within a natural setting or constructed (or technological) space. The gaze of the other robs us of intimacy, annihilates us from us (from being in itself). In other words, we turn here to Jean-Paul Sartre's ontology of the I-Other relation; he establishes the possibility of consciousness of Others from within the first person perspective. In the Other's eye, I am in-itself, while the Other is for itself, and vice-versa. The For-Itself, which is always the object of the look of another For-Itself, ascends to self-consciousness through its being and having been temporalized and spatialized by this other For-Itself (Mendieta 2001, 210). Thus, to be in the midst of the world is not something abstract—to be the object of someone else's look is to be a body that is seen. To take place in the world is to live out our bodies in that distended space that has become our place. To be in the world is to be the object of someone else's look; this means that the self and the other come into the world simultaneously.

> The structure of the world demands that we can not see without being visible. The intra-mundane references can be made only to objects in the world, and the seen world perpetually defines a visible object to which its perspectives and its arrangements refer. This object appears in the midst of the world and at the same time as the world. It is always given as an addition to some grouping of objects since it is defined by the orientation of these objects; without it there would be no orientation since all orientations would be equivalent. It is the contingent upsurge of one orientation among the infinite possibilities of orienting the world; it is this orientation raised to the absolute. But on this level this object exists for us only in the capacity of an abstract indication. (Sartre 1978, 317)

Sartre wrote on the city, city life, buildings, and city dwellers. He devoted a fair share of lines to description of the great European cities such as Berlin, Paris, Rome, London, and Naples, but the epitome of his deliberations on the city environment, it seems to me, is his essay "New York, the Colonial City" published in 1955. In contrast to the European city, the American counterpart represents an outpost in the wilderness. To him, this represented a very different experience, one based on open landscapes and technology, rather than on seclusion and history. "New York is a city for far-sighted people, a city in which you can only "adjust" to infinity. My glance met nothing but space. It slid over blocks of identical houses, with nothing to arrest it; it was about to lose itself in empty space, at the horizon" (Sartre 1962, 127). He described the feeling of space itself in his crossing through New York, "quickening and expanding it."

Based on these insights, Sartre developed a notion of pure space (or any place): "And suddenly pure space looms into view. I imagine that if a triangle could become conscious of its position in space, it would be terrified at the realization of the rigorousness of its defining co-ordinates, but that it would also be terrified to discover that it is merely any triangle, any place" (Sartre 1962, 129). The concept of any place points to the phenomenological experience of the human-constructed environment. "My body is co-extensive with the world, spread across all things, and at the same time it is condensed into this single point which all things indicate and which I am without being able to know it" (Sartre 1978, 318). But, at the same time, in the city my co-extended body co-exists with other co-extended bodies. David Seamon, in his book "Geography of the Lifeworld," describes the urban space in terms of place ballet: "The integral parts of place ballet are individual habitual bodies in synchrony with a supportive spatial configuration that generates animated streets and places" (Seamon 1979, 11). He offers to look into the matter by employing two contrasting modes of daily life—the triad of habituality and the triad of openness. The former refers to the typical ordinariness and humdrum of everyday life which involves unquestioned repletion and routine that could be changed through thoughtful design of the physical environment, whereas the triad of openness consists of those moments in everyday life and everyday life practices at which a person is suddenly alert to the world in a more sensitive, intense way and experiences a heightened encounter with the world.

> What I'm suggesting here is that a phenomenology of lively urban places indicates how a particular fit between people—specifically, habitual bodies—and world—specifically, particular pathway pattern—supports physical co-presence and potential encounter that may facilitate sociability and a sense of community and neighborhood. Hillier's demonstration of how a particular pattern of spatial configuration—the deformed grid—has the potential to found a nexus of lively pedestrian movement illustrates in a remarkably new way the basic phenomenological principle that people are immersed in world as world is immersed in people. (Seamon 2011, 239–40)

The latter phenomenon has been described also as the phenomenon of the *flaneur*. Experience, here, is much more than experience of an individual's encounters with a harsh reality; the reflecting consciousness becomes an integral part of the city experience. In other words, it is consciousness' reflection upon itself reflecting. Perhaps the best explanation of this phenomenon is given by Walter Benjamin in his "Arcades Project": "The street conducts the *flaneur* into a vanished time. For him, every street is precipitous. It leads downward—if not to the mythical Mothers, then into a past that can be all the more spellbinding because it is not his own, his private. Nevertheless, it always remains the time of a childhood" (Benjamin 1999, 416). He continues, "That anamnestic intoxication in which the *flaneur* goes about the city not only feeds on the sensory data taking shape before his eyes but often possesses itself of abstract knowledge—indeed, of dead facts—as something experienced and lived through" (Benjamin 1999, 417).

Developing the theme of *flaneur*, Michel de Certeau, in his turn, offers a distinction between two types of city experience—the first resting on distanced, elevated viewing (voyeurism), and the second on direct exposure to surroundings and to other people (*flanerie*).

> When one goes up there, he leaves behind the mass that carries off and mixes up in itself any identity of authors or spectators. An Icarus flying above these waters, he can ignore the devices of Daedalus in mobile and endless labyrinths far below. His elevation transfigures him into a voyeur. It puts him at a distance. It transforms the bewitching world by which one was "possessed" into a text that lies before one's eyes. It allows one to read it, to be a solar Eye, looking down like a god. The exaltation of a scopic and gnostic drive: the fiction of knowledge is related to this lust to be a viewpoint and nothing more. (de Certeau 1988, 93)

On the other hand, he writes,

> The ordinary practitioners of the city live "down below," below the thresholds at which visibility begins. They walk—an elementary form of this experience of the city; they are walkers, Wandersmänner, whose bodies follow the thicks and thins of an urban "text" they write without being able to read it. These practitioners make use of spaces that cannot be seen; their knowledge of them is as blind as that of lovers in each other's arms (de Certeau 1988, 94).

They strive to enjoy idle walking and taking in experiences of unexpected encounters with the unknown (therefore, their attitude can be characterized as rather passive), and they feel an urge to reflect upon those encounters and upon themselves as reflecting. This, according to Certeau, opens the path for phenomenological description of the city experience that can be viewed in the broader context of eco-phenomenology and of phenomenology of life.

## Works Cited

Benjamin, Walter. 1999. *The Arcades Project*. Trans. Howard Eiland and Kevin McLaughlin. Cambridge MA/London: The Belknap Press of Harvard University Press. Print

Brown, Charles, and Ted Toadvine, eds. 2003. *Eco-Phenomenology. Back to Earth Itself*. Albany: State University of New York Press. Print.

Cohen, Robert S., ed. 1998. *Philosophies of Nature: The Human Dimension. In Celebration of Erazim Kohak*. Dordrecht: Springer. Print.

de Certeau, Michel. 1988. *The Practice of Everyday Life*. Trans. Steven Rendall. Berkeley/Los Angeles/London: University of California Press. Print.

Heidegger, Martin. 1997. Building, Dwelling, Thinking. In *Rethinking Architecture. A Reader in Cultural Theory*, ed. Neil Leach, 95–114. London/New York: Routledge. Print.

Holl, Steven, Juhani Pallasmaa, and Alberto Gomes-Perez. 2006. *Questions and Perceptions. Phenomenology of Architecture*. San Francisco: William Stout Publishers. Print.

Kohak, Erazim. 1984. *The Embers and the Stars*. Chicago/London: University of Chicago Press. Print.

Mendieta, Eduardo. 2001. The City and the Philosopher: On the Urbanism of Phenomenology. *Philosophy & Geography* 4 (2): 203–218.

Naess, Arne. 1989. *Ecology, Community and Lifestyle. Outline of an Ecosophy*. Trans. David Rothenberg. Cambridge: Cambridge University Press. Print.

Norberg-Schulz, Christian. 1979. *Genius Loci. Towards a Phenomenology of Architecture*. New York: Rizzoli. Print.

Pallasmaa, Juhani. 2005. *The Eyes of the Skin. Architecture and the Senses*. Chichester: Wiley. Print.

Sartre, Jean-Paul. 1962. *Literary and Philosophical Essays*. Trans. Anette Michelson. New York: Collier Books. Print.

———. 1978. *Being and Nothingness*. New York: Pocket Books. Print.

Seamon, David. 1979. *A Geography of the Lifeworld*. New York: St. Martin's Press.

———. 1993. *Dwelling, Seeing and Designing. Toward a Phenomenological Ecology*. Albany: State University of New York Press. Print.

———. 2011. Seeing and Animating the City: A Phenomenological Ecology of Natural and Built Worlds. In *The Natural City: Re-envisioning the Built Environment*, ed. Ingrid Leman Stefanovich and Stephen Bede Scharper. Toronto: University of Toronto Press. Ch. 15. Print.

Sessions, George, ed. 1995. *Deep Ecology for the Twenty-First Century.* Boston/London: Shambhala. Print.

Tymieniecka, Anna-Teresa. 1988a. *Logos and Life.* Book 1. *Creative Experience and the Critique of Reason,* Analecta Husserliana XXIV. Dordrecht/Boston/London: Kluwer Academic Publishers. Print.

———. 1988b. *Logos and Life.* Book 2. *The Three Movements of the Soul,* Analecta Husserliana XXV. Dordrecht/Boston/London: Kluwer Academic Publishers. Print.

———. 1990. *Logos and Life.* Book 3. *The Passions of the Soul and the Elements in the Ontopoiesis of Culture, the Life Significance of Literature,* Analecta Husserliana XXVIII. Dordrecht/Boston/London: Kluwer Academic Publishers. Print.

———. 2000. *Logos and Life.* Book 4. *Impetus and Equipoise in the Life-Strategies of Reason,* Analecta Husserliana LXX. Dordrecht/Boston/London: Kluwer Academic Publishers. Print.

———. ed. 2005. *Phenomenology of Life. Meeting the Challenges of the Present-Day World,* Analecta Husserliana LXXXIV. Dordrecht: Springer. Print.

# Eco-Phenomenology of the Human Environment: The Case of Intercultural Dialogue

**Angela Ales Bello**

*Invisible Harmony is better than Visible Harmony*

(Heraclitus, fragment 14)

**Abstract** The historical moment in which we live creates an opening onto cultures and religions different from our own, and it does so not only for reasons of knowledge and spontaneous curiosity but also to establish comparisons and to strengthen connections, all in the hope of realizing the unity of humanity that goes beyond difference. Following this direction of inquiry, which makes possible an encounter and, therefore, a dialogue, both the theoretical and practical can be correlated.

Phenomenological inquiry is very "versatile," and it is successful at finding an originary common ground, which is useful for interpretation, especially in terms of analyzing the essence of what we call "lived experiences." By probing the cognitive-expressive structure of human beings, it becomes possible to understand the sense of cultural formations and visions of the world that characterize diverse peoples.

I find phenomenological archaeology the most adept instrument for probing the inside of cultures and religions. In this regard, I would like to demonstrate how phenomenological inquiry can investigate connections and differences among cultural and religious expressions as well as how this investigation can be useful for concrete and lived human encounters from an ethical perspective. My research is in the line traced by Anna-Teresa Tymieniecka when she inaugurated the book series "Islamic Philosophy and Occidental Phenomenology in Dialogue."

---

Translated by Antonio Calcagno

A. Ales Bello (✉)
Pontifical Lateran University, Rome, Italy

 
W. S. Smith et al. (eds.), *Eco-Phenomenology: Life, Human Life, Post-Human Life in the Harmony of the Cosmos*, Analecta Husserliana CXXI,
https://doi.org/10.1007/978-3-319-77516-6_39

The theme of my paper is a recurring one and, perhaps, even overworked. When I first discussed this theme at the beginning of the 1990s, it was still "new" (Ales Bello 1992). I propose, however, that we return to it (Ales Bello *The Divine in Husserl and Other Explorations*). I maintain that a philosophical-phenomenological investigation can significantly contribute to clarifying the theoretical presuppositions of intercultural and interreligious dialogue. We can remember on this topic Anna-Teresa Tymieniecka's project of prompting dialogue between Islamic Philosophy and Occidental Phenomenology in order to foster a New Enlightenment (Tymieniecka et al. 2014).

In elaborating my own phenomenology of religion, I have two points of reference: one is historical and the other theoretical. They are largely held to be distinct in contemporary research, ultimately constituting two separate disciplines. The phenomenological inquiry that I have undertaken holds, on the contrary, that these two points are connected. I present a phenomenology of cultures and religions that establishes a relation between the aforementioned points of reference on different levels. On one hand, my reflection focuses on what "religion" and "culture" mean. On the other hand, in order to carry out my study I have to move within particular cultures and religions that present themselves as phenomena whose "sense" needs to be analyzed while avoiding superfluous abstraction. I proceed, then, by following two directions: from bottom to top and from top to bottom; that is, from the plane of concrete history to theory and vice versa, but also from human experience to philosophical theorization and vice versa. We will discover that such planes of reference are, in reality, connected.

Here I will present the theoretical guidelines of my project. Phenomenological inquiry is very "versatile," and it is successful at finding an originary common ground, which is useful for interpretation, especially in terms of analyzing the essence of what we call "lived experiences." By probing the cognitive-expressive structure of human beings it becomes possible to understand the sense of cultural formations and visions of the world that characterize diverse peoples. Moreover, the historical moment in which we live creates an opening onto cultures and religions different from our own, and it does so not only for reasons of knowledge and spontaneous curiosity but also to establish comparisons and to strengthen connections, all in the hope of realizing the unity of humanity that goes beyond difference. Following this direction of inquiry, which makes possible an encounter and, therefore, a dialogue, both the theoretical and the practical can be correlated.

I find phenomenological archaeology the most adept instrument for probing the inside of cultures and religions. In this regard, I would like to demonstrate how phenomenological inquiry can investigate connections and differences among cultural and religious expressions as well as how this investigation can be useful for concrete and lived human encounters from an ethical perspective.

It seems impossible in our present day to hold that in order for various cultures to encounter one another it is necessary to pass through a religious moment. This passing through a religious moment or stage appears to contrast directly with our own western culture, which largely presents itself as a-religious. If western culture is examined more deeply, however, one can see that in certain distinct moments, for example, liberal democracy and its view of human rights, we find traces of Christianity, now, of course, secularized. In this sense, the recognition of the role played by Christianity decisively contributes to the understanding of our culture, which is also strongly influenced by a commitment to philosophical and scientific reason.

All this demonstrates the complexity of the western cultural world, a complexity that reintroduces itself, even if not in the same terms, in other cultures. We need to reenter the mindset of other cultures and, above all, to reenter the domain of individual and collective lived experiences that characterize these cultures. I would like, therefore, to pause and analyze the methodology of the phenomenologist Edmund Husserl. Proceeding from the analyses he left behind, he enables us to understand how it becomes possible to delineate a phenomenological archaeology, which is an apt tool for describing cultural formations.[1]

## The Life-World as the World of Culture

The entry way constituted by the life world as described by Husserl, especially in his work *The Crisis of European Sciences and Transcendental Phenomenology*, permits us to enter the human world as a cultural world as well as an historical one (Husserl 1970). We encounter human expressions that we call cultures; we observe operations that manifest themselves in the transformations of the given natural world and a variety of productions. At the base of such diversity, we can trace the creative and expressive capacities that constitute such a world. If we wish, however, to understand how this diversification happens, we cannot let ourselves be dragged down by the "Heraclitean river," as Husserl says, of those variegated productions. In reality, we note that there exist many disciplines, especially within western culture, which were formed to plumb the depths of the aforementioned river in order to trace out its most important currents: aesthetic disciplines that deal with artistic production, scientific disciplines that focus on the understanding of nature, historical disciplines that analyze human events, as well as sociological and psychological disciplines.

One can observe in western culture and even in other cultures – especially in Asian ones, albeit in other ways – a radical exigency, namely, to plumb the depths of reality in order to understand phenomena. This demand has been called philosophy in our culture. It continues to have meaning because philosophy demands that we not entrust our understanding solely to particular and specific domains of research; rather, it is necessary to suspend their validity, to bracket them, to subject them to an *epoché*. This is why Husserl maintains that he needs to insert himself within the furrows of western philosophy. In fact, phenomenological philosophy proposes to see itself as *erste Philosophie*, first philosophy, understood as a rigorous science, *strenge Wissenschaft*, which he saw as unfolding with the ancient Greek philosophers, including Plato and Aristotle. Husserl believed that the project of first philosophy was most valid and had to be brought to completion. Indeed, this is the "trade" of the philosopher, her existential calling, her vocation or *Beruf*. One must bracket everything, which does not mean negating everything but instead critically examining everything.

We are dealing with a difficult operation, for it could lead to an empty territory. One risks not finding any supporting way stations: one may head into a desert and lose sight of the very human productions one seeks. On the contrary, however, after bracketing the products of humans, we are faced with the human beings who developed these products within the march of history. Here, a problem arises: who are we that ask ourselves questions concerning the meaning of our cultural productions, and who are they that produced them? The personal pronoun "we" has a double meaning: we are the ones whom they seek but we are also the ones who produce. The first "we" reveals our insertion into an effective community of researchers, whom we remember and consider present in our own day. The second "we" is that of humanity and a group of humans understood as producers of culture.

The *epoché*, which opens up a pathway of backward-moving research, leaves as its residue, then, the "we" and the I, always situated within the tension of the two moments. We understand why Husserl, then, can speak of a reduction to the subject or even, at the same time, a reduction to intersubjectivity. But, the way he chooses to carry out his programmatic work, namely, the project of his *Ideas Pertaining to a Pure Phenomenology and Phenomenological Philosophy*, is one that passes through the residue of subjectivity and not the empirical subject (Husserl 1983). He focuses on that which emerges from the completed investigation of the fundamental structures that I find in myself and others. What do I find in myself and in others? The object of the discovery, suggested and indicated by the studies on psychic acts carried out by Husserl's teacher Franz Brentano, constitutes the original moment of Husserl's position. This discovery is based on a preliminary belief, which says that every phenomenon that presents itself can be grasped intuitively in its essence: we can understand the sense of the phenomenon. This grasping is what Husserl calls the eidetic reduction or the reduction to essence, and it constitutes the first step of his method. Which phenomena's sense am I able to grasp fully, if not those that are related to my very own interiority? It is here that the dimension of the acts that I am living discloses itself to me: this is the dimension of lived experiences or *Erlebnisse*.

This dimension is understood to be transcendental, for it is characterized by the presence of cognitive, affective, and valuing structures, which are common to all human beings and which also permit reciprocal exchange among humans. These structures are human. In reality, phenomena that permit full and intuited knowledge, insofar as they are objects, are *Erlebnisse*, which we translate in English as lived experiences. The term literally translates as "experiences that are lived by us in consciousness." It is easier to understand the sense, the essences, of lived experiences because the researcher lives them. This is highlighted not only by Husserl but also by his disciple, Edith Stein. By employing the term "lived experience," the notion of interiority comes to the fore, for *Erlebnisse* are experiences lived from within. Interiority, however, does not mean that we remain closed within ourselves:

we have to probe it, for in its depths we find an exit, a veritable opening that realizes itself through the lived experience of empathy or intropathy[2] (*Einfühlung*), which can be understood as an opening onto the alter-ego, the other.

## The I and Others: The Theoretical Basis of Dialogue

Intropathy is a lived experience present in all human subjects. It has a preeminently cognitive character. Intropathic acts sometimes achieve, in certain cases, the goal of living that which the other lives, but never is there an absolute identification between me and the other. Here, we face the real possibility and limits of intropathy.

To understand deeply the lived experience of intropathy from a phenomenological point of view we need to perform a double reduction: first, to my 'I' that 'feels' what the other is living and, second, to my 'feeling' of what the other is living. In this case, the content of my feeling does not belong to me as does my remembering an event. I present to myself something that is not mine and it, therefore, transcends me. The other's joy can be seized by me as joy, but the modality, intensity, its quality, all of these can only be given to me in adumbrated forms. I am unable to live it in the first person, even if I rejoice with the other for the very same reason that the other rejoices.

Many implications follow from the description just given, even though appears simple. First, as was said earlier, the possibility and limits of communication come to the fore. I can live what the other is living, and I can even reverse this very statement by saying that because I live what the other is living, I 'feel' (*fühlen*) that the other is a human being that is similar to me. I can complete this operation of 'analogy' between me and the other in a spontaneous and non-reflective way. I maintain that this is the profound core of an analogy that only secondarily can become an intellectual and reflective fact, that is, analogical reasoning upon which many philosophical arguments rest. In this way, we understand Husserl's insistence on distinguishing this type of analogy from all other types of theorizing that discursively lead to the establishment of analogy.

Insofar as the understanding of the other can be actualized to its depths, its transcendence *vis-à-vis* me is constitutive, but I will never identify myself with the other: his/her individuality and my own will always remain separate and different, even though they are also similar and communicable. Here, we are not dealing with a theory that needs to be confirmed by experience. Rather, we are confronted by an experiential fact that is read phenomenologically, that is, this fact is understood in its specificity.

The results of my analysis up until now are not foreign to the investigation of intercultural encounters nor to the recognition of a common humanity in particular, which is felt through the act of intropathy and which does not exclude cultural differences; rather, humanity remains despite these differences.

This being said, we still need to pause and investigate the 'why' behind cultural differences in order to understand them more deeply. Such understanding forms an indispensable part of any possible encounter. To this end, it would be useful to probe further the dimension of collective and personal lived experiences. Among them we find religious experiences.

## The Divine as "the Other": Toward a Phenomenological Archaeology of Religious Experience

Owing to the links between culture and religion – in my opinion religion is the core of every culture – we can proceed to trace the common base of religious experience, which, even though it manifests itself in diverse forms in the history of religions, is present in all religions. First, we turn to the phenomenology of religion, especially as Gerardus van der Leeuw develops it, in order to highlight the presence of an "existential" attitude that characterizes human beings. In fact, this attitude is not limited to accepting the life that is given; rather, it seeks the *Power* behind life. The human being does not find the source of this power within itself, and not finding it in itself, it "seeks to allow the Power one believes in to enter into its life; one seeks to elevate one's life, to let one's life grow, seizing for oneself a deeper and wider meaning for one's life" (van der Leeuw 1933, 536).

The ultimate sense, the highest meaning, is religious meaning, insofar as no other sense can supersede it in terms of depth and breadth. But it is also a sense that gives and occludes itself and that is always "beyond." The *homo religiosus* wishes to understand life in order to control it, and this is why he always seeks new ways to do so; but he is also aware that he can never ultimately transcend his own limits. Human being will never reach the greatest heights; rather, the highest reality has to reach down to him, and does so in an understandable and mysterious fashion. This is why human beings become aware that they are being led into a strange terrain. One is conscious of finding oneself in every instant surrounded by marvelous things, and one knows with certainty that something is *coming towards one along the path*: it is the *Other* that does not have a name, but who nevertheless inserts itself into one's life.

If we examine the history of religions, we note that when a religious experience occurs, human beings are, first, stupefied, then they are terrified, and often this is followed by faith. Differences among religions notwithstanding, all religions are linked through a common guiding thread. This does not mean that they are all of equal value; rather, they are all *religions*.

On the basis of van der Leeuw's analysis, it is possible to tackle two vital questions: (1) given that religious experience must have a permanent essential structure, is it possible, then, to speak of the religious phenomenon as being a unified phenomenon?; (2) is such an experience constitutive of the human being? The first question legitimizes the very possibility of the history of religions, whereas the second represents a response to theoretical atheism.

Religiosity is not accidental: it is not an experience that can give or not give itself, that can be traced to something other, say, a psychological attitude (as for Feuerbach, Freud), or social conditions (as for Marx), or human existence carried through to its extreme consequences (as for Sartre). The human being is able to claim for himself "the will to power," as Nietzsche invites us to do, but even he, in employing such a turn of phrase, reveals a desire for a "Power" that he sees operating in him. In fact, even though Nietzsche shows us the push for liberation from the Other through his proclamation of the "death of God," he is nevertheless tormented by the difficulty of achieving that very act of liberation, that is, the re-appropriation of power.

To address the first question mentioned above concerning the common element present in all religious expressions we need to deepen our analysis of religious consciousness and the lived experiences that lay at its base.

## Anthropological Presuppositions of Religious Experience

Let us probe more deeply, then, within ourselves to understand the presence/absence of the divine. By doing so, we can analyze the structure of the human being. I will proceed by employing the directives of the phenomenological school and, in particular, those of Edmund Husserl and Edith Stein. The investigation of interiority not only helps us understand the sense of religious experience but also assists us in dealing directly with the question: What is the human being? This question cannot be grasped form the "outside"; rather, it is seized from within through an analysis of lived experiences. It is through *Erlebnisse*, "lived experiences," understood in the phenomenological sense of Husserl and, therefore, grasped in their structure or "purity," that it becomes possible to describe in an essential way the human subject (Ales Bello 2013). The subject is extraordinarily – Husserl even says "paradoxically" – capable of presenting herself both as the subject and object of a phenomenological investigation. One discovers that one is capable of "perceiving," "remembering," "imagining," "loving," "suffering," "desiring," "thinking," "valuing," and "reflecting." The aforementioned capacities and potentialities are uncovered as transcendental experiential structures that are consciously recognized: they are the mirror of the complex and elaborate layers that constitute the human being.[3]

From their very beginnings, religions and philosophies have understood that the human being is both body and soul. But how do we validate or refute their respective interpretations after we have bracketed, performed the *epoché* of, the question of belief or interpretation? The qualitative analysis of lived experiences is not only useful for understanding how we arrive at knowledge of the external world but it is also a useful instrument for the uncovering of human dimensions through lived experiences.

Perception actualized through sensations leads us to corporeity, whereas desire, attraction, or repulsion, the living of impulses and instincts, lead us to that domain we call psyche. Finally, acts of valuing, reflection, and decision-making lead us to the spiritual dimension (Husserl 1990).

I employ this stratification and the complexity of the human being in order to describe the sense of religious experience. Where can one find the source of that idea of "that which no greater can be conceived?" In other words, where can one find the traces of the divine?

I maintain that it is precisely in the core (*Kern*), in which consists the source of our singular person. The core remains always identical and there we can discover the traces of the Presence of what is not transitory, that is the Power. As the source of human development, the core manifests itself – and, therefore, we can say that it indirectly develops – through the human characteristics of corporeity, psyche, and, in a particular way, the spirit.

The recognition of such a Presence is subject to the spirit. If the spirit is fundamentally decision, will, and intellect, and is known through its essential characteristics, namely the lived experiences of the spirit that manifest themselves as free acts, it is possible to accept or refuse such a Presence. But, given that all the strata of the human being exist and possess unity, this Presence "passes" also through the acceptance or refusal made on the part of the psyche, which offers to the spirit material that it can examine, gathering or correcting psychic impulses. Even corporeity is involved because acceptance allows the body to take on certain attitudes that manifest a relation with this Presence; rites and liturgical acts justify themselves in this way.[4]

At this point, we can hypothesize two paths. On one path, one who 'feels' the presence of the Divine accepts it at the level of the psyche, and also consciously welcomes it at the spiritual level. One also feels this presence at the corporeal level. On the other path, one who refuses such a presence because it is not acceptable at the psychic level will devise theoretical arguments at the intellectual level in order to show that such a presence is illusory. This is the path of atheism.

## Welcoming and Listening as Presuppositions of Dialogue

The philosophic-phenomenological point of view and its theoretical contributions along with its ethical implications are both important and preliminary. It can create an opening unto other cultures, which in turn may lead to the taking on of the practical position of welcoming different points of view.

It is clear that once one has seized the connection between unity and difference, one can then take on a position of evaluation. In fact, every human being, even one who assumes the phenomenological attitude of the "disinterested spectator," belongs to a life world, to a cultural dimension, all of which provide evaluative criteria.

Every culture and religion has always considered itself as the true religion and the best culture, and it is good that it does so, if it wishes to remain in "good faith."

But from such stances two attitudes can arise: one is of absolutization of one's own point of view even to the point of condemning other religions and mocking those who are "different," and the other is an attitude of welcoming.

The criteria of welcoming – a term that is a better one than tolerance – is not seen in all religions and it certainly does not cross all cultures. But, given the growing frequent contact among religions and cultures, a question has arisen, especially in the west in the last decades of twentieth century, concerning the possibility of intercultural dialogue, a question that was hardly conceivable in past ages.[5]

Given that nothing is ever guaranteed in the unfolding of history and that there are always risks of regression, theoretical investigation has as its task the bearing of the burden of investigating difference without absolutizing or eliminating it. Moreover, the taking on of the evaluative position and a coherent behavior toward such a position allows us to place ourselves in an attitude of listening to the reasons of others, not so that we abdicate our own respective points of view, but in order to examine them more closely. In comparing one religion with another, one can arrive at greater certainty about one's own religion and what one believes, always mindful that the other is like oneself and that it is necessary to account for this fact. The encountering of other religions and the comparison that follows does not mean that one cedes to the other. From an anthropological perspective, Christianity has given us fundamental directions concerning the need to respect other human beings: from listening, attention can arise in such a way that one's own behavior can become exemplary for the other, showing in a concrete way the validity of certain principles.

All of this ascribes to western culture, which is born from the uniting of the Greek *logos* to the Christian message, a historical project of great importance that points beyond cultural differences and that shows the lines of a common humanity: we must allow the coexistence of the singularity of cultural subjects with the universality of the structure of the human to thrive, without eliminating particularities. Rather, we must establish the deep ties that make possible a genuine living-together.

We cannot hide the difficulty of such an undertaking because of the obstacles created by cultural differences, which are often lived as "difference," "foreignness," and which sometimes, sadly, are absolutized such that they become barriers that cannot be crossed. The insistence on the communal aspects present in human beings, discovered through philosophical, phenomenological investigation, can indicate a useful pathway for ethical behavior, ultimately guiding a practice of welcoming, listening, and especially "care" for human beings belonging to diverse cultures that have been struck physically and psychologically by an encounter, which often can become a conflict where others appear "different" and, therefore, "foreign." We can quote at this point Tymieniecka's proposal of a New Enlightenment, which aims to discover the common roots that all philosophies and all cultures share and which is linked up with the primeval *logos*.

It is useful, therefore, to develop and explain the theoretical instruments of knowledge that permit us to grasp the unity of human beings that underlies difference: both unity and difference provide us with an interpretative key for recognizing a genuine plurality of cultural subjects, which can lead to the realization of an authentic encounter.

In my opinion, these are the theoretical presuppositions of a philosophy of dialogue that can help us to probe the religious and cultural complexity of people, through a philosophical anthropology that brings to light universal elements present in every human being.

## Notes

1. The description of the phenomenological project of Edmund Husserl that follows can only make reference to certain aspects of his comprehensive analysis. I cannot develop fully my own interpretation here. For a more lengthy and detailed treatment, please refer to my *L'universo nella coscienza* (2007—2nd edition).
2. I prefer the term intropathy because empathy has been identified with an emotive capacity, namely, sympathy, which is foreign to the meaning of the German term as employed in classical phenomenology.
3. To understand the essential elements of the phenomenological method, please consult my lectures constituting an introduction to phenomenology, which have been published in Brazil and in Italy (Ales Bello 2006, 2009a). Concerning the relationship between the phenomenological method and anthropology, please see my book on the sense of things (*Il senso delle cose,* Ales Bello 2013, Chap. IV).
4. Concerning the relationship between human being and the Divine please see my book (Ales Bello 2009b) and also a later one on the sense of the sacred *Il senso del sacro* (Ales Bello 2014).
5. Excellent examples of such an attitude can be seen in the person of John Paul II, who organized a meeting of all religious believers in Assisi in 1986. In 2000, the opening of the Holy Door was also occasion for interreligious dialogue. In October 2012, Pope Benedict the XVI called for interreligious dialogue. One must also not forget the homage paid to Mother Teresa of Calcutta on the part of the Indian Government at her funeral in 1997. Also, Gandhi, when asked why there are so many different religions, replied: "Like a tree that has only one trunk but many branches and leaves, there is one true and perfect Religion: it becomes many when it passes through human beings. Religion alone transcends every word" (Achuraparambil 1986, 239). John Paul II, when discussing the centrality of the work of the Trinity, notes that the true religion is connected to the work of the Holy Spirit through the "*semina Verbi* (the seeds of the Word)," which constitute a sort of radical soteriological community of all religions (John Paul II 1994, 89).

## Works Cited


Achuraparambil, Daniel. 1986. *La spiritualità dell'induismo*. Rome: Studium. Print.

Ales Bello, Angela. 1992. *Culture e Religioni: Una lettura fenomenologica*. Rome: Città Nuova. Portuguese Translation: *Culturas e Religiões. Uma leitura fenomenológica*. 1998. Trans. A. Angonese. Bauru: EDUSC. Print.

———. 2006. *Introdução a Fenomenologia*. Bauru: EDUSC. Print.

———. 2007. *L'universo nella coscienza. Introduzione alla fenomenologia di Edmund Husserl, Edith Stein, Hedwig Conrad-Martius*. 2nd ed. Pisa: ETS. Print.

———. 2009a. *Introduzione alla fenomenologia*. Trans. F. Rodrigues and A. M. Sciacca. Rome: Aracne Editrice. Print.

———. 2009b. *The Divine in Husserl and Other Explorations*, Analecta Husserliana XCVIII. Dordrecht: Springer. Print.

———. 2013. *Il senso delle cose. Per un realismo fenomenologico*. Rome: Castelvecchi. Print; Angela Ales Bello, ed., *The Sense of Things. Toward a Phenomenological Realism*. 2015. Analecta Husserliana CXVIII. Dordrecht: Springer. Print.

———. 2014. *Il senso del sacro. Dall'arcaicità alla desacralizzazione*. Rome: Castelvecchi. Print.

Husserl, Edmund. 1970. *The Crisis of the European Sciences and Transcendental Phenomenology*. Trans. D. Carr. Evanston: Northwestern University Press. Print.

———. 1983. *Ideas Pertaining to a Pure Phenomenology and Phenomenological Philosophy*, vol. 1. Trans. F. Kersten. Dordrecht: Springer.

———. 1990. *Ideas Pertaining to a Pure Phenomenology. Ideas Pertaining to a Pure Phenomenology and Phenomenological Philosophy*, vol. 2. Trans. R. Rojcewicz and A. Schuwer. Dordrecht: Springer. Print.

John Paul II, 1994. *Varcare le soglie della speranza* [Crossing the Threshold of Hope]. Milan: Mondadori. Print.

Tymieniecka, Anna-Teresa, Nazif Muhtaroglu, and Detlev Quintern, eds. 2014. *Islamic Philosophy and Phenomenology in Dialogue. The Logos of Life and Cultural Interlacing*, Islamic Philosophy and Occidental Phenomenology in Dialogue 7. Dordrecht: Springer. Print.

van der Leeuw, Gerardus. 1933. *Religion in Essence & Manifestation: A Study in Phenomenology*. Trans. John Evan Turner. New York: Macmillan.

# Digital Reason vs. the Modern "Metamorphosis of Man": From the Perspectives of the Philosophical Anthropology of Józef Bańka and Anna-Teresa Tymieniecka

**Jan Szmyd**

**Abstract** The paper contains a critical analysis of the *digital reason* understood as an antithesis for theoretical and practical reason and the *heart's reason*. The main thesis is as follows: *digital reason* performs a very important role in the field of quantitative cognizance of reality and acts as an accelerator of cognition, yet it is not capable of replacing natural theoretical and practical reason in the process of the quantitative cognition of reality, its essence, and development. It is not capable of understanding contacts with human subjectivity and personality, with the *humanum*. The domination of *digital reason* in a digital civilization would entail not only epistemological destruction but also the possibility of the inner disintegration of the subjective harmony and spiritual life of *Homo sapiens* and transformation of *Homo sapiens* into some kind of superficial humanoid *post Homo sapiens*.

**Keywords** Philosophical anthropology · Digital reason · Metamorphosis of man · Recentivism and Banka · Digital humanism · Tymieniecka and philosophical anthropology · Phenomenology of evolution

J. Szmyd (✉)
University of Krakow, Krakow, Poland



W. S. Smith et al. (eds.), *Eco-Phenomenology: Life, Human Life, Post-Human Life in the Harmony of the Cosmos*, Analecta Husserliana CXXI,
https://doi.org/10.1007/978-3-319-77516-6_40

## Changing Humanity Amid the "Contemporaneity" of Scientific and Technical Civilization: As Seen from the Perspective of Józef Bańka's Recentivism

### *Józef Bańka's Assumptionless, Intuitive Philosophical Approach to Humanity's Current Metamorphosis*

Cognitive clarification, performed in an in-depth and understandable manner, along with assessment, done in an evaluative manner, of the issue of the contemporary transformation/metamorphosis of man is not possible without the participation of philosophy, that is, the form of human thought that asks about the core, the sense, the main principle, and the objective of "Everything." This includes the core and the sense of development of the "human being," understood as a specific biological species (type) and as an individual. Obviously, not everything that is related to the potential of cognitive activity and evaluative philosophy may be, within the expected time, satisfactorily accessible. This is also related to the issue of the development of philosophy as such. Nevertheless, it can be assumed that philosophy can offer a "good beginning" for this weighing of the contemporary scene or, at least, provide a correct "key" for the right direction and even manner of effective investigation along this indispensable path of cognitive research.

It is clear that only an inventive and original philosophy is up to such a task. The best type of philosophy to be pursued here would be a philosophy without assumptions, unburdened by the ballast of the ready-made mottos and theoretical concepts of traditional systems and philosophical lines. This should be a philosophy that is "virginally" intuitive, interested, and inquisitive, intellectually non-schematic and spontaneous, and, as far as possible, maximally exploratory and inventive. Such particularly creative and explorative instances of philosophizing are relatively few and far between in modern times. Among these few are the so-called recentivism philosophy of Józef Bańka[1] and the phenomenology of life and man of the recently deceased Anna-Teresa Tymieniecka.[2]

It is necessary to briefly consider the contributions of these two original formulations of philosophical thought to the issue of the current transformation or metamorphosis of human being.

Let us start with Józef Bańka's recentivism. The starting point for this intellectual system and the main context for its philosophical descriptions is the character and the specific features of the thoroughgoing (however imperceptible it be to the "naked eye," or usually so) transformation of man in the present times, that is, the so-called "great metamorphosis" of the twenty-first century. This consists in the fact that, as a result of the exceptionally swift development of such features and properties of the modern stage of civilization as our burgeoning information technology, the development of the Internet, digitization, and the advent of so-called "digital reason" and "digital humanism," new qualities and modalities are emerging in people on the basis of these processes, which form a new man in a new era, so-called

*Homo digitalis*. A separate line for investigating man, one called *eutyphronics,* has been grown out of *recentivism*. It is aimed at saving man from the deformation and the evaluative regression that are threatening us. With this objective in mind, the concept of *Homo eutyphronicus* has been advanced and a manner by which to gradually attain this presently most optimal model of being human has been demonstrated. This is a person who cannot avoid a number of significant changes caused by contemporary technical, scientific, informatics, and medical progress, mainly in the form of the technization and informatization of a person's life and activities, and more particularly in the form of the newly honed and increasingly conquering *digital reason*. Nevertheless, this is a man who does not exceed the "human limit" and who remains, in its classic understanding, a human being, preserving his *humanum* and combining thinking reason ("clear reason") and feeling reason ("the heart's reason") with calculating reason (digital reason), which is useful indeed, but which has not been operationally absolutized into constituting the entirety of human functioning. According to the philosopher, only a harmonious combination within the mentality of the changing human being of these types of reason can allow avoidance of evolutionary deformation and possibly our complete destruction, which is to say, a transformation into some kind of post-human "techno-organizational hybrid" (Bańka 2014, 605–607).

The first step in providing such philosophical support for man, one embedded in *eutyphronics,* in the face of this evolutionary threat is a simple attitude towards protecting the human psyche from the various negative impacts of modern technique, including "protection of the psychic sphere of man from the impacts of the complete digitization of the contemporary information civilization" and postulating "the creation of a humanistic culture capable of generating technique in a manner that increases the value of personal life" and "warm" principles of human conduct, "appealing to the feelings and emotions of the individual"; in particular, this means setting a requirement of proper "measure and proportion between the spheres of *thymos* and *phronesis* in man," that is, the subjective, emotional, experiential, and moral sphere, on the one hand, and the intellectual, cognitive, functional, and practical sphere, on the other (Bańka 2014, 620, 625).

Another step along the path leading in the direction of the desired model for the man of the IT and digital age is the formulation of the ideal of straight-thinking, with its cognitive, axiological, moral, existential, and anthropological aspects. The cognitive aspect of the idea of straight-thinking consists, generally speaking, of an assumption that conceptual and discursive cognition – whose main expression is scientific, rational, and philosophical cognition – not only does not exhaust the cognitive potential of man but also does not provide natural and fully authentic cognition. Therefore, it requires supplementation by various types of direct cognition, namely, intuitive, emotional, commonly experienced, and via theoretically assumption-free thinking: "thinking for the first time" (*a recentiori*). However, the axiological and moral aspect of the idea of straight-thinking relies on the belief that the most important human values are "simple values of the spirit and the intellect," values that are experienced directly and personally, subjectively recognized, and valued most. These include such values as: truth, kindness, beauty, personal happiness, and dignity. The ethical thread as well of straight-thinking is manifested in the postulate of referring for our conduct to the moral feeling inborn in human nature, our natural sensitivity and moral feelings (Bańka 2000).

## *Preserving the Natural and Personal in the Human Without Falsely Opposing Quantification and Depth*

The summit of the idea of straight thinking, as it were its anthropological culmination, is personified in the "straight-thinking person," who constitutes, together with personal and subjective features, mental and behavioral features, the starting point for the desired model of man in our modern scientific and technical civilization, essentially, the ideal "eutyphronical man" (*Homo eutyphronicus*).

This ideal person embodies such characteristics as balance (*equilibrium*) between the emotional sphere (*thymos*) and the intellectual sphere (*phronesis*) in the structure of his or her personality and, on the intellectual level, a harmonious cooperation of reason, thought, and qualitative understanding ("clear reason") as well as calculating and information-fed reasoning ("digital reason"). The integral and very interesting personal features of this discussed model of man also include vitality and cultivation in one's attitude toward life, natural morality (straight-thinking), and acknowledgement of the priority of spiritual, auto-telic, and personal values over material, utilitarian, and impersonal values. Generally speaking a eutyphronical person who fully preserves a naturalness and all that is personal in the sphere of his or her predispositions and interests, and does not hand over cognitive autonomy to machines; he or she does not discard his or her cognitive powers and potential and, in particular, does not surrender his or her analytical and comprehensive powers for the advantages offered by technique. Through this, one remains a *Homo naturalis* and *Homo sapiens* without being reduced to *Homo digitalis* (Bańka 2014, 22–25).

The author assumes, contrary to the dominant tendency regarding this issue in the literature on the subject, the possibility of preserving, even within the perspective of a digital and information-fed civilization, such a natural model of man and basic standard of classically understood humanity and the humanistic mental paradigm related to it.

However, it is not necessary to in desperation entrench oneself in the culture of the classical humanities; one should be open to enrichment by new values, which may be created by the further development of scientific and technical civilization.

> A man cannot be analyzed in enlargement; in the best case, he can be offered stilts to be taller, but he always has to be accepted in cognitively natural dimensions corresponding to his nature. Thence, the situation becomes paradoxical: some notice an emptiness of form without content, which characterizes digital humanism; others, who are beyond such worthless (in their opinion) and empty network formalisms, trumpet a return to traditional non-quantitative humanism (copying objects that have already been shaped). In this very situation, the correction proposed by eutyphronics goes in the direction of fulfilling the ideal of straight-thinking, i.e., of replacing the choice between quantity and depth with the

choice between complexity and simplicity. Within these very bounds, there are beautiful ideals of freedom and democracy, omnipresent affluence, and science guarding everything as the bulwark of rational thought, guarantor of unlimited progress and peace in the world. (Bańka 2014, 607)

## A Critique of "Digital Reason"

The main strategy for protecting the personal and mental status of *Homo sapiens*, our identity and naturalness, straight-thinking, and all of those features constituting humanity, involves making a thorough and – as it seems – effective critique of "digital reason" and, in consequence, of *Homo digitalis*, who potentially represents a bridge and is a probable springboard by which to transfer to the side of *post-Homo sapiens*, as forecast by many observers. Below are the most important elements of this critique:

- "Digital reason" is only able to capture but one dimension of actual and virtual reality, by its capturing it quantitatively. It is only able to procure and transfer digital information. It can indeed calculate and so analyze reality much faster and "further" than can the natural mind. However, it is not capable of performing the reasoning that is indispensable for maintaining a deeper, rational, and evaluative orientation in the world through thinking, meditating, and the intuiting of cognitive insights. It lacks what only "natural reason" and the "heart's reason" can accomplish, the definitive level of reasoning, wisdom, feeling the reality, having emotional and intuitive contact with it, qualifying it from the point of view of good and evil, beauty and ugliness, friendly closeness and alienness, kindness and wrongdoing, and the like (Bańka 2014, 435–548);
- Digital reason is not capable of comprehending and expressing the sphere of the spiritual values of man, the core and specific nature of man's spirituality, everything that is unique and extraordinary in his personality and individuality. Digital reason is also not capable of comprehending, in a cognitive and understanding manner, what is subjective and personal, for example, the sphere of personal predispositions and interests, desires and longings, dreams and beliefs, experiences and existential problems, metaphysical and religious problems, aesthetic preferences and creative acts, internal movements and moral reactions, ethical and humanistic "straight-thinking," and so forth (Bańka 2014, 528–603);
- Digital reason's "sphere of incapacity" is particularly clearly marked in reference to the area of human desires and the will, as opposed to the capacity of the natural mind and the "heart's reason." This area includes the choice of original human objectives and so-called cognitive values, that is, values "which offer man a surplus beyond the digital in his life and which humanity shows the utmost interest in relating to. This is to point to *the values which make it "good" to live, even if they prove difficult, and without which life is worthless, even if it be easy"* (Bańka 2014, 609); (emphasis added).

## *Spotting the Dysfunctional Repercussions of the Digital Advantage*

The displacement of the "heart's reason" by "digital reason" in the personal and mental structure of man that is taking place at the present stage of development of our scientific and technical civilization, this age of "technotronics," is resulting in basic changes in our subjective identities, species, human nature, and humanity. It is the beginning of the formation of a new type of person; this man, as he is called by the author, is "incomplete," "quantified" and he is losing the balance between the "values of personal life," the sphere of "warm" feelings and experiences, spontaneous stances and kinds of behavior, "straight-thinking" references to others and to oneself and the sphere of "cold," calculating, clearly pragmatic, and "scheming" reason. Reason is guided, in the world of digits and digital concepts, by the idea of countability and utility and not by any form of ethical and social law, such as the rule of selflessness and seeking the common good (Bańka 2014, 609).

*Recentivism* not only exposes the main transformations of man in the "huge metamorphosis of the 21st century" but also projects the trends within humanity into a more or less predictable future, foreseeing an approaching border, i.e., a transition from the human to the post-human. This transition, whereby "*Homo sapiens* ceases to be himself and becomes *Pianthropus digitalis*, becoming an 'incomplete man,'" one "endowed only with a capacity to receive information and that only in one form: the digital form"; in still further transformations, we would become some sort of post-human, "techno-organic hybrid" (Bańka 2014, 15).

This may be considered to a certain extent a both probable and ominous vision, one of the end of "natural man" (*Homo naturalis*). But it is sketched only out of criticism of "digital civilization" (which, according to the author, "in spite of doubtless advantages, does not perform well," because "it is not the proper manner of implementing the strictly humanistic values of man," from which the also projected ideal of *Homo eutyphronicus* has been derived (Bańka 2014, 614). Eutyphronic man is a spiritually "complete" person, one whose personality has been harmoniously shaped (with a balance between emotional, individual experience, and all the axiological sphere and the rational, practically functional and information-fed sphere), one who then behaves without detriment to self and others and who may even regain human identity and proper humanity so threatened or fundamentally deconstructed under the pressure of our digital civilization. Currently, considering the constantly growing role and intensifying impact of "digital reason" on the life of modern man, "it is necessary to talk more often about the "heart's reason," instead of the "digital reason." Digital reason is, in principle, a mathematical mode, whereas the heart's reason is something that offers identity, which is a liminal fact for a human being; identity is man's "starting point's riddle" (Bańka 2014, 616). It is quite easy to agree with this very justified statement and this very valid, humanistically-oriented advice. However, the question arises as to whether it may still count on broader acceptance by "post-modern man" and recognition of its practical, utilitarian ultimate purposefulness?

## Phenomenology of the Evolution of Life and Reason: From the Perspective of the Metaphysics Advanced in Anna-Teresa Tymieniecka's *Logos and Life*

### *Advancing the Metaphysical Vision amid Information Overload and Civilizational Confusion*

The "post-modernist" age is characterized, according to Anna-Teresa Tymieniecka, in the first place by an accelerated transformation of human life in all its areas and manifestations: social, cultural, intellectual, customary, moral, aesthetic, practical and life-related, existential. The main reason for this transformation, according to her, is scientific and technological development, an unprecedented "display of scientific and technological knowledge" so great that "we are facing a *real upheaval in our view of the world and in our approach to life and its conditions*" (Tymieniecka 2010, 7); (emphasis added). The unprecedented events made possible by this progress such as, for example, sending space probes to other planets, unique inventions that change human life in time and space, quick telecommunication, and numerous devices that facilitate and accelerate the speed of everyday life "have not only transformed in numerous ways our existence but also have us on the alert for further wonders and shocks. All humanity simply expects*, and is in some dread of a never-ending, advancing transformation of life*" (Tymieniecka 2010, 7); (emphasis added). The uncertainty of today and tomorrow is intensifying; paradoxically, the rapid development of scientific knowledge is contributing to it. "Expanding knowledge of nature, the world, the cosmos, of human beings too, *keeps humanity in perpetual incertitude*" (Tymieniecka 2010, 7); (emphasis added). The situation is also aggravated by a rapid increase in the variety of new ideas, experiences, customs, and intuitions, which require a lot of time and effort to comprehend, to understand them, and, sometimes, to apply them in practice. Generally speaking, "we remain *lost in the mass of the ever changing...cannot come to terms with and embrace the ever fresh, even startling appearance of reality*" (Tymieniecka 2010, 7); (emphasis added).

Among the characteristic properties of the "unique times" in which we happen to live is the fact that we have to more broadly confront people from other cultures (owing to the increased migration of people). We have to cope with various and often very difficult social problems, customs, and inter-human problems amid this historically-unprecedented process, even as there is a loosening of the value criteria of own cultural identity and a more and more acute lack of clear guiding points in life and valuating measures (ethical, worldview, aesthetic) and worrying signs of the warping and crisis (and possibly even a clearly marked fall) of own culture and the regress of culture into the abyss of "barbary" (Tymieniecka 2010, 7). These negative cultural transformations are also accompanied by a radical intellectual, cognitive, and philosophical breakthrough; essentially, "old, fossilized chains of terms, theories and pre-determined concepts concerning the human character, nature, moral standards, rules of ethics" are becoming loose and weakened in their validity and "the strength of conviction they carry with themselves has yielded to new perspectives opened by scientific progress" (Tymieniecka 2010, 9). In particular, this refers to the traditional concepts by which the sense of the world around us has been apprehended and the basic significance and intuition of human experience relayed (Tymieniecka 2010, 7). This shakes classic rationality (Aristotle, Descartes, Bacon, Hobbes, Locke et al.) and, in consequence, causes "disorientation within the fluctuating and rapidly advancing waves agitating with ever new perspectives opening upon reality" (Tymieniecka 2010, 8).

However, in the growing confusion in the area of man's orientation in and comprehension of the world, humanity's classic dream of a metaphysical vision does not disappear, according to the author. In this vision, the deepest riddles of reality and human cognition are discovered and clarified, along with final causes and principles, perspectives of human reason in the examination of life, the world, and humanity. And in the exploration of the evolutionary development of reason and skills, manners, and techniques of domesticating the forces of nature and controlling them, understanding the sense and the core, the genesis, and the transformations of the reality surrounding man, she discerns the driving and causative force: the Logos of All-being, and especially the particular reality that is Life as such and the *Logos of Life*.

Therefore, amid our contemporary civilizational confusion, information overload, and cognitive disorder, philosophy, in its metaphysical sense, does not wither and cannot wither. Only traditional and fossilized forms of philosophy can wither or lose their significance, but in its modern forms philosophy is revived and gains new directions for development.

However, modern metaphysics faces new tasks not dealt with before, without even mentioning the necessity of taking the new situation into account in the area of scientific and technological progress. However, what the author writes about the new cognitive tasks of philosophy is particularly interesting. She says:

> There is to be considered not only our more fundamental understanding of our fabric, of the human mind in its evolutionary course, but also the contemporary clarification of the nature of language in framing reality's interpretation. There are being elaborated stricter postulates of reasoning, criteria of certainty that call for a critical assessment of conceptions hitherto accepted in philosophical inquiry (e.g., subject and object, individual and community, essence and existence, substance and accident). Furthermore, there is to be appreciated the significant new insights we have into the associative links, communicative threads, etc. that lead to a more adequate picture of the real. (Tymieniecka 2010, 8)

Such a philosophy is possible and needed at a time when "humanity, after further periods of human barbarism and despair, is apparently plunging into further chaos as disorientation about everything and the "deconstruction" of all footholds in life proceeds. We cannot be struck by the seeming failure of hope, but equally by the profound misunderstanding therein of the present situation of humankind." (Tymieniecka 2010, 9); (emphasis added).

However, we need something more than a renewal of philosophy that is maximally and metaphysically oriented; in these times of great transformation caused by sudden scientific, technological, social, and civilizational changes, a *new critique of reason* is possible and needed, according to the author of the *Logos and Life* volumes. This critique should lead to its renewal. Tymieniecka declares, "A vision of reason that breaks out from the narrow traditional framework and opens up creatively toward appreciation of the host of new rationalities now expounded is need in order to deal with the changeable currents of existence, to generate criteria of validity, predictability, prospects, measure." (Tymieniecka 2010, 10).

This fundamental renewal of reason is already taking place. It is visible in modern scientific studies and in the development of the skills of human reason that conduct such studies. This may be, according to the author, the announcement of something greater: the "renewal and repair of man," a "New Enlightenment of humanity."

The *credo* of Anna-Teresa Tymieniecka, with respect to this important issue which determines the fate of humanity, is as follows:

I am claiming that in fact, beneath under the present-day mood of disarray and our feeling that we lack a compass, there is a deeply brewing flux of renewal, growth, and the perfecting of humanity. As Voltaire, the herald of Enlightenment, voiced it, the progress of humanity depends upon the renewal of reason. It is, indeed, from a rebirth of reason proper that we are heading toward a New Enlightenment, which I herald. In a situation comparable to that of the Eighteenth Century, we are, indeed, ready to launch A NEW ENLIGHTENMENT OF HUMANKIND. (Tymieniecka 2009, xxiii-xxiv)

## *Weighing Tymieniecka's Diagnosis of Our Contemporary Plight and the Fix She Tenders in a New Critique of Reason*

From the point of view of the issue that is of major interest to us, that is, the transformation and the development/regression of humanity in an era of great and historically unprecedented transformation of the "human world" – civilizational, social, cultural, scientific, technological, psychological and mental transformation – there are several remarks and general conclusions to be made here.

First of all, metaphysical determinism and creationism (the stimulating, ordering, and creative role of the Logos in All-being: the main demiurge, the driving force, and the "sense of the senses" of everything that exists, lives, and changes and develops in the Cosmos, the perpetrator and the guide of the evolutionary process and the objective/objectives of the changing human and para-human reality) is strongly emphasized and displayed in the philosophy Tymieniecka worked out in her *Logos and Life* volumes, which does not mean that she excluded temporary inhibition and turning back, regression, and various disruptions and perturbations in this technological and progressive process. Her philosophy also does not exclude temporary regression of the human being, the "receding of man and his culture into the abyss of barbarity" and certain regions of moral and existential disorder, axiological chaos and commotion; "slipping" into the feeling that we lack a compass and "deep misunderstanding of the current location," even a fall of rationality and hope.

Secondly, the evident dynamic intellectual development and quick growth of the practical skills of man in an epoch of great transformation is not accompanied by any confirming important development in other spheres of life and activity: spiritual, personal, moral, artistic, and the like. Failure to perceive any positive changes in this very important sphere from the point of view of complete humanity, with simultaneous ascertainment of such achievement only in the circle of narrow elites, for example, researchers, inventors, and scientists, and not among average people, seems to indicate not only a diagnosis of a quite narrow and one-sided development of a few representatives of modern *Homo sapiens* but also, indirectly, the suppression, standstill, and even regression of many other, equally important sectors of humanity, such as intellect, "theoretical reason" and "practical reason" which gives us indirect, but very acute, confirmation of manifest anthropological regression.

Thirdly, ascertaining significant "wear and tear" in modern rationalism (that of Descartes and Bacon) today, along with its clear cognitive and methodological malfunctioning, is accompanied by indications of an even more burdensome and unfortunate situation, the appearance, among various alternative genres of rationalism of such varieties that have nothing to do with their own rationality and even contradict it (in this case, an example may be provided by certain concepts of rationality adopted in the radical trend of postmodern philosophy such as so-called deconstructionism). And there is a rising tide of various influential types of irrationalism related to philosophy, worldview, life, and religion.

A "new rationalism" and a "new vision of reason" worked out by some philosophers and contemporary researchers, and cognitively practiced by Anna-Teresa Tymieniecka, one which would allow for finding one's bearings in the "thicket of constant changes" and keeping up with the pace of a constantly new and surprising reality, which would be creatively open to the multiplicity of currently examined rationalities, does not constitute a breakthrough in the grounding of philosophical awareness and scientific "modernity." It still leaves unanswered the question of whether a modern man can be open to being a rational being (Szmyd 2012, 223–243).

Fourthly, there is the announcement of a "New Enlightenment of humanity" according to the author of *The Fullness of the Logos in the Key of Life*, Book I: *The Case of God in the New Enlightenment*. That we have a "chart of renewal, growth and repair of humanity," which is becoming realized via a "revival of reason" is not so much a statement of specific empirical, and historical fact, but a desired idea and noble intent and, indirectly, a confirmation of the general bad and impoverished spiritual and mental condition of the humanity of the "post-modern" age, along with being a postulate for its renewal and revival (Tymieniecka 2009, xxiii).

Here, we are dealing with an accurate and in-depth philosophical diagnosis of our condition, and even though this vision is not unique in the contemporary panorama of philosophical and anthropological thought, with its humanistically-involved, pro-human intellectual perspective and desired vision – projecting and practical – at the same time, balancing all the advantages and shortcomings of such a vision, its dramatic qualities and hope, unverified realism, and utopia, it cannot be granted unanimous and unqualified endorsement.

## Notes

1. Józef Bańka, *Ja teraz. U źródeł filozofii człowieka współczesnego*. Katowice: 1983; *Świat poręczenia moralnego. Medytacje o etyce prostomyślności*. Katowice: 1988; *Metafizyka zdarzeń*, Katowice: 1981; *Filozofia wieczysta. Krótka lekcja ludzkiego losu* Vols. 1 and 2, Poznań: 2010–2011.
2. Cf. Anna-Teresa Tymieniecka, *Logos and Life: Creative Experience and the Critique of Reason*, Book 1, Analecta Husserliana 24 (Dordrecht: Kluwer, 1988); *Logos and Life: The Three Movements of the Soul or The and Creative in Man's Self-Interpretation-In-The-Sacred,*" Book 2, Analecta Husserliana 25 (Dordrecht: Kluwer, 1988); *The Passions of the Soul and the Elements in the Ontopoiesis of Culture: the Life Significance of Literature*, Book 3, Analecta Husserliana 28 (Dordrecht: Kluwer, 1990); *Logos and Life: Impetus and Equipoise in the Life-Strategies of Reason,* Book 4, Analecta Husserliana 70 (Dordrecht: Kluwer, 2000).

## Works Cited


Bańka, Józef. 1983. *Ja teraz. U źródeł filozofii człowieka współczesnego*. Katowice: Śląsk.

———. 1988. *Świat poręczenia moralnego. Medytacje o etyce prostomyślności*. Katowice: Wydawnictwo Uniwersytetu Śląskiego.

———. 1991. *Metafizyka zdarzeń*. Katowice: Wydawnictwo Uniwersytetu Śląskiego.

———. 2000. *Główne zasady etyki prostomyślności*. Poznań: Wydawnictwo Psychologia i architektura,"Print B."

———. 2010 and 2011. *Filozofia wieczysta: Krótka lekcja ludzkiego losu*, Vol. 1: *Nadmonada wieczysta świat równoległy jako nierzecaywisty*; Vol. 2: *Czlowiek—"niemowle wszechświata" i Bog kwantowy*. Poznań: Wydawnictwo Psychologia i Architektura.

———. 2014. *Krytyka rozumu cyfrowego. Recentywizm i wielka metamorfoza XXI wieku*. Katowice: Śląsk Sp. z o.o., Wydawnictwo Naukowe.

Szmyd, Jan. 2012. *Myślenie i zachowanie nieracjonalne. Tradycyjne i współczesne wymiary. Z psychologii i filozofii irracjonalizmu*. Katowice: Śląsk Sp. z o.o. Wydawnictwo Naukowe.

Tymieniecka, Anna-Teresa. 2009. *The Fullness of the Logos in the Key of Life, Book I: The Case of God in the New Enlightenment. Analecta Husserliana C*. Dordrecht: Springer Print.

———. 2010. The New Enlightenment. In *Phenomenology and Existentialism in the Twentieth Century, Book 3: Heralding the New Enlightenment*, ed. Anna-Teresa Tymieniecka and C.V. Analecta Husserliana, 7–15. Dordrecht: Springer Print.

# Part IX
# Flesh, Body, Embodiment/Space and Time

# From the Archeology of Happenings … the Matter of Corporeality

**Aleksandra Pawliszyn**

**Abstract** I would like to propose a new outlook on philosophical hermeneutics as an archeology of happening. Thus, the hermeneutics described here will be a method, one worked out in the framework of the phenomenological tradition. This method could be helpful for contemporary humanity in our facing the task of understanding the essence of situation constellations in life, which are always unexpected.

The human being, encompassed here as a consciousness meditating on the world, attempts to grasp the sense of the world's occurrences. Thus, an archeology of events (the archeology of happening) will be applied to study occurrences, which are indispensable for expressing human existence in a proper way, occurrences such as: corporeality in the context of death, eroticism, creativity, human freedom.

In considering these interesting and intertwining corporeal aspects of the mystery of life and death, I draw from some reflections of Maurice Merleau-Ponty, Martin Heidegger, and Hans-Georg Gadamer.

In this paper it is presented that corporeality, death, the work of art, and eroticism are experiences that measure the profoundness of life, by which the human being is able to fully and freely participate in the existence—a dispersed corporeality especially could be treated as the basis of various transformations in the circle of existence. The work of art, in this context, appears as an expression of human infirmity facing an unknown future.

**Keywords** Corporeality · Life · Creativity · Eroticism · Death · The work of art · Human freedom · The circle of existence

---

This text (here reedited) has appeared in: Pawliszyn, Aleksandra. 2014. *An Archeology of Happenings Inspired by Death. The Philosophical Word Before That Which Is Inevitable.* Saarbrűcken: LAP LAMBERT, Academic Publishing, 26-39. Print.

A. Pawliszyn (✉)
Academy of Arts, Gdańsk, Poland
e-mail: aleksandra.pawliszyn@asp.gda.pl



W. S. Smith et al. (eds.), *Eco-Phenomenology: Life, Human Life, Post-Human Life in the Harmony of the Cosmos*, Analecta Husserliana CXXI,
https://doi.org/10.1007/978-3-319-77516-6_41

## Life in Front of the Mystery of Death

When daily we step out into the whir of traffic on a public road, we can expect that at some instant our soft body, consisting mostly of water, could be torn by steel, could brutally collide with the hard asphalt, so that we will lose our life. But then, how our feeling for existence increases when we are successful in crossing the road, reaching the other side of the street without injury. It is as if to having had the experience of the possibility of death brings with it the effect of a sort of holiday, by which existence manifests its force.

Thinkers in the European philosophical tradition have discerned the human way of experiencing the world as occurring within a sphere—a sphere of distance—which frees the human being from the pressures of an eternally changing reality. Since Heraclitean times, we, the people of Europe, have been astonished at the pendulum of life and death in the river of reality, what was later named the dialectic of the entity. However, the death of a human being is special because not only is it the ordinary end of a life, but it demonstrates that human life is exceptional.

I will here consider some interesting and intertwining aspects of the mystery of death, drawing on reflections from Merleau-Ponty, Heidegger, and Gadamer.

## A Few Words About Method

Who among us does not stand before the task of having to interpret happenings occurring in life? However, not many people are interested in the methodological constitution of interpreting these events. In our culture we use language formulated in philosophical disputations about the riddle of relationships, such as that of the human being with the world and, also, in philosophical considerations trying to solve the problem of knowledge beyond doubt. Therefore, of interest to us is the zone of human learning that embraces these epistemological, oft called transcendental, themes, which have been highly developed in the modern period of European philosophy.

When we look at our philosophical tradition, we can say that it was René Descartes who encouraged us to follow a meditative route to learning. On the one hand, Descartes grasped reality as being based on two elements of reality that are irreducible to one another, *res extensa*, and *res cogitans*. On the other hand, he noticed that the sphere of *ego cogito* is the spring of evidence and truth. However, it is to Immanuel Kant's merit that he noticed that the acts and processes of authentic learning cannot be reached out of the human sensorium. According to the philosopher from Königsberg—in opposition to the naivete of English empiricism—our sensing is to be apprehended critically in a richer way than that of empiricist schemes, namely in their being denominated *a priori* forms of human sensing (visuality). Let us note here that Kant also wrote of a sensate intuition that as it were provides concepts with "objective reality,"[1] and in this, he also transcended the naivete of English empiricism.

The transcendental motif of philosophical reflection promoted by Descartes and Kant was continued by Edmund Husserl, and Descartes' accent on meditation yields, as a result, phenomenological meditation, which is crucial for contemporary philosophy. In the framework of phenomenological activity, the learning subject rises to the rank of consciousness, giving total sense to the world. Husserlian considerations are the basis from which to speak about events, and from which archeology escapes the phenomenological method. It is this kind of archeology that Emmanuel Levinas seeks to employ in order to consider the mystery of time. He sees that, to enter into the question of what time as time is, it is necessary to move outside of phenomenology.[2]

It may well be necessary to pay attention to a new kind of philosophical narration, irrespective of whether we consider the possibility of the realization of Levinas' task to be outside of phenomenology. For the European epistemological tradition is enriched by Levinas' moral reflection. The French philosopher treats the transcendental ability of a learning consciousness as a kind of transformation, as something proceeding from the world's "wound" into the rational game of The Same. Among French thinkers, Georges Bataille also accents the corporeal side of human existence, revealing it as pain, suffering, delight, decay, putrefaction, and so on.

However, it may also be necessary to observe that analysis of the world's occurrences happens in the framework of a learning model setting the possibilities of human learning. In that case, every expression, as a yield of the human sensibility of the world, finds its place in description within epistemological borders.

Thus, one can assert that transcendental reflection, starting with Descartes' philosophy, developed in Kant's critical analysis, and culminating in Husserlian phenomenology, describes, in substance, the framework of the learning potential of the human being. On the one hand, one cannot transcend this framework, but on the other, it guarantees, theoretically, infinite interpretations of the events that humans can meet in the world.

Into the context just sketched, we can introduce a new view of the hermeneutical interpretation as a kind of archeology of occurrences. This kind of hermeneutics, developed from a phenomenological background, gives to the astonished human being a hint of how to understand the essence of the state of affairs that is revealed on the path of human existence. The human being, as the meditating power over the world of consciousness, tries to grasp the sense of that which is happening in the world. The world is here as the Heraclitean logos (Logos), drawing into conversation the searching sense of the human being. Let us take note that Gadamer's project, which grasps philosophy as a kind of notional history, is a kind of meditating entry into the meanings of notions, and can be denominated a notional archeology.

So, in this sketch, we use hermeneutics as an archeology of happening to consider the crucial characteristics of human existence, such as: corporeality in the context of language gestures, death, human creativity, and sovereignty (freedom).

## Archeology of the Body

Since we exist in perfect harmony with *the corporeality of the world*, we can only experience it in the way *our bodies* allow. To some extent, the corporeality of existence as a whole 'lines' the private worlds of every one of us, turning them into something familiar to us all. That given, one should not forget that the body can constitute a source of incomparable bliss but, when hurt, can also be a mass of suffering tissue (See Bataille; Maurice Merleau-Ponty).[3] When we look at the theme of the body from yet another perspective, it would appear that it can also serve as a language of meaningful gestures and, hence, a source of preverbal meanings.

Human body movements can themselves be meaningful because, as Merleau-Ponty shows, the human body is a sort of condensation or clot of *scattered corporeality* (*la chair*), which, in his view, is the essence of existence, a rule transformed into action, also described as a "clump" or "pulp of body tissue".[4] Therefore, *la chair* here would appear to be a kind of scattered corporeality of the universe, a certain manner adopted by the world, also incorporating the impulses of logos, possibly even being the logos itself, in the ancient meaning of the word.

The joy that accompanies comprehension of the world logos would also significantly connect with the joy experienced in a corporeal manner.

Merleau-Ponty's philosophical thought introduces us to the atmosphere of existence in the world of scattered corporeality, a corporeality that is streaked with meaning, as it were, to the atmosphere of meaning-laden corporeality which, by 'padding' the world, becomes a world in itself. Our body in this world is a sort of eruption, a temporary condensation of this scattered vibrant element and for us, here, it is also the measure of this world. It is, therefore, our body which gives the measure of events in the porous horizon of corporeal existence.

The body is a symbol of finiteness, an end sign with the stench of corruption (Bataille), a token of death in which the qualities of the corporeal world, similar to the seeds mentioned by Anaxagoras (of blood, hair, muscles, bones), undergo the process of decomposition, thereby becoming visible signs of finiteness.

The mere reading of the body itself as a sign evokes a perception and a concept of the world, revealing it in its primordial state of unity. At the ontic level one may speak of an ontic metaphor (Jacques Derrida) "expressing motion in terms of what it facilitates"[5] and, as such, *combining* the temporal with the spatial, external and corporeal. An ontic metaphor can also be treated as a manifestation of the motion of *dissimilarity*, a dissimilarity between two orders: the order of temporariness and the order of objects facilitated by temporariness. But, then, each of these terms accentuates man's indirect access to the motion of existence itself, which becomes obscured by the span of spatial entities and their corporeality.

The body can also be construed as a revelation, a symptom, a disclosure of the invisible, yet one which constitutes the motion of time itself. In the motion of the metaphor described above (whether unifying or differentiating), the invisible can be called the lining (to use Merleau-Ponty's term) of that which is visible, tangible, and generally discernible. Therefore, from the context outlined above, it becomes clear that without a body and without the corporeal emergence of the 'happening' of time we would be totally unable to experience the invisible pulse of existence or to take part in the mystery of being. Contemplation would serve as a trap for catching the scattered germs of meaning offered to man by the corporeal logos of the world.

## Delight and Suffering Written in the Body

It can be said that the human body is an open wound, as nothing can compare to physical pain and, as Levinas indicates, no means exist by which one's own perception of pain could be transferred to another human being.[6] Physical pain is torn tissues and the suffering that constitutes the ultimate fate.

One should not forget, however, that the body is also, as previously-mentioned, a source of elation and the joy of existence, a fact which, then, seems all the more miraculous. To say this is not only to acknowledge a dialectic supplement, however, but also to indicate the struggle to identify the nature of corporeality itself, taking us as it does towards the experience of pain and bliss. It is about recognizing that only thanks to the body can we experience being at all.

All writing and talking about pain is simply writing and talking, because through the medium of language (be it spoken or written) the indirectness of sensation therein undergoes a significant metamorphosis. Through this metamorphosis, a knot of mysterious unification reveals itself: the unification of what exists now with what has passed in time. This is how the matter is perceived at the level of the form of temporal change. By contrast, in the realm of the real events permeating this form, we encounter a dimension where the sternness of the inevitability of universal change will dominate, bringing with it a boundless, almost barbarous pain, which will eventually end.

## The Subtle Corporeality of Language, Tissue and the Scar of a Word

Let us now examine the level of the form of worldly events, which can also constitute the level of language, and which is, therefore, a level on which a subtle corporeality will dominate—'the less weighty' corporeal tissue, as Merleau-Ponty chooses to express it, namely, the corporeal tissue of language.[7] Let us consider yet again that the merging of dimensions—the form and the events contained within the form—contains in itself a mystery of a kind that accompanies every transformation of the scattered corporeality of the world from a given body's experience of pain into flashes of meaning captured within words. The bond between words and experience seems fundamental to an appropriate interpretation of the human condition.

For it seems that a gesture can be a word, as it emerges from the environment of the corporeal logos, as if saturated by seed-germs of meaning that possess the power to free man from the conditioning of change, but not the power to eliminate this conditioning. Therefore, the creation of the dimension of the world of subtle corporeality (effectively, language) becomes a scar from the laceration of intense corporeality and a special memory of the pain of ripped muscles. The scars are no longer the pain itself but a memory of the pain and, as such, can be viewed as words describing pain but not as the pain per se.

## The Distance of a Word vs. Freedom

Creating a relative distance from the pressure of eternal, universal change allows the possibility of escaping the bounds of the stream of change. It would seem that, in the light of Merleau-Ponty's corporeal ontology, the chance to acquire distance is rooted in the condition of the corporeal world, based on the aforementioned germs of meaning yielded by the corporeal element. Only in the context of still-pulsating bodily fabric do these germs have any power to incorporate a niche for peace of thought into the fluidity itself, a niche that is also a word zone and a kind of ideal in relation to the things that pulse with changeability. Escape outside changeability is only relative, but appears to introduce a new quality into human existence in a corporeal world.

That is to say it brings new words to the world, which narrate the world itself and constitute a source of freedom for the one who speaks against the backdrop of that about which he speaks. The very act of spinning a tale about the world makes the changeability of reality unreal to the extent that the tale acquires somewhat ineffable undertones. It should be stressed here that the disassociation of the word itself, whilst generating the freedom from being crushed by inevitable change, simultaneously signifies a loss of direct contact with the fertile 'soil' of corporeal existence.

## Reigning by Corporeality

The human world of temporality is, here, a homely feeling turned into corporeality, revealed by bodily ageing, which has emerged from the otherness of that which is out at the edge, and beyond this edge will come back.

The eternity of that which is different and not changing, not ageing, namely, the mysterious difference of immortal death, the eternal death corpse, is, here, a stigma of the cosmic element running through the world's corporeality. It has a part in every entity, tempting every human to attempt to be a creator, to espouse a clarity of the body and of understanding—a clarity in opposition to an all-absorbing darkness which fetters the darkness fettering the miracle of the happening of existence.

## The Gradual Coldness of the Body...

Corporeal proximity to the grand universal change is so intimate that every act of detachment is a heart-rending cry of sorrow; it is a lack of the beloved body which pains most sorely. Therefore, the other side of the unspeakable joy that corporeal proximity to existence brings with it, as an expression of its own self-generating power, proves to be a maddening pain after the loss of a warm, tender body which is growing cold.

Then arises the otherness of the forever-untamed coldness, which deals a blow to the pulsating corporeal rule over existence only, in the next instant of change, to paralyze with its longing for peace and stillness, while at the same time leading man into a time lapse, into a zone of thought or creation that, for the human being, turns out to be a relative, yet liberating, escape beyond the realms of time.

## Creativity in Front of Death

Death, as the essence of the human world, frightens and, at the same moment, like a spur tempts the human being into life, to take risks to enter into the strange dimension of time. It is as if, carrying within himself death, the human being can call on the power of existence that carries him into life and forces him to reveal himself, in order, in a twinkling, to make clear the flickering sense of life, which is sometimes forced through an accumulation of happenings (so Heidegger impresses on us).[8]

Breaking into the uncanny dimension of the time of the work of art is an incitement to life, an attempt to save the truth about human existence. Human existence, as sailing upon the ocean of life and death, is a call to our endowment for learning, to grasp that which is inexpressible, but also that which is the salt of existence. Balancing on the edge of life and death, the human being attempts to learn, as it were through the architectonic sculptures of his world, that world itself, washing away into every day changeability as if demanding human apprehension, which can rely on bewitching the truth of the world into a crystal work of art.

## Creativity as a Life and Death Symphony

Therefore, in a work of art, the effect of an attempt at human creativity, we are dealing with the revealing of the still-hidden sense of human existence—a burning fire, a continuous life and death symphony. Thus, the work of art is situated in a realm of continuous transcendence of a diametrically different state: life and death. It is as if a fruit of transgression has pushed the human being into this realm. It also seems as if the work of art brings a new type of temporality into the game of realizing existential time—stopped "portions of eternity," as it were, a kind of quantum temporality that, grasping and stopping, drops away from the sensibility of the existing world.[9]

## The Consolidated Transgression of the Mystery of Passage—The Work of Art

It is possible for the human being to creatively penetrate the tissue of existence pulsating through death and life, as a result of which the time of human death is concealed. Considering a reinterpretation of the myth of Prometheus as told by Aeschylus, Gadamer observes: "Before Prometheus brought the gift of concealment, his people lived in a miserable and passive way in caves, not creating any consolidated works of culture that distinguished them from among other living creatures."[10] So, balancing on the border of exclusive spheres of reality, spheres separated from each other by the secret mystery of passing, is found the real way that humans realize existence: existence distinguished by the attempt at creativity, by which the human being measures up to, on the one hand, the unusual impetus and impulse of life, and on the other hand, the dark abyss of nonexistence.

## Grief, the Heroism of Extorting from Existence an Enclave of Immortality

In this context, the work of art appears as an expression of human infirmity facing an unknown future. This infirmity causes an expected loneliness amid such an unresponsive world, as well as promotes the heroic deed of tearing out of existence an enclave of immortality. The tearing out of a work of art is a contraction against the rim of space and an efflorescence in the center of time; it is a human being crying out in pain, when she tears out of uninhabited spaces a shelter of fulfilled sense, a household retreat, where, after the exertions of battle, a wearied warrior can rest.

## Notes

1. As we read in Immanuel Kant, "… all concepts must be capable of resting upon an intuition of some sort, to provide them with objective reality; but all our intuition is sensuous." (Kant 2002, 385).
2. Levinas (1998).
3. Bataille (1976).
4. Merleau-Ponty (1964).
5. Derrida (1967).

6. Levinas (1998).
7. M. Merleau-Ponty (1964).
8. M. Heidegger (1987).
9. In this case see: A. Pawliszyn, Krajobrazy czasu (Pawliszyn 1996).
10. In this case see: H.-G. Gadamer, "Die Erfahrung des Todes" (Gadamer 1983). [Our translation is after the Polish translation by A. Przyłębski, *Gadamer*, Warszawa: 2006, p. 243.]

## Works Cited

Bataille, Georges. 1976. *L'Histoire de l'érotisme*. Paris: Gallimard. Print.

Derrida, Jacques. 1967. *La voix et la phénomène. Introduction au problème du signe dans la phénomènologie de Husserl*. Paris: PUF. Print.

Gadamer, Hans-Georg. 1983. Die Erfahrung des Todes. In *Neuere Philosophie I-II*. Gesammelte Werke, Bd. 4, 288–294. Tübingen: Mohr. Print.

Heidegger, Martin. 1987. *Einführung in die Metaphysik*. Tübingen: Niemeyer. Print.

Kant, Immanuel. 2002. *Theoretical Philosophy after 1781*. Cambridge: Cambridge University Press. Print.

Levinas, Emmanuel. 1998. *Le temps et l'autre*. Paris: PUF. Print.

Merleau-Ponty, Maurice. 1964. *Le Visible et l'Invisible suivi de notes de travail*. Paris: Gallimard. Print.

Pawliszyn, Aleksandra. 1996. *Krajobrazy czasu. Obecne dociekania egzystencjalnej wartości czasu*. Gdańsk: Uniwersytet Gdánski. Print.

———. 2014. *An Archeology of Happenings Inspired by Death. The Philosophical Word Before That Which Is Inevitable*, 26–39. Saarbrűcken: LAP LAMBERT, Academic Publishing. Print.

# Multi-layered Time and the Unity of the Unfolding Logos of Life

**Kamil Łacina**

**Abstract** A careful analysis A.-T. Tymieniecka's theory of time in light of her *New Enlightenment* uncovers the incompleteness of the account. This paper tries to rectify this incompleteness by introducing an additional element of temporalization to the theory. Later, the three elements of *kāla*, *chronos*, and *kairos* are combined to form a new multi-layered model of time – achieving the temporalization of the *Logos of Life*. Further comparison with other theories of time provides a deeper understanding and appreciation of the unfolding of the *Logos of Life*, the harmony of the *Cosmos*, and of the Unity-of-everything-there-is-alive.

**Keywords** Tymieniecka · Ingarden · Time · Chronos · Kairos · Logos · Dynamic existence · Self-individualization

Time is an ever-present philosophical problem. From the earliest period of philosophical reflection, the problem of time occupies a fundamental place, among such issues as the good, the true, and existence. While many a great philosopher has attempted to tackle this problem, very few have managed to make an unquestionably important contribution to the field.

Much like Edmund Husserl, their master, Anna-Teresa Tymieniecka and Roman Ingarden, while approaching the problem of time from different vantage points, use a method characteristic of phenomenological enquiry. Ingarden focuses on the problem of existence in time, paying little heed to the overall metaphysical conundrum. Tymieniecka, on the other hand, takes a wide perspective, looking for time and temporalization within the progression of the Logos of life. In the culmination of her analysis, chronos and kairos become two interwoven arteries of becoming. The theory is, however, incomplete for the progression of the Logos is not exhausted by the two modes of temporality.

K. Łacina (✉)
Institute of Philosophy, Jagiellonian University, Kraków, Poland

 
W. S. Smith et al. (eds.), *Eco-Phenomenology: Life, Human Life, Post-Human Life in the Harmony of the Cosmos*, Analecta Husserliana CXXI,
https://doi.org/10.1007/978-3-319-77516-6_42

In this paper, I attempt to complete the model by reconciling the relevancies, universal harmony, and scientific perspectives with the already-present elements of Tymieniecka's analysis of time and temporality. The resultant model and its subsequent analysis reveal an overarching Unity-of-plurality – echoing the Husserlian Ego – by bringing together the transcendence of the Logos and the intentional, yet simplifying, apprehension of human consciousness.

The analysis of time and the subsequent building of Multi-Layered Time starts by delineating Ingarden's account of time. Then, Tymieniecka's more robust theory from will be taken up. Afterward, in order to provide us a basis for a new interpretation, a theory advanced by Dirk J. Pons et al. will be presented for inspirational purposes only. While this theory itself is not viable (Łacina 2014), its structure will help us to build a new theory. Synthesizing the three approaches will allow us to see time as a multi-layered construction following the *Logos of Life* in its progress in the unity-of-everything. A comparison of our model with two other theories of time will bring our analysis full circle, allowing for a deeper understanding and appreciation of the unfolding of beingness guided by the *Logos*.

## Ingarden's Existence in Time

The considerations of this presentation, with regards to Ingarden's theory of time, are informed by his opus magnum, *The Controversy over the Existence of the World*, and his short essay "Man and Time". These two works seem to approach the problem of time from two different perspectives: with what exists within and outside of time and the diachronic identity of the former being addressed in *The Controversy*; and man's relationship with time being addressed in "Man and Time". Yet, in fact, they both address the same subject – what it means to be temporal. While Ingarden's main work is focused on the metaphysics of temporal existence (what entities can be distinguished as being temporally determined, and just how they exist within the stream of change) and the essay is devoted to the way humans interact with and experience time – the two perspectives subjected to a rigorous analysis are seen to focus, first, on man as a lasting and unchanging unity of self-identity and, second, on man as a victim of change and the flow of time.

The subject of time and temporality interests Ingarden only with regard to the modes of existence. Time itself (its existence, metaphysical status) is soon of little relevance to this philosopher. He does, however, distinguish three separate understandings of time: i) abstract time – physical time, time as mathematically described and determined; ii) common time – the comparative time between the different particular times of particular objects; iii) concrete (or particular) time – a time 'filled' by what is happening within it. The definition of understanding *iii* brings to mind the relationist definition of time, while understanding *i* is similar to the substantivalist account. Ingarden dismisses, somewhat offhandedly, the problem of time itself by writing that the questions "is this particular time homogeneous or heterogeneous, without qualities or qualitatively determined – these are all questions that can only be properly formed, answered by formulating a general theory of time" (Ingarden 1962, 214–5). Instead, the philosopher focuses on the entities whose existence is temporally determined.[1]

Here, a quick remark is in order. The reason I decided to focus on accounts such as Ingarden's and Tymieniecka's (ones focusing on what exists in time, or ones that undergo temporalization) is because, contrary to some theories, we have no direct access to what we call time; we perceive the flow of time by observing changes in our surrounding world. While there were contemporary attempts at defending the substantivalist theory of time, most notably a thought experiment by Shoemaker, they ultimately fall short of the staggering phenomenological evidence we have to the contrary.

Within *The Controversy…*, Ingarden distinguishes three entities that have an intimate, metaphysical relationship with time – events, processes, and enduring objects. An event is defined as a state of affairs or an objectual situation that obtains (Ingarden 1962, 216). Contrary to the other two, an event has no temporal breadth – it is not extended in time. "An event is characterized by not enduring. They obtain, and thus they cease to exist" (Ingarden 1962, 216). While an event has no temporal extension, it still is considered a temporally-determined entity by virtue of its relation to some process (which is temporally extended), and also by its being located at some instant of time *t* (for temporal sequences see McTaggart and others). The former is more relevant since, while being qualitatively different from processes, events serve to initiate and finalize each process. Ingarden does not claim that events are point-like (which would be suggested by a mathematical-geometric idea of time): "this question [of a point-like nature of events – K.Ł.] may only be answered by establishing if moments are just places in a one-dimensional temporal continuum, or, on the contrary, are particular units of time that mark their distinctiveness in the flow of time" (Ingarden 1962, 217–8). Additionally, he notes that "an event does not span further than the extent of one specific now" (Ingarden 1962, 218). Thus, we have here one type of temporal entity – a self-insufficient event.

Second, are processes. In contrast to events, processes are temporally extended – they last for a period of time. While acknowledging that a process is a "continuous sequence of phases", the philosopher distinguishes between the "continuous entirety of phases, which grows in a continuous fashion while the process is occurring, and, on the other hand, the object being constituted in these phases, as a particular subject for qualities, which the process is" (Ingarden 1962, 211). The bulk of investigation here is devoted to the mode of existence of a process – an issue with which we are not concerned at present. What should be mentioned, however, is the fact that Ingarden does not equate a process with an ordered sequence of independent events. Admittedly, only the present enjoys a robust existential standing. Nevertheless, the past and the future, as the 'already realized' and the 'to be realized', are not existentially vacuous. All three temporal periods are equally important for the unity-within-itself of the process. All phases are interconnected, self-insufficient, and dependent. In his own words, "what is future, is up front determined by what is present, but not always by this alone, as sometimes it is determined also by what has already become past" (Ingarden 1962, 225). In this theory, the hallmark of actuality is a *direct* activeness, agency, allowing for causality. The continuous nature of the process, as well as its temporal breadth spanning all three modes of time (past, present and future) assures direct activeness. Moreover, the co- influential nature of the unity of times blurs the identity of past, present, and future phases that comprise a process. Further distinction between processes and events (and the impossibility of identifying one with the other) arises from the fact that, while each event is separate and self- contained, the phases of a process are not – they are existentially and causally interlocked with one another, thus constituting a greater whole, which is a process.[2]

The third type of temporal entity described by Ingarden is objects enduring in time. This type is by far the most problematic, both in terms of the metaphysics of time and in this present analysis. Temporally enduring objects are considered different from processes on the basis of a metaphysically grounding priority – according to Ingarden, it is enduring objects that serve as the existential grounding for processes. While it is true that processes influence objects (creating, changing, and even destroying them), they are nevertheless metaphysically dependent on the existence of objects. Ingarden's theory of enduring objects is, in fact, an endurantist theory with all the merits and flaws of such (Balashov 2011). The flaws referred to concern the identity of an object. Endurantists claim that an object is present as a whole in all the moments of time. This raises several questions regarding properties, change, and issues described as problems of constitution. Since objects change with time, we need to subscribe to some form of a substrate theory in order to maintain that an object is, in fact, the same through different moments of time (by virtue of its *substantia, haeceitas*, or some other element grounding all its qualities). Ingarden seems to be subscribing to some version of substrate theory, when he highlights a difference between an object and its state, in order to solve the problem of a changing yet identical object (Ingarden 1962, 246). Ingarden's example of Napoleon I actually darkens the issue instead of illuminating it. A living organism, which incidentally qualifies as an enduring object, causes even more problems for this theory – problems that, in light of the work of Tymieniecka's phenomenology of life, may have a viable solution. What Ingarden claims is that identifying an enduring object with a process stems from a too broad a definition of an aforementioned state. If we see the state of an object as the totality of its qualities (too broad a definition, according to Ingarden), then an object becomes a process. Surely, however, the metaphysical difference between an object and a process should not be so flimsy as to admit any blurring of boundaries based on semantics. Leaving this problem aside, there is still an issue of what Ingarden designates the "core of an object" (Ingarden 1962, 252). What that core might be, for now, we do not know.

The problematic nature of the living enduring entity is also elaborated on in his earlier essay – "Man and Time". The essay is focused on the human being's dialectic experience of time. One the one hand, man feels surprisingly impervious to the passage of time. We feel that our early childhood, our adolescence, our present state as well as all our future moments have one thing in common – us. There is an undoubted feeling of continuity and identity that accompanies our reflection on our own lives. While this is less so with regards to others (for we sometimes say things like “Mary’s changed; she’s not the girl I used to know”), when it comes to ourselves we have little doubts about the consistency of the protagonist of our first-person narrative. At the same time, however, we feel “carried by the stream of time”, slaves to its destructive power. Not only do we not stay the same, we are being slowly devoured in the process of existence. This contradiction seems to stem from the dual nature of our existence, which is exemplified by *individuality* and the *unity-of-life*.

## Tymieniecka and the Unfolding of the *Logos of Life*

Anna-Teresa Tymieniecka takes a rather different – although, as will be shown – complementary view of time. With her enquiry into, and interrogation of, the *Logos of life*, the Polish-born philosopher uncovers a structure of time that, while at the outset seems far more complex, is in itself a much more metaphysically parsimonious theory that that of her master. Her metaphysics of the *New Enlightenment* no longer speaks of time. Rather, and in a way somewhat similar to Heidegger’s approach, it is concerned with temporalization – the process of the progression of the logos within being in becoming.

Tymieniecka refuses to absolutize time. Instead, she sees it as life-emergent. Time, according to her, is an intrinsic process of self-measurement of a living being. Within this self- measurement of life framework, she distinguishes two modes of time: *chronos* and *kairos*. To quote directly, “Time, in both of its modes – chronos and kairos, as differentiated long ago in Antiquity – is the grand, infinitely complex, flexible artery-in-progress of the constructive advance of life, encompassing its relevant cosmic links, on the one hand, and its reach for a portal to the transcendent aspirations of the human being, on the other” (Tymieniecka 1997, 19). This definition, and Tymieniecka’s theory of time, is strongly influenced by Aristotle. While in broad strokes I agree with her, the basis for the theory is disputable at best. For one thing, Aristotle’s definition of time is de facto circular – a fact often passed over by philosophers. Tymieniecka even states this explicitly, saying that “time and motion remain in reciprocal relation to each other when it comes to measurement: ‘we measure not only motion by time, but also time by motion, because they determine each other mutually: because time determines motion of which it is the number, and movement determines time’” (Aristotle 2008 bl4–22).

This account is clearly circular: we measure movement with time and time with movement. A similar mistake is that which destroys the recently published theory of time of Pons et al. Nevertheless, an ontopoietic grounding of time can be successfully defended – which, incidentally, is precisely the goal of this presentation. Before continuing, let us take a brief look at both these modes of time. This will also allow us to see what is in fact missing from the account presently being delineated.

Let us first look at the modus of time denominated *chronos*. Chronos represents the somewhat familiar measurable time that spans, allowing processes to unfold. The difference that stands out in Tymieniecka’s account of chronos is that time “is not … added to the constructive operations from the 'outside' as if it 'existed' or is 'just there'" (Tymieniecka 1997, 9). It is the constructive movements of bios that, by measuring themselves from within, create this particular modus of time. "[T]heir advance marks and measures temporal progression in itself. ... Life proceeds and temporalizes itself without it" (Tymieniecka 1997, 9–10). The constructive tendency of life, the *entelechial principle of life*, is present and governs these constructive, self-individualizing movements of bios. "Chronos is the order and sequences of life" (Tymieniecka 1997, 11).

Thus far, we have one modus of time. According to the ontopoietically grounded theory, processes, such as 'an opening of a flower petal', are not somehow situated within an externally imposed, or added, time. They, themselves, produce time through the living soul's measurement of its constructive motions. Moreover, when comparing Ingarden with Tymieniecka, a discord can clearly be seen. While processes are clearly present in both accounts, they are not so conceived that they can be identified and separated from one another. Were we to agree with both of these theories, we would be forced to recognize a type of monism and there would be only one process, that of bios unfolding – the progress of the logos of life. Within this monoprocessual universum one could identify various sub-processes, those being the self-individuating instances of life. Nevertheless, there is hardly any possibility of comprehending these individualized beings (living things) other than by a "simplifying human mind" (Tymieniecka 1997, 4). This is a problem of discernibility, which cannot be addressed here owing to the scope and spatial limitations of this essay.

Next is the mode of time denominated *kairos*. According to Tymieniecka, "life-constructive fulfillments marking ontopoietic progress, and their occurrence within the play of favorable and contrary conditions, are the moments of kairos" (Tymieniecka 1997, 11). In accord with the classical meaning of the Greek word, Kairos denotes a propitious moment for the achievement of a goal set forth by the Human Being through *imaginatio creatrix*. It is only this particular self-individualized instance of life that is able to set out for itself goals and priorities other than those dictated by animality. These goals' "timing is their partaking of shades of significance that vary in the infinite modalities and qualitative nuances of the aesthetic, moral, and intellectual sense- giving factors brought into the progress of life by the Human Condition" (Tymieniecka 1997, 15). And, further, as

> Chronos diversifies into innumerable streamlets of occurrences, there emerge conundrums of propitious conditions that allow the tying of the knot of accomplishment. It is by the tying of these knots of accomplishment that the human personal – and social – self-interpretive course of existence not only proceeds but is also measured. Indeed, along the path of human creative self individualizing, kairos is concurrently the timing of the propitious circumstances and forces leading toward the realization of constructive projects, as it is their accomplishment, and finally their measure. (Tymieniecka 1997, 15)

Another important factor is the distinction between "the inner" and "the outer" life. The former is an arena of preparation, planning, and goal-setting, while the latter is the space of relations and realizations – the interweaving of various unfolding self-individualized sub- processes of the grand *process majeur* of the progress of *Logos*. The inner life, often identified with experience, or consciousness, exemplifies a timing of its own – both chronos and kairos. For, according to Tymieniecka, "[w]e do not experience stagnation and inertia. We experience a "continuity" of the "present" of our actual state" (Tymieniecka 1997, 16). Thus, the theory advanced here, at least on the level of the 'inner life', resembles Bergson's account of *durée*.

Now, neurobiological evidence regarding human time perception[3] and both progressive and episodic memory[4] seem to suggest that Tymieniecka's theory is 'on the right track'. This 'right track' is the continuous dynamic influence that the transcendent world and consciousness have on our perception and cognizance of the process that we refer to as "time". The two modalities of time, the two *arteries*, pass through one another, intimately connected, constituting, on the one hand, the continuity and flux of becoming, and on the other, teleological drive. To sum up this account of the two modalities of time, "Thus, human life's inner experience and external activity is carried by two interwoven arteries; that is, it is timed into chronos, the everyday tacit carrying on of repetitive assignments for life's maintenance, and the kairic rhythms of urgency, promise, expectation, ecstatic hope, and final attainment of goals" (Tymieniecka 1997, 16).

With regard to so-called cosmic time, Tymieniecka states that we, through the operations of the simplifying human mind, simply project 'a mere skeleton' of living time onto the world to arrive at a measurable, mechanic account of cosmic events. Considering this together with life-derived time is quite problematic. For one thing, we are aware that life appeared on Earth (we have no certainty of any other appearance) approximately 3.5 billion years ago (Schopf, et al. 2007), with accounts ranging from 4.25 billion years ago (Tenenbaum 2002) to 4.4 billion years ago (Steenhuysen 2009). At the same time, we know that the Universe is approximately $13.798 \pm 0.037$ billion years old (Planck Collaboration 2014). How would we then conceive the progress of any processes or events "unfolding" before life began? Here is where we need to "build" time.

## "Building" a Multi-layered Time

The ultimately erroneous theory of time advanced by Dirk J. Pons, Arion D. Pons, and Aiden J. Pons will help us construct our multi-layered time. In their paper "Time: An Emergent Property of Matter," they stipulated, on the basis of the non-local hidden variables (NLHV) interpretation of quantum mechanics, that time is multi-layered, proceeding from the frequency of oscillation of active ends of a non-local particle, and arriving, gradually, through decoherence, at the physical time of complex structures, i.e., chemical time, biological time, and, finally, human (consciousness related) time. The theory, while interesting, suffers from a number of errors (already discussed in Łacina), the most serious of which is defining time as frequency of oscillation. Nevertheless, the layered structure is quite inspiring and can be applied to our present theoretical synthesis. We already have certain layers in place, specifically chronos (corresponding to bios), and kairos (corresponding to the human condition).

Chronos, as already told, is the self-measurement of the creative motions of Beingness, the realization of the progression of the *Logos of Life* in self-individualized processes identified by us as living beings (viruses, plants, animals, and the like). This sequenced and ordered progression is just one of the layers of time. Just as life, in its complex, interwoven, and immeasurable complexity of existential relations is complex in each of its instances (as each one of its instances is intertwined with the unity-of-everything-that-is-alive), so too is time, being emergent from life, soul, a complex structure. This complexity is, in fact, the result of the very unity-of-everything-that-is-alive.

Both chronos and kairos constitute different layers of our time-structure. They are, however, in no way separate and, much as in the theory of Pons et al., each succeeding layer emerges from the former. Kairos emerges from chronos. Without chronos, the progression, self- individualization, and constructive movements of bios, there would be no Human Condition and, thus, no kairos. An interdependence of temporal layers is essentially inscribed in the interdependence of all life. All processes within the unity-of-everything-that-is-alive have numerous multifarious connections (for example, the simple opening of a flower petal). The same principle prevails across all realizations of life – time included.

There is, however, a missing piece of the puzzle. Time, to be a coherent notion, needs more explication than that given it by the apprehension of chronos and kairos. As previously mentioned, life emerged at a certain point in the evolution of the universe. What we need is something that will account for everything that occurs pre-life. Additionally, life emerges in the evolution of the universe. It emerges not *ex nihilo*, but out of non-living matter. How that happens is still a mystery – the fact that it happens is not. Thus, since the layer of chronos emerges from life, and life emerges from pre-life conditions, there is something prior. Tymieniecka claims, with regards to cosmic time, that we simply project a skeletal derivative of living time (chronos) onto cosmic events. While this, in itself, is not controversial, the pre-human and pre-life period is. Nevertheless, the progression of the logos of life, its entelechial telos-oriented progress, allows us to unravel this issue.

According to Leslie's theory of a fine-tuned cosmos, even a minute change in any one of the fundamental forces and ratios would result in the universe's not existing. Everything seems to be literally fine-tuned for life to emerge (a good example here is the *anthropic principle*). Furthermore, the decoherence theory of Zeh suggests that the world-structures emerge (become individualized) through interactions of particles with their environment. It is because of these interactions that quantum effects (non-locality, entanglement, and superposition) are not observable in our 'macro' world of classical physics. Following this trail, one soon realizes that what plays a fundamental part in these interactions are, in fact, the fundamental forces that were already mentioned. Without this fine balance (Cosmic harmony), familiar material structures never would have emerged from quantum chaos. The balance seems to be both the source (in causal terms and in terms of specifically understood boundary conditions) and *telos* of the evolution of the Universe. While decoherence theory has its problems, as detailed by Fields, it is, nevertheless, the best theory we currently have that explains, at least partly, the transition from the quantum world to the classical world.

On the basis of this theory, others have developed a theory of quantum Darwinism (Zurek 2009), where a 'survival-of-the-fittest-like' situation is posited to occur, with particles competing with one another for "survival" – in the constitution of objects and the preservation of their qualities.[5] Now, regardless of whether we subscribe to quantum Darwinism or not, the early evolution of the universe is precipitously close to mirroring the evolution that is the progression of the logos of life embodied in the creative movements of self-individualized living beings. As already told, it is the soul's measurement of these very movements that gives rise to chronos. Moreover, any pre-life events need to be properly sequenced and ordered for life to appear – for the *Logos of life* to be realized within an embodied life form. However, within the sphere of pre-life there is no soul to measure and "time" its movements, even though movements seem to occur. It is this pre-life space of unfolding that is missing from Tymieniecka's original analysis of time (Tymieniecka 1997).

In order to rectify this, we need to be able to grasp and define (at least partially) the space of the preestablished harmony of the *entelechial* unfolding of the Logos of life within its preparatory processes. We require a space of Unity-of-everything-within-the-unfolding-of-Logos to comprise a harmonious telos-bound order of the approaching-unfolding Life-to-be. What we have then is the first layer of time, which I have decided to call *kāla*. (*Kāla* is a Sanskrit word meaning "Time" – more specifically, "a fixed or right point of time, **a space of time**, time … destiny, fate … death".) This is a space of unmeasured movement, understood holistically as a monolithic sphere of ordered sequence. It is the most basic layer, from which chronos, as measures, emerges through the fact of being measured by the living soul. The ordered sequence is a direct and intimate preparatory progression of the not yet embodied, but already *telos*-bound, logos. For the logos of life to be realized, its initial conditions have to be ordered and sequenced. The *entelechial* principle governing life needs to be established as a form of a Leibnizian preestablished harmony. Further, this harmony, this order is still present within life as – what Tymieniecka describes as – 'relevancies'.

*Kāla* completes the tapestry of the progression of the Logos of Life, where time (of Bergsonian persuasion) is abstracted from life/change, which is framed by the *entelechial* principle. While time proper – chronos – emerges from and with life, for the realization, coming into beingness of the *Logos of Life*, the stage must first be set in and through the preestablished order of the unfolding.

How does Ingarden figure into all of this? His "events" become part of time itself. No longer to be considered as something immersed in time, they become *kairoi* – propitious moments propelling life towards its telic unfolding. His processes, rather than being realized in time, realize time. The constructive movements of bios, as interwoven processes, measure themselves, thus, creating time (*chronos*). Objects enduring in time no longer endure in time – they become instances of self-individualization. What, then, of the dialectic of the contradictory experience of time that we face. Man feels above and beyond time because, through self-individualization and the power of *imaginatio creatrix*, man in his Human Condition creates a new layer of time – kronos. As such, man feels himself to be something more than time. At the same time, man feels bound to the flow of time, being both created and destroyed by it. This part of her experience stems from the unity-of-everything-that-is-alive. Man is still part of bios and, so, measures and is measured with regard to chronos. All this is bound by the preestablished harmony of the ordered sequence of kāla – the unmeasured space of the order teleologically imparted to the world by the *Logos of Life*.

Multi-layered time – comprised of kāla (a preestablished, unmeasured, ordered sequence), chronos (the measured movements of bios), and kairos (propitious moments of the Human Condition propelling the progression of Logos) – creates an intricate artery of life's progress towards its divine telos. Thus, the complexity of time reflects the complexity of life.

## Remarks on the Philosophy of Time

It is widely known that the field of philosophy of time is a multi-faceted minefield, one densely salted with a wide array of problems and approaches. While most of the debates taking place in this field of research are obsessed with language and existence, focusing mostly on Anglo-American analytic approaches having little to do with eco-phenomenology, still, the above delineated model finds confirmation in some recent works. See Craig Callender, ed., *The Oxford Handbook* and his article "Time's Ontic Voltage" (Callender 2011, 2012).

Two works are of particular interest here. The first is Broad's theory of time defined as *absolute becoming*, found in his *Examination of McTaggart's Philosophy*. Broad's theory was to be a retort to McTaggart's famous argument against the reality of time. Time can be defined as neither movement, nor change. Any definition making use of the former notion automatically evokes a question of the rate of said movement – the famous 'How fast does time pass then?' question. The usual answer, 'One second per second', is clearly circular, and gets one into more trouble, rather than providing any coherent solution. Similarly, any definition referring to the latter requires time for the change to occur in, as well as for the thing that is subjected to said change to already exist in time. Faced with these difficulties, and with several versions of infinite regress, Broad proposed to define time through absolute becoming – an undefinable notion of pure simplicity. It is through absolute becoming that time is defined and constituted, with the becoming itself requiring it, and admitting no definition whatsoever. For Broad, absolute becoming was confined to the domain of instantaneous events (for fear of being forced to admit the reality of the past and future – in this theory, only the present exists). While we could plausibly present Tymieniecka's understanding of time as well as the above-given layered model of time in Broad's terms, as instantaneous events of the unfolding of Life, a better theory still can be found in a recently published monograph by her compatriot Jerzy Gołosz.

In his *Upływ czasu i ontologia*, Gołosz proposes to define time not in terms of absolute becoming but, rather, in terms of dynamic existence. Objects, dynamically existing and enduring, constitute moments of time by dynamically and continuously transporting their existence into these moments. The concept of dynamic existence frees us from Broad's instantaneity of events and transports us into the realm of objects and their fluid dynamism of being. Time becomes the mode of existing, of coming into object's own beingness. Gołosz, both a philosopher and a physicist, demonstrates that dynamic existence can be easily reconciled with relativity by stating that objects create their individual histories, which are measured in their own, individual time (Gołosz 2011, 20). This account – *dynamic existence* – mirrors the dynamic unfolding of bios exemplifying the progression of Logos. It is through the dynamic unfolding of life that time is measured and created (with movement being understood metaphysically, rather than physically). Each living being creates its own history – its own evolution of beingness – measuring it in its own individual time. These individual histories are all instances of the progression of the Logos of Life which, while determining the telos of unfolding, remains transcendent to the individual processes derived from its totality.

How can our model of a multi-layered time be reconciled with these theories? In order to properly understand time, we need to go back to the seemingly naive monistic understanding of time. We initially criticized the commonsense approach, stating that time is a monolithic structure. We have divided time into three proper parts: *kāla, chronos*, and *kairos*. However, upon making a closer examination, and confronting both Broad and Gołosz, it should be stated that the appropriate structure for time is more complex than initially anticipated. When considered from the perspective of the unfolding of the *Logos of Life*, time is a tripartite unity of kāla, chronos, and kairos, all separate, yet unified. While each layer has its purpose in the unfolding of beingness, ultimately, time is the emanation of the progression and realization of the *Logos of Life* – one process, progress, simplistically bound in three distinct layers by the simplifying grasp of our investigations. This three-in-one structure of time-as-it-is-for-us echoes the tripartite structure of the Husserlian Ego, which brings our understanding of the transcendent and of consciousness full circle. The *Logos of Life*, finding the fullness of its realization in the Human Condition, allows us to gain insight into its structure through the phenomenological analysis of its ultimate achievement – human consciousness in its ontopoietic creativity.

## Conclusion

Tymieniecka's almost mystical understanding of time can be reconciled not only with the more conservative Ingardenian thought on the subject but also with some interesting theories currently being discussed within the broad field of philosophy of time. The complete model presented above allows us to gain new insights into both (what Tymieniecka calls) a simplifying grasping performed by the human mind and the broader structure of the unfolding of the *Logos of Life* – in a *New Enlightenment*. The trinitarian characteristic of unity and divisibility provides us with grounds for examining time from a new perspective, while not withdrawing from our interrogation of the *Logos of Life* in the Unity-of-everything-there-is-alive and the harmony of Cosmos.

## Notes

1. For a further, and explicit, account of support given to the relationist theory of time, see Ingarden's *Controversy* (Ingarden 1962, 220).
2. The problematic nature of processes has recently been unfolded by Dolev. He shows that processes and events (like a football game) are usually complex, consisting of various stages. The entire process need not be present for parts of it to be present. Nevertheless, we commonly say that we are presently "watching a game". Such an approach gives some credence to the stage theory of existence in time.
3. Grush, following the original Husserlian account of the structure of time consciousness, utilizes information technology – specifically the Kalman Filters – to reconcile the protention/attention/retention account of Husserl with contemporary cognitive science. This original approach models the constant dynamic process of existence, which influences the Bergsonian flux of the time we experience.
4. For details, see studies by Hassabis, Kumaran, and Maguire, and by Mullaly and Maguire. These studies provide us with an insight into the intimate connection between what in the phenomenological tradition have been referred to as protention and retention. While these studies focus greatly on memory and anticipation – more precisely, on Episodic Future Thinking – the results are very much relevant to our understanding of the phenomenology of time. The near structural identity in neural correlates between the way we conceive the past and the future offers support to an idea of a dynamic existence, where the constant flux is a function of the unity of a living process.
5. For an excellent philosophical analysis of hidden metaphysical assumptions and the problems they generate for scientific theories, please see an excellent paper by Fields, "A Physics-Based Metaphysics Is a Metaphysics-Based Metaphysics" (Fields 2014).

## Works Cited

Aristotle. 2008. *Physics*. Trans. R. Waterfield, ed. D. Bostock. Oxford: Oxford Paperbacks. Print.

Balashov, Yuri. 2011. Persistence. In *The Oxford Handbook of the Philosophy of Time*, ed. Craig Callender, 13–40. Oxford: Oxford University Press. Print.

Broad, Charlie D. 1938. *Examination of McTaggart's Philosophy*. Cambridge: Cambridge University Press. Print.

Callender, Craig, ed. 2011. *The Oxford Handbook of the Philosophy of Time*. Oxford: Oxford University Press. Print.

Callender, Craig ed. 2012. Time's Ontic Voltage. In *The Future of the Philosophy of Time*, ed. Adrian Bardon, 73–98. New York: Routledge. Print.

Dolev, Yuval. 2012. Perceiving Transience. In *The Future of the Philosophy of Time*, ed. Adrian Bardon, 56–72. New York: Routledge. Print.

Fields, Chris. 2014. A Physics-Based Metaphysics Is a Metaphysics-Based Metaphysics. *Acta Analytica* 29: 131–148. Print.
Gołosz, Jerzy. 2011. *Upływ czasu i ontologia*. Kraków: Wydawnictwo UJ. Print.
Hassabis, Demis, Dharshan Kumaran, and Eleanor A. Maguire. 2007. Using Imagination To Understand the Neural Basis of Episodic Memory. *The Journal of Neuroscience* 27.5: 14365–14374. Print.
Ingarden, Roman. 1945. *A Book on Man*. Warszawa: PWN. Print.
———. 1962. *The controversy over the existence of the world*. Warszawa: PWN. Print.
Łacina, Kamil. 2014. It is not what it looks like – A response to D.J. Pons, A.D. Pons, and A.J. Pons Theory of Time. *Applied Physics Research* 6.1: 26–34. Print. McTaggart, John. The Unreality of Time. *Mind* ns 17.68 (1908): 457–474. Print.
Leslie, John. 1988. How to Draw Conclusions from a Fine-Tuned Cosmos? In *Physics, Philosophy and Theology: A Common Quest for Understanding*, ed. Robert J. Russell, William R. Stoeger, and George V. Coyne, 297–312. Vatican City: Vatican Observatory. Print.
Mullally, Sinéad L., and Eleanor A. Maguire. 2013. Memory, Imagination, and Predicting the Future: A Common Brain Mechanism? *The Neuroscientist* XX.X: 1–15. Print.
Planck Collaboration. 2014. Planck 2013 Results. XVI. Cosmological Parameters. *Astronomy & Astrophysics*. arXiv: 1303.5076. Web. 05 September 2014.
Planck reveals an almost perfect universe. 2014. *Planck ESA* 21 March 2013. Web. 03 September 2014.
Pons, Dirk J., Arion D. Pons, and Aiden J. Pons. 2013. Time: An Emergent Property of Matter. *Applied Physics Research* 5.6: 23–47. Print.
Schopf, J. William, Anatoliy B. Kudryavtsev, Andrew D. Czaja, and Abhishek B. Tripathi. 2007. Evidence of Archean Life: Stromatolites and Microfossils. *Precambrian Research* 158: 141–155. Print.
Shoemaker, Sidney. 1969. Time Without Change. *The Journal of Philosophy* 66.12: 363–381. Print.
Steenhuysen, Julie. 2009. Study Turns Back Clock on Origins of Life on Earth. *Reuters*, 20 May 2009. Web. 23 July 2014.
Tenenbaum, David. 2002. When Did Life on Earth Begin? Ask a Rock. *Astrobiology Magazine* 14 October 2002. Web. 23 July 2014.
Tymieniecka, Anna-Teresa. 1986. The First Principles of the Metaphysics of Life. *Analecta Husserliana* XXI, 3–9. Dordrecht: Kluwer. Print.
———. 1997. Life's Primogenital Timing: Time Projected by the Dynamic Articulation of the Ontogenesis. In *Phenomenology of Life as the Starting Point of Philosophy*, ed. Anna-Teresa Tymieniecka, 3–22. Dordrecht: Springer. Print.
———. 2009. *The Fullness of the Logos in the Key of Life*. Analecta Husserliana, vol. C. Dordrecht: Springer. Print.
Zeh, H. Dieter. 1970. On the Interpretation of Measurement in Quantum Theory. *Foundations of Physics* 1: 69–76.
Zurek, Wojciech H. 2009. Quantum Darwinism. *Nature Physics* 5: 181–188.

# The Question of Placeness

**Carla Danani**

**Abstract** Returning to a long and fruitful anthropological tradition which has ripened from phenomenological-hermeneutical and metaphysical thought, we talk of the human being as "bodily consciousness" or "embodied consciousness." Therefore, human beings, like things, have to be seen as entities that must "always be somewhere" and, so, it makes sense for them to question about the place they are in. But human beings and things have different relationships with the places that they occupy. Indeed, we must say that human beings *inhabit*: they are not in the world in the way things are, because the relationships they have with space are intrinsic to their existence. "Place consciousness" is therefore also one of the conditions necessary for building individual identities and communities that can establish long-lasting relationships with nature and humans. These can only be based on self-sustainability and awareness of our relationship with the environment we inhabit.

The renewed place consciousness that comes from a new understanding of what places are, from renewed practices of inhabiting, and from the strengthening of the experience of links, connections and relations in which the conditions of life and the reproduction of life are given – also made possible by technologies of communication, too – requires an innovation of horizon.

The paper aims to offer a contribution in that direction.

**Keywords** Anthropology · Place · Space · Dwelling · Sustainability · Smart city

## The Human Being: *An Inhabitant, in Transit*

Returning to a long and fruitful anthropological tradition which arises from phenomenological-hermeneutical and metaphysical thought, we can speak of the human being as "bodily consciousness" or "embodied consciousness"

C. Danani (✉)
Università degli Studi di Macerata, Macerata, Italy
e-mail: carla.danani@unimc.it



W. S. Smith et al. (eds.), *Eco-Phenomenology: Life, Human Life, Post-Human Life in the Harmony of the Cosmos*, Analecta Husserliana CXXI,
https://doi.org/10.1007/978-3-319-77516-6_43

(Melchiorre 1987). Through means of the body, existing is "staying with" other human beings and things in the world even before thought can conceptualize it and reflect on how one relates to it.

Being a human is an existence within and through space, simply because it is a life lived in, and through, a conscious body. This body perceives, acts, and moves with a certain grasp on space, thanks to which it can locate and orient itself. So human beings, like things, have to be seen to "always be somewhere," and it makes sense for them to have questions about the place they are in. But human beings and things have different relationships with the places that they occupy. In fact, it must be said that human beings *inhabit*: they are not in the world in the way things are because the relationships they have with space are intrinsic to their existence.

For humans, to be in the world is to always exist in a "here" whose frame is the connection with all the "there" as compared to that which it is "here." Thus, living is also a constant act of familiarization, that is to say, a dynamic of orientation between that which is usual and that which is still unknown. This way of existing in the world can be summarized by saying that the human being "inhabits"; space and time are intrinsic to it. The body applies itself to these and embraces them. The familiarity with the world is "older than thinking" as Maurice Merleau-Ponty taught.[1] More specifically, the human being is an inhabitant *en passant* (Nancy 1999): one inhabits, *in transit*. In fact, within the dimension of space, time is lived, through which human existence is equally constituted. Thus, the logic of existence is also always a topology.

Therefore, it is a matter of understanding the ways of that familiarity, and of discovering how it manifests itself, within the multitude of human beings' experiences and practices. It is about grasping the meanings that appear in it, to interpret them; and thus, to understand how thinking works, which actions are generated according to it, and what kind of responsibilities emerge. This is very important for an anthropological elaboration, but also for ethics, which concerns good living in the world, and for politics, which is the construction of space for living together. Space is something so intimate to humans that each culture creates orientation systems that facilitate the development of positive environmental images – that is to say, a conscious interrelation – because "a distinctive and legible environment not only offers security, but also heightens the potential depth of human experience" (Lynch 1960, 75). Anthropological studies have highlighted the relevance of this.

## Place and Relationship

The complex structure of what we call "places," as well as their resistance to any sort of categorisation or characterisation, consists of both objective and subjective elements, which are inextricably linked to the practices and procedures that built them. All this tells us that a place, which can be seen as "marked *spatiality*," cannot be understood as a coherent, limited, and stable entity (Massey and Jess 1995). This is not true only of our era, one that particularly challenges any fixation because of its dynamics. By looking at the nature of places in depth, we can appreciate their intrinsic relationality.

The "identity of place" can only be interpreted as the result of a long sequence of connections between places, which all are peculiar inter-relations in a wider field: this reveals the possibility and the duty not only to appreciate any specific originality, but also to recognise the various bonds. Hence, places shall be seen as open and porous, as the result of connections rather than closed, exclusive, and separate entities: it is because of these intersections that a place acquires not only its uniqueness but also its commonality with what is elsewhere.

If the quest for closed and purified entities results in a geography of rejection, an open and interactive paradigm of identity results in a possible geography of reception (Young 1990). The latter is more faithful to the frame and life of places and human beings. *"Inhabiting" means being allocated, being somewhere, which is at the same time an experience of displacement*; each "where" comes as a relationship of many places, and living implies a constant decentralisation. An authentic existence is a continuous moving out from our quiet place. On the one hand, we should recognise limits and boundaries, because this is a condition that makes life possible; we cannot do without them. We lose our sense of direction in the desert, for example. On the other hand, delimitations are constantly crossed, redefined, and overstepped. Therefore, each "identity of place" does not have a static and *a priori* definable content. It is not even a fixed essence to be understood. As Norberg Schulz theorizes, we can speak of a *genius loci*, but it must be seen as the set of place-related values that define a place in history and as a co-evolution of natural and anthropic elements.

We can describe places as a geographical and historical "taking shape": they express history, characters, and structures over a long period, giving life to territorial "types" and individualities. They convey meaning. In order to understand places, functional considerations are not enough; we need descriptions that take into account all their heritage: nature, history, society, aesthetic elements and their perceptions, archives of past knowledge, traditions, memories and plans for the future. It is a matter of understanding the interwoven tissue between persistence and new potentialities.

According to Heidegger, we build because we inhabit. So he reverses the obvious and usual axiom according to which we think about building and farming; he conveniently highlights a hermeneutic circle at work among them (Heidegger 1971, 141–160). If it is true that inhabiting is the human way of existing on which all the practices are based and grow, practices develop and open new horizons to different places. This is not a mere relationship between goal and means; building and farming are a way of inhabiting, and inhabiting happens through building and farming.

The "identity of place" reveals itself as a kind of extremely concrete reality, but it cannot be reduced to any objectification; it is built collectively and its meaning should be constantly interpreted; it sprouts from "stones that tell stories" and from tales that have become shapes and stones. In this sense we can explain "place consciousness" as:

> The awareness of the asset value of common territorial goods (material and relationships), acquired through a process of cultural transformation of the inhabitants. Such goods have been recognised as essential elements for the reproduction of individual and collective life, biological and cultural life. This becoming conscious of the above-mentioned value is achieved through an individual and collective path that underlines the importance of community and of open and supportive ways. (Magnaghi 2007, 9)

Becoming "place conscious" is, therefore, also one of the conditions for building individual identities and communities that, aware of their constitutive relationship with the place they inhabit, could establish long-lasting interactions with nature and humans based on self-sustainability.

This does not imply the idea of place as a safe haven, a culturally coherent space. From such an inadequate perspective comes the opposite interpretation of contemporary transformation as disarticulation, as the emergence of a space of flux that supersedes the space of places. These ideas, combined with the modern outlook of the fluidity of all certainties, provide the illusion of justifying, with good reason, a reactive defence of the only possibility of stability. Hence the inappropriate mottos defending the legitimacy of being "ruler in one's own home." However, each place is the result of a long lasting relationship with other places, and its intersections define both its uniqueness and interdependence on other places (Massey and Jess 1995).

"Place consciousness" relates to one's personal moral life, since it is related to actions and behaviours that are attributable and estimable on the scale of good and evil, for which the human being is the agent and responsible one (Alici 2011, 11). Social practices and political decisions, together with individual actions, do not take place in a homogeneous isotropic space, but in a space characterised by qualitative differences. We inhabit different places so much that it is said that a house should be protective, the church solemn, the office functional. It should be noted that the peculiarities do not depend, on the utility that can be gained from the place to which they refer; they express, rather, their significance for the human beings living in the world. A road is not a square and not a yard, a school is neither a house nor a church. The experience of being "out of place," or words and acts that are "out of place," are negative possibilities in the "taking place" of existence which is always drawn to suitability. Each meaning that has to be interpreted is, without doubt, a social construction and has historical importance; it is part of the dialogue where individuals and societies encompass and comprehend each other. In any case, in the capability to be oneself and have a style, place consciousness is also involved. It is being able "to stay with" the world so that we can be informed as to what the appropriate behaviour is for each occasion, meaning for a specific moment at a specific place, that is, the ability to acknowledge the specificities of the environment we are in. The disciplinary aspect, which Michel Foucault underlines[2], although relevant to this, can only be built on a more original adherence that the individual and collective existence have the responsibility to interpret.

Nothing benightedly conservative or politically reactionary[3] need inevitably connote discourse on the "sense of place," on place consciousness, or on "identity of place" although they contain reference to the duration of a place, a sort of stability. The places offer themselves as written texts, read and interpreted over and over again. In this sense, as I have said, that objectivity cannot be reduced to any objectification, and we can say that "time is read in space" (Schlögel 2003). We can also point out that only a well-cultivated place consciousness can support a correct relationship to the natural environment. This concerns both individual behaviours and collective rules. Only if we refrain from considering places as mere commodities, instruments of wealth and "resources," will the stress on the so-called "green economy" (or more generally on the continuous call for renewable energy sources, rules on differentiated waste collection, attempts to reduce emissions and to preserve water pureness), have a deep and long-lasting effect. Only in this sense will it represent an alternative to the consumerist and economistic mentality. The environmental crisis, as Norberg-Schulz says, reveals a human crisis and it can be faced effectively only by basing our actions on the understanding of human bonds to places. Modernity long believed that science and technology would set people free from their dependence on places. This proved to be an illusion. Pollution and environmental chaos appeared out of the blue as a frightening curse and put the problem of "place" under the spotlight once again (Norberg-Schulz 1980, 11). We cannot properly address it since places are considered mere sites, mere objects, or mere contexts in which life happens. Thus, if ethics and politics want to deal with the environmental question radically, then laws, technological innovations, and virtuous attitudes are not enough. A deep reflection about the transcendental allocation of human beings has fundamental relevance: this is the horizon where questions are to be found and addressed. We know: beliefs change practices.

## The Place of the Web

The so-called "Web social revolution" is a recent phenomenon to take into consideration: it seems to suggest, in fact, unexpected connections between *online* and *offline*.[4] If Web 1.0 can be described as "impersonal," consultancy-oriented, based primarily on a one-way flow of information, Web 2.0 is, instead, bottom-up and participatory, interaction-oriented. The rapid spread of such a phenomenon as Facebook over the past decade and its overtaking of Google as the most visited website seem to be indicative of a significant step. At the same time, sociologists record the decreasing popularity of the virtual universes one may enter as theatres in which to stage parallel lives, such as Second Life, in favour of social networking sites where you normally sign in with your real name and you tell about your *offline* life, past and present.

In social networks, what matters are the contacts and being connected; the semantics is subordinate or incidental. "Talking" here is not just "saying something" but it is, first of all, "saying something to someone." As pointed out by Chiara Giaccardi, being in a relationship with someone, even if it comes to nothing, is of major importance compared to what one has to say; the phatic function seems to be dominant, and it seems to play a social role rather than convey information. Marcel Mauss has shown that the phatic function – such as chatting about anything around the campfire in the evening – in traditional villages,[5] has always been an important "glue" among people, and it seems to have the same role in the digital village (Malinowski 1923, 296–336). Surveys of sociologists have shown that in social networks the transition from "contacts" to "friends" can certainly be read as a "trivialization" of friendship, but perhaps – and probably more appropriately – as expressions of a way of building relationships, as well. Moreover, each contact, even if communication effectively does not take place at that time, is a "potential relationship." That relationship, despite all the skepticism, and all the critical observations, seems the fundamental category for understanding the dynamics of the next-generation network.

These surveys also revealed that the apparently immaterial and incorporeal space of social networks seems to intertwine seamlessly with the real life of those who access it. The social network can be understood as a real place where people try to live a dimension of "community," that "being in touch" which existence needs in order to share issues and problems. This attitude is not free of danger. If, for example, it can lead to conformity, another risk is what has been called "all in": the network is a total field without a center, one which incorporates everything. A closer look, however, will show that this type of disease is lurking in every experience of place: when perception of the "outside" is muffled, people build a magic illusionary circle.

As underlined by Chiara Giaccardi in her comment on the results of a recent research in the field, there are many elements that belie both the tendency to individualism and the prevalence of utilitarian attitudes. The web world is cooperative and "horizontal": digital practices are based on sharing, exchange, the building of materials and knowledge from the bottom, and an open and scattered knowledge which people participate in, which is not apparently subject to preset directions. Of course, being connected should not be confused with communicating, and everything could become an exploitable field for non-virtuous logics, but this must not lead to considerations governed only by a fear of possible negative outcomes. We cannot simply say that the virtual constitutes a withdrawal from placeness, although Levy insists that virtualization creates a nomadic culture and when a person, a community, a place, a question, becomes virtual, "ils se mettent 'hors-là', ils se detérritorialisent."[6] He says that a sort of release pulls all these away from the usual physical and geographical space and also from the usual time of the clock and the calendar. Levy is, however, aware of the impossibility of totally denying space and time, but he insists that virtualization represents a way to escape them. The extolling of estrangement from the corporeal, embodied dimension, which characterizes many idolaters of cyberspace, does not take into account that escape from body and spatiality does not lead to a destiny of freedom, but to the weakening of that grip on the world that allows people to really make it a "world." It involves the *Welt-Armut* that Heidegger ascribed to animals and is an impoverishment of the human being.

Against all this, the effort to proceed, not to a reduction of humanity, but to a strengthening through communicative layers that do not neglect location, promises an enrichment of our world. For this we must equip ourselves with competences and virtues. Cultural training has an important responsibility here; it means educating one's self-awareness also in its corporeal dimension, and it means taking into account "awareness of place" in order to understand oneself in the world as a transcendentally allocated being.

## Views for Future Projects

It is important to think about the horizons that can be opened by a renewed place consciousness, one able to use the new means of social communication wisely and virtuously. In *L'Invention du Quotidien,* Michel De Certeau makes a distinction between the city of "seeing" and the city of "doing." Urban reality can be seen from above, transposed on a map, planned, and organised, but there is a strangeness in everyday life. De Certeau explains: "escaping the imaginary totalisations produced by the eye, the everyday has a certain strangeness that does not surface, or whose surface is only its upper limit, outlining itself against the visible" (de Certeau 1984, 93). If you go to the street level and become a walker, a pedestrian: "a *migrational, or metaphorical, city thus* slips into the clear text of the planned and readable *city." The walker is blind and cannot see the city as a whole, but he lives its everyday reality with spatial and social relations. De Certeau adds: those who* "live down below, [live] below the thresholds at which [an everyday] visibility begins. They walk – an elementary form of this experience; they are walkers, *Wandersmänner*, whose bodies follow the thicks and thins of an urban "text" they write without being able to read it. These practitioners make use of spaces that cannot be seen. Their knowledge of them is as blind as that of lovers in each other's arms.... It is as though the practices organizing a bustling city were characterized by their blindness" (de Certeau 1984, 93). We must, therefore, be aware of the many ways in which we perceive the world and orient ourselves, because to do so today offers an important opportunity.

A renewed place consciousness can find a new alliance with communication tools, to create and delineate geographies of hospitality and plural identities: building spaces of differentiation without exclusion. While it is becoming increasingly clear that everyone depends on the interaction of thousands of other people and enormous organizational resources to achieve their individual goals, we understand that we have a number of problems and interests in common. If it is true that this does not in itself give life to a community in the sense that people share ultimate ends and reciprocity, it still brings to light something common that constitutes the premise of that community. This is today, the condition of the possibility of a compromise, that is "the form that clothes mutual recognition in situations of conflict and dispute" (Ricoeur 2005, 210). On the basis of a place consciousness together with the potential for new means of social communication through the web, electronics, and IT devices, this difficult moment of disorientation can also offer the opportunity to think of a new way of living places, one capable of deep awareness of layers, relationships, and interconnections. Even the public social space is expanding and becoming more accessible. An unprecedented horizon for the articulation of the link between singularity and multiplicity is provided.

We can show at least two thorny issues on the agenda. First is the instance of an overcoming of the modern perspective in order to finally understand urban life according to the hermeneutical key of the territory. Second is to recognize that every individual has a responsibility to participate in debates and decisions regarding the place in which he or she dwells. The agenda therefore requires a change in perspective, that understands places as content and not as context for the question of living together.

Let us try to put our reflection to work, then, taking into consideration a perspective of great importance in European social planning, that covers an increasingly relevant place for the human community – the city – and emphasizes the value of communication and information technologies. I am referring to the emerging importance of the theme of *smart cities*. Today the UN estimates that about half of the world's population lives in urban centers; from about 750 million residents in 1950, the urban population grew to 3.6 billion in 2011, and it is expected to reach 60% of the world population by 2030. UN-Habitat (the United Nations human settlement program) warns us: today a third of the world's urban population lives in slums, and if it continues to increase, in 2020 the number of people who are in a similar condition will come to about 1.03 billion. The cities of the future will have to be smart, they say.

Now, we must go back to the 1990s to find the first occurrence of the term "smart city," a usage linked to two large multinational companies: IBM and Cisco. These giants of the digital developed a vision of an ideal city focused on automation, information, and communication technology as propulsive tools for urban "intelligence" (Zanirato 2012, 22–23).

In 2007 an important report was published; the fruit of a research led by Rudolf Gillinger of the Vienna University of Technology and conducted in collaboration with the University of Ljubljana and the Technical University of Delft. The report represented an early stage of the later evolution of the "smart city," one marked by a transition from understanding it as a digital city to the smart city as a more livable and more inclusive city. More specifically, the research, which is aimed at measuring the level of "smartness" of seventy European cities of medium size, defines smart cities as those which pursue the improvement of their performance in meeting six strategic objectives: smart economy, smart government, smart environment, smart living, smart mobility, and smart people. We shall look carefully at this approach, which has become the mainstream of reference and influences European politics on the revision of its priorities and allocation mechanisms; it also influences many regional policies. We need to watch this perspective closely, so that it does not run the risk of a simplified reading of the city and its needs (Granelli 2012). We should acknowledge that the ranking suggested by Rudolf Gillinger was able to go beyond the definition of smart city in which "intelligent" has the reductionist meaning of digital or technologically advanced. An additional merit of this model is that it does not focus on the economic side only, being aware that the reduction of everything to mere economic factors puts other factors at risk of deteriorating.

But the question of "what a smart city is" remains partially unanswered. A city is an interweaving of *civitas*, *urbs*, *polis*, and land. It is not enough to consider single aspects. This difficulty brings to light the real question: when do we declare that a city is smart? On the basis of quantitative data, or on performance, or on the basis of the perception of its inhabitants? First of all, we should understand how to develop appropriate methods of assessment. On the one hand, this is a matter of understanding whether the adoption of a data-based "performances" criterion is enough; on the other hand, it is a matter of being aware that "satisfaction" involves "adaptive preferences". In fact, people tend to adjust their desires to their means, that is, those desires are themselves determined by a criterion that predetermines the choice. On the hedonism scale of values, the satisfaction induced by resignation may be indistinguishable from the satisfaction of individual wills (Elster 1983). Moreover, the comparison made among the levels of satisfaction of different individuals is problematic.

We must not only ask ourselves how to develop adequate methods of evaluation, but also discuss the criteria by which the evaluation is based. The term "smart" is used for various aspects ranging from smart city as an IT district, to a smart city as regards the education or the intelligence of its inhabitants. When referring to economy or jobs, the term describes the city as an industry, or a business park, which mainly involves the field of information and communication technologies (ICT), both as products and as used in the manufacturing processes. The term is also used with regard to the level of education of the citizen, and in this case, smart city means a city with inhabitants who have a high level of education. In other literature the term smart city refers to the relationship between the city government (that is, its administration) and the citizen. It is also used to describe modern technology in everyday urban life, including not only ICT but also modern technologies of transport. Finally, various aspects of life in a city are mentioned, such as security, greenness, energy, efficiency and sustainability. Therefore, there are different fields of activity which are described in the literature as being related to the term smart city: industry, education, participation, technical infrastructure, and other "soft" factors.

However, neither determining efficiency and efficacy, nor researching what is useful in terms of advantages and objectives achieved, nor adopting a concept of "welfare" according to which a situation is chosen only on the basis of its usefulness, seems sufficient. It is a matter of clarifying what "well-being" means, and a new semantics of this term is probably necessary, not only from the perspective of the person, but also of the community. A critical theory would take into account both the distribution structures and the processes and relationships that these structures produce and reproduce. This also applies to access to tools of communication. The distribution of goods and resources is a fundamental issue of social justice, but equally important are issues concerning power and decision-making processes, as well as culture and division of labor. Justice is to be seen as a constitutive factor of well-being: each person must be – and feel – as a subject and not as an object in the politics of living together.

Social planning has become explicitly aware of the need to go beyond purely quantitative and efficiency-related parameters. This is a valid premise for good practices within social and natural environments. But, we have to be careful that the factors of smartness are outlined in non-neutral contexts: they could only be an attempt to define the ranking of procedures in a competitive horizon, that reduces everything to a means of wealth and does not consider the existence of "values" as wealth.

Competition may be an important factor, but we should not accept its meaning as obvious, and indeed a new semantics is again very urgent in order to replace the obsolete model of positional competition based on Hobbes' anthropological assumption *mors tua, vita mea*. It is not necessary for us to defeat others to exalt ourselves. Moreover, we now know that the cooperation between stakeholders is as important as competition between them. We have learned from extensive empirical evidence that there is a critical threshold in competitive intensity, beyond which the benefits associated with increase in performance levels fall below the level of the disadvantages following from the lack of motivation, and the loss of identity among those who remain out of or are marginalized by market competition.

Moreover, theoretical debate and social planning should be pushed to go beyond the horizon of smart cities,[7] in the direction of the wider question about *smart lands*. In fact, the renewed place consciousness that comes from a different understanding of what they are, from renewed practices of inhabiting, from the strengthening of the experience of links and relations in which the conditions of life and reproduction of life are given, requires an innovation of horizon: one must not only think of cities but also of agricultural areas, which do not take their identity from urban connotations. Rural areas - and the third landscape - are not places resulting from the city, but they do safeguard the conditions of life and reproduction of life albeit in an increasingly precarious balance with the urban context and its infrastructures.

Philosophical reflection can make a contribution, with other disciplines, to the construction of new representations and new narratives, to the development of new practices. It is important to reinforce critical understanding and to think in depth; this is a matter of challenging obvious meanings, of rethinking some categories. New technologies can make a great contribution in this direction. Philosophical reflection, proceeding from the transcendental placeness of humankind, can be a very fruitful perspective: from a theoretical and a practical point of view.

## Notes

1. See M. Merleau-Ponty, *Phénoménologie de la perception*, Paris: Gallimard, 1945; English trans. *Phenomenology of Perception*, London and New York: Routledge & Kegan Paul, 1962, p. 162: "I belong to them [space and time], my body combines with them, and includes them", and "Space and perception generally represent, at the core of the subject, the fact of his birth, a perpetual contribution of his bodily being, a communication with a world more ancient than thought" p. 296. We refer to the phenomenological research of Husserl too. See Edmund Husserl, *Studien zur Arithmetik und Geometrie*, (Husserliana XXI), ed.

I. Strohmeier, Den Haag: Nijhoff, 1983; *Ding und Raum. Vorlesungen 1907*, Husserliana XVI, ed. U. Claesges, Den Haag: Nijhoff, 1973. Space is pertinent to kinaesthetic spontaneity and not to receptivity. Vincenzo Costa illustrated clearly that for Husserl sensation is the bearer not only of color, sound, touch, form, and of the characteristic of the thing, but also of a system of orientation. Kinaesthetic consciousness is a condition for experience, for the manifestness of things and space. In sensitivity there is "una certa attività, una certa spontaneità, non si tratta della spontaneità di un io del pensiero, bensì di un io incarnato, dell'attività cinestetica, attraverso cui l'io è nel mondo, lo abita"; the question here is not about an anthropological feature of the individual, but the structure of manifestness as such (see V. Costa, "La questione della cosa e il realismo," his "Introduction" in Edmund Husserl, *La cosa e lo spazio*, Soveria Mannelli: Rubbettino, 2009, pp. xliii and xlv; see also V. Costa, *Il cerchio e l'ellisse. Husserl e il darsi delle cose*, Cosenza: Rubbettino, 2007).

2. M. Foucault refers to the organization of space in barracks, at school, and in hospital.
3. I do not intend, however, to accept the current ideology that affirms that everything new is always good and what is preserved as such is negative.
4. I refer to the research that was led by the Università Cattolica di Milano in partnership with the UCS (Ufficio Nazionale per le Comunicazioni Sociali) of the CEI (Conferenza Episcopale Italiana). The integral report is available here: www.testimonidigitali.it/ricerca, and it was published in C. Giaccardi (ed.), *Abitanti della rete*, Milan: Vita e Pensiero, 2010.
5. Malinowski pointed out that the action done when "chatting about nothing" is the fundamental action for the constitution and reconstitution of the community.
6. P. Lévy, *Qu'est-ce que le virtuel?*, Paris: La Découverte, 1995, especially pp. 17–19, English trans. *Becoming Virtual: Reality in the Digital Age*, New York: Plenum Trade, 1998, pp. 27–29: "The fact of not being associated with any 'there' of clinging to an unassignable space (the one in which telephone conversations take place) – none of this prevents us from existing" (28); "they are not totally independent of a referential space-time since they must still bond to some physical substrate and become actualized somewhere sooner or later" (29).
7. The very interesting reflections about "*sense/able cities*" (see Carlo Ratti, etc.) do not concern a radical new paradigm in our sense, for they refer only to an enhancement of human capabilities through technology. Regarding smart cities, see also: T. Campbell, *Beyond Smart Cities. How Cities Network, Learn and Innovate*, Routledge, 2012; E. Riva Sanseverino, R. Riva Sanseverino, and V. Vaccaro, *Atlante delle smart cities. Modelli di sviluppo sostenibili per città e territori*, Milan: Franco Angeli, 2012; M. Vianello, *Smart cities. Gestire la complessità urbana nell'era di Internet*, S. Arcangelo di Romagna: Maggioli Editore, 2013; C. Zanirato, *Ricreare la città. Smart cities*, San Francisco/Bologna: Edizioni Blurb/Pamphlet, 2012; L. Matteoli, *Cityfutures. Verso la città del futuro*, Milan: Hoepli, 2010; L. Hatzelhoffer, K. Humboldt, M. Lobeck, C. Wiegandt, *Smart City in Practice. Converting Innovative Ideas into Reality*, Berlin: Jovis, 2012.

## Works Cited

Alici, Luigi. 2011. *Filosofia Morale*. Brescia: La Scuola.

Campbell, Tim. 2012. *Beyond Smart Cities. How Cities Network, Learn and Innovate*. London: Routledge.

Costa, Vincenzo. 2007. *Il cerchio e l'ellisse. Husserl e il darsi delle cose*. Cosenza: Rubbettino.

———. 2009. La questione della cosa e il realismo. Introduction. Edmund Husserl. La cosa e lo spazio. Soveria Mannelli: Rubbettino..

De Certeau, Michel. 1984. *The Practice of* Everyday Life. Trans. S. Rendall. Berkeley/Los Angeles: University of California Press. *A translation of L'invention du quotidien*, Vol. 1: Arts de faire. Paris: Gallimard, 1980.

Elster, Jon. 1983. *Sour Grapes*. Cambridge: Cambridge University Press.

Giaccardi, Chiara, ed. 2010. *Abitanti della rete*. Milano: Vita e Pensiero.

Granelli, Andrea. 2012. *Città intelligenti? Per una via italiana alle smart cities*. Rome: Luca Sossella Editore.

Hatzelhoffer, Lena, Kathrin Humboldt, Michael Lobeck, and Claus-Christian Wiegandt, eds. 2012. *Smart City in Practice. Converting Innovative Ideas into Reality*. Berlin: Jovis.

Heidegger, Martin. 1971. Building, dwelling, thinking. In *idem, Poetry, Language and thought*. New York: Harper & Row. A translation of Bauen, Wohnen, Denken (1951) In idem, *Vorträge und Aufsätze*. Stuttgart: Neske, 1994.

Husserl, Edmund. 1973. *Ding und Raum. Vorlesungen 1907* (Husserliana XVI), ed. U. Claesges. The Hague: Nijhoff.

Husserl, Edmund. 1983. *Studien zur Arithmetik und Geometrie* (Husserliana XXI), ed. I. Strohmeier. The Hague: Nijhoff.

Lévy, Pierre. 1995. *Becoming Virtual: Reality in the Digital Age*. New York: Plenum Trade, 1998. Translation of *Qu'est-ce que le virtuel?*. Paris: La Découverte.

Lynch, David. 1960. *The Image of the City*. Cambridge, MA: MIT Press.

Magnaghi, Alberto, ed. 2007. *Scenari strategici: visioni identitarie per il progetto di territorio*. Florence: Alinea.

Massey, Doreen, and Pat Jess, eds. 1995. *A Place in the World?: Places, Cultures and Globalization*. Oxford: Oxford University Press.

Malinowski, Bronislaw. 1923. The problem of meaning in the primitive languages. In *The Meaning of Meaning*, ed. C.K. Ogden and I.A. Richards, 296–336. New York: Harcourt, Brace & World.

Matteoli, Lorenzo, and Roberto Pagani, eds. 2010. *Cityfutures. Verso la città del futuro*. Milan: Hoepli.

Melchiorre, Virgilio. 1987. *Corpo e persona*. Genoa: Marietti.

Merleau-Ponty, Maurice. 1962. *Phenomenology of Perception*. London/New York: Routledge/Kegan Paul. A translation of *Phénoménologie de la perception*. Paris: Gallimard, 1945.

Nancy, Jean-Luc. 1999. *La ville au loin*. Paris: Fayard.

Norberg-Schulz, Charles. 1980. *Genius Loci. Towards a Phenomenology of Architecture*. London: Academy ed.

Ricoeur, Paul. 2005. *The Course of Recognition*. Trans. D. Pellauer. Cambridge, MA: Harvard University Press. A translation of *Parcours de la reconnaissance*. Paris: Stock, 2004.

Riva Sanseverino, Eleonora, Raffaella Riva Sanseverino, and Valentina Vaccaro, eds. 2012. *Atlante delle smart cities. Modelli di sviluppo sostenibili per città e territory*. Milan: Franco Angeli.

Schlögel, Karl. 2003. *Im Raum lesen wir Zeit*. Munich: Hanser Verlag.

Ufficio Nazionale per le Comunicazioni Sociali di Conferenza Episcopale Italiana. www.testimonidigitali.it/ricerca. Web. 1 March 2015.

Vianello, Michele. 2013. *Smart cities. Gestire la complessità urbana nell'era di Internet*. S. Arcangelo di Romagna: Maggioli Editore.

Young, Iris Marion. 1990. The ideal of community and the politics of difference. In *Feminism/Post-Modernism*, ed. L. Nicholson. London: Routledge.

Zanirato, Claudio. 2012. *Ricreare la città. Smart cities*. San Francisco/Bologna: Edizioni Blurb/Pamphlet. www.smart-cities.eu/download/smart_cities_final_report.pdf. Web. 1 March 2015.

# Holographic Memory of Life Situation

**Salahaddin Khalilov**

**Abstract** Every phenomenon is rich and meaningful against its background and environment. The lives of a human being are not limited to their thoughts. The background, life situation, and *ontopoiesis* accompanying these thoughts are more important. Indeed, Husserl rightly tries to reduce our emotions and subjective features. However, there is hazy and vague content (substance) in the fine structure of a phenomenon. The goal is not to get rid of it but rather to illuminate it more and discover the background and substructure of this structure. In my opinion, it is possible to find a hologram of the noumena in the phenomenon. Our goal is to discover the method by which to find it. This event cannot be expressed fully by the term "intuition." Intuition is the acceptance of the truth of "things-in-themselves" not from their manifestation or phenomenon, but from an unknown cosmic source. Actually, the illuminated front part of consciousness or a phenomenon should not be isolated from the context but, rather, should be taken in with its background and context. A pure consciousness or phenomenon isolated from the world and bordered on all sides cannot allow any precise thinking. Extreme precise extraction from conditions, mathematical or logical modeling of the world, the construction of the world from pure phenomena – all of these are idealization and, in fact, are a distortion of the reality. The mathematization of life is impossible. Life is unique and unrepeatable, having endless wealth. The desire to specify events on the level of mathematical figures is too idealistic. Hence, there is a need in phenomenology to go beyond formal and mathematical logic and to apply the idea of *fuzzy logic*.

**Keywords** Holographic memory · Mathematical modeling · Fuzzy logic · Husserl · Tymieniecka

---

S. Khalilov (✉)
Azerbaijan University, Baku, Azerbaijan

 
W. S. Smith et al. (eds.), *Eco-Phenomenology: Life, Human Life, Post-Human Life in the Harmony of the Cosmos*, Analecta Husserliana CXXI,
https://doi.org/10.1007/978-3-319-77516-6_44

## Eco-Phenomenology and Holographic Memory

My research is on the relation of the phenomenology of life and more particularly eco-phenomenology to holographic memory. I have tried to show how the phenomenology of life and Anna-Teresa Tymieniecka's concepts of memory provide a sound methodological base for holography. Indeed, this philosophy is the most appropriate methodology to apply in framing scientific theories about the living and non-living realms, establishing a common denominator and offering scientific explanations of the creation of the living world. Synergetic principles, for example, cannot be explained by applying traditional scientific methodology. Ilya Prigogine has written, "Today the 'vitalist principle' has been superseded by the succession of improbable mutations preserved in the genetic message 'governing' the living structure" (Prigogine and Stengers 1984, 84).

Tymieniecka's phenomenology of life not only serves as a methodological base for this worldview, it also explains some related problems. She has instilled new ideas into the foundation of strict scientific theories about the creation of life. We must go beyond traditional approaches to take the measure of her philosophy. Her thought is a bold new step challenging existing paradigms and creating a new terminology. I believe the solution to this problem of the holographic memory to be found in phenomena has a place among the purposes and praxis of her philosophy's followers.

## Object and Habitat

Every living organism exists in accord with its material environment. From the scientist's viewpoint, a habitat is the mean interrelation of a knowable object and external factors affecting that object. In other words, no object can be truly known in isolation. Max Scheler writes: "That which is seen and experienced is *given* only *in the seeing and experiencing act itself*, in its being acted out [*Vollzug*]; it appears in this act and only in it. It does not simply stand there and let itself be observed so that now this feature, now that, stands out in relief without any alteration in the object" (Scheler 1973, 138). Therefore, various situations offer different angles of view or contexts. Cognition of an object depends on its situation. For example, a flower can change in its appearance from sunny to dark days or in cold or warm weather. The situation or environment of living beings is known as the lifeworld.

Perhaps this variability is why Tymieniecka considered knowledge about all living beings within an identified ecosystem to be so important. She writes, "As was already clearly brought out in the twentieth century, and has glaringly come to light in contemporary life and science, it is not only the circumambient sphere encircling singular, living beingness that is affected by its growth, development, and becoming, but the entire system of life in the range of life that that living beingness participates in, in what is today called the 'ecosystem'" (Tymieniecka 2003, xxviii).

Indeed, this is a very important condition. Along with the living beingness, its habitat and lifeworld, its knowable and known sides should be borne in mind to establish a more comprehensive understanding. Sartre writes:

> Let us suppose that my imaging consciousness aims at the Pantheon. As it is knowledge, it aims at the Pantheon in its sensible nature, which is to say as a Greek temple, of grey colour, with a certain number of columns and a triangular pediment. On the other hand, in a certain manner, the Pantheon aimed at is present: it is given in its affective reality. On this affective presence, my knowledge intentionality apprehends the qualities just cited. (Sartre 2004, 87)

In my view, phenomenology is to be directed toward and focused on human consciousness, including the background of an event in the mind. The unveiling of the phenomenon in Husserl's phenomenology assumes a void environment, where an image is composed on a Lockean *tabula rasa*. However, that ideal of a perfectly blank slate does not exist. Focusing consciousness on only one object in isolation is also unachievable. First of all, human consciousness is fuzzy and unclear. Certain images may leave a deep impression which can never be completely removed. Therefore, new images are drawn over previous ones, even those that have faded significantly over time. Husserl's phenomenological reduction serves to merge these ideas, but this, in our opinion, is unacceptable. True reality requires attention to both the object actualized by our consciousness and that image which expresses the initial state. This is of paramount importance in the apprehension of a phenomenon by living beings, especially human beings. It is not by chance that Anna-Teresa Tymieniecka especially considered the phenomenon of living beings, which she saw "as a filigree, a microcosmic counterpart of the great macrocosmic horizon."

## Holographic Apprehension

The initial state, the *tabula rasa* of consciousness, is closely related to memory. Thus, a human being must yield everything in his consciousness to the passive fund of memory during the process of a new cognition. If previous knowledge and impressions are not completely cleared, consciousness will remain unclean and fuzzy. Then the phenomenon created as a result of focusing one's consciousness on an object will result not in an accurate representation but in an image displayed against this unclean background.

Every phenomenon is rich and meaningful in combination with its background and environment. Similarly, the lives of human beings are not limited to their thoughts. A person's background, situation, and the *ontopoiesis* accompanying these thoughts are also important. Husserl rightly tries to reduce emotions and subjective features in the apprehension of phenomena. However, the goal is not to get rid of the hazy and vague content or substance within the fine structure of the phenomenon; rather, the aim is to illuminate the content and reveal its background and substructure.

In modern science, taking a "photo" of such complex systems is possible through a new method – holography. In contrast to two-dimensional photos, holograms make it possible to see an object in three-dimensional space and from different angles. The purposeful illumination by a laser corresponds to the intentionality of consciousness in phenomenology and to *ishrak*, or enlightenment, in ishraqism. The inner information a thing carries is called holographic memory. Michael Talbot explains:

> Pieces of holographic film containing multiple images also provide a way of understanding our ability to both recall and forget. When such a piece of film is held in a laser beam and tilted back and forth, the various images it contains appear and disappear in a glittering stream. It has been suggested that our ability to remember is analogous to shining a laser beam on such a piece of film and calling up a particular image. (Talbot 1991, 21)

The consideration of such situations using traditional logic and mathematical models is impossible, which explains the need for fuzzy logic. In our opinion, Tymieniecka's phenomenology of life also provides a methodological basis for fuzzy logic. It was not an accident that she focused on the problems of memory and remembrance. Consequently, here is the method for uniting two separated paths in philosophy.

## Learning and Feeling

A paramount tenet of traditional philosophy and psychology is that the thought of the human being is related not only to his inner world and experiences but to the reflected, external, and knowable world. In Husserl's phenomenology, intentionality or the focusing of consciousness on an external thing is taken as a base. Moreover, Husserl considers being free of all subjective factors when apprehending objects, avoiding psychologism, to be an important condition for making philosophy a strict science. And indeed, the idea of an object cannot depend on the feelings of particular individuals or their circumstances. Certainly, feelings and senses are isolated and remain only pure ideas. However, until the unveiling of a phenomenon during the process of cognition, the human being is in the stage of sensory experience, variously connected to the object of cognition, and is experiencing different feelings as well. Consequently, this process cannot reach resolution without leaving an impression on the inner world of the human being. In other words, although an idea is beyond feelings, feelings are affected. Sartre writes:

> We have seen, in the second part of this work, that one of the essential factors of the imaging consciousness is belief. This belief aims at the object of the image. All imaging consciousness has a certain positional quality in relation to its object. An imaging consciousness is, indeed, consciousness of an *object as imaged* and not consciousness *of an image*. But if we form on the basis of this imaging consciousness a second consciousness or reflective consciousness, a second species of belief appears: the belief in the existence of the image. (Sartre 2004, 86)

Knowledge is beyond subjectivity. However, subjects do change, not because a person knows anything but because his feelings become richer during the process of learning. It is not coincidence that the well-read, educated, and intelligent people differ from the non-educated people even in their appearance. This is the non-learning side of the visible process in traditional phenomenology.

## The Eye's Mark

On the other side, does the process of learning affect the object? In other words, does a cognized and known thing differ from a thing never seen or learned of by human beings? Furthermore, upon discovering a new object, the viewer naturally wonders if it has previously been unveiled by anybody else. The consideration of being "unveiled" or "veiled" calls to mind the idea of virginity. An Azerbaijani song refers to this visible change:

> *There is an eye's mark on your face,*
> *Who has looked at you, my soulmate?*

Similarly, Pierre Bezukhov, a character in Leo Tolstoy's *War and Peace*, addresses the eye's mark when he describes the beautiful Helene Kuragin, the jewel of many parties and balls. He declares that a "thousand glances have turned the beauty of Helene into a cold and dead marble bust."

What is this eye's mark? As the seen object is a human being, perhaps he or she has a subconscious reaction (inner reflection) to the alien glances, which leave their mark on the inner world. However, this concept only applies to glances known and recognized by the observed human being. From a greater perspective, all living and nonliving beings that are the object of cognition incur some mark. Everything changes owing to external influences such as direct light, wind, or rain. Indeed, this is indisputable fact. However, why are changes to objects of cognition ignored? In fact, the sensory experience is enabled by the material interrelation.

According to Heisenberg's uncertainty principle we cannot measure simultaneously the position (x) and the momentum (p) of a particle with absolute precision. The more accurately we know one of these values, the less accurately do we know the other. This means that no experience, even a sensory experience, fails to make an impact. Thus, in order for us to see something, it must be illuminated. And, of course, illuminated and never before illuminated objects are different. The extrapolation of this universal principle leads to the idea of holographic memory, meaning that we can also speak of memory in the material world.

It is necessary for us to reconsider the relation between past and present and to create a new concept of memory. According to Einstein's theory of relativity, it is impossible to return to the past, because the past is the actual image of an event and its physical-optical visibility is moving away from us at the speed of light; for reaching and catching it we would need to go faster than $c$ – and that is impossible.

A past local event is "kept" in our memory as video tape and the like, as it were. It is possible to review these: to revive in the memory, to re-watch as a video, and so on. But this is not simply repetition, because both sides that meet have now been changed. On the one side, the memory, the tape, etc. are not a complete copy of the reality but manifest it only from a certain angle. And, then, the subject is also different from its previous state. Our capabilities for assessing the event have now changed. For example, new details actually caught on a tape but not disclosed through the technology of that time may be discovered now. Our own experience also improves, and when we remember that event now, some details not clear to us before and so not brought to the fore or not focused upon stand illuminated and brought to the fore. Of course, one might then wonder what remembered part of the phenomenon was precisely cognized by us during the first contact.

## Tymieniecka's Phenomenological Advance

The phenomenon itself is alive. It has its own relatively independent life. In our opinion, the phenomenon continues to live not only during our direct contact with the real event, when we focus our attention on something, but also after that immediate contact vanishes. It passively remains alive in the background, certainly, and this is not a phenomenon that Husserl's methodology is suited to apprehending, for this is a potential phenomenon holding a chance to revive. That is why Anna-Teresa Tymieniecka made a special place for "memory." She wrote, "Living beingness appears as carrying within itself the prolife schema and life requirements that mother earth possesses. Through this essential existential network, the project of life appears subtended by the earth's participation in the forces of the universe, the cosmos. Hence, we may see living beingness as a filigree, a microcosmic counterpart of the great macrocosmic horizon" (Tymieniecka 2003, xxix). Regarding the issue that we here pay attention to, it is to be put forth that a reminiscence's passive and potential life is not limited purely by memory, but continues in real life too. Moreover, this "second life" continues in the form of a hologram and takes part in the revival process of the passive form in our memory in due course of time. In addition, this new phenomenon could have more reach, more impact than did the initial and "real" phenomenon.

## Holographic Eternity

The difference between the passive and active state in the physical realm is that an active event occurs in a concrete space-time continuum. In the passive state an event turns into the "memory of nature," "natural history," "eternity, world," in other words, to a life "after a death," and exists in the form of a hologram.

When and under what circumstances can the passive phenomenon remaining in our consciousness and the hologram remaining in physical reality spring back to life? This is a virtual continuation of the past in the present. However, it occurs not in photos, tapes, books, nor finally in a person's memory, but in a virtual cosmic reality – in the world of opportunities. It cannot affect the environment, and it is not possible to affect it either. We cannot consider its interrelation with the current (actual) physical world. Yet, we cannot refute the possibility of such an interrelation, or influence, because the passive phenomenon has a chance to be actualized again. Of course, this can be realized only through the divine will, but it is also possibly a potential of the human being's consciousness. Consciousness has a power to bring into focus of a certain fragment of physical reality and to illuminate it, as well as to revive the virtual past. How could it happen?

The ability of consciousness to remain in passive form in the memory and to actualize and restore "information" is obvious fact and has been researched sufficiently in philosophy and psychology. However, we are considering now not the information in memory summoned in remembrance, but the "restoration" and revival of the existence of the virtual cosmos and the ability to return the past and bring it to the same level as the present.

## Conclusion

Husserl saw the world through the human being's consciousness. However, today we are at a level of technological development where we sense the existence of a virtual world consciousness as if our consciousness is a part of a common and universal consciousness. Along with our individual memory and remembrance, Nature has its own memory. The actualization and illumination of our consciousness in a certain direction occurs under a coordination with a "world consciousness" or "world memory." And new developments in science and technology, especially holography, give new opportunities for philosophical assessment.

## Works Cited

Prigogine, Ilya, and Isabelle Stengers. 1984. *Order out of chaos. Man's New Dialogue with Nature*. New York: Bantam Books Print.

Sartre, Jean-Paul. 2004. *The Imaginary: A Phenomenological Psychology of the Imagination*. Translation and Philosophical Introduction by Jonathan Webber. Taylor & Francis e-Library. Web.

Scheler, Max. 1973. *Phenomenology and the Theory of Cognition*. Selected Philosophical Essays. Evanston: Northwestern University Press. Print.

Talbot, Michael. 1991. *The Holographic Universe*. London: HarperCollins Publishers. Print.

Tymieniecka, Anna-Teresa. 2003. The Unveiling and the Unveiled. In *The Passions of the Soul in the Metamorphosis of Becoming*, ed. A-T. Tymieniecka. Islamic Philosophy and Occidental Phenomenology in Dialogue 1, xxi–xliv. Dordrecht: Kluwer Academic Publishers. Print.

# The Chronotopic Content of Esoterism and the Models of Thought

**Sadaqat M. Aliyeva**

**Abstract** This article considers the form and content of the models of thought in scientific and literary creativity. When the "atomic" content of thought is conveyed with its literary content, a philosopheme or a philosophical concept can be understood as a model of thought. Consequently, the structural content of the ideal construction of a phenomenological thought does not obliterate its literary essence. Herein the literary content of phenomenological thought is considered as its second phenomenological feature. For example, if the phenomenon "death" gives one the chance to interpret life as common "destiny," the human being is able to include the essences of the "death" and "destiny" in one thought and one model of thought. Re-modeling occurs as an act of reconstruction in philosophical thought. The esoteric essence is considered through comparison of the two models of thought. The concept of the "space of thought" developed by Salahaddin Khalilov gives us an opportunity to understand philosophical space as a phenomenological space. And the "passion of the Earth" expounded by Anna-Teresa Tymieniecka relates this primary idea with cosmic harmony. Anna-Teresa Tymieniecka shows not only the logical perception of place but also the abilities of literary imagination. On this basis, there is generated an instinct of place or "nest" and the model of the "passion of the Earth" and the primary and divine homesickness turns toward an esoteric place.

**Keywords** Chronotope · Cosmos · Model · Esoteric space · Thought

Relation to time and space is the simplest form of relation to being. Therefore, the chronotope realizes existence and restores relation between being and non-being. The connection of being and non-being at first appears as an ideal connection and, then, non-being defines the being of our imagination and realization.

---

S. M. Aliyeva (✉)
Azerbaijan University, Baku, Xatai region, Azerbaijan Republic

 
W. S. Smith et al. (eds.), *Eco-Phenomenology: Life, Human Life, Post-Human Life in the Harmony of the Cosmos*, Analecta Husserliana CXXI,
https://doi.org/10.1007/978-3-319-77516-6_45

The essence of any relation subsists in the fact of its obtaining a new space or realization of space. But this is not always so, and directs realization of physical space. What we call connection is the opening of a new field, another place in space, and is perceived, in fact, as an extension of space. Space always implies form and always tries to change a shape, but space, at the same time, simulates this form in time. Space itself can only form in time. The act of existence outside of a chronotope creates in us an amorphous representation.

Esoteric space is the base condition for esoterism. This esoteric space is entirely a correlator of cosmic space, and it is transformed into the main consideration that bases the existence of cosmic space for esoterism. Just as the graded refraction of light rays justifies the existence of light, esoteric space also proves the existence of our common space and simplifies the conception of common space. That is to say the presence of any constitutive expression is the indicator of its existence, as well as of the constitutive intentionality of consciousness that helps us to realize its existence.

The shortest and the simplest form of the essence of an esoteric chronotope is discerned in the way in which all the ideal processes of esotericism occur in the mental field. Ceaseless ideal modeling comes with esoterism, and the ideal corporeal, space network, is formed in local esoteric thought. Within the microclimate of esoterism, tracking this process is probably not possible because of very high speed of qualifications, which express real meanings even as they pass out of sight.

Therefore, in esoterism only the specific essence of a chronotope is felt. Models of thought arise in a local esoteric environment. In fact, they are cosmic models, and here microcosm is the continuation of the macrocosm and individual consciousness passes in submission to a spatial order. At the same time, the real results of the time and place remain beyond the esoteric environment, as do the internal processes and external relations of consciousness, and thinking is charged with the elements of esotericism. Somewhere, the cosmic network is always formed, knitted, and, from another place, the cosmic network proceeds and arises in an absolutely other context. Esoterism always and in all circumstances, in all geographical distributions, investigates the mental maintenance of this difficult network. This interlacing, which was overshadowed in esoteric space, essentially consists of this. This internal ideal connection between order and chaos displays cosmic thinking. Anna-Teresa Tymieniecka explains this spatial thought as internal potential. For example, the germination of a seed requires, along with such conditions as light and heat, the internal possibility of germination. Just this internal ability to germinate directs the attainment of cosmic space.

In the Middle Ages, Eastern esoterism experienced its renaissance. The relationship of time and space assumed a special place in medieval philosophy. A distinctive meaning of chronotope appeared. For example, in the atomistics of An-Nazzam (811–845) the discreteness of space was raised as an issue. An-Nazzam, with his concept of "tafra" (a leap), explains that in its becoming discrete the material essence of matter is to share and to merge infinitely (Zackuev 1961, 11). According to An-Nazzam, overcoming space is possible only with a "leap" because, if the space is discrete, to get continuous action matter needs to be passed directly over a void.

Just as much as being, non-being also acts as a prime condition for the emergence of a chronotope; also, the concept of the integrity of time consists of imaginations of stable sequences of intervals of time. That said, just as in the apparatus of sight, the aperture plays the role of a cosmic space for light reflection, even so the void between the fragments of space and time creates a special microenvironment for their realization – the cosmos. Thus, the simplest way to understand the attitude to adopt vis-à-vis time and space can be found in the essence of esotericism that is the chronotope. However, in a philosophical sense, analysis of a chronotope cannot be limited within esotericism. The phenomenological approach to space and time allows for explaining a chronotope not only in an ideal but also in a real form. At this point, the ontological essence of a chronotope is investigated not as a basis for time and space but as the attitude of existence. Here, the cognitive level of knowledge stands first rather than the common presence of action. In her cognitive approach to a chronotope, Anna-Teresa Tymieniecka's thoughts on an esoteric gravitation toward space helps the understanding of space as a special phenomenon. Here, space is presented as aspiration to self-realization, a special element. This element receives a certain frame in the creative values of art, aesthetics, and by that it shows the essence of human existence. Unlike Sigmund Freud, Tymieniecka produces individual subconscious tendencies out of the submission of individual consciousness. She, like Freud, does not submit existence to consciousness; to the contrary, Tymieniecka explains consciousness as a part of cosmic consciousness. For her, consciousness is always under the influence of a specific esoteric gravitation. Perhaps it is under the influence of the chaosmos, spoken of by Deleuze and Guattari. Edmund Husserl's intersubjectivity expresses the chaosmos of Deleuze and Guattari, but Anna-Teresa Tymieniecka's esoteric gravitation expresses space; the main difference here consists in this. According to Tymieniecka, poetic space is the cosmic space. However, in all cases this esoteric gravitation does not consist of local gravitation only, which exists in the esoteric environment. The esoteric gravitation to space is the cosmic force aspiring to their harmony. The person with this esoteric gravitation can transmit internal creative energy in various directions. If the esotericism created for local environment is a spherical, mystical opportunity, esoteric gravitation is a cosmic opportunity having ontopoietic content. This cosmic opportunity proceeds from the material and spiritual borders of existence. Tymieniecka connects literary thought, the relation to beauty, and aesthetics with just this cosmic opportunity. The esoteric gravitation to space receives attention as an important factor which, being the presence connecting to the social-public environment, gives to it a public-social essence. But according to Tymieniecka, here is an internal decay that, unlike Kantian internal decays, is regulated not by analytical consciousness, and not by public and social relations, but directly by the cosmic order. In this case, internal addiction, the internal tendency to growth, carries the global essence as spiritual decay. The person is identified with the cosmos. Therefore, Tymieniecka, within this cosmic space, can treat as equal weeds, fruit trees, and the person.

If cosmic space expands constantly, that means that it was put there beforehand to allow possibility in the universe for all living forms. That is to say that presence was created as the act of potential presence. Therefore, Tymieniecka considered the initiative of instinctive life naturally; she saw this initiative as the ontopoietic essence of life in artistic creativity. For example, the constant turning of a sunflower, always toward the sun, no matter how much it seems to be an artistic, aesthetic event, proceeds from the flower's internal genetic essence, which relate to its cosmic beginning. Literature connects it, in the metaphorical way of literature, to a beautiful legend, and that legend with its artistic content caresses the soul of the person. But aspiration to space, going towards light, taking a place under the Sun is a natural instinct of a living creature. Gaston Bachelard analyzed water as an organic cosmic space in his "water reveries," and he showed the elements to be a primary protospace for all mankind. Water, being the protospace giving the initial start for any living being, does not meet the requirements of the "instinctive shelter" posited by Tymieniecka. For her, the "instinctive nest" is connected with the element of the soil, which by an elemental instinct turns into space, which can be managed space. According to Gaston Bachelard, the force of water elementally and in its quintessence engenders connective abilities: the union of water and earth produces clay, dough, all that is pasty and is in continuity with our work on the clay, the dough (Bachelard 1998, 151).

Anna-Teresa Tymieniecka probes the part played by the elements in human consciousness and the covering gravity of space: "this passion, which I call 'esoteric passion', finds its projection and crystallization in space, in an imaginary, extraordinary, dream place. This is the 'esoteric passion for place'" (Tymieniecka 2008, 74). Consciousness of elements cannot be managed like other elements; only the creative act plays the role of organic space for consciousness. In creativity the elements of consciousness turn organic space into poetic space. Thus, all the universe, always and in all versions, tries to express life and find reflection in the struggle for space. This is the most important aspect of the creativity that gives ontopoietic content to this struggle. Fighting for space expresses not only the instinctive content of creativity, but also the harmonious connection regulated by the order of space. Tymieniecka in expounding her phenomenology dwells on these harmonious contents. The ontopoietic content of philosophical thought is the expressive poetic content. The opportunity for artistic modeling is used in the content of a chronotope. Artistry extends the internal possibility of thought and brings in semantic content. The artistic-poetic content of Anna-Teresa Tymieniecka's vital phenomenology and her aesthetic conceptualizing providing the universe existence through continuous expansion consist in this. In the Bible, the continuous expansion of the Universe is the given primary potential opportunity seen as a protoforce, like the force by which yeast increases the mass of dough several times, or like the way a mustard seed can become space for birds to perch upon. Internal decay, a primary possibility of expansion, impels a constant attempt to acquire new space and the aspiration to realize the protophenomenological features inhering in time. The expansion of the universe is a process of pulsation and is part of the desire to live; at the same time, it helps us to understand the physical content of a chronotope.

Sometimes Tymieniecka's thought, which linked an esoteric gravitation to space and the phenomenological thought of Edmund Husserl, made the link through intersubjectivity in such a way as to seem to be extrapolating from inside to outside. The only difference is that in Tymieniecka's phenomenological thought, cosmic space is more external space, a transcendental space; but, according to Edmund Husserl, cosmic space makes an internal, immanent space impression. Both thinkers focus on cosmic space, but left unaddressed is the distance between internal and external space. The difference appears in the perception of their ontological content. Internal space is endless, subjective, but external space is objective, limited at the level of consciousness. Therefore, transcendental space is both phenomenological and psychological space, but immanent space is at the same time cognitive space, and here existence is an abstract existence. Husserl's idea of space involves how much distance there is for a person from itself to itself. Husserl, in his concept of the "other I," sees the person in an internal microcosm that is located in its mental network. The main feature of the concept of the "other I" is that here cosmic connection is internal connection. The import of this phenomenologically-based relation is that it exists directly. That is to say that the existence of this relation is not dependent on whether we perceive it or not. Besides, it forms a special system operating us and the social environment. In Husserl's phenomenology, the internal space relation is an ideal connection that has not yet been polluted by time and space and the loss of real belonging; it is only the "pure" essence of that which consciousness created in imagination – possibly to accept its "purity." However, this "purity," as Alfred North Whitehead stressed, depends "on the degree of purity of our eyes." In Husserl's phenomenology intersubjectivity turns itself for itself into a subject and, having also attached phenomenal significance to ideal communication, turns the person into society and strangles the person's instinctive freedom. According to Edmund Husserl, consciousness aspires to initial ontopoietic contents, but this space is limited within the frame of intersubjectivity.

Therefore, as with the cosmic space of Tymieniecka, ontopoietic space cannot be shared with all. Edmund Husserl frames the idea of the "other I" in such a way that the human being at a distance from itself looks to itself to be very problematic and at the same time magnificent; there is no place for weeds, however, for he sees, "'another' in the phenomenological sense, as a modification of my egoism" (Husserl 2000, 467). Husserl places the other internally – the global content of a person is in the special framework of the philosophical value of essence. Here, society cannot be the beginning of the public and social person. Society is a result of public consciousness, and is also its philosophical result. Analyzing Husserl's phenomenological approaches, Herbert Spiegelberg declared, "As many phenomenologists as exist, there are as many phenomenologies." That is to say, sometimes phenomenological thought is to be identified with one's conceptual creativity, which becomes simpler and is transformed into a scientific method. However, Husserl's phenomenology expresses an immanent chronotope in an internal, spiritual polemic. Probably, therefore, his ideas of the "other I" and the intersubjective amount to a categorical version of Sartre's moral and spiritual searches in *Nausea*. Yet, it does not prove to be solipsism, as the person turning himself into a subject, insofar as it separates itself from society, finds himself only as a member of society.

The main difference between Edmund Husserl's and Gustav Shpet's phenomenology proceeds from here. According to Husserl, society is a result, but according to Shpet, it is society – historical activity. Husserl's intersubjective does not require society; according to Shpet, the subject is more being than it is an object. Addressing just this difference, Tymieniecka, by positing the force of an esoteric gravitation towards a space, bases Edmund Husserl's intersubjectivity in an ontopoietic way because the person who does not find himself does not see the Universe.

A chronotopic approach and the concept of "model of thought" are helpful in disclosing the value of esotericism. Salahaddin Khalilov, who has put forward the concept of the "space of thought," analyzes the concept "space of thought" as a special space connected with evocative speech, and points out that the "space of thought" is formed at the expense of time duration. Discussing time, Martin Heidegger said that "time is casual, and a case is temporary." Connecting time with a case considers only temporal time. Continuous time is cosmic time; therefore, it covers the cosmic contents of the space of thought.

Khalilov's concept of the "space of thought" does not have a logical sequence, for in its compressed, logarithmic structure, it hides potential chaos. In the "space of thought" the sentence is not 'the atomic fact,' it is one of the elements of a difficult model of the majority of numerous cases. The "space of thought, " which, being analyzed by Khalilov as cosmic space, investigates the main forms of thinking, leads to a conclusion that "the space of logical thinking" is isotropic, that is, here all directions of thought are equivalent (Khalilov 2008, 172). The thought that all directions are equivalent is abstract thought put in a structural frame, this because "isotropy is correct for unidirectional coordinate systems." Owing to the fact that these models of thought have no transitional points, it is possible to create only their isotopes. For while in logical thought there is perhaps the formation of isomorphism, since one logical thought does not intersect with another, one's space of thought is limited. While explaining the value of chronotopes, Khalilov based the model "space of thought" not on isotropy, but on anisotropy. He points out that logical models are isotropic and logical foundations of cognitive acts; additionally, spatial unidirectional cases develop gradually and do not go beyond a certain structure. In the one-dimensional "space of thought" there are concrete borders. In two-dimensional and three-dimensional spaces the distance for transitional points is formed, and the "space of thought" as poetic space extends and becomes boundless.

In Khalilov's model "space of thought, "the fact of an inclination of space substantiates the similar content of the refraction and inclination of light. Khalilov's model "space of thought" allows for understanding of different models of thought and uniting these models in their synergetic content. In the idea of "the inclination of the space of thought" philosophical thought models do not equate to an act of consciousness, nor to the information received from the outside, as well as to cosmic consciousness. This cosmic consciousness contributes the phenomenon of artistry to the essence of philosophical thought while more and more complicating its value. In Khalilov's account of creativity, the phenomenon of artistry in philosophical thought is opposed by logical knowledge, and artistry relates to philosophical thought. This in turn points to the uniqueness of the artistic content of Khalilov's model of philosophical thought. Here is an apprehension considered mandatory by

Edmund Husserl, an example of an inter-subjective experience trying to acquire poetic space as a way of thought:

> "...What connects us beyond the existing natural passions?" If you always exist in my thoughts, I mean, living in one life as I live another? Am I a little you? In these relations I am confident only in the accuracy of three things:
>
> – For some reason, I could never lie to you...
>
> – For some reason, I could never move away from your truth completely...
>
> – For some reason, I could never present this world without you...
>
> Perhaps spiritually, morally, we combine these "never possibles" and in this timelessness we are the same? Is it just so?

## Works Cited

Bachelard, Gaston. 1998. *Water and Dreams. (French philosophy of the twentieth century). The experience of the imagination of matter*. Moscow: Humanities Publisher Print.

Husserl, Edmund. 2000. *Logical Investigations. Cartesian Meditations. The Crisis of European Sciences and Transcendental Phenomenology. Philosophy and the Crisis of European Man. Philosophy as a Rigorous Science*. Moscow: AST Print.

Khalilov, Salahaddin. 2008. The Doctrine of the 'space of thought' in Abu Turhan. *Philosophy and Socio-Political Sciences* 4:159–185. Print.

Tymieniecka, Anna-Teresa. 2008. Esoteric addiction to 'place'. *Philosophy and Socio-Political Sciences* 5:73–78. Print.

Zackuev, Achmed Kharadi. 1961. *Philosophy An-Nazzama*. Baku: AN SSR. Print.

# Phenomenology of the Virtual Body: An Introduction

**Roberto Diodato**

**Abstract** The problem of perception in a virtual environment could be reformulated thusly: what can the philosophy of perception gain from a theory of "perception in virtual environments," given the specific nature of that environment? This discourse goes in circles, because the analysis by which we frame the difference always proceeds from theories elaborated in the field of the so-called "real," but it is a typically philosophical process, one that can make sense, however, only if it can be shown that the virtual is an existent being that has an ontological structure of its own. It is thus distinguished by asking these elementary questions: what are the elements that make it possible for one to perceive a virtual environment? How are difference and the subject-object relationship constituted in virtual environments? What does it mean, in short, to perceive a virtual object? The answer to these questions may emerge from a phenomenology of the virtual body.

**Keywords** Virtual · Body · Presence · Perception · Phenomenology · Environment · Aesthetics

## Presence in Virtual Environments

Reflection on the concept of *presence* in virtual environments has been unfolding for years, and has seen contributions from scholars representing many different disciplines in order to allow the construction of environments that can better simulate the complex sense of presence in a way that is perceptually faithful. Therefore, the objective of the research is a definition of presence that is a function of efficacy: the sense-feeling, so to speak, of presence in a virtual environment is all the more interesting the more it is able to compete with the same "feel" in non-virtual environments or, as it is wrongly put, in "real" environments. The degree of illusion induced by the device counts, therefore, and research is aimed precisely at the reproduction

R. Diodato (✉)
Università Cattolica del Sacro Cuore, Milan, MI, Italy
e-mail: roberto.diodato@unicatt.it

 
W. S. Smith et al. (eds.), *Eco-Phenomenology: Life, Human Life, Post-Human Life in the Harmony of the Cosmos*, Analecta Husserliana CXXI,
https://doi.org/10.1007/978-3-319-77516-6_46

of the pretense of reality, and so on; obviously, this purpose guides the research. This desire is infinitely lost, as is well known by the designers of virtual environments, and therefore the quest must be limited to an analysis of the degree of user attention: the more the virtual environment loses contact with reality, the more pervasive and effective it will be. Thus, what is interpreted is a situation in which a user is set in a double environment, virtual and real, and is consistently able to focus his attention on one or the other. The aim is to encourage immersion in the virtual primarily by means of relevant stimuli that serve to bind, to constrain, attention. It is suggested in this regard that we consider presence in virtual environments to be an illusion of non-mediation (a perceptual illusion of non-mediation), and, relatedly, that we understand the non-mediation as disclosive of the degree of presence. But, of course, the quality of presence, taken as an indicator of the quality of the immersion and also of the interactivity, involves the possibilities of action in the environment, expectations and adaptation, comprehension and dispositional attitudes, and thus outlines a field of remarkable theoretical density: "The experience of presence is a complex, multidimensional perception, formed by an intersection of (multi) sensory raw data and a variety of cognitive processes – an experience in which the factors of attention play a crucial role" (Ijsselstein and Riva 2003, 5). It is therefore necessary to enter into the speculative quality of this interweaving, into a conceptual depth that proceeds from perception to culture, taking on the inescapable background layers of the complexity in order to evaluate reflections on what will ultimately amount to a theoretical proposal.

## Credible Environment as an Objective Environment

The illusion of non-mediation can be a starting point because it requires, basically, the position that presence is immediately perceptible: the subjective feeling or sensation of being present in an environment that is characterized by a feeling of perception. The question then is to analyze perception in the virtual environment such that it is different, or that it proposes to differentiate itself from a mediated environment and therefore from the perceptual qualities that overlap. On a second level, the perception of the environment, such as mediation in film, television, photography, and interactions, mediate the reading of texts or hypertextual navigation. In some respects, then, the question is simplified and the investigation may focus directly on perception. In this way we place the question in the context of the simulation and assume that the task of the builders of virtual environments is that of a perceptual adequation, for which a virtual environment achieves its objective if it can trick the user as to its status as "virtual." Now, the fact that the user does not distinguish, at the level of perceptual interaction, a virtual body from a "real" body, certainly does not per se entail "that virtual objects are to be faithful copies of real objects. Emphasis is placed significantly on the interaction: it is the motor and perceptual conditions that must be similar, and not the properties of the objects represented, so that they can also be created worlds very different from the real, provided that they maintain an appropriate level of interaction" (Pasquinelli 2003, 480). In this way, the problem of mimesis as true copy according to the middle-line of perceptual similarity is overcome. The issue becomes one of objectivity as credibility: what makes a credible environment like an objective environment? More generally, what is it that causes something (object, event, world) to be seen as objective? The question is relevant, if one can manage to loosen the perception-objectivity structure, because it permits an extension of semantic space for the question of "presence." The answer may not, in fact, cover only the sensory qualities of the environment or simply the evidence of the senses as the criterion of credibility, particularly as a criterion for the justification of perceptual fidelity. It will exceed the physical conditions of the environment as conditions sufficient for the justification of belief in objectivity, involving factors such as the success of recognition in social interaction, and therefore the symbolic and cultural components related to communication and cooperation.

Following the same strategy and working from the results that mark a return to the question of perception, we can then introduce the *affordances* of the ecological theory of visual perception to complete the picture.[1] This allows us to obtain an overall classification of the environment as a place of complex subject-object experience, one at least relatively consistent with expectations (which are always to some extent not simply private but also dependent on personal "history") and with the projects of consumers, and one that can adapt to the exceptions that confirm the rule.

## How Can We Perceive a Virtual Object?

Now, the results of this approach are practically and technically effective for the construction of virtual environments; however, they presuppose the homogeneity of the ontological structure, that is, they consider the virtual to be a possible being and therefore as having a relation of essential similarity and existential difference with respect to what is called real. In this way, none of the semantic fields are complicated (real, actual, possible, potential, virtual, etc.) or revealed in relationship to one another, which makes the discussion less interesting from the speculative standpoint.

The problem of perception in a virtual environment could be reformulated, then, as: what can we learn in the philosophy of perception from a theory of "perception in virtual environments," given the specific nature of that environment? It is obvious that the discourse goes in circles, because it always by proceeding from theories elaborated in the field of the so-called "real" that we develop the difference. This is a typically philosophical procedure, one that makes sense, however, only if it can be shown that the virtual is an existent being that has an ontological structure of its own. It is thus distinguished by asking the elementary question: what are the elements that make it possible for one to perceive a virtual environment? How are difference and the subject-object relation constituted in virtual environments? What does it mean, in short, to perceive a virtual object?

The answer to these simple questions may emerge from an ontology of the virtual, which necessarily includes a discussion of the concept of the virtual. Note: this avoids the obvious consideration that the concept of objectivity, related to that of object, has the same meaning for beings regarded as "real" as it does for virtual beings. If, for example, one thinks that an objective entity or event is "external," in the sense of existing independently of the (perceptual) experience that a subject can undergo, then the virtual "object" is not objective, even if one can say that this object exists beyond experience, or does not exist solely through experience. What, then, is the degree of re-identifiability of such an object? It will be partial and belong mainly to the part unidentifiable by spatial coordinates: not to the recurrence and constancy of form in diverse places, but to the partition of the object, that is, to the computer script. If we believe, however, that we can distinguish between perceptions or, more extensively, representations that show up as being "external" and stable, and representations that show up as being "internal" in continuous adjustment through feedback mechanisms, then our experience of the virtual body denies the possibility of the distinction. Indeed, the multisensorial multiplicity of the input of virtual bodies is coherent and unitary for the constancy of the link between a property and a place, as one cannot disregard the principle of non-contradiction (for which one single element from every class of perceptual property belongs in time $t$ to the object in location $l$ – note that one must assume the hypothesis to be demonstrated, namely, that there exists a definable time-instant). Yet, at the same time, the virtual body, through its interactivity, dissolves the difference between distal and proximal stimulus since it is, so to speak, both ontologically near and perceptually distant. A first step to consider, then, is the possibility that certain consolidated analyses do not work for virtual objects because of their ontologically different status. Finally, it is clear that the answer to the question of *presence* requires a clarification and deepening of the notion of virtual body-environment so that we can, circularly, demonstrate its peculiarities in terms of perceptual theory. I hope to arrive at this point, but it will take some time.

## Problems of the Phenomenological Approach

Now, when we speak of presence we can still assume the banality so well-expressed in Husserl's phenomenology: every given is the result of a complex of intentional operations, and all intentional unities come from a deliberate genesis, are 'constitutive' unities that have relationship with our past, with the history of our perceptual life, with the legacy of experience. Phenomenological description (the "static" aspects of phenomenology) has taught us to take into account, in our study of the structure of the sense of *presence*, both the noematic and the noetic points of view. So, for example, from the noematic standpoint, the relations will be explained, and in particular the relations of precedence, in terms of validity, between the system of phantasms and the system of movements, and therefore of deformations, of partitions, and so on. From the noetic point of view, there will be inspected, in contrast, the differences between retention, recollection, and expectation, to name a few – in short, those modes of consciousness that are, for the phenomenologist, conditions of subjectivity. The phenomenological perspective belongs, of course, to the philosophical analysis of perception, but the former is normally an apprehension that intentionally embraces all those layers together that the analysis distinguishes. From here there emerges a difference – obviously beyond that between the empirical and the transcendental – between the psychological approach, following from studies of the notion of presence in virtual environments, and the philosophical approach, a difference that an aesthetics of the virtual can waive, at least on a preliminary basis. Now the question might be: in confronting a virtual body, what changes, if anything, from the perspective of a phenomenology of perception? This is certainly not the only question one can ask in relation to the theme of presence in virtual environments, inasmuch as the phenomenological point of view is one of many possible philosophical ways of understanding *presence*, but it is my first question.

This operation, which could now be carried out at least partially on the basis of experiences and case studies in terms of the theory of perception in virtual environments, could provide results that permit another question, which regards, circularly, the sense of the phenomenological method and its purpose. Well known is the difficulty, not unlike the one that in Platonic theory concluded in the doctrine of recollection, regarding the relation between the historicity of the transcendental subject and the pretense of truth delivered by the eidetic method: if in all perception there is present, and can never be absent, the inheritance of a history of practical experience, almost like the passively established *a priori* essence that is a condition of the possibility of capturing an *eidos*, then we have a projection of the empirical onto the transcendental, of historically constituted subjectivity onto the apprehension of world. Now "if the passively pre-constituted *eidos* that serves as a guide in the eidetic variation is, however, made up from a world with a specific ontological structure, *how can it claim validity for a possible world in general?*" (Costa 1999, 30–31). This is a problem involving the basic constituent of any representation from spatiality onwards (and, more radically, temporality), as one can imagine, or at least not prejudicially exclude, cultural variations of the feeling of space. Moreover, the phenomenologist is necessarily conditioned by the fact that he took himself as his starting point: "Transcendentally he finds himself as the ego, then as generically an ego, who already has (in conscious fashion) a world – a world of our universally familiar ontological type" (Husserl 1999, 76).

## The Opacity of the Cogito and Its Implications

He is therefore always to start from "a world of our well-known ontological type," from an ontology both obvious and necessary, that produces fiction and eidetic variation, and will thus be limited in his freedom.

Derrida, in *Edmund Husserl's* Origin of Geometry, has made the point that, beginning from the insuperability of limit, the phenomenologist will develop his program of research in the direction of "quasi-transcendental" figures, but certainly with an awareness of the limit of freedom from variation and therefore in substance the limit of the imagination, proper to Husserl and proceeding on its way towards a consideration of genesis, one that gradually enhances awareness of an essential perspectivalness. Of course, the transcendental ego is not empirical, but is a functional center of intentional acts, in and of itself prior to any worldly egoism, and one that puts aside the natural attitude linked to the psychological. This is very important, as we shall see, even for an approach to virtual reality. It still allows one to understand the particular curvature that the term "ontology" assumes within phenomenology and its relationship with the term "world"; the world, be it real or possible, as the theme of transcendental phenomenology, is never an already given, a being present in the form of phenomena that contain an essence to be revealed, but is an eidetic complex that emerges through intentional operations. And ontology is, relatedly, the attempt to locate and describe layers of meaning in this world. Yet the project is not only that of phenomenological description and so aims not only at identifying the essential differences between intentional acts but also becomes a search for "what works in subtle, subliminal ways, what is absorbed and metamorphosed in the passive synthesis" (Ghiron 2001, 178). When that synthesizing correspondingly emerges as being internal to the genesis, the power of the imagination, then there emerges also the question of the limits of that power: "Can I imaginatively make it so that my body is transformed into that of my childhood, and my mind into that of a child? Isn't this by all rational accounts a kind of nonsense? With these questions we see that the problem of the changing of perspective [*umdenken*] has not been properly placed or treated" (Husserl, ms. E III 9/49a, quoted in Costa 1999, 35). It is no coincidence that the issue of "changing perspective" is central in Sartre and Merleau-Ponty, that is, in those authors who have developed their own approaches in phenomenological research, approaches that devote great attention to the dimensions of opacity, and implication, to the lack of purity of the cogito – in short, that give renewed attention to the body, and therefore to the historicity, affectivity, emotions, and, conclusively, the perceptual materiality of the ego. And so phenomenological research, which typically concerns the invariable structures that *should* be found to be common to the different conceptions of world, while not interrupted, becomes complicated and problematized.

## Is a Virtual Environment a Simple Simulation?

At issue is the actualization of a revision of transcendental subjectivity in the direction of corporeality as a condition of the possibility for any objectual apprehension, an emphasis present in Husserl's own work, but made here with important additional stress on the further development of the open question with respect to the complexity and novelty of the virtual field. If, in other words, we hold fast to the methodological lesson of phenomenology that leads to the exercise of the imagination as a variation of the aim of capturing the transitory from the emergence of forms, from the variety of images, to the identity of objects and to their essential meaning, and if we think of this experience not only as a typology or empirical description of the field but as an analysis of its stratification, and finally, if this account is only possible at the level of intentionality, then, in the case of its own experience of virtual images, the body of the subject is necessarily prosthesis-equipped and the body of the object is an eminently interactive hybrid, all of which may have consequences that are reflected in the claims of the phenomenological method to transcendentality. Can one still speak of an "ego in general, that is already a consciousness of world, a world of our well-known ontological type," where there no longer exists a "well-known ontological type," or where the ontological type is no longer so "well-known?" Abstractly, one can *ex parte* also ask this of the object: What happens when the intentional object is not a body, not an image, but is a hybrid body-image precisely in the same way as a painting, a photograph, a picture, a film, a television image, not least because it has the same intersubjective quality, a quality that is always connected to the degree of interaction? I will return to this point below, but for now note that the degree of interactivity implies a difference that is reflected in the very possibility of experience. And what point of view, what perspective ought we assume for the inquiry: the perspective of a split egoic experience, both internal and external to the virtual environment? Again, it is clear, we are close to the situation of the impossibility of dream analysis, and this may call into question the possibility of any "common sense." Obviously, this leads us to consider the question from the point of view of the constitution of the aesthetic object, where imagination and image play in an inextricable circularity. But for now I have only posed a problem: the challenge is to enter into the question of "presence" in the virtual environment holding firm, in a manner not too naïve, to the essential phenomenological advantage that renders an account of the complexity of the processes possible. The question that then arises is this: how important is the analysis of presence in the virtual environment, and the *gap* or *variation* that must be compared to the perfect simulation? It should also be noted that the analysis of the field (a field of interacting forces, in our case consisting of users and virtual bodies) cannot be a merely verbal description of the content of a user's experience, even if that user is the subject making the description, and that is so precisely because the use of language attributes to the contents of experience an awareness that might not be legitimate. Such contents are, rather, the unfolding of the layers of a history that led up to that experience, a regressive demonstration of the development, a grasping of the dynamics of constitution, but without the pretense of universality that is negated at the root in the development of technologies.

## Who or What Is the Subject?

Equally abstractly, the subject must take into account the inorganic prostheses that come to constitute the "body" of the operator-consumer-actor in the virtual environment. This is relevant for the phenomenological tradition, and to me it seems to point in the direction of its further deepening, and not to a denial of its method. Let us take up the question from some distance: Who or what is the "subject" that "feels," perceives, and understands? Not a disembodied mind, it seems, but an inextricable mind-body complex. To be sure, this is what Merleau-Ponty has progressively taught, first stressing the factor of the body, and then developing the strange and complex notion of flesh as an environment of organic-inorganic participation. But the simplest – or at least most primary – approach to the question arises from the consideration of the user's body as a body with inorganic connections that make it possible to perceive and, in part, to constitute virtual bodies. This body is a cybernetic organism, a hybrid of machine and organism; post-human theory indicates "new ways of living out identities through a mutating body, never finite or definite, a hybridization of the organic and inorganic, between biology and technology, between flesh and circuits" (Combi 2000, 119). We can understand it conceptually as a figure of the limit and surpassing of the limits between organic and inorganic and, more generally, between nature and artifice. But for our purposes the body in a virtual environment is simply a "structural coupling," a body-prosthesis that belongs to the project of comprehending the "functions of the organism within a machine's encoding," that enhances the body's ability on the one hand, and on the other develops the skills of the machine with the implant of "an organic body" (Boccia Artieri 2000, 231). One will need to ask oneself at the phenomenological level, and also at the level of ontology, precisely what are the consequences of the artificial.

## The Living Body in a Virtual Environment

It is now necessary to reflect with tact on the following point: we cannot simply hypothesize that the body endowed with inorganic prostheses corresponds, in a virtual relationship, to the living body or to the body proper as described by the phenomenological tradition. Perhaps, as Nancy writes, "we shouldn't think the 'ontological body' except where thinking *touches* on the hard strangeness of this *body*, on its un-thinking, unthinkable, exteriority" (Nancy 2008, 17). It is true that bodies "take place at the limit, *qua limit*: limit – external border, the fracture and intersection of anything foreign in a continuum of sense, a continuum of matter.

An opening, discreteness" (Nancy 2008, 17). It is also true that the interval between bodies is where they will take place as image. And in particular, "The *image* (that [the body] is), has no link to either the idea or, in general, to the visible (and/or intelligible) 'presentation' *of* anything. The body is not an image *of*' (Nancy 2008, 63). Everything functions, and we therefore exploit it, as a description of the immersive and interactive virtual field.

Yet the issue is not to think of the negation of the living body as much as it is to think of the body that comes from inorganic integration *as* a living body. It is not to think of the non-living as organic, but of the living as being inorganic without being inanimate; we are working here in a deliberately non-obvious modality. Certainly deficient is the notion of a hierarchy of mere body-objects, on the one hand, which would exist *partes extra partes*, and the living body, on the other hand, as a condition for "having a world." The idea of the body as a space of appropriation, as that which can be inhabited and directed from within, is also lacking. Any internal-external difference more profound than that essential internal-external involvement, one which only the view of the living body allows to be thought, is therefore to be abandoned.

Moreover, the possibility is also indicated of an overcoming of the univocal relation between consciousness and identity that developed from the theory of the living body as an organon or schema of self-consciousness. In sum, self-consciousness is consciousness that both derives from the world and restores the world. A case in point, it seems to me, an experiment that goes in this direction, are the body-modifying operations of Stelarc, and perhaps the last mechanotronic performance of Marcel•lí Antúnez Roca.

## The Complexity of the Body's Experience in a Virtual Environment

Since making the distinction is a matter of perspective or intention, the organic-inorganic body synthesis can be considered at the same time to be a living body and to be a body-object, an aggregate of extended parts that carry out biological functions, thinkable and describable in physical terms. The notion of body *qua* machine is no longer a mark of reductionism. The cyborg is a complex made up of the psycho-physiological and the physical, the mental and the mechanical, the natural and the artificial, the organic and the inorganic. As such, it is a vehicle for expressive possibilities in the virtual environment; the cyborg is precisely "our general medium for having a world" in the virtual domain (Merleau-Ponty 1989, 146). Now, after having indicated the direction in which to think of the body endowed with technological prostheses, we can rethink it in terms of the *Phenomenology of Perception*. Structurally an intentional openness, taking in the world as it presents itself in the opening movement, such a body is no mere object and cannot be grasped in its complexity from the objectifying perspective; and so it is always imminent to any scientific approach and stands rather as a condition for the possibility of our having a descriptive language: this body also is "the horizon latent in all experience and itself ever-present and anterior to every determining thought" (Merleau-Ponty 1989, 92). It is a preobjective view able to perceive its own movement as an event within the environment, a complex of affective and kinesthetic sensations, a "body schema," a space-temporal complex *in situ* that determines the very existence of the environment's space-temporal characteristics. This is a body that, because it is a condition for the possibility of experiencing an environment, itself constitutes that environment. It is an embodied consciousness that arises primarily not as an "'I think that' but as an "I can," one that takes the environment as both possibility for action and possibility for expression that constitutes qualitative horizons (Merleau-Ponty 1989, 137).

## The Paradoxical Nature of the User-Body

In the virtual environment, the user-body inhabits the space and time that its interaction constructs in the specificity of a virtual time-space and the phenomenalization of the interaction (making up the environment itself) between virtual body-objects. The technological prosthesis seems to confer upon the body an extraordinary power, which in non-virtual environments is the effect of habit, of repetition, "the power of dilating our being-in-the-world, or changing our existence by appropriating fresh instruments" (Merleau-Ponty 1989, 143). Here is the power to assimilate otherness into one's body, making it an expansion of the expression of one's own time-space, so that the body's perceptual and cognitive activities constitute the field of presence as a simultaneous field of meaning. In other words, in the dynamics of a virtual relation or the constitution of a virtual environment, the peculiar relationship between a body equipped with prostheses and a virtual body is such that meaning is internal to the perception, or to that encounter that we continue to call "perception." This is a delicate point, but supposing that in the virtual environment, understood in the strong sense that we have defined, relations between users by means of avatars is possible, then it will be the case in such an environment that "the parts of my body together comprise a system, so my body and the other's are one whole, two sides of one and the same phenomenon; and the anonymous existence of which my body is the ever-renewed trace henceforth inhabits both bodies simultaneously" (Merleau-Ponty 1989, 404). The *other* user-body, being virtual, will be, strictly speaking, hybrid and paradoxical. I realize that this body is hardly conceivable in its perceptual and imaginative dynamic, inasmuch as it is difficult to think of it outside the gaze that is the condition of objectivity. But this is the most complex case because it supposes the possibility of a relationship between identity and difference in which one relation to three positions is reduced to one relation to two positions by assimilation, in an avatar, of a body equipped with prostheses and a virtual body: an identity among diverse elements that at the same time remain diverse. Perhaps at this level the perceptual environment most similar to that of the virtual, at least in terms of complexity and the immediate feeling of presence, is that of the dream.

## Note

1. See James J. Gibson, *The Ecological Approach to Visual Perception*. Boston: Houghton Mifflin, 1979.

## Works Cited


Boccia Artieri, Giovanni. 2000. *Lo sguardo virtuale*. Milano: Franco Angeli. Print.

Combi, Mariella. 2000. *Corpo e tecnologie*. Roma: Meltemi. Print.

Costa, Vincenzo. 1999. *L'estetica trascendentale e fenomenologica. Sensibilità e razionalità de Edmund Husserl*. Milano: Vita e Pensiero. Print.

Ghiron, Valeria. 2001. *La teoria dell'immaginazione di Edmund Husserl. Fantasia e coscienza figurale nella "fenomenologia descrittiva."* Venezia: Marsilio. Print.

Gibson, James. 1979. *The Ecological Approach to Visual Perception*. Boston: Houghton Mifflin. Print.

Husserl, Edmund. 1999. *Cartesian Meditations. An Introduction to Phenomenology*. Trans. D. Cairns. Dordrecht: Kluwer Academic Publishers. Print.

Ijsselstein, Wijnand, and Giuseppe Riva. 2003. Being there: the experience of presence in mediated environments. In *Being There: Concepts, Effects, and Measurement of User Presence in Synthetic Environments*, ed. Giuseppe Riva, Fabrizio Davide, and Wijnand Ijsselstein, 1–14. Amsterdam: IOS Press. Print.

Merleau-Ponty, Maurice. 1989. *Phenomenology of Perception*. Trans. C. Smith. London: Routledge & Kegan Paul. Print.

Nancy, Jean-Luc. 2008. *Corpus*. Trans. R. A. Rand. New York: Fordham University Press. Print.

Pasquinelli, Elena. 2003. Oggetti e presenza in realtà virtuale. *Sistemi intelligenti* XV 3:475–491. Print.

The manufacturer's authorised representative in the EU is Springer Nature Customer Service Centre GmbH, Europaplatz 3, 69115 Heidelberg, Germany. If you have any concerns regarding our products, please contact ProductSafety@springernature.com

Printed and bound by CPI Group (UK) Ltd, Croydon, CR0 4YY
01/10/2026
02226550-0002